**Table B**  Most Common Positive Oxidation Number for Each Element

| | 1 | 2 | 3f | 4f | 5f | 6f | 7f | 8f | 9f | 10f | 11f | 12f | 13f | 14f | 15f | 16f | 3 | 4 | 5 | 6 | 7 | 8 | 9 | 10 | 11 | 12 | 13/IIIA | 14/IVA | 15/VA | 16/VIA | 17/VIIA | 18/VIIIA |
|---|---|---|---|---|---|---|---|---|---|---|---|---|---|---|---|---|---|---|---|---|---|---|---|---|---|---|---|---|---|---|---|---|---|
| 1 | H +1 | | | | | | | | | | | | | | | | | | | | | | | | | | | | | | | | He |
| 2 | Li +1 | Be +2 | | | | | | | | | | | | | | | | | | | | | | | | | B +3 | C +4 | N +5 | O | F | Ne |
| 3 | Na +1 | Mg +2 | | | | | | | | | | | | | | | | | | | | | | | | | | | | | | |
| 4 | K +1 | Ca +2 | | | | | | | | | | | | | | | | | | | | | | | | | | | | | | |
| 5 | Rb +1 | Sr +2 | | | | | | | | | | | | | | | | | | | | | | | | | | | | | | |
| 6 | Cs +1 | Ba +2 | La +3 | Ce +3 | Pr +3 | Nd +3 | Pm +3 | Sm +3 | Eu +3 | Gd +3 | Tb +3 | Dy +3 | Ho +3 | | | | | | | | | | | | | | | | | | | |
| 7 | Fr +1 | Ra +2 | Ac +3 | Th +4 | Pa +5 | U +6 | Np +5 | Pu +4 | Am +3 | Cm +3 | Bk +3 | Cf +3 | Es +3 | | | | | | | | | | | | | | | | | | | |

ents as to the most common oxidation number are based on data given in M. Pourbaix,
*Environmental Chemistry of the Elements*, Academic Press, New York, 1979, p. xi.

# Principles of Descriptive
# Inorganic Chemistry

# Principles of Descriptive Inorganic Chemistry

GARY WULFSBERG

Middle Tennessee State University

UNIVERSITY SCIENCE BOOKS
Sausalito, California

**University Science Books**
55D Gate Five Road
Sausalito, CA 94965
Fax: (415) 332-5393

Printed in the United States of America

10   9   8   7   6   5   4

ISBN 0-935702-66-0

Library of Congress Catalog Card Number: 90-72042

*On the cover:*
Sodium thiosulfate, better known as "photographers' hypo." The crystals were
formed by evaporating a supersaturated solution of sodium thiosulfate on a glass
microscope slide. The microscope was an Olympus model BH research stand, using
crossed polarizing filters and a first-order red compensation filter. The photo was
taken with an Olympus computerized automatic-camera system, PM-10, AD. The
magnification was 100X, and Kodak Ektachrome, 50 ASA, tungsten film was used.

To those who have given me
so much personal support—
Marlys, my mother,
my father,
and my children, Joanna and Paul

# Preface

## Appropriate Courses for this Text

This textbook covers the basic principles of inorganic chemistry and the main properties and reaction chemistry (*descriptive chemistry*) of inorganic elements and compounds. It is written at a level appropriate for students in their *first full semester* of undergraduate inorganic chemistry, which may be in any one of three places in the curriculum of American colleges today.

■ First, it may follow general chemistry but precede physical chemistry and a second advanced inorganic chemistry course. This text covers both the theoretical and the descriptive inorganic topics recommended for such a course by the American Chemical Society Division of Inorganic Chemistry. It also includes some topics that are recommended for the advanced course but that also fit in well with the first-semester course. A background in physical chemistry is not presupposed.

■ Second, the first *full* course may be called advanced inorganic chemistry and may follow physical chemistry and a general chemistry course that contains only a small amount of descriptive inorganic chemistry. The same basic inorganic principles are those as covered, for example, in the first eight chapters of Huheey's *Inorganic Chemistry: Principles of Structure and Reactivity* (3d ed., Harper and Row, New York, 1983), although at a pre–physical chemistry level. But, in addition, this text exposes students in such courses to enough descriptive inorganic chemistry to allow them to function safely and effectively in the lab as professional chemists.

■ Third, a full semester of the general chemistry course may be devoted to inorganic chemistry. This text might also be appropriate for such a course provided that the students have been well prepared in high school and have already had, in general chemistry, a good exposure to the concepts of equilibrium and equilibrium constants, electrochemistry and the Nernst equation, and free energy and entropy.

Over the past 35 years, the field of inorganic chemistry has expanded its body of factual and theoretical knowledge manyfold and has also developed large new subfields such as bioinorganic chemistry and transition-metal organometallic chemistry. Consequently, the full breadth of inorganic chemistry cannot even be adequately surveyed in a one-semester course or text such as this one. But not only has inorganic chemistry grown in scope and magnitude, it also has become more and more *fundamental* to the proper understanding of other areas of chemistry or allied sciences. This textbook thus tries to emphasize the *fundamentals* of inorganic chemistry that the student will need to succeed, not only in any further course in inorganic chemistry but also in other areas of chemistry or in allied fields. The descriptive chemistry included is mainly that of the elements and their most frequently encountered classes of compounds (halides, oxides, hydrides, hydrated ions, oxo anions, and the like) rather than the latest in cluster organometallic compounds.

The principles of inorganic chemistry emphasized are those that are most useful for predicting and explaining this descriptive inorganic chemistry. This emphasis allows the students to learn the maximum amount about both descriptive chemistry and principles of inorganic chemistry in the minimum amount of time. Indeed, I found in the early days of developing this material that after as few as 13 lectures, students who took the Graduate Record Exam as part of the now-defunct Undergraduate Assessment Program scored, on the average, about two standard deviations above the normative mean on the inorganic chemistry portions of the exam.

Because most students taking this inorganic course will not become inorganic chemists, this text contains many optional sections that apply inorganic principles to phenomena in nontraditional areas of chemistry, such as geochemistry, materials science, metallurgy, chemical safety, medicinal chemistry, and biochemistry. The use of these sections could reveal additional career options about which undergraduate chemistry majors might not have known. This course could also be valuable for students already majoring in these auxiliary fields; I have been told, for example, that this material was of invaluable help for a student who went on to take a materials science course as part of an engineering curriculum.

## Organization of the Text

This textbook goes far beyond the much-touted but seldom-successful "integration" of descriptive chemistry and principles: Insofar as possible, it *unifies* the two by removing the distinction. Descriptive chemistry is first observed by the student (in labs or demonstrations), who then devises principles to explain the observations. The principles are then amplified and explained in the text and finally applied to predict more facts. Gone is the necessity of learning facts about the compounds of one element, or one group of elements, in isolation from theory or from the very similar facts about the compounds of the next element or group of elements. Instead, the text focuses at one time on *one particular reaction or property type of one class of inorganic compounds* (for example, hydrolysis reactions and hydrolysis products of the halides of the elements begin Chapter 2). This organization is modeled after that used so successfully in organic chemistry to organize descriptive organic chemistry, which can no longer be distinguished from the principles of organic chemistry.

Chapter 1 reviews the basic periodic properties of atoms, ions, and molecules that will be used to explain and predict inorganic chemical properties throughout the text—i.e., valence-electron configurations, Lewis dot structures, trends in Pauling electronegativities, ionic radii, and common oxidation numbers. Similar review topics not needed until the latter part of the text are introduced later. For example, see net ionic equations in Chapter 3, electrochemistry in Chapter 5, ionization energies and electron affinities in Chapter 6, and VSEPR in Chapter 7. In my experience, teaching at three quite different schools, students often fail to grasp these topics when they are presented in the general chemistry course, either because not enough time is devoted to them or because students really need to have a better background in inorganic chemistry than they can have in their first year. Of course, these sections may be skipped if the students at a particular school already demonstrate a satisfactory understanding of this material.

The principles used in Chapters 2 through 10 to predict and explain descriptive chemistry are as versatile and simple as possible and are based on periodic variations in the fundamental properties reviewed in Chapter 1. In Chapter 11, we take the theory to a deeper level of sophistication, explaining these periodic variations of fundamental atomic properties in terms of orbital shapes, screening effects, Slater's rules, and even relativistic effects. Instructors who prefer to cover all the theory before proceeding to the study of chemical reactions and properties may wish to proceed directly from Chapter 1 to Chapter 11; I have written the text with this possibility in mind. Other instructors (for example, those who teach a lab with the course) may prefer to get to chemical reactivity first and thus may prefer to take the chapters in numerical order. In this case, Chapter 11 should provide a bridge to subsequent, more theoretical courses in physical or inorganic chemistry; it relates theory and actual chemical reactions more thoroughly than do the typical introductory chapters of texts in those courses.

Chapters 2 and beyond cover the familiar types of chemical reactions in the following order: acid-base and precipitation reactions in Chapters 2 through 4, oxidation-reduction reactions in Chapters 5 and 6, and coordination (Lewis acid-base) chemistry in Chapters 7 and 8. The different classes of inorganic compounds are introduced gradually, the more common forms first: hydrated ions, oxides and hydroxides, and oxo anions in Chapter 2; soluble and insoluble salts in Chapter 3; the elemental forms in Chapter 5; complex ions in Chapter 7; halides, nitrides, and sulfides in Chapter 9; and hydrides and simple organometallic compounds in Chapter 10. I have endeavored to avoid the out-of-sequence introduction of principles and compound and reaction types (for example, statements such as "As will be seen in later chapters ..."); thus, only the acid-base reactions of hydrated ions, oxides and hydroxides, and oxo anions are discussed in Chapter 2. The precipitation reactions of these species are left to Chapter 3, redox reactions to Chapter 5, and coordination chemistry to Chapters 7 and 8. This kind of logical development of concepts is impossible, of course, in a text that presents the chemistry of the elements group by group or period by period.

In Chapter 2, we develop a very fundamental skill—that of classifying the relative acidity and basicity of different positively and negatively charged species. This skill is discussed quite carefully and thoroughly at this point, because its use is important throughout the remainder of the book. The student may find that Chapters 2 and 3 are the most difficult in the book. On the other hand, the effort spent mastering these concepts will be amply rewarded in later chapters.

Chapter 12 is a brief one, covering one miscellaneous but significant topic (the abundance and costs of the elements) and helping the student to sum up, integrate, and apply the principles found throughout the book.

Many educators have noted that descriptive inorganic chemistry is most interesting when observed in the laboratory, because many inorganic reactions are very spectacular and colorful; hence *optional* experiments are included in Chapter 13. My former students strongly recommend that these be performed, but this may not be possible at many schools. Alternatively, most of these experiments can be done as lecture demonstrations to be discussed immediately afterward by the class. If neither of these approaches is possible, the experiments can be read by the student after completing the appropriate chapter. If the principles of the chapter are truly understood, the student should be able to predict what will happen in the experiment (i.e., to dry-lab it). These experiments, of course, can be completely omitted.

As is usual for modern textbooks, many instructors will find that not all topics in this book can be covered. The chemistry of the alkoxides, sulfides, nitrides, hydrides, and organometallics is covered in the last half of Chapter 9 and in Chapter 10; the coverage, however, is somewhat less fundamental than that found earlier in the book and thus can be omitted. As mentioned before, Chapter 11 is a bridging chapter that could profitably be covered at the beginning of a theoretical inorganic course. There are a few sections that are not closely related to the others; they have been included principally because they include topics suggested for coverage by the American Chemical Society. These could be omitted with no loss of continuity (details are provided in the instructor's manual).

## Acknowledgments

The approach in this textbook has been carefully developed and tested over an eight-year period, both at a selective liberal-arts school (St. John's University in Minnesota) and at a regional state university that includes graduate students (Middle Tennessee State University), with equal success found with the very diverse types of students at the two institutions. I wish to thank Thomas Jeffries, who class-tested the approach at Campbellsville College in Kentucky. Given the diversity of setting in which inorganic chemistry is taught in this and other countries, I know that further improvements are both possible and desirable and would very much appreciate hearing from teachers and students who use this book.

Special acknowledgments are due two individuals. One is a former student at St. John's University, Tim Sloan. While teaching from a theoretical text, we happened to cover the hard and soft acid-and-base theory. Tim exclaimed in astonishment, "You mean that you can actually predict what will happen in an *inorganic* reaction?" This alerted me to the great fallacy of traditional teaching of both theoretical and descriptive inorganic chemistry, a fallacy that keeps the two far apart. The other individual is a former teaching colleague at Northland College, Dave Whisnant (now at Wofford College). His innovations in laboratory and lecture teaching provided me with the idea on how the unification of theoretical and descriptive inorganic chemistry might be carried out.

I am also indebted to the following reviewers who provided useful and insightful comments on an earlier draft of the manuscript: John Alexander (University of Cincinnati), Alan Bates (Southeastern Massachusetts University), Henry Bent (North Carolina State University), Derek Davenport (Purdue University), Owen Faut (Wilkes Colleges, Pennsylvania), James Finholt (Carleton College, Minnesota), John Frey (Northern Michigan University), Russell Geanangel (University of Houston), Forrest Hentz (North Carolina State University), Albert Herlinger (Loyola University, Pennsylvania), Thomas Kallen (State University of New York—Brockport), Stanley Manahan (University of Missouri), and Wayne Wolsey (Macalester College, Minnesota).

I also wish to thank Middle Tennessee State University for providing me with time for writing; Dan Scott for his support as my department chairman; Kevin Benner, Khalid Shadid, Kathy Shelton, and Terry Hammonds for critically checking certain sections; Bob Jones and Geoffrey Hull for assistance with licensing questions; and Sue Ewing of Brooks/Cole for her constant encouragement. And finally, I thank my wife, Marlys, and Stacey Jones for their help in proofreading.

*Gary Wulfsberg*

# Contents

Contents

Contents

# CHAPTER 6                                                              180

# Properties of the Elements Themselves                                  180

# CHAPTER 7                                                              226

# Coordination Compounds and the Lewis Acid-Base Concept                 226

## CHAPTER 8                                                                266

## The Hard and Soft Acid-Base (HSAB) Principle and Its Applications        266

## CHAPTER 9                                                                306

## The Halides, Nitrides, and Sulfides of the Elements                      306

# APPENDIX C 441

# Frequently Used Tables 441

# Inorganic Chemistry: The Periodic Table and the World We Live In

## 1.1

## Why Study Inorganic Chemistry?

About one in ten professional chemists is an inorganic chemist, but *all* chemists and many other scientists must work with inorganic compounds—in the laboratory, in the field, or in theory. This book is intended to provide the fundamental facts and principles of inorganic chemistry needed by chemists and other scientists. It is also intended to show some applications of these principles in the different fields of chemistry and in several allied sciences.

Among chemists, organic chemists have always relied on inorganic reagents to carry out syntheses; this trend is increasing as organic synthesis turns more and more to the use of specific transition-metal catalysts (Chapter 7) and organometallic compounds (Chapter 10). Analytical chemists are often concerned with the detection and quantification of elements other than carbon and often use chelating ligands (Chapter 7) of appropriate hardness or softness (Chapter 8) to concentrate and detect metallic elements. Physical and theoretical chemists are concerned with measuring or calculating the fundamental properties of inorganic and organic substances. Modern biochemists are becoming increasingly aware of the critical role played in living systems by metal ions (Chapter 7). And there are many times when any chemist must make up a solution of a new type of inorganic reagent, modify a synthesis, detect an element in a new form, or study the properties of a different type of inorganic compound. At this point the chemist needs to have some ability to anticipate the properties of inorganic compounds he or she has not dealt with in the past. This book is designed principally to enable any chemist to develop an understanding of and an ability to anticipate the chemistry of the elements and their most common compounds.

Academic chemists are not the only ones who deal in a nonroutine way with the compounds of elements other than carbon. Most of the largest-volume industrial chemicals (Chapter 12) are inorganic compounds. Manufacturing processes are discussed in several chapters: oxo acids and glass in Chapter 4; nonmetals and metals in Chapter 5; inorganic polymers in Chapters 9 and 10. We need to know something about the processes for manufacturing these chemicals to understand their pollution aspects, for example, or their dependence on a raw material produced only in some politically unstable or hostile country. We need to know inorganic chemistry to be able to invent manufacturing processes that use other raw materials or that produce fewer pollutants (such as the air pollutants discussed in Chapter 4 and the water pollutants discussed in Chapter 8). Industrial organic chemistry is currently investing heavily in the study of organometallic transition-metal catalysts (Chapter 7), which offer the promise of producing industrial organic chemicals in high yield and purity, thus eliminating unwanted and often toxic byproducts that have often been dumped into the environment. All chemists need to be concerned with the fire or explosion hazards presented by some inorganic compounds (Chapter 5).

Many scientists and engineers who do not even consider themselves to be chemists must also deal with inorganic chemicals in a safe and insightful way. Biologists may have to make up a solution of an inorganic reagent and may need to anticipate whether the reagent will be toxic to their organism (Chapter 8). Certainly they need to know whether the material will explode on contact with water, or will fail to dissolve, when they try to make up the solution! Environmental and aquatic chemistry (Chapters 2, 3, 4, 5, and 8), toxicology and medicinal chemistry (Chapter 8), industrial chemistry (Chapters 4, 5, 7, 9 and 10), chemical safety (Chapter 5), geochemistry (Chapters 4 and 8), and materials science and solid-state physics and chemistry (Chapters 3, 4, 6, 9, and 10) all deal with a wide variety of inorganic compounds and need ways to anticipate the properties of new inorganic compounds. Chemists who develop a good fundamental understanding of the facts and principles of inorganic chemistry not only make better chemists, but they also may find opportunities to contribute to the fields allied with traditional chemistry.

## 1.2

## Understanding and Predicting Descriptive Inorganic Chemistry

This book deals with **descriptive inorganic chemistry**: the physical properties and chemical reactions of the 108 known elements and their countless compounds. Rather than trying to memorize this vast body of information, inorganic chemists refer constantly to the **periodic table** and try to understand chemical reactions and properties by relating them to the way a small number of atomic properties of the elements vary throughout the periodic table. Many of the most important properties of the elements and their compounds can be adequately understood and often predicted using atomic properties that are already familiar to you from General Chemistry: electronegativity, ionic charge, atomic radius, electron configuration, and the like. In the remainder of Chapter 1 we will review these properties and their periodic trends. Some of these topics are treated in slightly different ways in different general chemistry textbooks; hence we suggest that you look through this chapter even if you have had a good background in these topics in your general chemistry course.

Ultimately all the familiar chemical properties of the atom, as well as the trends in electronegativities, atomic radii, electron configurations, and so forth, should be explainable in terms of our quantum-mechanical model of the atom. This will be the objective of Chapter 11 of this book, which we hope will serve as a bridge to the more extensive treatments of the atom you will encounter in your physical chemistry course and in subsequent inorganic courses.

We feel that it is difficult to understand physical explanations of chemical phenomena that a person has never observed. Chapter 13 contains laboratory experiments that your instructor may have you try or may demonstrate and have you discuss before you begin many of the chapters. This will provide you with an opportunity to observe several examples of the type of chemistry that will be discussed in that chapter. Or the instructor may want to substitute other experiments or demonstrations, refer you to inorganic experiments that you performed in your general chemistry class, or simply let you read the experiments on your own and try to imagine the results. In order to enhance your enjoyment of the experiments, we have written them in such a way that you may be able to discover a periodic trend to the reaction tendency being investigated before this tendency is discussed in lecture or in this text. Rest assured that there are often many ways of interpreting periodic trends and organizing and explaining them; if your discovery does not match the interpretation given in the text it may just mean that you are more creative than the author. (The test would be, of course, in seeing how many other physical and chemical properties you could predict using your idea.) When it comes time to take an exam over the actual physical and chemical properties of the elements and their compounds, we hope that you will discover to your pleasure that you are able to reason your way to the answers, perhaps using principles that you discovered yourself.

## 1.3

## The Periodic Table

The periodic table was originally invented in a form such that if the elements were listed in order of an increasing fundamental property (originally atomic weight, now atomic number—i.e., the number of protons in the atom), very similar chemical properties would recur periodically through the list. When the properties first recurred, the horizontal listing of the elements would be disrupted, and a new **period** of elements would be begun, with the chemically similar elements listed below each other in a **group**.

To do this with the known elements, it is necessary to set up seven horizontal periods of elements. As it turns out, however, these periods become considerably longer as atomic numbers get larger, so that the periodic table becomes inconveniently wide (32 groups at the bottom); hence one usually sees a shorter form of the periodic table, with the last groups to appear (those of the "lanthanides" and "actinides") separately listed at the bottom. We will usually resort to this shorter form, too, but in Table 1.1 we present a fully expanded version of the periodic table [1].

As we can see by its shape, the periodic table has four regions; some of these are named differently by different authors. Both the two groups at the far left and the six groups at the far right are often called the representative elements. To distinguish them we will call the two groups at the far left the **s-block elements**. According to the latest recommendations of the International Union for Pure and Applied Chemistry (IUPAC), these groups should be numbered 1 and 2. The six groups at the far right we shall call the **p-block elements**. The IUPAC recommends that these be numbered 13 through 18, since there is much confusion resulting from the American tradition of numbering them IIIA through VIIIA while Europeans number them IIIB through VIIIB. At the time of this writing, however, it is far from certain that the IUPAC recommendation will become practice, so we will number them using both the IUPAC and the American systems: i.e., the boron group will be numbered 13 (IIIA).

The low, long region next to the s-block elements has a plethora of names: "rare-earth elements," "inner transition elements," or "lanthanides and actinides." For simplicity we will call these elements the **f-block elements**. They are not usually numbered at all, but for figuring valence electron configurations it is useful to do so; for this purpose we will number them from 3f to 16f. Although the third block of elements is universally recognized as the "transition elements," for consistency we will call them the **d-block elements** and will number their groups from 3 to 12, as is recommended by the IUPAC.

Note the general shape of the periodic table: Every *two* periods a new block of elements is introduced, just after the s-block. This is due to occur again in the eighth period—between the s- and f-blocks there should be a new block of elements that would be called the **g-block elements**. Note also that each new block of elements is wider than the ones introduced before: The s-block of elements is 2 elements wide, the p-block is 6 elements wide, the d-block is 10 elements wide, the f-block is 14 elements wide, and the g-block is expected to be 18 elements wide.

## 1.4

## Characteristic Valence Electron Configurations

Most of the electrons of most of the elements are found in orbitals of such low energy that they cannot be shared, lost, or otherwise altered during chemical changes. These **core electrons**, as we will see at the end of the book, are not entirely without influence on chemical properties, but we will find it most useful to concentrate our attention on the electrons in the one or two highest-energy occupied orbitals of atoms. These **valence electrons** are the ones symbolized by the dots in Lewis (electron-dot) symbols used in general and organic chemistry (Figure 1.1).

**1.** *Count the total number of valence electrons in the molecule or ion.* To do this, add up the group numbers for each atom in the molecule; add one extra electron for each unit of negative charge, and deduct one electron for each unit of positive charge. Thus in $PCl_3$, P, coming from Group 15/VA, contributes 5 valence electrons. Each Cl (Group 17/VIIA) contributes 7 valence electrons, for a total of 26 valence electrons. In $SO_3^{2-}$, both the S and O atoms are in Group 16/VIA, so these four atoms contribute 24 electrons. But to this we must add the two electrons that are responsible for the $-2$ charge on the ion, to give a total of 26 valence electrons. Similarly, $SO_3$ has 24 valence electrons, $XeO_4^{2-}$ has 34, $SbCl_5$ has 40, and $SbCl_5^{2-}$ has 42 valence electrons.

**2.** *Draw a skeleton structure of the molecule or ion, joining atoms with lines to represent shared electron pairs,* i.e., single bonds. Unless we know otherwise, we will assume that the unique atom in the formula is the central atom of the molecule or ion. Now *deduct two electrons for each bond drawn.* This gives us the number of electrons still available to complete the Lewis structure. Thus (a) $26 - 6 = 20$ electrons are still available for $PCl_3$, and (b) $24 - 6 = 18$ electrons remain for $SO_3$. The count of valence electrons remaining in the other examples are (c) 20, (d) 26, (e) 30, and (f) 32.

**3.** *Draw a circle around each nonhydrogen atom, including all bonds to it. See how many electrons each atom still needs to achieve an octet of electrons; total these numbers.* (We will say that an atom such as P in $PCl_5$ that starts with more than an octet of electrons needs *zero* additional electrons.) Thus, as shown in Figure 1.2, the total number of electrons *needed* by the atoms of each structure are (a) 20, (b) 20, (c) 20, (d) 24, (e) 30, and (f) 30.

**4.** *If the number of valence electrons needed (step 3) matches the number of electrons available (step 2), only unshared electron pairs are needed to complete the octets of each atom; fill these in.* This is true in examples (a), (c), and (e). In each of these the outer Cl or O atoms gets three additional unshared electron pairs; the P atom in $PCl_3$ also needs and gets one unshared electron pair (Figure 1.3). (The circles should now be omitted.)

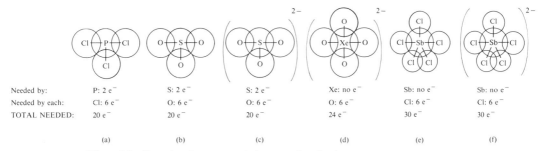

| | | | | | |
|---|---|---|---|---|---|
| | (a) | (b) | (c) | (d) | (e) | (f) |

Needed by:    P: 2 e⁻    S: 2 e⁻    S: 2 e⁻    Xe: no e⁻    Sb: no e⁻    Sb: no e⁻

Needed by each:    Cl: 6 e⁻    O: 6 e⁻    O: 6 e⁻    O: 6 e⁻    Cl: 6 e⁻    Cl: 6 e⁻

TOTAL NEEDED:    20 e⁻    20 e⁻    20 e⁻    24 e⁻    30 e⁻    30 e⁻

**Figure 1.2** Counting electrons needed to complete Lewis structures.

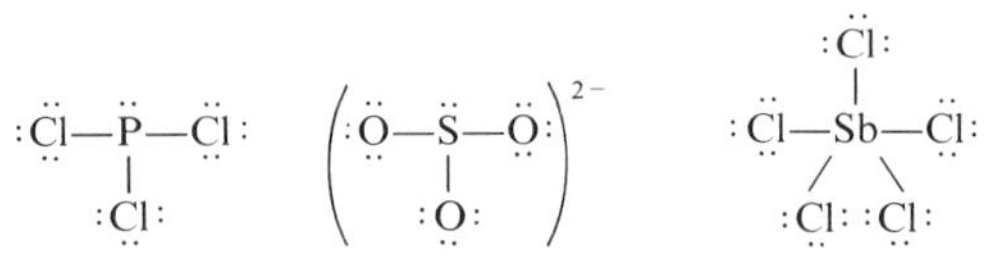

**Figure 1.3** Completed Lewis structures for $PCl_3$, $SO_3{}^{2-}$, and $SbCl_5$.

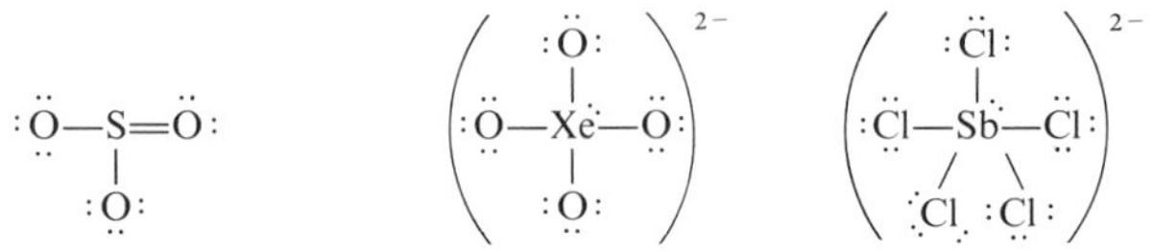

**Figure 1.4** Completed Lewis structures for $SO_3$, $XeO_4{}^{2-}$, and $SbCl_5{}^{2-}$.

**5.** *If more electrons are needed (step 3) than are available (step 2), make multiple bonds by adding one extra line between pairs of atoms for each two electrons you are short.* If there is a choice of different outer atoms to which the multiple bonds may be drawn, it is best to draw them to O, N, C, or S; they cannot be to H. *Then complete the placement of the remaining electrons as unshared electron pairs.* $SO_3$ needs 20 more electrons but 18 are available; hence one S=O double bond is drawn in and the remaining 16 valence electrons are distributed on the oxygen atoms as unshared electron pairs (Figure 1.4).

**6.** *If there are too many electrons to locate, put the extra electrons on the central atom,* which will then have to violate the octet rule. The central atom must be from the third period of the periodic table or below. Thus the dot structures of $XeO_4{}^{2-}$ and $SbCl_5{}^{2-}$ are completed by adding one pair of unshared electrons on Xe and one on Sb.

**7.** Finally, *check your answer!* Be sure that (a) the total number of electrons shown agrees with the number calculated in step 1 and (b) each atom has an octet of electrons (except hydrogens with two, and, if necessary, central atoms from the third period or below, which may have more than eight).

## 1.6

## Periodic Trends in Electronegativities of Atoms

When atoms of different kinds form a covalent bond, we expect that the attraction of the two nuclei for the shared electron pair will not in general be the same and that the electron pair will, on the average, be closer to one of the nuclei. Pauling sought a measure of the relative abilities of the different atoms to attract such bond electrons to themselves, which he called the atom's **electronegativity**. It is not clear how such a property of atoms is to be measured experimentally; different people have proposed different measures of electronegativity. Pauling himself examined bond energies of molecules (we will look into this later in the book) and obtained a table of what we now call **Pauling electronegativities**; an updated version of these values is included as Table A. (Throughout this text we will refer to Table A to find the Pauling electronegativity of an element in order to make predictions about the chemistry of that element. Consequently, Table A and three other similarly useful tables have been placed inside the front and back covers for convenient reference.)

Despite the ready availability of the tables inside the covers, it is still worthwhile to note the general periodic trends in electronegativities so that we can quickly predict trends in chemical reactivity even when the tables are not available. We recall that the overall trends in any scale of electronegativity are (1) electronegativities of atoms generally *increase* from left to right as we go *across* the periodic table, and (2) electronegativities generally *decrease* as we go from top to bottom of the periodic table.

Before reading further, it is worthwhile to examine Table A period by period and row by row to see how exactly these statements seem to hold in practice. (The method of measuring these values has an uncertainty of $\pm 0.05$ units, so changes of less than this may not be meaningful.) There are parts of the periodic table in which these trends disappear or even reverse. If Pauling electronegativities strongly influence chemical properties, we can expect to find some anomalous chemical trends in these areas. Some trends we note on going *across periods* are that (1) periods 2 and 3 are "well behaved"; (2) electronegativities rise more slowly and sometimes decrease across the *d*-block elements in Periods 4 to 6; (3) electronegativities scarcely vary at all across the *f*-block elements.

Going *down groups* of the *s*-block elements, we see that all goes well; but going down groups of the *f*- and *d*-block elements, we often note significant increases in electronegativites. Going down Groups 13/IIIA and 14/IVA, we notice very unusual "zigzag" trends that seem to dampen out and disappear in the later *p* groups. In the final chapter of this book we will try to find explanations for these anomalies; for now we will take them as given and see if they correlate with observed chemical properties other than bond energies.

## 1.7

## Oxidation Numbers of Atoms and Charges of Ions

### 1.7.1    Numerical Rules for Calculating Oxidation Numbers

The following set of rules for calculating oxidation numbers works as long as the specified elements do not have unusual oxidation numbers and provided the species does not have more than three elements in it, or if it has more than three, that it can be separated into ions of three or fewer elements each. The rules are meant to be applied *in order*, with any contradictions being resolved by the earlier rule overriding the later rule.

**1.**    The sum of the oxidation numbers of all the atoms in a molecule equals zero; the sum in an ion equals the charge on the ion. (Oxidation numbers are on a per-atom basis, and before being summed must be multiplied by the number of atoms of that element present in the molecule or ion.)

**2.** The oxidation number of the least electronegative element may be set if it is in Group 1, 2, 3*f*, or 3, or is Al (in Group 13/IIIA). The oxidation number is numerically equal to the group number ($+1$, $+2$, or $+3$, respectively). (These elements follow a noble gas by no more than three elements and predictably form ions with the above charges).

**3.** The oxidation number of the most electronegative element may be set as $-1$ if it is in Group 17/VIIA or $-2$ if it is in Group 16/VIA.

**4.** The oxidation number of the remaining element is called $x$. The oxidation number of each element is multiplied by the number of atoms of that element present and is summed as indicated in step 1. The resulting equation is then solved for $x$.

EXAMPLE    Compute the oxidation numbers of each element in the ion $H_2IO_6^{3-}$.

Solution    The sum of all oxidation numbers in this ion must equal the ion charge, $-3$. First, we assign the oxidation number of the least electronegative atom, H, as $+1$. Then we assign oxygen the oxidation number $-2$, and we assign I the unknown oxidation number $x$. Taking into account the number of atoms of each type, we have $2(+1) + x + 6(-2) = -3$. We solve the equation for $x$: The oxidation number of I must be $+7$.    □

This procedure gives the average oxidation number of all atoms of a given element in a given species but will not differentiate atoms of an element in different bonding environments. Consequently it can give fractional oxidation numbers. Also, especially in the realms of coordination and organometallic chemistry, these numerical rules are inadequate for assigning oxidation numbers. Hence we also need a more versatile approach to the assignment of oxidation numbers.

### 1.7.2    Oxidation Numbers from Lewis Structures

The degree to which a bond electron pair is unequally shared between two atoms of differing electronegativity is difficult to measure. If the two atoms are of extremely different electronegativities, the electron pair may reside completely on the more electronegative atom, giving a positive and a negative ion (**ionic bonding**). If the two atoms are identical, the electrons should be equally shared in a pure **covalent bond**. Most cases fall in between, however, with a partial negative charge being assignable to the more electronegative atom and a partial positive charge to the less electro-negative atom. In such cases, rather than attempt to measure the actual magnitude of the partial charges in the molecule, for bookkeeping purposes we may decide to assign the bond pair of electrons completely to the more electronegative atom—i.e., we pretend that the bonding is ionic. After doing this to all the bonds in the molecule, we have produced a number of (sometimes imaginary) positive and negative ions. The imaginary charge that we have produced on each of the atoms of the molecule may be assigned as the oxidation number of that atom.

To determine oxidation numbers in this manner, we first draw the Lewis dot structure of the molecule. Second, we move the electrons of each bond onto the

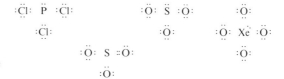

**Figure 1.5** Assignment of bond electrons to the more electronegative atom.

**Figure 1.6** Original Lewis symbols of elements appearing in Figure 1.5.

more electronegative atom of that bond (if the two atoms are equal in electronegativity, we move one electron onto each atom). Third, we compare the remaining number of valence electrons on each atom with the number of valence electrons found on a free atom of that element (i.e., its group number). For each electron that has been lost in forming the molecule or ion, we add 1 to the oxidation number of the element. For each electron that has been gained, we contribute a $-1$ to its oxidation number.

Let us take as examples the first four molecules for which we drew Lewis dot structures in Section 1.5. Since Cl is more electronegative than P, and O is more electronegative than either S or Xe, we artificially assign the bond electrons to the outer atoms, as shown in Figure 1.5.

Now we can compare the number of valence electrons left on each atom with those on the original Lewis symbol of each atom, shown in Figure 1.6. Comparing the two, we see that oxygen has gained two electrons in each case and thus has an oxidation number of $-2$ in each of its compounds. Likewise, each Cl in $PCl_3$ has gained one electron, so Cl has an oxidation number of $-1$. The other atoms have lost electrons and thus have positive oxidation numbers (for P in $PCl_3$, $+3$; for S in $SO_3$, $+6$; for S in $SO_3^{2-}$, $+4$; for Xe in $XeO_4^{2-}$, $+6$).

The oxidation numbers we get from operations like these represent another periodic variable that may reflect chemistry of the compounds of the elements. Most elements are capable of exhibiting more than one oxidation number. In Chapters 5 and 6 we will be concerned with analyzing in detail the periodic trends in the various possible oxidation states of the elements and with the chemical consequences of that variation. Until then we will usually assume that, in our reactions, the oxidation number of an element does not change. In the early chapters we will mainly be concerned with elements having *positive* oxidation numbers. In Table B (inside the front cover) for each element we list one of its most common positive oxidation states when in aqueous solution. Many oxidation numbers correspond to the preferred positive ionic charges of the elements, another periodic variable that is useful in understanding descriptive chemistry.

### 1.7.3 Periodic Trends in Common Oxidation Numbers

For later use in predicting trends in chemical reactivity, it is useful to note some of the periodic trends in common oxidation numbers. For example, negative oxidation numbers occur for *p*-block elements when their atoms acquire enough electrons to form anions (or form molecules with a less electronegative atom such as hydrogen) with an octet of valence electrons. For the elements of Group 14/IVA and beyond, *the minimum oxidation number equals the (old style) group number minus eight—e.g., oxidation number −4 for Group 14/IVA.*

A necessary trend is that the *highest oxidation number of an element cannot exceed its group number*, since it cannot, by the definition of valence electrons, lose or share anything beyond its valence electrons. A second trend is that *no oxidation number in any known compound exceeds +8* (in fact, no common oxidation number exceeds +7). Evidently, any element is electronegative enough to resist pulling away or sharing more than seven or eight of its own electrons with other atoms. As a consequence, *toward the right in any block of elements, the maximum oxidation number is less than the group number*—exceptions being the *s*-block elements and sometimes Cl, Br, I, and Xe in the *p*-block elements.

Toward the right in blocks, we observe the following common positive oxidation numbers: (1) In the *p*-block, we often find an oxidation number equal to the group number minus two. This corresponds to the retention of one unshared pair of electrons by the element, as we observed in the Lewis dot structures of $PCl_3$ and $SO_3^{2-}$. (2) In the *d*-block, we often find oxidation numbers (or ionic charges) equal to +2, +3, or +4. (3) In the *f*-block, only at the far left does the oxidation number reach the group number. The most common oxidation number for most of the *f*-block elements is +3.

Generally speaking, common positive oxidation numbers do not change as we go down a group of the periodic table. But when they do change, we find the following vertical trends: (a) They are highest in the *middle* of a group in the *p*-block of elements; (b) they are highest at the *bottom* of a group in the *d*- and *f*-blocks.

### 1.7.4 Oxidation Numbers in Nomenclature

Since most of the elements outside of the *s*-block have more than one accessible positive oxidation number (oxidation state), when we name compounds of that element we need to designate the oxidation state involved. We do this by using the corresponding Roman numeral after the name of the element. Thus we refer to $BiCl_3$ as bismuth(III) chloride; the $Bi^{3+}$ ion is called the bismuth(III) ion. Since the elements of the *s*-block (or others that follow the noble gases by three or fewer positions in the periodic table) have only one positive oxidation state, the number in parentheses is unnecessary for them and is not used in practice.

## 1.8

## Periodic Trends in Ionic Radii

Another important periodic variable that is useful in explaining and predicting the chemistry of the elements is the **radii** of their atoms and ions. The radii of atoms vary depending on the type of bonding: Thus there are covalent radii, metallic radii, van der Waals radii (when no bonding is occurring between atoms that touch), and ionic radii. In the next few chapters we will principally be concerned with the chemistry of ionic species, so for now we will only examine the periodic trends in cationic radii. The best available measurements of ionic radii, the Shannon-Prewitt values, are tabulated for cations in Table C, inside the back cover of the book.

Horizontal trends in ionic radii may be examined in either of two ways. We may select a series of ions that are **isoelectronic**—that is, a series of ions of differing atomic number that have the same electron configuration. To examine such a trend, we may start with a Group 1 cation (with one more proton than its total number of electrons), then add protons, producing cations in later groups with increased charge due to increased numbers of protons but with the same number of electrons. For example, we may compare the cationic radii of the following isoelectronic ions in Period 5: $Rb^+$, $Sr^{2+}$, $Y^{3+}$, $Zr^{4+}$, $Nb^{5+}$, $Mo^{6+}$, $Tc^{7+}$. We note that as the positive charge on the nucleus increases, the attraction for the 36 electrons becomes stronger, resulting in a strong *contraction of the ion with increasing positive charge*.

We observe a second horizontal trend by holding the charge on the cation constant while we move from left to right on the periodic table. The most extensive such series of cations is in the $f$-block of elements, among the $+3$ ions of the lanthanides (elements of Period 6). We see that even though the overall surplus of protons over electrons remains constant, the ions still *decrease in radii*, albeit much more slowly. The same trend may be seen in a given period among the $d$-block elements of constant charge (i.e., $+2$, $+3$, or $+4$), although some anomalies may be noted.

Among ions that form more than one cation, we find that with increasing positive charge, the radius of the cation decreases (see, for example, the data for V, Cr, or Mn in Table C). By removing electrons, we increase the surplus of protons over electrons and reduce the tendency of electron clouds to expand due to repulsion of their like charges.

Finally, taking some group such as Group 1, we may examine the vertical trends in atomic radii. Since the principal quantum number of the valence orbital steadily increases down a group, we expect to and do find that the radius of the ion increases in size. This trend holds quite well among the $s$-block, $p$-block, and (to a lesser extent) the $f$-block elements. It is worth noting the anomalous trend among the $d$-block elements: In most $d$-block groups, there is a normal increase in size on passing from Period 4 to Period 5, but surprisingly there is little or no increase on passing from Period 5 to Period 6. We will explore the reasons for these trends and anomalies in Chapter 11; for now we will take them as given and proceed to investigate what chemical consequences they have in different types of reactions and properties of different classes of compounds, beginning in Chapter 2 with elements in positive oxidation states (i.e., as real or imagined cations).

## 1.9

## Thermodynamic Measures of Tendencies to React

In the remaining chapters of this book we will look at certain types or classes of reactions (such as of metal ions with water, or of compounds that melt at room temperature). We generally observe that with some of the 108 elements, the reaction in question does proceed to give (an equilibrium mixture favoring) products, while with others it does not. It will help us in understanding the causes of the periodic trends we observe if we can characterize these tendencies quantitatively, using chemical thermodynamics.

A chemical reaction that does tend to proceed to give a preponderance of products is said to exhibit a *negative* **Gibbs free energy change, $\Delta G$** ($\Delta G < 0$). If we characterize this tendency in terms of the concentrations of products and reactants, we find that the reaction exhibits an **equilibrium constant, $K_{eq}$**, that is *greater than one* ($K_{eq} > 1$). If we characterize this tendency in an electrochemical cell, we find that a reaction that tends to proceed generates a positive voltage or **electromotive force, $E$** ($E > 0$). Naturally, since all three of these parameters measure the same tendency, they can be interconverted by equations (which will be given later in the text as needed).

A chemical reaction that does *not* tend to proceed to products shows the opposite characteristics, that is, $\Delta G > 0$, $K_{eq} < 1$, and $E < 0$. Such a reaction tends to go in the *reverse direction*: If we mix the compounds listed as products on the right side of the equation, they will tend to react to give the compounds shown on the left side of the equation.

In the special case of a mixture of products and reactants in which each product and reactant is present in its standard state and at unit activity (or concentration), we add a superscript to two of the symbols, namely, $\Delta G°$ and $E°$. These are the standard conditions at which it is convenient to compare the compounds of different elements, for example. The standard state of an element or compound is the phase that is usual at room temperature and atmospheric pressure (gas, liquid, or solid, in its most common allotrope or other modification). We will discuss the difference between activities and concentrations later; generally we will ignore the difference and refer to concentrations. The standard concentrations of substances are taken as pure for solids and liquids, at a pressure of 1 atmosphere (atm) for gases, and at a concentration of 1 molar ($M$) for solutes. Reaction tendencies generally depend on the concentrations of the species involved, so our comparisons may not hold strictly—for example, under environmental conditions, pollutants are seldom if ever present at anywhere near these concentration levels. However, there are equations (such as the Nernst equation) by which the comparisons can be adjusted to apply to more realistic concentrations.

How *quickly* a reaction proceeds to the favored side of the equation is something we cannot predict from these measurements. It often happens that a reaction that shows a strong tendency to go to products is hindered by a very high *activation energy* that must first be provided; in the absence of this initiating energy, the reaction may take a very long time to occur.

A very important question for us to ask ourselves as we observe these varying reactions is, Why is the tendency to react greater in some cases than in others? This question can be answered on many levels. On the level of chemical thermodynamics two factors are connected with the tendency of a reaction to proceed to products (to be spontaneous) or not to proceed to products (to be nonspontaneous).

The first of these factors is that *products are more likely to be favored if they are lower in energy than are the reactants.* For example, if the products of a reaction contain stronger covalent bonds than the reactants, the reaction will be more likely to favor products. If the reaction is carried out in an open container (at constant pressure), the energy change can be measured as an evolution of heat by the reacting system and is known as the **enthalpy change** of the reaction, $\Delta H$ or $\Delta H°$. A system that evolves heat (and is thus more likely to proceed to products) is said by convention to have a *negative enthalpy change* ($\Delta H < 0$).

However, not all spontaneous reactions involve the liberation of heat by the reacting system. For example, ice tends to melt at room temperature, but a melting ice cube certainly does not heat its surroundings! There must be another reason for reactions to proceed, one that sometimes can override the tendency to go to lower-energy products. This is the tendency of reactions to proceed in the direction of products that are more *disordered* or *random*. It is measured thermodynamically as the **entropy change**, $\Delta S$, of the system. A system that proceeds toward greater disorder or a more random state is said by convention to exhibit a *positive entropy change* ($\Delta S > 0$).

We find that the amount of heat evolved by a reaction (its enthalpy change) and the degree of disorder produced (its entropy change) are not very much affected by changes in temperature, but reaction tendencies are affected. In particular, at low temperatures the tendency toward heat evolution usually determines whether the reaction will proceed, while at higher temperatures the tendency (or lack of it) toward increasing disorder becomes more and more significant. Thus the relationship between the two tendencies involves not only $\Delta H$ and $\Delta S$ but also the temperature $T$ (on the Kelvin scale). In particular, the free energy change, which indicates whether the reaction tends to proceed or not, is related to these three parameters by the equation

$$\Delta G = \Delta H - T\Delta S \qquad (1.1)$$

When we observe varying tendencies of a reaction type to occur for different elements or compounds, it will often be useful to determine whether the sign of $\Delta G$ is governed predominantly by the sign of the $\Delta H$ term or by the sign of the $-T\Delta S$ term. This can give us clues to answers to the question of why the reaction tends to go or not at a different level—that of a physical picture of molecules forming more stable bonds or being more randomly distributed in space, and the like. We usually (but not always) find that the $\Delta H$ term is larger in magnitude than the $-T\Delta S$ term at room temperature, so we usually concentrate our attention on factors such as the strengths of the covalent bonds involved. However, if we find the $-T\Delta S$ term to be larger in magnitude at room temperature, we are alerted to the fact that something is going on in the reaction that causes an unusual degree of disorder (such as the production of a large number of gas molecules).

On the other hand, the fact that we observe that a reaction tends to go in a certain direction at room temperature does not mean that it will show the same tendency at all temperatures. Indeed, at high enough temperatures the $-T\Delta S$ term must predominate because all complex gaseous molecules are unstable there and tend to decompose to large numbers of smaller gaseous molecules or atoms, which spread out over a larger volume of space (at constant pressure).

## 1.10

## Study Objectives

1.  In this (and subsequent chapters) know the definitions of the terms given in **boldface**. Many of these are found in the glossary, Appendix B.
2.  Know the long form of the periodic table so that given a short-form table and the symbol (or atomic number) of an element, you can write its characteristic valence electron configuration.
3.  Given the charge of a monoatomic cation or anion, write its valence electron configuration.
4.  Write the Lewis dot structure of a specified molecule or ion.
5.  Without referring to Table A, describe the characteristic horizontal and vertical periodic trends in Pauling electronegativities.
6.  Calculate the oxidation numbers of atoms in specified molecules or ions (a) working from Lewis dot structures and (b) using numerical rules. Give the name of an element or ion in a specified positive oxidation state.
7.  Without referring to Table B, describe the main trends in common oxidation numbers in different blocks of the periodic table.
8.  Without referring to Table C, describe the two main horizontal and the main vertical periodic trends in radii of ions.
9.  Extend the periodic table to include additional elements; extrapolate plausible valence electron configurations, electronegativities, ionic radii, and common positive oxidation numbers for these elements.

## 1.11

Exercises   Answers to questions preceded by an asterisk ($*$) are found in Appendix A of this book. Questions preceded by a dagger ($\dagger$) are answered in part in Appendix A.

*1.  Write the characteristic valence electron configurations of the following atoms or ions: **a.** Fr;   **b.** As;   **c.** Pt;   **d.** Dy;   **e.** $Bi^{3-}$;   **f.** $Bi^{3+}$;   **g.** $Ba^{2+}$;   **h.** $Pt^{2+}$;   **i.** $Dy^{3+}$;   **j.** Ge;   **k.** $Ge^{2+}$;   **l.** $Ge^{4-}$.

2.  Write the characteristic valence electron configurations of the following atoms or ions: **a.** Sr;   **b.** At;   **c.** W;   **d.** Bk;   **e.** $S^{2-}$;   **f.** $Tl^+$;   **g.** $Cs^+$;   **h.** $W^{4+}$;   **i.** $Bk^{3+}$;   **j.** Mn;   **k.** $Mn^{4+}$;   **l.** Pa.

*3.  Write the Lewis dot structures of the following molecules or ions:
    **a.** $CO_2$;   **b.** $NO_2^-$;   **c.** $NO^+$;   **d.** $H_2O_2$;   **e.** $CH_4$;   **f.** $H_2SiBr_2$;   **g.** $IF_5$;   **h.** $OsO_4$.

4. Write the Lewis dot structures of the following molecules or ions:
   **a.** HCN; **b.** $SO_3^{2-}$; **c.** $NO_2^+$; **d.** $XeF_4$; **e.** $H_2N—NH_2$;
   **f.** $SiH_4$; **g.** $ICl_4^-$.

†5. Assign oxidation numbers to each atom in the molecules and ions of
   **a.** question 3; **b.** question 4.

6. Working from Lewis structures, assign oxidation numbers in the following
   molecules or ions: **a.** $HCCl_3$; **b.** $NSF_3$; **c.** $C_2H_6$; **d.** $C_3H_7Cl$;
   **e.** $Fe(CO)_5$; **f.** $TeO_6^{6-}$.

*7. Working from Lewis structures, assign oxidation numbers in the following
   molecules or ions: **a.** HOF; **b.** $Si_2H_6$; **c.** $Si_3H_8$; **d.** $CSe_2$;
   **e.** $V(CO)_6^-$; **f.** cyclo-$(HN—BH)_3$; **g.** $ClCH_2—CH_2Cl$.

8. Without referring to Table B, describe the most common positive oxidation numbers to be expected in the following regions of the periodic table:
   **a.** $s$-block; **b.** left side of the $p$-block; **c.** right side of the $p$-block; **d.** left side
   of the $d$-block; **e.** right side of the $d$-block; **f.** left side of the $f$-block; **g.** right
   side of the $f$-block.

*9. List all the known chemical elements that have six valence electrons (and no
   more). Then, without referring to Tables B or C, circle those elements that
   commonly give up or share all six of those valence electrons with more
   electronegative elements to achieve an oxidation number of six. Give the
   atomic numbers of the next five elements that can be expected to have six
   valence electrons.

10. Without referring to Table C, arrange the following sets of atoms and ions
    in order of increasing radii:
    **10.1** $Cr^{6+}$; $Cr^{4+}$; $Cr^{2+}$; $Cr^{3+}$.
    **10.2** $Ra^{2+}$; $Mg^{2+}$; $Be^{2+}$; $Sr^{2+}$.
    **10.3** The ions of valence electron configuration $6s^0$ between Cs and Re.
    **10.4** The $f$-block $+3$ ions of Period 7.

11. Without looking at the relevant tables, describe: **a.** the main horizontal
    trends in Pauling electronegativities; **b.** the main horizontal trends in ionic
    radii; **c.** the main horizontal trends in common oxidation numbers; **d.** the
    main vertical trends in Pauling electronegativities; **e.** the main vertical trends
    in ionic radii; **f.** the main vertical trends in common oxidation numbers.

12. In what regions of the periodic table are there anomalous vertical trends
    in Pauling electronegativities? Are there corresponding anomalous vertical
    trends in these regions in ionic radii? Common positive oxidation numbers?

*13. Make reasonable predictions of the following:
    **13.1** The valence electron configurations of the elements of atomic number
          126, 144, and 162.
    **13.2** The Pauling electronegativities of the elements *directly below*
          **a.** U; **b.** Au; **c.** At.

**13.3** The common positive oxidation number for Rn; the element number 121; the elements directly below Np, No, Os, Pb.

**13.4** Using Table 1.4, the ionic radii for the cations corresponding to the common positive oxidation numbers and the elements in the previous question.

---

**Notes**

[1] There are a few elements that have such similar chemical and electronic properties that their placement into groups of the periodic table is somewhat arbitrary. Such a case is that of the pair of elements lanthanum (La) and lutetium (Lu), and the pair actinium (Ac) and lawrencium (Lr), each of which has an equal claim to resemble the early elements in Group 3, scandium (Sc) and yttrium (Y). As pointed out by Jensen (*J. Chem. Educ.*, 59, 634 [1982]), the metallurgical resemblance is much stronger for lutetium than for lanthanum, so we have adapted the metallurgist's convention of listing Lu (and by extension Lr) below Sc and Y. An important additional advantage of this is that the periodic table becomes more symmetrical, and it becomes easier to predict electron configurations.

[2] Perhaps it is best, in the *p*-block, just to think of the "1" in 13, 14, and so forth as serving the same function as the *A* or *B* in the old group numbers. Arguably, this could also be done for Group 12.

[3] In isolated (gaseous) atoms of 23 of the *d*- and *f*-block elements, there are slight discrepancies between our predictions and the observed electron configurations. Among these elements the valence $ns$, $(n - 1)d$, and (sometimes) $(n - 2)f$ orbitals are of quite similar energy, so frequently one or two of the valence electrons will "slip" from one orbital to another. This occasional irregularity is of no consequence whatsoever for the ordinary chemistry of the *d*- and *f*-block elements, which involves *ions* in solution, the solid state, or the metallic form of the element; these anomalies disappear when we consider the electron configurations of the atoms in these forms.

# Metal Cations and Oxo Anions
# in Aqueous Solution

Most of the elements are found in nature not as free elements but in the form of ions; hence we begin our study of the chemistry of the elements with some chemical properties of their ions. In this chapter we will investigate the interaction of some common ions of the elements (cations and oxo anions) with water and see how the periodic trends in these reaction tendencies can be related to the atomic properties reviewed in Chapter 1. We will then construct a physical model of what is happening during these reactions. Finally, we will apply these periodic trends and this model to a practical situation, predicting the forms that the different elements will take in unpolluted natural waters and those that are polluted with acid rain.

Your instructor may choose to have you begin your study of the chemistry of the ions of the elements with a laboratory investigation (or classroom demonstration and discussion) of the process of dissolving the cations of the elements (in the form of their chlorides) in water. This may sound trivial, but you will find some unexpected excitement in the process. You will find, upon analyzing the results, that even so simple a reaction involves some important chemistry.

You will find that the principles you derive apply to far more than just the reaction of a cation or an anion with the humble water molecule; ions react similarly with many other chemical species. The ways in which you begin looking at positively and negatively charged species in this chapter will be useful in subsequent chapters.

## 2.1

## Hydration of Cations

When writing chemical equations for reactions of ions in solution we often write ions as if they were simple particles in solution—e.g., we may write the sodium ion as $Na^+$ or perhaps as $Na^+(aq)$. But there are definite reactions between ions and the solvent water that produce what we call **hydrated ions**. Hydrated ions arise as a consequence of the polar nature of the water molecule. Since the oxygen atom

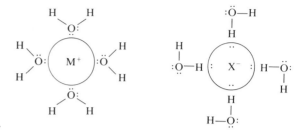

**Figure 2.1** A hydrated cation and a hydrated anion.

of the water molecule is much more electronegative than the hydrogen atoms, each H—O bond is a polar covalent bond in which the bond electrons are (on the average) closer to the oxygen atom than the hydrogen atom, giving rise to a partial negative charge on oxygen and a partial positive charge on hydrogen. Since $H_2O$ is not a linear molecule, it has a partially negatively charged end (its oxygen end) and a partially positively charged end (the hydrogen end). Since opposite charges attract, a positive ion (cation) placed in water surrounds itself with water molecules, with the oxygen ends inward toward the ion (Figure 2.1). Conversely, a negative ion surrounds itself with water molecules, hydrogen ends inward. These surrounded ions are what we call hydrated ions.

The attraction of opposite charges is really quite a strong force. If we were to plunge gaseous cations into water, they would form hydrated ions and release large amounts of energy, which we call the **hydration energy** of the cation. (This experiment is quite impossible to perform, but the energy released can be determined indirectly. Note that in the experiment that you just did, you added cations to water from the solid state and also added anions. This makes a large difference, as we will see in the next chapter, and consequently you did not usually detect a lot of energy being released.) Hydration energies of a number of cations are listed in Table 2.1; by any normal chemical standard these are large energies.

The data in Table 2.1 show that the hydration energy of a cation depends upon the charge and the radius of the cation, as expected qualitatively from Coulomb's law, and also depends upon the electronegativity of the element. Latimer [1] observed that if the electronegativity of the metal is not too great, the hydration energies of metal ions are given approximately by the equation

$$\Delta H_{\text{hyd}} = -\frac{60,900Z^2}{(r + 50)} \text{kJ/mol} \tag{2.1}$$

where $Z$ is the charge on the cation and $r$ is the cationic radius (in pm). (The constant added to the radius of the cation we can loosely equate with the radius of the oxygen in the water.)

No attempt is made in Latimer's equation to include the effects of electronegativity, but examination of the data for metals of Pauling electronegativities greater than 1.5 (on the right side of Table 2.1) shows that their hydration energies are substantially higher than those of ions of comparable radius and charge on the left

**Table 2.1**  Hydration Enthalpies of Metal Cations (kJ/mol)

| Electronegativity $\leq 1.5$ | | | Electronegativity $\geq 1.5$ | | |
|---|---|---|---|---|---|
| ION | RADIUS | $\Delta H_{hyd}$ | ION | RADIUS | $\Delta H_{hyd}$ |
| **+1 Ions** | | | | | |
| Cs | 181 | −263 | | | |
| Rb | 166 | −296 | Tl | 164 | −326 |
| K | 152 | −321 | | | |
| Na | 116 | −405 | Ag | 129 | −475 |
| Li | 90 | −515 | Cu | 91 | −594 |
| H | | −1091 | | | |
| **+2 Ions** | | | | | |
| Ra | | −1259 | | | |
| Ba | 149 | −1304 | | | |
| Sr | 132 | −1445 | Pb | 133 | −1480 |
| No | 124 | −1485 | Sn | | −1554 |
| Ca | 114 | −1592 | Cd | 109 | −1806 |
| | | | Cr | 94 | −1850 |
| | | | Mn | 97 | −1845 |
| | | | Fe | 92 | −1920 |
| | | | Co | 88 | −2054 |
| | | | Ni | 83 | −2106 |
| | | | Cu | 91 | −2100 |
| Mg | 86 | −1922 | Zn | 88 | −2044 |
| | | | Be | 59 | −2487 |
| **+3 Ions** | | | | | |
| Pu | 114 | −3441 | | | |
| La | 117 | −3283 | | | |
| Lu | 100 | −3758 | Tl | 102 | −4184 |
| Y | 104 | −3620 | In | 94 | −4109 |
| Sc | 88 | −3960 | Ga | 76 | −4685 |
| | | | Fe | 78 | −4376 |
| | | | Cr | 75 | −4402 |
| | | | Al | 67 | −4660 |
| **+4 Ions** | | | | | |
| Ce | 101 | −6489 | | | |

Ionic radii are from Table C; hydration enthalpies are taken from J. Burgess, *Metal Ions in Solution*, Ellis Horwood, Chichester, England, 1978, pp. 182–183.

side of the table. Such metals have electronegativities within about two units of that of oxygen, which suggests that for these metals there is not just an electrostatic attraction between the metal ion and the negative end of the water molecule, but that there also may be some degree of covalent bond formation, in which an unshared electron pair on water is shared with the metal ion.

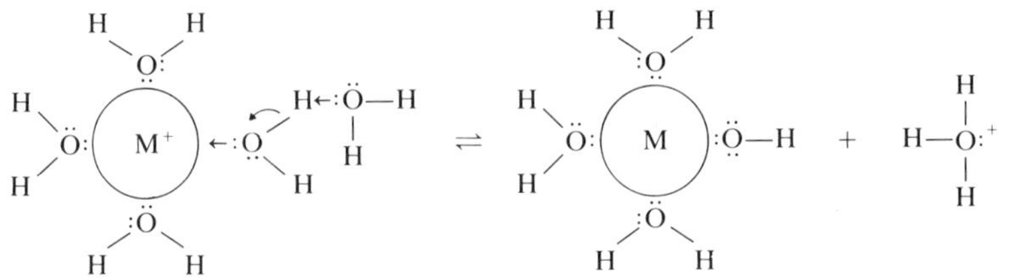

**Figure 2.2** Hydrolysis of a hydrated cation.

## 2.2

## Hydrolysis of Cations: Hydroxides, Oxides, Oxo Acids, Oxo Anions

If the attraction of the metal ion for the negative end of the water dipole is strong enough, the water molecule itself is affected (Figure 2.2). As the unshared electron pairs of the water molecule are pulled closer to (or even shared with) the metal ion, the electrons in the H—O bonds move closer to the oxygen to compensate some of its loss of electron density. Consequently the hydrogen ends up with an increased positive charge, which makes it more closely resemble a hydrogen ion. Eventually it may dissociate completely, attaching itself to solvent water molecules to make a hydronium ion and leaving a hydroxide group attached to the metal. We may represent this equilibrium by equation (2.2):

$$[M(H_2O)_6]^{z+} + H_2O \rightleftharpoons [M(H_2O)_5(OH)]^{(z-1)+} + H_3O^+ \tag{2.2}$$

The aluminum ion, for example, readily undergoes this reaction:

$$[Al(H_2O)_6]^{3+} + H_2O \rightleftharpoons [Al(H_2O)_5(OH)]^{2+} + H_3O^+ \tag{2.3}$$

The equilibrium constant for this process may be measured, although there are many experimental difficulties. One commonly finds tabulated [2] the negative logarithm of the equilibrium constant, $pK_a$, for reaction (2.2); selected $pK_a$ values are listed in Table 2.2. (An important consequence of taking negative logarithms, as in pH's, is that *lower* values of $pK_a$ and of pH correspond to a *greater* extent of hydrolysis and a *higher* acidity of the solution.) Note the similarity of equation (2.2) to the equation for the equilibrium process of ionization of a weak acid such as acetic acid:

$$HC_2H_3O_2 + H_2O \rightleftharpoons C_2H_3O_2^- + H_3O^+ \tag{2.4}$$

The $pK_a$ for a hydrated ion is thus analogous to the $pK_a$ (negative logarithm of the acid dissociation constant) of a species such as acetic acid.

Some of the $pK_a$ values are the averages of different measurements that may differ by more than one $pK_a$ unit, so we should not attempt to interpret small differences; but it is clear that the extent of this process, commonly called **hydrolysis** of the metal ion, increases with increasing charge and electronegativity of the metal and increases with decreasing radii of the metal ion. This process has a number of important practical consequences that we will investigate in this chapter and

**Table 2.2** Hydrolysis Constants for Metal Cations

| Electronegativity < 1.5 | | | | Electronegativity > 1.5 | | | | |
|---|---|---|---|---|---|---|---|---|
| ION | RADIUS | (a) | pK$_a$ | ION | RADIUS | (a) | (b) | pK$_a$ |
| **+1 Ions** | | | | | | | | |
| K | 152 | 0.007 | 14.5 | Tl | 164 | 0.006 | 0.016 | 13.2 |
| Na | 116 | 0.009 | 14.2 | Ag | 129 | 0.008 | 0.049 | 12.0 |
| Li | 90 | 0.011 | 13.6 | | | | | |
| **+2 Ions** | | | | | | | | |
| Ba | 149 | 0.027 | 13.5 | | | | | |
| Sr | 132 | 0.030 | 13.3 | Pb | 133 | 0.030 | 0.066 | 7.7 |
| | | | | Sn | | | | 3.4 |
| Ca | 114 | 0.035 | 12.8 | Hg | 116 | 0.034 | 0.082 | 3.4 |
| | | | | Cd | 109 | 0.037 | 0.055 | 10.1 |
| | | | | Cr | 94 | 0.043 | 0.043 | 10.0 |
| | | | | Mn | 97 | 0.041 | 0.046 | 10.6 |
| | | | | Fe | 92 | 0.043 | 0.075 | 9.5 |
| | | | | Co | 88 | 0.045 | 0.082 | 9.6 |
| | | | | Ni | 83 | 0.048 | 0.088 | 9.9 |
| Mg | 86 | 0.047 | 11.4 | Zn | 88 | 0.045 | 0.060 | 9.0 |
| | | | | Be | 59 | 0.068 | 0.074 | 6.2 |
| **+3 Ions** | | | | | | | | |
| Pu | 114 | 0.079 | 7.0 | | | | | |
| La | 117 | 0.077 | 8.5 | Bi | 117 | 0.077 | 0.127 | 1.1 |
| Lu | 100 | 0.090 | 7.6 | Tl | 102 | 0.088 | 0.140 | 0.6 |
| Y | 104 | 0.086 | 7.7 | Au | 99 | 0.091 | 0.191 | −1.5 |
| Sc | 88 | 0.102 | 4.3 | In | 94 | 0.096 | 0.123 | 4.0 |
| | | | | Ti | 81 | 0.111 | 0.115 | 2.2 |
| | | | | Ga | 76 | 0.118 | 0.148 | 2.6 |
| | | | | Fe | 78 | 0.115 | 0.147 | 2.2 |
| | | | | Cr | 75 | 0.120 | 0.135 | 4.0 |
| | | | | Al | 67 | 0.134 | 0.145 | 5.0 |
| **+4 Ions** | | | | | | | | |
| Th | 108 | 0.148 | 3.2 | | | | | |
| Pa | 104 | 0.154 | −0.8 | | | | | |
| U | 103 | 0.155 | 0.6 | | | | | |
| Np | 101 | 0.158 | 1.5 | | | | | |
| Pu | 100 | 0.160 | 0.5 | | | | | |
| Ce | 101 | 0.158 | −1.1 | | | | | |
| Hf | 85 | 0.188 | 0.2 | Sn | 83 | 0.193 | 0.222 | −0.6 |
| Zr | 86 | 0.186 | −0.3 | Ti | 74 | 0.216 | 0.220 | −4.0 |

SOURCES: Values of hydrolysis constants (pK$_a$) taken from C. F. Baes and R. E. Mesmer, *The Hydrolysis of Cations*, Wiley-Interscience, New York, 1976; and when not available there from J. Burgess, *Metal Ions in Solution*, Ellis Horwood, Chichester, England, 1978, pp. 264–267.

NOTE: (a) $Z^2/r$ ratio for the cation; (b) $Z^2/r + 0.096\,(\chi_P - 1.50)$ for the cation, as in equation (2.9).

the next, including the one suggested in part 8 of Experiment 1, and a number of environment consequences that we will see later.

Many of these consequences arise because equation (2.2) is only the first of several reactions that may occur. A second, then a third water molecule in the hydrated ion may hydrolyze, producing more hydronium ions and giving rise to cations containing more than one hydroxy group. These **hydroxy cations** undergo a bewildering variety of chemical reactions: Often they polymerize—and sometimes they lose a molecule of water from two hydroxy groups to give an **oxo cation**—but the most significant reactions are those generating usually insoluble hydroxides:

$$z[M(aq)(OH)]^{(z-1)+} \rightleftharpoons M(OH)_z(s) + (z-1)[M(aq)]^{z+} \tag{2.5}$$

The equilibrium constant for this reaction for most metal ions is rather large (approximately $10^{5.6}$) [3], so the hydroxy cations of most metals tend not to persist over wide pH ranges but instead give rise to precipitated **metal hydroxides**, as observed for several of the metal ions in Experiment 1. Thus the partly hydrolyzed aluminum ion produced in equation (2.3) readily gives rise to a gelatinous precipitate of aluminum hydroxide:

$$3[Al(aq)(OH)]^{2+} \rightleftharpoons Al(OH)_3(s) + 2[Al(aq)]^{3+} \tag{2.6}$$

It can be shown [4] that the metal hydroxide will precipitate at a pH that is roughly equal to the $pK_a$ value of the metal ion:

$$pH = pK_a - \left(\frac{1}{z}\right)\log[M^{z+}] - \frac{5.6}{z} \tag{2.7}$$

This equation tells us that the lower the $pK_a$ of a given metal ion, the less basic the solution need be for the metal hydroxide to begin precipitating. Certainly the concentration of the metal ion does have some effect on this pH, but we can easily calculate that moderately acidic ions such as $Al^{3+}$ or $Fe^{3+}$ will form insoluble hydroxides even in solutions that have quite high hydrogen-ion concentrations (low pH's).

Often the insoluble metal hydroxides will lose molecules of water to give insoluble oxides, as represented by equation (2.8):

$$M(OH)_z(s) \rightleftharpoons MO_{z/2}(s) + \frac{z}{2}H_2O \tag{2.8}$$

Since it is difficult for us to tell when this has happened, we will not attempt to distinguish metal hydroxide and metal oxide precipitates.

Even this may not be the end of the story: If the attraction of the metal ion for the pair of electrons on oxygen is strong enough, the hydroxy groups of the metal hydroxide may start to lose their remaining hydrogens as hydronium ions:

$$M(OH)_z + z\,H_2O \rightleftharpoons MO_z^{z-} + z\,H_3O^+ \tag{2.9}$$

Thus the metal hydroxide may begin to act as a weak **oxo acid**, which finally may

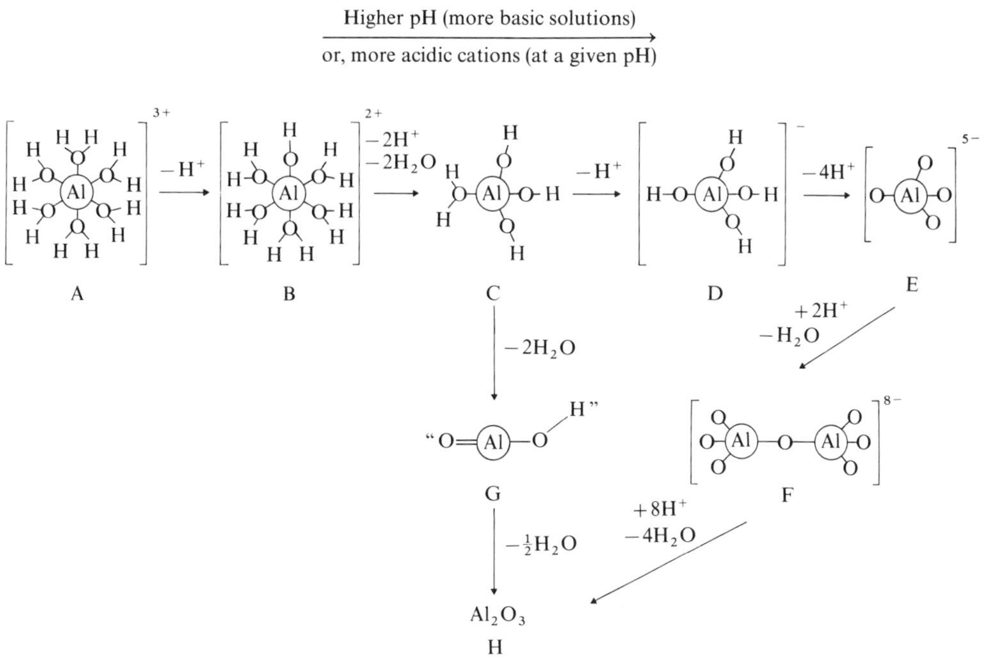

Higher pH (more basic solutions)
or, more acidic cations (at a given pH)

**Figure 2.3**  Main species arising from the hydrolysis of cations ($Al^{3+}$ is used for purposes of illustration) in aqueous solutions and the sections in which they are discussed. (A) a hydrated ion, Sec. 2.1; (B) a hydroxy cation, Sec. 2.2; (C) a metal hydroxide, Sec. 2.2; (D) a hydroxo anion, Sec. 2.7; (E) an oxo anion, Secs. 2.6, 2.8. 2.9; (F) a polynuclear oxo anion, Secs. 4.7–4.9; (G) an oxo acid (hypothetical in the case of aluminum), Sec. 2.10; (H) an oxide, Secs. 4.1–4.6.

ionize to give an **oxo anion**, thus completely dismembering the water molecules that were originally attached in a hydrated ion. Although we will not look at oxo acids and oxo anions until later in the chapter, we can see that a remarkable variety of chemical species can arise out of the "simple" interaction of a positively charged ion and its water molecules of hydration. The main species that we will be studying in Chapters 2, 3, and 4 are summarized in Figure 2.3.

## 2.3

## Predicting the Degree of Hydrolysis of Cations

### 2.3.1    Effects of Charge and Radius on Hydrolysis

There are important practical consequences of the acidic properties and hydrolysis reactions of compounds of elements with high oxidation numbers or small radii or high electronegativities. The suggested laboratory experiment hints at some of these: A compound with an innocent-looking formula such as $TiCl_4$ may react quite violently with water. It may react with the water vapor in the air and fill the laboratory with choking fumes of the acid HCl. You may be trying to prepare a solution of a metal ion for some experiment in biology, only to find a precipitate forming. You may need to know the form a metal-ion pollutant takes in a lake, to know whether it will end up as an insoluble sludge of oxide at the bottom of the lake or whether it will remain in solution as a cation or as an oxo anion—which may be taken up or rejected by quite different mechanisms by living organisms. Thus it is important for us to be able to gauge the approximate acidity of a given cation so that we can anticipate (even in the laboratory without a calculator) how violently a given compound will react with water or atmospheric humidity and whether its hydroxide or oxide will precipitate or whether an oxo anion will be produced.

Rather accurate equations have been presented for predicting the extent of hydrolysis of metal ions [5,6], but these are not simple enough to be used for our purposes of quick estimation of approximate acidity. To obtain such a relationship, we have graphed (in Figure 2.4, using solid circles) the $pK_a$ values of metal cations of Pauling electronegativities ($\chi_p$) of 1.5 or below versus the $Z^2/r$ ratios of these ions. Equation (2.10) expresses the relationship we find empirically.

$$pK_a = 15.14 - 88.16\frac{Z^2}{r} \tag{2.10}$$

Normally we will find it quicker not to solve equation (2.10) to obtain exact values of $pK_a$, but instead we will evaluate $Z^2/r$ in order to place it in a certain range or category of acidity, as outlined in Table 2.3. We find experimentally that the ions in a given category of acidity share several important chemical properties. These properties include not only the degree to which the ions react with water but also (as we will see in subsequent chapters) the solubility or insolublity of salts formed by these ions and the properties of compounds formed by these ions (such as their oxides and halides) in the absence of water.

Clearly the smallest ratios of $Z^2/r$ correspond to the least hydrolysis and belong to the cations that we can most accurately call *nonacidic cations*. We can predict that if their electronegativities are below 1.5, ions of $Z^2/r$ *less than 0.01* should show such negligible hydrolysis reactions that their acidity does not manifest itself in any important way. Included are such ions as $Cs^+$ and $Rb^+$, whose hydrolysis constants ($pK_a$ values) are absent from Table 2.2 because the hydrolysis of these ions is too slight to be measured. Consequently we find that the hydroxides of these elements do not precipitate from solution; on the contrary, the solid hydroxides **deliquesce** in humid air, removing water from the air to form a solution!

The hydrolysis of ions such as $Li^+$, $Ba^{2+}$, $Sr^{2+}$, and $Ca^{2+}$, which have $Z^2/r$ values between 0.01 and 0.04, can be measured in sensitive experiments and occasionally has significant consequences. We will refer to these ions as *feebly acidic cations*. Their hydroxides also do not normally precipitate from water, but they are less soluble than the hydroxides of the nonacidic cations.

Acidity becomes an important part of the chemistry of ions of low electronegativity and $Z^2/r$ ratios between 0.04 and 0.10, although this acidity is not apparent

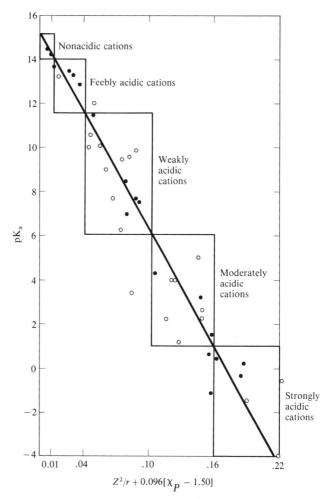

**Figure 2.4** pK$_a$ values of cations as a function of charge, size, and electronegativity. Closed circles equal metals of electronegativity($\chi_P$) less than 1.5, plotted as a function of $Z^2/r$. Open circles equal metals of electronegativity greater than 1.5, plotted as a function of $Z^2/r + 0.096(\chi_P - 1.50)$.

In the figure:
- y-axis: pK$_a$
- x-axis: $Z^2/r + 0.096[\chi_P - 1.50]$
- Regions labeled: Nonacidic cations, Feebly acidic cations, Weakly acidic cations, Moderately acidic cations, Strongly acidic cations

**Table 2.3** Relationship between $Z^2/r$ Ratios and Acidity of Metal Ions

| $Z^2/r$ Ratio | $\chi_P$ | Category | pK$_a$ Range | Examples |
|---|---|---|---|---|
| 0.00–0.01 | <1.8 | Nonacidic cations | 14–15 | Most +1 ions of the s-block |
| 0.00–0.01 | >1.8 | Feebly acidic cations | 11.5–14 | Tl$^+$ |
| 0.01–0.04 | <1.8 | Feebly acidic cations | 11.5–14 | Most +2 ions of the s- and f-block |
| 0.01–0.04 | >1.8 | Weakly acidic cations | 6–11.5 | Most +2 ions of the d-block |
| 0.04–0.10 | <1.8 | Weakly acidic cations | 6–11.5 | All +3 ions of the f-block |
| 0.04–0.10 | >1.8 | Moderately acidic cations | 1–6 | Most +3 ions of the d-block |
| 0.10–0.16 | <1.8 | Moderately acidic cations | 1–6 | Most +4 ions of the f-block |
| 0.10–0.16 | >1.8 | Strongly acidic cations | (−4)–1 | Most +4 ions of the d-block |
| 0.16–0.22 | <1.8 | Strongly acidic cations | (−4)–1 | |
| 0.16 and up | >1.8 | Very strongly acidic cations | <(−4) | |
| 0.22 and up | <1.8 | Very strongly acidic cations | <(−4) | |

NOTE: The electronegativities of the p-block elements vary too greatly to allow their inclusion in one category of "cation."

in simple pH measurements because the solution normally also contains dissolved carbon dioxide, which is at least as strong an acid as these ions. Included in this group of *weakly acidic cations* are many important $+2$ ions such as $Mg^{2+}$ and the $+2$-charged $d$-block ions whose biological functions, as we will see later, revolve around their acidity. These ions show enough acidity to react with moderate concentrations of the (strong base) hydroxide ion to precipitate insoluble metal hydroxides in neutral or just slightly basic solutions.

The solutions of cations with $Z^2/r$ ratios between 0.10 and 0.16, such as $Al^{3+}$ and the $+3$-charged $d$-block ions, are unmistakably acidic, with quite low pH's and sometimes the appearance of some cloudiness due to formation of some insoluble metal hydroxide. If the solutions of these ions are not kept highly acidic, metal hydroxides will precipitate even though the pH may still be below 7. We will call these ions *moderately acidic cations*, as their $pK_a$ values are comparable to the $pK_a$ values of typical organic acids such as acetic acid.

Cations with $Z^2/r$ ratios between 0.16 and 0.22, such as $Ti^{4+}$, react violently and nearly completely with water, giving strongly acidic solutions and copious amounts of precipitate of insoluble metal oxide or hydroxide. It is appropriate to call these *strongly acidic cations*. But we may also note that their reaction with water is usually reversible—the precipitate will often redissolve in quite concentrated hydrochloric acid, at least if it has not aged too much.

Cations with $Z^2/r$ values over 0.22 may be expected to react irreversibly with water, generating an oxide or hydroxide. The higher the $Z^2/r$ ratio, the more likely that this hydroxide will act as a weak acid, or even a strong acid, ionizing to give an oxo ion. We will call this group of "cations," which really cannot remain as cations in water, the *very strongly acidic cations*. Later in this chapter we will find methods for distinguishing which hydroxides from this group will act as weak acids, which as strong acids, and which will be neutral—the use of $Z^2/r$ is not the best approach to this question. It is worth noting now, however, that the "cations" of the purely nonmetallic elements of the periodic table have $Z^2/r$ ratios in this range, which is to say that they cannot exist in water as cations. (This is one of the characteristics by which we differentiate metals and nonmetals.) Hence the practical use of these calculations lies with the metal ions.

## 2.3.2    Effects of Electronegativity on Hydrolysis

So far we have not attempted to take into account the effects of high electronegativity of the cation. From Table 2.2 we can see that the cations of metals with Pauling electronegativities ($\chi_p$) over 1.5 are more acidic than other metal ions of similar charge and size. We have derived a rough relationship between the "excess" Pauling electronegativity of a cation and its "excess" acidity, which allows us to modify equation (2.10) to include the effect of electronegativity:

$$pK_a = 15.14 - 88.16\left[\frac{Z^2}{r} + 0.096(\chi_p - 1.50)\right] \qquad (2.11)$$

Note that this equation should be used *if and only if* the Pauling electronegativity of the metal exceeds 1.50.

In Figure 2.4 we have used open circles to show the relationship of the $pK_a$ of cations to their "modified" $Z^2/r$ ratio, $Z^2/r + 0.096 \, (\chi_p - 1.50)$. There is more scatter in the data for these more electronegative cations than there was for the "simple," less electronegative cations, which suggests that we have oversimplified the relationship to electronegativity. But our main purpose is to quickly categorize the relative acidity of these cations; the scatter is not so serious as to prevent us from suggesting an even simpler rule of thumb: *If the electronegativity of the metal ion is 1.8 or greater, move it up one category in acidity.* The suggested categories of acidity, their corresponding $pK_a$ values, and their relationships to $Z^2/r$ and $\chi_p$ are summarized in Table 2.3.

EXAMPLE    Classify each of the following cations and describe their reactions with water: $Eu^{2+}$, $B^{3+}$, $W^{6+}$.

Solution    We begin by finding ionic radii in Table C; then we compute $Z^2/r$. For $Eu^{2+}$, this ratio works out to be $2^2/131 = 0.031$; for $B^{3+}$, $3^2/41 = 0.220$; for $W^{6+}$, $6^2/74 = 0.487$. Checking the table of electronegativities, we find that the latter two elements have electronegativities in excess of 1.8. Hence, using Table 2.3, we can classify $Eu^{2+}$ as a feebly acidic cation, which should be present largely unchanged (as a hydrated ion) in water. The latter two "cations" must be classified as very strongly acidic and will not actually be present at all in water. Halides of these two cations will react violently with water to generate hydrohalic acid and the oxides, oxo acids, or oxo anions of these elements. ☐

## 2.4

## Halides That Fail to Undergo Hydrolysis

According to the rules we have developed, carbon tetrachloride should be one of the most reactive of all the chlorides of the elements, since $Z^2/r$ for $C^{4+}$ is quite impressive (0.533). The predicted explosive reaction of $CCl_4$ and water fails to materialize, however. Indeed, $CCl_4$ is used as an inert solvent that is insoluble in and unreactive with water. The key to this seeming violation of our concepts lies in the *very* small size of the carbon atom, which is so crowded in by the four large chlorine atoms that the water molecule cannot get in to start the reaction.

This nonreaction (at low temperatures) illustrates an aspect of the radius in the central atom in a molecule: There is a limit to the number of atoms that may be bonded to it. This number is called the **maximum coordination number** of the element, and although it depends on the size of the atoms around the central atom, it obviously *increases down the periodic table*. Elements of Period 2 almost never have more than four atoms bonded to them in compounds; this is an important basis of the **octet rule**. Hence we say that the maximum coordination number for an element in Period 2 is four. But (see Table C) the ions of Periods 3 and 4 are

31

substantially larger than those of Period 2; hence they can bond to as many as six other atoms. In Periods 5 and 6 the atoms are larger still, and cases of atoms bonded to more than six neighboring atoms are known.

Another case of "unexpected" nonreactivity of a halide is that of sulfur hexa-fluoride, $SF_6$, which resists reacting with steam at $500\,^{\circ}C$ and resists molten KOH. The atoms get larger as we go down Group 16/VIA, however; thus $SeF_6$ is unreactive with water at $25\,^{\circ}C$ but is hydrolyzed in the respiratory tract (with quite undesirable consequences); $TeF_6$ reacts slowly with cold water.

Although the rate of reaction of water and carbon tetrachloride is negligible at room temperature, it is still true that the end products of their very slow reaction ($CO_2$ and $4\,HCl$) are more stable than the reactants ($CCl_4$ and $2\,H_2O$) by 52 kJ. At high temperatures the reaction does indeed occur. An intermediate product that forms in this reaction (which may happen when a carbon tetrachloride fire extinguisher is used along with water on a very hot fire) is carbonyl chloride or phosgene, $COCl_2$, a deadly poison. It has been suggested that the main reason that life can exist based on carbon chemistry rather than silicon chemistry is that the small size of carbon causes the reactions of carbon compounds to be much slower, so that many more carbon compounds that are not inherently thermodynamically stable can exist in the presence of reactive compounds such as water and oxygen [7].

Oddly enough, for some ions near the center of Periods 5 and 6 (e.g., $Ag^+$, $Sn^{2+}$, $Hg^{2+}$, $Au^+$) the preferred coordination number is often quite small—as low as two. As you might expect, there is some anomalous chemistry in this part of the periodic table. Thus the acidities of the hydrated $Hg^{2+}$ and $Sn^{2+}$ ions are much higher than expected from equation (2.11), which may be attributed to the fact that there are unexpectedly few water molecules coordinated to the metal ion, so the electrostatic force pulling on the electrons of each water molecule is stronger than in more normal hydrated ions. Similarly, the smallest hydrated ion, $Be^{2+}$, is also more acidic than one would predict [6].

## 2.5

## Some Details of the Chemistry of Hydrated Ions

### 2.5.1    Oxo Cations

Although we wish generally to ignore the complex chemistry of the intermediate species that result after a metal ion hydrolyzes but before it precipitates as a hydroxide, in some cases this species is persistent and important enough to merit mention. Sometimes when two or four hydroxide groups are attached to a (highly charged) cation, they may lose molecules of water and form **oxo cations**. Such species are particularly persistent for the $+5$ and $+6$ oxidation states of the $f$-block elements U, Np, Pu, and Am; the most common form in which you are likely to encounter uranium is in salts of the yellow uranyl ion, $UO_2{}^{2+}$. Similar species are relatively common in the chemistry of molybdenum (V) and (VI), and vanadium (IV) and (V). Many nonmetal halides, on partial hydrolysis, form species that resemble these in formula—e.g., $POCl_3$. These generally hydrolyze further with more water, however. (In the suggested experiment, the precipitates formed by antimony and bismuth are not the oxides, but antimonyl chloride, SbOCl, and bismuthyl chloride, BiOCl.) Oxo cations are commonly named by replacing the -ium ending of the metal by -yl.

## 2.5.2    Color and Metal Ions

One of the most attractive features of inorganic chemistry is the numerous and varied colors of many inorganic compounds. Examination of a collection of hydrated metal ions, however, would be largely disappointing in this respect; the hydrated ions of the *s*- and *p*-block elements are uniformly colorless as are (in general) those hydrated ions of the *d*-block and *f*-block elements that have *empty* or *completely filled* sets of valence *d* or *f* orbitals. But those hydrated ions with partially filled subshells of *d* or of *f* orbitals have much more to recommend themselves; most of these ions have attractive colors (Table 2.4), although these colors are not particularly intense. *Crystal field theory* explains (among many other things) how the *d*-block ions produce their many colors, although the study of this theory will have to await a further course in inorganic chemistry.

**Table 2.4**    Colors of Hydrated Metal Ions

| *n* | Period 4, *d* block | Period 6, *f* block | Period 7, *f* block |
|---|---|---|---|
| 1 | $Ti^{3+}$, violet | $Ce^{3+}$, colorless | $Pa^{4+}$, colorless |
| 2 | $V^{3+}$, blue | $Pr^{3+}$, green | |
| 3 | $V^{2+}$, violet | $Nd^{3+}$, lilac | $U^{3+}$, red-brown |
|  | $Cr^{3+}$, violet | | $Np^{4+}$, yellow-green |
| 4 | $Cr^{2+}$, blue | $Pm^{3+}$, pink | $Np^{3+}$, purple |
|  | | | $Pu^{4+}$, tan |
| 5 | $Mn^{2+}$, pale pink | $Sm^{3+}$, yellow | $Pu^{3+}$, blue-violet |
|  | $Fe^{3+}$, pale purple | | $Am^{4+}$, red |
| 6 | $Fe^{2+}$, pale green | $Eu^{3+}$, pink | $Am^{3+}$, pink |
|  | | $Sm^{2+}$, red | $Cm^{4+}$, yellow |
| 7 | $Co^{2+}$, pink | $Gd^{3+}$, colorless | |
|  | | $Eu^{2+}$, colorless | |
| 8 | $Ni^{2+}$, green | $Tb^{3+}$, pink | |
| 9 | $Cu^{2+}$, blue | $Dy^{3+}$, yellow | |
| 10 | | $Ho^{3+}$, yellow | |
| 11 | | $Er^{3+}$, lilac | |
| 12 | | $Tm^{3+}$, green | |
| 13 | | $Yb^{3+}$, colorless | |

SOURCE: F. A. Cotton and G. Wilkinson, *Advanced Inorganic Chemistry: A Comprehensive Text*, 4th ed., Wiley-Interscience, New York, 1980, Chapters 21 and 22 and pp. 985 and 1019.

NOTE: *n* = number of electrons in valence *d* or *f* orbitals.

## 2.6

# The Hydrolysis of Oxo Anions

Before you go further in this chapter, your instructor may wish to have you perform or observe and discuss Experiment 2 (found in Chapter 13).

As you may have read in Experiment 2, when we deal with nonmetals in positive oxidation states in aqueous solution, we are not dealing with cations, but with species such as oxides or hydroxides, which may react as weak oxo acids or finally as strong oxo acids, ionizing completely to hydronium ions and **oxo anions**, $MO_x^{y-}$. These forms are sufficiently different from cations that our rules based on the ratio $Z^2/r$ of cations are no longer useful enough in predicting their chemical properties; hence we will develop new approaches. It is most convenient in doing this to begin at the end of the series, looking first at the oxo anions. In an oxo anion of formula $MO_x^{y-}$ we will be particularly interested in the effects of the number of **oxo groups**, $x$ (oxygen atoms bonded to M and to no other atom [8]), and in the charge on the oxo anion, $-y$.

Just as metal cations do not exist "bare" in aqueous solutions, neither do anions such as oxo anions. They also attract water molecules to form hydrated ions, although in this case we would expect to (and do) find that the *positive* end of the dipolar water molecule—that is, the end with the hydrogens—is attracted to the unshared electron pairs on the oxygen atoms of the oxo anion. Substantial hydration energies also result from this interaction (Table 2.5). As we would expect, these energies increase with increasing charge and decreasing size of the anion.

Once again the interaction between an ion (in this case, a negatively charged oxo ion) and its waters of hydration frequently is great enough that one or more of the water molecules may be "pulled apart" in the process of hydrolysis. In this case

**Table 2.5** Shannon-Prewitt Radii and Hydration Enthalpies (kJ/mol) of Some Anions

| Anion | Radius | Hydration Energy | Anion | Radius | Hydration Energy |
|---|---|---|---|---|---|
| $F^-$ | 119 | −497 | $OH^-$ | 119 | −453 |
| $O^{2-}$ | 126 | | $CN^-$ | 177 | −334 |
| $N^{3-}$ | 132 | | $N_3^-$ | 181 | −290 |
| $Cl^-$ | 167 | −355 | | | |
| $S^{2-}$ | 170 | −1356 | $SH^-$ | 193 | −328 |
| $Br^-$ | 182 | −328 | $BF_4^-$ | 215 | −215 |
| $Se^{2-}$ | 184 | | $ClO_4^-$ | 226 | −227 |
| $I^-$ | 206 | −287 | $I_3^-$ | | −160 |
| $Te^{2-}$ | 207 | | | | |

SOURCES: Radii for monoatomic anions are taken from R. D. Shannon and C. T. Prewitt, *Acta Crystallogr.*, B25, 925 (1969) and R. D. Shannon, ibid., A32, 751 (1976). "Radii" for polyatomic anions (which of course are not truly spherical) are thermochemical radii, taken from J. E. Huheey, *Inorganic Chemistry*, 3d ed., Harper and Row, Cambridge, 1983, p. 78.

NOTE: Hydration enthalpies are from M. C. Ball and A. H. Norbury, *Physical Data for Inorganic Chemists*, Longman, London, 1974.

the partially positively charged hydrogen atom of the water may bond to a partially negatively charged oxygen atom of the oxo anion, releasing the remainder of the water molecule—a hydroxide ion— and producing a basic solution. This process, studied for many anions (such as the acetate ion) in general chemistry, may be represented as follows:

$$MO_x^{y-} + H_2O \rightleftharpoons [MO_{(x-1)}OH]^{(y-1)-} + OH^- \qquad (2.12)$$

A familiar example of this process is the hydrolysis of carbonate ion to give bicarbonate:

$$CO_3^{2-} + H_2O \rightleftharpoons HCO_3^- + OH^- \qquad (2.13)$$

An equilibrium constant can be written for the above reaction; this is often called a hydrolysis constant, symbolized by $K_b$ in recognition of the fact that the oxo anion acts like any other base. As before, we will not tabulate these exponential numbers but rather will use their negative logarithms, which are symbolized by $pK_b$ or $pK_{b1}$. (This last designation recognizes that reaction (2.12) may be followed by second, third, and subsequent steps of hydrolysis.) A large value of $pK_{b1}$ corresponds to a case in which reaction (2.12) goes very little to the right, so that the oxo anion is acting as a weak base, producing a solution with a pH not much above that of distilled water.

## 2.6.1 Calculating the Basicities of Oxo Anions

Again we will try to develop simple rules that will enable us to predict just how basic the solution of an oxo anion should be, and that will enable us to predict some practical consequences. (For example, any solution that is highly basic is also corrosive to many materials, including human tissue!) With cations we looked at three variables: charge, size, and electronegativity of the metal ion. With oxo anions we will neglect one of these: the size of the nonmetal atom (or metal atom in a high oxidation state). These are inherently much smaller than the several oxygen atoms in the oxo anions, so oxo anions with different nonmetal atoms but with the same number of oxygen atoms should all have very nearly the same size.

Just as increasing positive charge of a cation increases its tendency to undergo hydrolysis and give rise to acidic solutions, we would expect to (and do) find that increasing negative charge on an anion increases its tendency to hydrolyze and give rise to basic solutions. Referring to the (partial) list of oxo anions in part 5 of Experiment 2, we see that the negative charges of oxo anions can be quite substantial indeed, so it is not surprising that the solutions of many of them are very basic. Comparing the $pK_{b1}$ values for a series of oxo anions of increasing negative charge leads us to the conclusion that *the $pK_{b1}$ of an oxo anion decreases by* 10.2 *units for each ( additional ) negative charge on it* [9]. This translates into a substantial increase in basicity with increasing negative charge.

Most of the nonmetals show more than one oxidation number and can form oxo anions that differ in the number of oxo groups that are attached to the nonmetal atom. These different oxo anions differ substantially in basicity. The most complete series of oxo anions is that of chlorine, which forms four different oxo anions with the following $pK_{b1}$ values: $ClO^-$, 6.5; $ClO_2^-$, 12.1; $ClO_3^-$ and $ClO_4^-$, unmeasurable (these are such weak bases that their hydrolysis cannot be detected). Examination of the $pK_{b1}$ values for a number of such sets of oxo anions shows that, on the average, *each additional oxo group in an oxo anion increases its $pK_{b1}$ by 5.7 units.*

Why should the number of oxo groups have an effect on the basicity of an oxo anion? Students contemplating this question in the past have devised a number of explanations that fit the facts and are chemically reasonable.

■ Those who have already studied organic chemistry have recognized the fact that the more oxo groups there are in an oxo anion, the more resonance structures can be drawn by which the negative charge of the anion is delocalized; hence there is greater resonance stabilization of anions with more oxo groups.

■ Oxygen, of course, is a very electronegative atom. When we add an oxo group to an anion such as $ClO^-$, we attach O to an unshared pair of electrons that was on the Cl atom of $ClO^-$. The new oxygen atom strongly withdraws these electrons from the Cl, which compensates by attracting electrons more strongly from the other oxygen atom(s); hence there is less effective negative charge at each oxo group for attracting the hydrogen atom of the water of hydration.

■ A third explanation focuses on the fact that when we attach an additional oxygen atom, the oxidation number of the nonmetal atom increases by 2. In effect, it acts as a "cation" with a positive "charge" increased by 2, which will attract the electrons on all its oxo groups more strongly, reducing their attraction for the hydrogen atom of the water of hydration.

■ A fourth explanation looks at the Lewis dot structures of oxo anions and asks the question, On which atoms of this oxo anion are the negative charges located? Of course, they are not really on any one atom, since oxo anions are (internally) covalently bonded species, but knowing that oxygen is more electronegative than the other atom in the oxo anion, we can assume that they are on oxygen. But in an oxo anion such as $ClO_4^-$, there is no reason to assume that the negative charge is on one particular oxygen any more than on any of the others, so we may suppose that each oxygen atom bears a *partial charge* of $-\frac{1}{4}$. Likewise in $ClO_3^-$, we suppose a greater partial charge of $-\frac{1}{3}$; while the oxygen atom in $ClO^-$ is stuck with the whole charge of $-1$. Although these numbers are fictitious, there should indeed be partial negative charges on oxygen, which should decrease as the number of oxo groups increases, resulting in reduced basicity.

It is not infrequent in science to have more than one explanation of a phenomenon such as this. If the explanations are all satisfactory, we may continue to use more than one, depending on the situation or on personal preference. Of course, if you think about the preceding four explanations, you will see that they are far from being four completely different explanations—to a large degree they are different ways of saying the same thing.

The electronegativity of the nonmetal atom also influences the basicity of an oxo anion. Comparing the $pK_{b1}$'s of $ClO^-$, 6.50; $BrO^-$, 5.3; and $IO^-$, 3.4, we find, as we may expect, that reducing the electronegativity of the halogen atom increases the basicity of the oxo anion. For $d$-block metals of much lower electronegativity than $p$-block elements, this effect is greater: $pK_{b1}$ for $SO_4^{2-} = 12.1$; $pK_{b1}$ for $CrO_4^{2-} = 7.5$. Insufficient data, however, are available to allow a quantitative estimate of the effect of electronegativity; qualitatively it appears to be a smaller effect than that of the charge or the number of oxo groups in the oxo anion, so for purposes of developing approximate rules we will ignore the influence of electronegativity.

We may incorporate the effects of the number of oxo groups and of the negative charge of an oxo anion into one equation that allows a reasonably accurate calculation of the constant for the first hydrolysis of an oxo anion:

$$pK_{b1} = 10.0 + 5.7x - 10.2y \qquad (2.14)$$

Here $x$ is the number of oxo groups and $y$ is the number of units of negative charge in the oxo anion of formula $MO_x^{y-}$. Keep in mind the physical reasoning behind this equation: (1) Higher $pK_{b1}$'s correspond to weaker basicity (lower pH); (2) additional oxo groups weaken basicity, hence add to $pK_{b1}$; (3) additional negative charges increase basicity, hence subtract from $pK_{b1}$.

In Table 2.6 we have tabulated the calculated values of $pK_{b1}$ for the important

**Table 2.6** Suggested Classification of Oxo Anions

| Classification | Type | Calculated $pK_{b1}$ | Examples with Known $pK_{b1}$'s |
|---|---|---|---|
| Nonbasic anions | $MO_4^-$ | 22.6 | M = Cl, Br, Mn, Tc, Re |
| | $MO_3^-$ | 16.9 | M = N, Cl, Br, I(13.2) |
| Feebly basic anions | $MO_4^{2-}$ | 12.4 | M = S(12.1), Se(12.0), Xe, Cr(7.5), Mo(9.9), W(9.4), Fe, Ru, Os |
| | $MO_2^-$ | 11.2 | M = N(10.7), Cl(12.1) |
| Moderately basic anions | $MO_6^{4-}$ | 3.4 | M = Xe, Os |
| | $MO_4^{3-}$ | 2.2 | M = P(2.0), As(1.5), V(1.0) |
| | $MO_3^{2-}$ | 6.7 | M = C, S(6.8), Se(7.4), Te(6.3) |
| | $MO^-$ | 5.5 | M = Cl(6.5), Br(5.3), I(3.4) |
| Very strongly basic anions | $MO_6^{5-}$ | −6.8 | M = I, Np |
| (Exist as hydroxo anions in solution) | $MO_4^{4-}$ | −8.0 | M = Si, Ge |
| | $MO_3^{3-}$ | −3.5 | M = As, Sb |
| | $MO_6^{6-}$ | −17.0 | M = Te |
| | $MO_4^{5-}$ | −18.2 | M = B, Al, Ga |
| | $MO_3^{4-}$ | −14.7 | M = Sn |

SOURCES: Known $pK_{b1}$ values are calculated from the appropriate $pK_a$ values given in F. A. Cotton and G. Wilkinson, *Advanced Inorganic Chemistry: A Comprehensive Text*, 4th ed., Wiley-Interscience, New York, 1980, p. 235; R. C. Weast, ed., *Handbook of Physics and Chemistry*, 50th ed., Chemical Rubber Publishing Co., Cleveland, 1969; and J. A. Dean, ed., *Lange's Handbook of Chemistry*, 13th ed., New York, McGraw-Hill, 1985.

simple oxo anions of the elements. Note that we have chosen the names and $pK_{b1}$ ranges of our categories to match those used for cations, as much as possible. (Fewer categories are needed to describe the properties of oxo anions, however.) As before, nonbasic ions show no detectable tendencies to undergo hydrolysis and do not alter the pH of their aqueous solvent. Feebly basic anions only occasionally manifest any basic properties. The category of moderately basic anions includes the ones on which equilibrium calculations were done in general chemistry, probably under the heading of "salts of weak acids"; their solutions are distinctly basic. The very strongly basic anions, however, are normally found only in anhydrous solids, since they hydrolyze completely or nearly completely in water according to equation (2.12).

### 2.6.2    Quick Estimates of Basicities

A quicker but rougher estimate of the basicity of an oxo anion may also be made by direct inspection of its chemical formula. We may note that in equation (2.14) the effect of one additional charge is roughly twice as great (and opposite in sign to) the effect of one additional oxo group; hence it is possible to estimate relative basicities by matching single negative charges and pairs of oxo groups and deleting them together from the formula of the oxo anion. Thus for a nonbasic anion such as $MO_4^-$, we can cancel out the effects of the unit of negative charge and the effects of two of the oxo groups, to be left with "$MO_2$." If all the negative charge cancels out but oxo groups remain, we may expect to (and do) find that the original oxo anion was *nonbasic*.

For the feebly basic anions $MO_4^{2-}$ and $MO_2^-$, we find that this cancellation process removes *both* the negative charge and the oxo groups, leaving us only with "M."

Species that end up with negative charges but no oxo groups we might expect to be appreciably basic. We find that oxo anions with $-\frac{1}{2}$ or $-1$ charges after cancellation are *moderately basic*. (For example, $MO_3^{2-}$ becomes "$M^{-1/2}$," and $MO_6^{4-}$ becomes "$M^-$.") Species that end up with charges that are *more* negative than $-1$ (e.g., $MO_3^{4-}$, which becomes "$M^{-5/2}$") are *very strongly basic*.

## 2.7

# Hydroxo Anions

The product of equation (2.12), the once-protonated oxo anion $[MO_{(x-1)}OH]^{(y-1)-}$, is still a base, but it now has one less negative charge and one less oxo group (one oxo group having been converted to a "hydroxo group"). If we calculate the basicity of this monoprotonated oxo anion using equation (2.14), we find that its $pK_{b2}$ is 4.5 *units more than its* $pK_{b1}$. Likewise, the resulting diprotonated oxo anion is still a base, but of further diminished basicity due to a $pK_{b3}$ that is yet another 4.5 units higher. In water this process continues until there results an equilibrium mixture of protonated oxo anions that have reasonable $pK_b$ values. In aqueous solution, then, the very basic oxo anions hydrolyze through several steps. Thus although the silicate anion $SiO_4^{4-}$ is found in several minerals, it is not really present in a solution of sodium silicate due to its very strong basicity (calculated $pK_{b1} = -8.0$). Instead the main species in solution is apparently the diprotonated ion $SiO_2(OH)_2^{2-}$, with a calculated $pK_{b3}$ of 1.0. (The formula of this ion is often written as if it had lost a molecule of water and become the so-called metasilicate ion, $SiO_3^{2-}$, but there is no evidence for the existence of such a species.)

When the negative charge on one of the very strongly basic oxo anions exceeds the number of oxo groups present, it is possible to transfer protons to each of the oxo groups and still have an anion. The resulting species, which no longer has any oxo groups, is perhaps better called a **hydroxo anion**. Thus the "impossibly" basic antimonate ion, $SbO_6^{7-}$, with a calculated $pK_{b1}$ of $-27.2$, is actually found in aqueous solution as $Sb(OH)_6^-$, with a much more moderate calculated $pK_b$ of $-0.2$. Likewise, the anionic species in solution for Sn(IV) and Pb(IV) are actually $Sn(OH)_6^{2-}$ and $Pb(OH)_6^{2-}$; those for B, Al, and Ga have the formula $M(OH)_4^-$ (M = B, Al, or Ga). The formulas of these species are also written as if they were oxo anions—e.g., $AlO_2^-$ instead of $Al(OH)_4^-$, but notice that such a formula, although it includes the correct charge, implies an absurdly low coordination number for the central atom.

Since the hydroxo anions are much less basic than the corresponding oxo anions, they can form with some metal atoms that have too low a $Z^2/r$ ratio (and too low an electronegativity) to form oxo anions. Thus in strongly basic solutions, hydroxo anions are known for several metals that have only moderately acidic cations: Examples include $Be(OH)_4^{2-}$, $Cr(OH)_6^{3-}$, $Fe(OH)_6^{4-}$, $Fe(OH)_6^{3-}$, $Ag(OH)_2^-$, $Au(OH)_4^-$, $Zn(OH)_4^{2-}$, and $Cd(OH)_4^{2-}$. These hydroxo anions are often formed by dissolving the corresponding metal oxide or hydroxide in strongly basic (NaOH) solutions. Since these same metal oxides or hydroxides also dissolve in strongly acidic solutions to give hydrated cations, such oxides or hydroxides are capable of acting either as acids (reacting with NaOH) or as bases (reacting with an acid). Oxides or hydroxides that can react both ways are termed **amphoteric**.

## 2.8

## The Formulas of Oxo Anions

The number of oxo groups in an oxo anion $MO_x^{y-}$ can be 1, 2, 3, 4, or 6 and depends on two properties of the **central atom M**: its size and its oxidation number. Looking first at the effects of the size of the central atom, we recall from Section 2.4 the concept of the **maximum coordination number** of a central atom—i.e., only so many other atoms can be packed around and bonded to an atom of a given size. This number depends in part on the size of the atoms being put around the central atom but must clearly be a function of the period in which the central atom is located, since atoms generally become larger, and can accommodate more neighbors, going down the periodic table. This trend can be seen from the listing of common oxo anions in Table 2.7.

**Table 2.7**   Names and Formulas of the Important Oxo Anions

Oxo anions in which the central atom oxidation number equals the group number

| | | | | | |
|---|---|---|---|---|---|
| $BO_3^{3-}$ borate | $CO_3^{2-}$ carbonate | $NO_3^-$ nitrate | | | |
| $AlO_4^{5-}$ aluminate | $SiO_4^{4-}$ silicate | $PO_4^{3-}$ phosphate | $SO_4^{2-}$ sulfate | $ClO_4^-$ **per**chlorate | |
| $GaO_4^{5-}$ gallate | $GeO_4^{4-}$ germanate | $AsO_4^{3-}$ arsenate | $SeO_4^{2-}$ selenate | $BrO_4^-$ **per**bromate | |
| | $SnO_6^{8-}$ stannate | $SbO_6^{7-}$ antimonate | $TeO_6^{6-}$ tellurate | $IO_6^{5-}$ **per**iodate | $XeO_6^{4-}$ **per**xenate |
| | $PbO_6^{8-}$ plumbate | $VO_4^{3-}$ vanadate | $CrO_4^{2-}$ chromate | $MnO_4^-$ **per**manganate | |
| | | | $MoO_4^{2-}$ molybdate | $TcO_4^-$ **per**technetate | |
| | | | $WO_4^{2-}$ tungstate | $ReO_4^-$ **per**rhenate | $OsO_6^{4-}$ **per**osmate |
| | | | | $NpO_6^{5-}$ **per**neptunate | |

Oxo anions in which the central atom oxidation number is two less than the group number

| | | | | | |
|---|---|---|---|---|---|
| | | $:NO_2^-$ nitrite | | | |
| | | | $:SO_3^{2-}$ sulfite | $:ClO_3^-$ chlor**ate** | |
| | | $:AsO_3^{3-}$ arsenite | $:SeO_3^{2-}$ selenite | $:BrO_3^-$ brom**ate** | |
| $:SnO_3^{4-}$ stannite | $:SbO_3^{3-}$ antimonite | $:TeO_3^{2-}$ tellurite | $:IO_3^-$ iod**ate** | $:XeO_4^{2-}$ xen**ate** | |
| | | | | $:FeO_4^{2-}$ ferr**ate** | |
| | | | | $:RuO_4^{2-}$ ruthen**ate** | |
| | | | | $:OsO_4^{2-}$ osm**ate** | |

The second factor that affects the number of oxo groups present is the *oxidation number* of the central atom. As discussed in Chapter 1, Section 8, among the p-block elements there are two common oxidation numbers, the *group number* and the *group number minus two*. Let us picture an oxo anion as being assembled from a "cation" of charge equal to its oxidation number plus the requisite number of oxide ions, $O^{2-}$. (The bonding within an oxo anion, however, is not ionic but is covalent.) If the p-block cation has an oxidation number equal to the group number minus two, it also has two remaining valence electrons that occupy space around the central p-block atom—at least as much space as an oxo group does. Consequently, p-block oxo anions in which the central atom has an oxidation number two less than the group number will have *at least one less oxo group* than the oxo anion in which the central atom has the group oxidation number. (This can be seen in Table 2.7 by comparing oxo anions of the same element with different oxidation numbers.) For *d*-block elements the oxidation number is not so predictable, and valence electrons remaining on the central atom do not generally take up the space that would otherwise be occupied by an oxo group, so the number of oxo groups does not usually depend on the oxidation number.

Thus larger central atoms are capable of accommodating greater numbers of oxo groups and/or unshared pairs of *p* electrons. (The term **total coordination number** has been devised [10] to indicate the total number of atoms *and* unshared *p*-electron pairs around a central atom.) We can summarize the trends in numbers of oxo groups in oxo anions as follows:

**1.** The smallest central atoms, those of the *p*-block of the second period, have ionic radii of only 27 to 41 pm. In *oxo anions* these central atoms have a *maximum total coordination number of* 3: They can accommodate three oxo groups or two oxo groups and one unshared pair of *p* electrons.

**2.** Central atoms of the third and fourth periods in the *p*-block, and the fourth and fifth periods in the *d*-block, have somewhat larger ionic radii of 41 to 73 pm and have a maximum total coordination number of 4 in oxo anions. Thus if the (non-*s*) valence orbitals of the central atom are 3*p*, 3*d*, 4*p*, or 4*d* orbitals, the central atom can accommodate *four* oxo groups and unshared *p*-orbital electron pairs.

**3.** Central atoms of the fifth and sixth periods in the *p*-block and the sixth period in the *d*-block have still larger ionic radii of 66 to 133 pm and have a maximum total coordination number of 6 in oxo anions. Often the maximum is not reached, especially if the central atom has unshared *p*-electron pairs. Thus if the (non-*s*) valence orbitals of the central atom are 5*p*, 5*d*, 5*f*, 6*p*, or 6*d* orbitals, the central atom can accommodate *four to six* oxo groups and unshared *p*-orbital electron pairs. Some of these oxo anions can exist in either of two forms; e.g., periodate is usually $IO_6^{5-}$ but is sometimes found as $IO_4^{-}$.

The other element of the formula of an oxo anion $MO_x^{y-}$ is its *charge*, $-y$. If we imagine an oxo anion as being assembled from a cation of charge equal to its oxidation number plus $x$ oxide ions, each of $-2$ charge, adding these individual charges gives the charge of the oxo anion.

EXAMPLE       Predict the formulas of the oxo anions formed by the following elements, each with the specified oxidation number: N(III), P(V), As(III), Se(IV), I(VII), I(V), Mn(VI).

Solution      ■ In Figure 2.5 we pictorially assemble these oxo anions from central atoms, oxide ions, and unshared pairs of *p* electrons: The smaller the central atom, the fewer oxo groups and unshared *p*-orbital electron pairs we expect to fit around it. Thus the smallest atom, nitrogen, which uses 2*p* orbitals, is expected to accommodate three unshared electron pairs and oxo groups; we represent this with three large empty

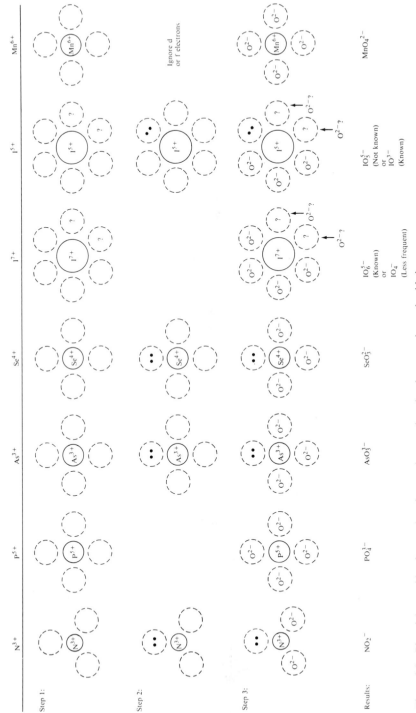

**Figure 2.5**  Pictorial assembly of oxo anions from cations, unshared *p*-electron pairs, and oxide ions. For steps and examples see the text.

circles around the small nitrogen atom in the figure. Since P uses $3p$, As and Se use $4p$, and Mn uses $3d$ valence orbitals, these are represented by larger atoms in the figure, around which there is room for four oxo groups or electron pairs. Around the very large iodine atom ($5p$ orbitals) there is room for four (to six).

■ Next we need to know how many unshared valence electron pairs to place on each $p$-block atom. To do this we compare the oxidation number of each element with its maximum possible oxidation number (group number). In this manner we find that there is one unshared pair of valence $p$ electrons on N(III) and one each on As(III), Se(IV), and I(V); P(V) and I(VII) have none. *This step is omitted for d- and f-block elements.* Thus although Mn(VI) has a single unshared $3d$ electron, it is ignored since it does not compete for space with oxo groups.

■ We fill in each large circle not already occupied by unshared $p$-electron pairs with an oxide ion, $O^{2-}$.

■ We compute the charge on each oxo anion: This is equal to the number of oxo groups times $-2$, added to the oxidation number of the central atom. We finally come up with the predicted formulas: $NO_2^-$, $PO_4^{3-}$, $AsO_3^{3-}$, $SeO_3^{2-}$, $IO_6^{5-}$, $IO_3^-$ (?—prediction unsure), $MnO_4^{2-}$.  □

## 2.9

### Naming of Oxo Anions

#### 2.9.1  Deriving the Name from the Formula

Having determined the formula of an oxo anion, we have to name it. A modern naming system sanctioned by the International Union of Pure and Applied Chemistry (IUPAC) has the following steps: (1) Start the name by indicating the number of oxo groups; (2) give the name of the element, replacing its final suffix (e.g., -ium) with -ate; then (3) indicate, in parentheses, either the oxidation number in Roman numerals or the charge on the ion with a negative number. Thus $NO_2^-$ can be named dioxonitrate(III) ion or dioxonitrate ($-1$) ion. Unfortunately these names have not caught on at all. Oxo anions have been so important in so many fields for so many centuries that an older, more difficult naming system is entrenched in usage.

This older but far more common naming system uses suffixes (and sometimes prefixes) to indicate *relative* oxidation numbers. All names replace the final suffix of the name of the element (e.g., -ium) with either the suffix -ate or the suffix -ite. In addition, a prefix (per- or hypo-) may be added. The most common oxo anion is given the name ending in -ate (e.g., arsenate for $AsO_4^{3-}$). The oxo anion in which the central atom has the next lower oxidation number is given the name ending in -ite (e.g., arsenite for $AsO_3^{3-}$). Any oxo anion in which the central atom has a still lower oxidation number is given the prefix hypo- and the suffix -ite. Finally, if there is a central-atom oxidation number higher than the most common one, the prefix per- and the suffix -ate are used.

For this system of nomenclature the key question is, What is the most common

**Table 2.8**  Relationship between Names of Oxo Anions and the Oxidation Number of the Central Atom

| Prefixes | Suffixes | In Most Groups | In Groups 17(VIIA), 18(VIIIA), 7, 8 |
|---|---|---|---|
| per- | -ate | Does not occur | Oxid. no. = group no. |
|  | -ate | Oxid. no. = group no. | Oxid. no. = group no. $-2$ |
|  | -ite | Oxid. no. = group no. $-2$ | Oxid. no. = group no. $-4$ |
| hypo- | -ite | Oxid. no. = group no. $-4$ | Oxid. no. = group no. $-6$ |

oxo anion? The general pattern is that the most common oxo anion is the one in which the central-atom oxidation number equals the group number, *provided that this does not exceed* 6. Oxidation numbers of 7 and 8 occurring in Groups 17(VIIA), 7, 7*f*, 18(VIIIA), and 8 are considered uncommon and hence are designated "per-ate," as in perchlorate. The oxidation number two less than the group number is generally considered common in these groups and is designated by the suffix -ate.

A complication occurs in the *d*-block, where the oxidation number can be one less than the group number. For clarity it may be best to use the IUPAC system, the suffix -ate plus the oxidation number in Roman numerals in parentheses; but one often sees $MnO_4^{2-}$ called "manganate" and $RuO_4^-$ called "perruthenate." Rather than to attempt to follow the above reasoning, perhaps it will be easier simply to learn the pattern of names given in Table 2.8, which is valid for *p*-block elements and in most cases for *d*-block elements.

EXAMPLE  Give the names of the oxo anions in the previous example.

Solution  ■ It is necessary to check two things: how the oxidation number of the central atom compares with the group number and the group number itself. If the latter is 17(VIIA), 18(VIIIA), 7, or 8, the naming system slips into "high gear." For $NO_2^-$, $PO_4^{3-}$, $AsO_3^{3-}$, and $SeO_3^{2-}$, the element is in a group numbered 6 or less, so if the central-atom oxidation number equals the group number, the suffix -ate is used: $PO_4^{3-}$ is phosphate. For the other three the oxidation number equals the group number minus two: $NO_2^-$ is nitrite, $AsO_3^{3-}$ is arsenite, and $SeO_3^{2-}$ is selenite.

■ For the iodine oxo anions, the names move up one category: $IO_6^{5-}$ is *periodate*, and $IO_3^-$ is iodate. The manganese oxo anion cannot be named directly from Table 2.8, since the oxidation number of manganese is *one* less than the group number. But in Groups 7 and 8, an oxidation number of 6 is given the -ate name, so $MnO_4^{2-}$ is named manganate.  □

## 2.9.2  Deriving the Formula from the Name

To go from the *name* of an oxo anion to its *formula*, somewhat the reverse process is suggested. (1) Identify the group number of the element and the prefixes and suffixes in its name. From this, using Table 2.8, determine the oxidation number of the central atom. Once the oxidation number of the central atom has been determined, the problem is just like the examples in Section 2.8, and the same process is used. (2) From the valence orbitals of the central atom, determine the total coordination number of the central atom. (3) If the element is $p$-block, determine the number of unshared $p$-electron pairs on the central atom; subtract this number from the total coordination number to get the number of oxo groups. (4) To determine the charge on the oxo group, mentally assemble the oxo anion from the central atom as an ion with positive charge equal to its oxidation number, plus one $O^{2-}$ ion for each oxo group. Add all the charges to find the charge on the oxo anion.

EXAMPLE  Give the formulas of the following ions: perxenate, vanadate, chlorite.

Solution  ■ First, V is in a group numbered 5, which is named normally. The suffix -ate tells us that the oxidation number of V equals its group number, so it is $+5$. Xe and Cl are in groups numbered 7 and 8; their names are "geared up." The per- and -ate with xenon tell us, in this case, that the oxidation number equals the group number, $+8$. The -ite suffix with chlorine tells us to drop down two steps (four oxidation numbers) from the group number of Cl to get its oxidation number, $+3$.

■ Second, from their valence orbitals we determine that chlorine and vanadium can each accommodate a total of four oxo groups and unshared pairs of electrons about the central atom, while xenon can take a total coordination number of 4 to 6.

■ Third, we look for unshared pairs of electrons on $p$-block elements. These occur only for chlorite, for which we had to subtract four from the group oxidation number, which means that chlorine has two unshared pairs of electrons; hence chlorite ions come with $4 - 2 = 2$ oxo groups. The Xe has no unshared electrons; thus it may have 4 to 6 oxo groups (6 is more likely in the absence of unshared electron pairs). This step is irrelevant for vanadium, which will have four oxo groups.

■ Finally, we mentally merge cations and oxide ions to generate the oxo anions. Combining $Xe^{8+}$ with six $O^{2-}$, we obtain $XeO_6^{4-}$. Combining $V^{5+}$ with four $O^{2-}$, we obtain $VO_4^{3-}$. Combining $Cl^{3+}$ with two $O^{2-}$, we obtain $ClO_2^{-}$.  □

## 2.10

## Naming and Writing Formulas of Inorganic Salts

The process of naming or writing the formula of an inorganic salt is mainly one of naming or writing the formulas of its ions, which we have already done. A brief summary of how these are brought together for the salt follows.

### 2.10.1  Naming Inorganic Salts

1. From its formula identify the oxo anion and write its name.
2. Identify the charge of the oxo anion.

3. Using the fact that the salt must be electrically neutral, derive the charge of the cation.
4. Name the cation, using its oxidation number in Roman numerals if needed. List the name of the cation before the name of the anion.
5. For strongly and very strongly acidic cations the nonmetal naming system may also be used. For example, $TiO_2$ is normally named titanium(IV) oxide, but it is also often known as titanium dioxide; however, $Mg(NO_3)_2$ is known only as magnesium nitrate (no oxidation number need be specified), never as magnesium dinitrate.

### 2.10.2 Writing Formulas

1. Write the formula of the cation.
2. Write the formula of the anion.
3. Find the electrically neutral combination. For example, calcium arsenate contains the $Ca^{2+}$ cation and the $AsO_4^{3-}$ anion. For the salt to be electrically neutral, the total charges of all cations and anions must add up to zero. The simplest formula results when the smallest possible number of each is used (three $Ca^{2+}$ ions and two $AsO_4^{3-}$ anions). Dropping charges gives us a formula of $Ca_3(AsO_4)_2$.

## 2.11

### Strengths of Oxo Acids; Nomenclature

As we continue to add protons to an oxo anion $MO_x^{y-}$ (either by adding acid and lowering the pH of the solution or by letting hydrolysis occur), we will ultimately arrive at an **oxo acid** of formula $H_yMO_x$, which will be expected to have a structure corresponding to $MO_{x-y}(OH)_y$, still possessing $(x - y)$ oxo groups. Similarly, protonation of a hydroxo anion, followed by the loss of some water molecules, will lead to a metal hydroxide $M(OH)_y$, which can be considered to be an oxo acid with no oxo groups. These oxo acids will have various tendencies to ionize:

$$H_yMO_x + H_2O \rightleftharpoons H_3O^+ + H_{y-1}MO_x^- \tag{2.15}$$

We can characterize the strength of these oxo acids by their $pK_a$'s (the negative logarithms of their acid ionization constants). If necessary these can be calculated from the $pK_b$ of the conjugate base on the right side of equation (2.9), using the relationship:

$$pK_a = 14 - pK_b \tag{2.16}$$

However, from the calculations we find that, at our level of approximation, the *strength of the oxo acid depends only on the number of oxo groups present*:

$$pK_a = 8.5 - 5.7(x - y) \tag{2.17}$$

(Writing the alternative formula for the oxo acid, $MO_{x-y}(OH)_y$, shows the number of oxo groups directly.) Since existing oxo acids of this type have from zero to three oxo groups, this allows us very readily to categorize the oxo acids:

**1.** Oxo acids with three oxo groups (in practice, $HMO_4$) are expected to have $pK_a$'s of about $-8.6$—i.e., to be *very strong acids*. In practice the concentrations of unionized $HMO_4$ left in equilibrium in solutions of such acids are too small to be measured at all accurately, so $pK_a$ cannot be measured in water solution; but there is no doubt that acids such as $HClO_4$ do ionize nearly 100%, and are thus very strong acids. Note that the very strong acids are the conjugate acids of nonbasic anions.

**2.** Oxo acids with two oxo groups ($HMO_3$, $H_2MO_4$, and $H_4MO_6$) are expected to have $pK_a$'s of about $-2.9$—i.e., to be *strong acids*. Again, exact measurements cannot be made, but these acids (such as $HNO_3$ and $H_2SO_4$) are known to ionize extensively in solution. (The second ionization of an acid such as $H_2SO_4$ has a $pK_{a2}$ that is about 4.5 higher than the first ionization and thus is a positive number. This indicates that in its *second* ionization, sulfuric acid does not act as a strong acid.)

**3.** Oxo acids with one oxo group ($HMO_2$, $H_2MO_3$, $H_3MO_4$, and $H_5MO_6$) are expected to have $pK_a$'s of about 2.8—i.e., to be *moderately acidic*. $pK_a$ values such as this can be measured, and have been for a number of these acids: They are found to be within a standard deviation of $\pm 0.9$ of this value [11].

**4.** Oxo acids without oxo groups (hydroxides of the nonmetals) are expected to have $pK_a$'s of about 8.5—i.e., to act as *weak acids*. The measured values for these acids fall within $\pm 1.0$ of this value. Note that the ultimate conjugate bases of most of these weak acids are classified as strongly basic oxo anions.

**5.** The second acid dissociation constants of these oxo acids can be estimated by adding 4.5 to the first ionization constant, and so on for the third and fourth ionizations, if they occur. Each successive ionization is thus less extensive than the previous one, since the remaining protons are attracted to the negative charge left behind by the earlier ionization(s) of one or more protons.

There are, of course, some oxo acids that do not behave as we expect. Notable among these are the oxo acids of phosphorus, $H_3PO_4$, $H_3PO_3$, and $H_3PO_2$, all of which have about the same $pK_a$ (about 2), which would suggest that all have one oxo group. In fact, there is independent evidence that this is the case and that the structures of the latter two have P—H instead of only O—H bonds (see Figure 2.6). Thus on neutralizing $H_3PO_3$ with strong base, only two of the three protons can be removed; with $H_3PO_2$ only one can be removed.

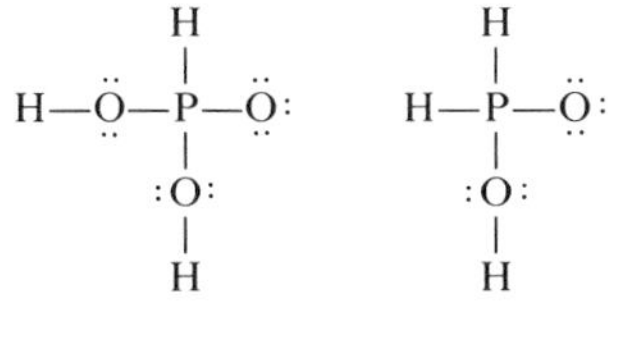

**Figure 2.6** Structures of $H_3PO_3$ and $H_3PO_2$.

The measured $pK_a$ for carbonic acid, 6.38, also seems to be out of line. But it is found that in a solution of carbonic acid, most of the carbon is in the form of hydrated $CO_2$, not $H_2CO_3$. If a correction is made for this, it is found that the $pK_a$ for the $H_2CO_3$ in the solution is 3.58, which is in the expected range.

The nomenclature of oxo acids is based on that of their ultimate conjugate bases, the oxo anions. Put quite simply, if the name of the conjugate oxo anion ends in -ate, the ending of the oxo acid becomes -ic. If the name of the conjugate oxo anion ends in -ite, the ending of the oxo acid becomes -ous. Thus the acid obtained by protonating the perchlor*ate* ion is called perchlor*ic* acid; that obtained from the phosph*ate* ion is called phosphor*ic* acid; that obtained from the sulf*ite* ion is called sulfur*ous* acid; and that obtained from the hypochlor*ite* ion is hypochlor*ous* acid.

## 2.12

## Most Common Forms of the Elements in Water

In this section we sum up the concepts in this chapter by using them to predict the forms in which the different chemical elements are likely to be found in water—say in a natural water such as a lake. The chemistry of the elements in natural waters is actually quite complex and involves types of chemistry that we have not yet dealt with, such as precipitation of salts, complexation by ligands, and oxidation by air or reduction by pollutants. At this time we will have to confine ourselves to discussing a rather unnatural body of natural water, one that contains no oxidizing or reducing agents or complexing agents, and in which no salts (except hydroxides) will be allowed to precipitate. (Distilled water fits this description better than any natural bodies of water, of course.) Since no oxidation-reduction reactions will be allowed to occur, we will assume that each element will be in the positive oxidation state specified in Table B. We will allow only those reactions that we have discussed in this chapter: the hydrolysis of hydrated cations to yield hydroxides or oxides of the elements, oxo acids, hydroxo anions, and oxo anions of the elements. We will allow the pH of the natural water to vary—as it may due to natural buffering on one hand or acid-rain deposition on the other—to see how this will affect the availability of the different elements as either nutrients or pollutants.

One point that we emphasize at the beginning is that the processes we have discussed in this chapter are equilibria, so there will generally be more than one form of an element present at any given pH of the water. This is often expressed (as you will see in the analytical chemistry course) by some form of distribution diagram that shows the relative concentrations of the different forms of an element as a function of pH and perhaps also as a function of the total concentration of all forms of that element. Figure 2.7 indicates (as a function of pH) the fraction of all phosphorus present that is in the form of phosphoric acid ($H_3PO_4$), the phosphate ion ($PO_4^{3-}$), and the two partially protonated phosphate ions $HPO_4^{2-}$ and $H_2PO_4^{-}$.

The concepts developed in this chapter are not designed to produce this detail

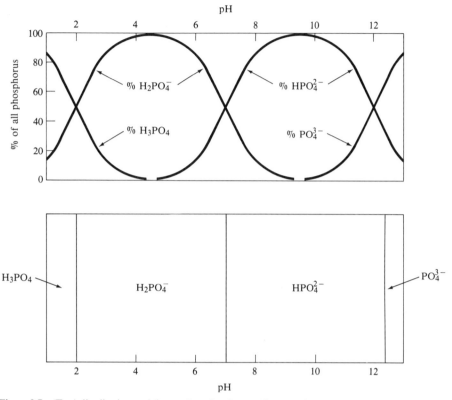

**Figure 2.7** (Top) distribution and (bottom) predominance diagrams for phosphate species.

of knowledge, however; their purpose is to allow us, with a minimum of calculation, to predict which of these species will be the most abundant at about what pH. This kind of information can be expressed in the form of *predominance diagrams* [12] such as those given for phosphorus at the bottom of Figure 2.7 and for other selected elements in Figure 2.8.

Natural waters usually have pH values between 6 and 9. Thus we can see from predominance diagrams such as those in Figure 2.8 that the elements that we might expect to find in such waters predominantly *in solution as hydrated cations* are the nonacidic, feebly acidic, and weakly acidic metal ions (since charge is the most important single variable determining acidity)—i.e., mostly $+1$- and $+2$-charged cations (plus large $+3$ ions, but not very small $+2$ ions). The elements we might expect to find predominantly *as insoluble oxides or hydroxides in the bottom sediments* (or perhaps in colloidal form) are those that give the moderately and strongly acidic cations—small $+3$ cations, most elements with $+4$ oxidation numbers, and metals with $+5$ oxidation numbers. We might also expect to find the elements with high oxidation states in solution in the lake as oxo anions, perhaps partially hydrolyzed (protonated).

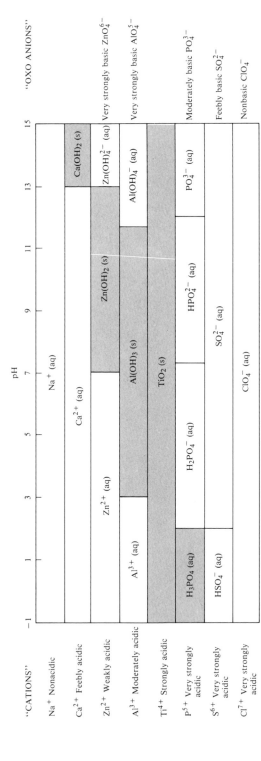

**Figure 2.8** Predominance diagrams for approximately $10^{-3}\,M$ solutions for selected elements in fixed oxidation states, showing relationships of predominant species to categories of acidity of cations and basicity of anions.

If we combine these predictions with our information on the most common oxidation number for each element and exclude the possibility of oxidation-reduction reactions, we can finally predict the probable species and location (hydrated cation in solution, oxide or hydroxide as a precipitate, or oxo anion in solution) for each of the elements. Such a series of predictions is embodied in Table 2.9. Please note, however, that these are not actual results obtained by environmental chemists; they are only rough predictions made under constraints that do not much resemble a real body of water!

EXAMPLE    The following radioisotopes (as fallout from an atomic bomb blast) are being deposited in a lake of pH 5.5 to 7: $^{99}Tc$, $^{90}Sr$, $^{244}Pu$. Without reference to Table 2.9, predict the chemical forms these elements will take in this lake.

Solution    ■ First, we need to predict the oxidation states each element will take. Table B allows us to postulate the presence of the following "cations": $Tc^{7+}$, $Sr^{2+}$, $Pu^{4+}$.

■ Next, we need to evaluate the acidity of each cation to determine whether it can remain as a hydrated ion in neutral water. Recalling that the pH of precipitation of a hydroxide is roughly equal to its $pK_a$, we would expect to find hydroxides precipitating as indicated in Table 2.10. Of the ions in question only $Sr^{2+}$ would be expected to stay in solution as a hydrated cation.

■ For the very strongly acidic cations, we need to evaluate the possibility of forming an oxo anion. In the example only $Tc^{7+}$ qualifies. From the valence orbitals of Tc we predict its oxo anion to have the formula $TcO_4^-$. We calculate the $pK_{b1}$ of $TcO_4^-$ to be 22.6; this ion is nonbasic and would thus be expected to persist in neutral water. Similarly, feebly basic anions would be expected to persist in solution. Partially protonated forms would be expected in natural waters for moderately basic anions (e.g., $HCO_3^-$ for the carbonate ion). Very strongly basic oxo anions are likely to be protonated even beyond the hydroxo-anion state—in such a case the insoluble hydroxide or oxide is the likely species.    □

One consequence of this division is that, by and large, the middle group of elements—those with moderately and strongly acidic cations—are seldom available for biological activity, either in natural waters or in the digestive system or in body fluids, unless chemical processes other than acid-base reactions can somehow make them available. These *unavailable* elements thus are seldom either essential for life (despite the abundance of titanium and silicon, most forms of life do not make use of them) or toxic to life. The biochemically more significant elements are generally present in solution (i.e., either as cations or as oxo anions). (Note, however, that if an element is *exceedingly toxic*, as is $Pu^{4+}$, the fact that most of it may end up in the sludge of a lake may be of little consolation, since even the small amount of the element the equilibria allow to remain in solution may be harmful.)

But if a body of water becomes highly polluted with acidic or basic pollutants,

**Table 2.9  Main Forms of the Elements in Moderately Aerated Water of pH 5.5 to 7**

| 1 | 2 | 3 | 4 | 5 | 6 | 7 | 8 | 9 | 10 | 11 | 12 | 13/IIIA | 14/IVA | 15/VA | 16/VIA | 17/VIIA |
|---|---|---|---|---|---|---|---|---|----|----|----|---------|--------|-------|--------|---------|
| $H_2O$ | | | | | | | | | | | | | | | | |
| $Li^+$ | $Be(OH)_2$ | | | | | | | | | | | $B(OH)_3$ | $CO_2$ / $HCO_3^-$ | $NO_3^-$ | $H_2O$ | $F^-$ |
| $Na^+$ | $Mg^{2+}$ | | | | | | | | | | | $Al(OH)_3$ | $SiO_2$ | $H_2PO_4^-$ / $HPO_4^{2-}$ | $SO_4^{2-}$ | $Cl^-$ |
| $K^+$ | $Ca^{2+}$ | $Sc(OH)_3$ | $TiO_2$ | $H_3V_2O_7^-$ / $H_2VO_4^-$ | $Cr(OH)_3$ | $Mn^{2+}$ / $MnO_2$ | $Fe(OH)_3$ | $Co^{2+}$ | $Ni^{2+}$ | $Cu^{2+}$ | $Zn^{2+}$ | $Ga(OH)_3$ | $GeO_2$ | $H_2AsO_4^-$ / $HAsO_2^-$ | $SeO_4^{2-}$ | $Br^-$ |
| $Rb^+$ | $Sr^{2+}$ | $Y^{3+}$ / $Y(OH)_3$ | $ZrO_2$ | $Nb_2O_5$ | $MoO_4^{2-}$ | $TcO_4^-$ | $Ru(OH)_3$ | $Rh_2O_3$ | $Pd(OH)_2$ | $Ag^+$ | $Cd^{2+}$ | $In(OH)_3$ | $SnO_2$ | $Sb_2O_3$ | $HTeO_3^-$ | $IO_3^-$ |
| $Cs^+$ | $Ba^{2+}$ | $Lu^{3+}$ | $HfO_2$ | $Ta_2O_5$ | $WO_3$ / $WO_4^{2-}$ | $ReO_4^-$ | $OsO_2$ | $IrO_2$ | $PtO_2$ | Au metal | $HgO$ | $Tl^+$ | $Pb^{2+}$ | $Bi_2O_3$ | $HPoO_3^-$ ? | |

| 3 | 4 | 5 | 6 | 7 | 8 | 9 | 10 | 11 | 12 | 13 | 14 |
|---|---|---|---|---|---|---|----|----|----|----|----|
| $La^{3+}$ | $Ce^{3+}$ | $Pr^{3+}$ | $Nd^{3+}$ | $Pm^{3+}$ | $Sm^{3+}$ | $Eu^{3+}$ | $Gd^{3+}$ | $Tb^{3+}$ | $Dy^{3+}$ | $Ho^{3+}$ | $Er^{3+}$ | $Tm^{3+}$ | $Yb^{3+}$ |
| $Ac^{3+}$ | $ThO_2$ | $Pa_2O_5$ | $UO_2^{2+}$ | $NpO_2^+$ | $PuO_2$ | $Am^{3+}$ / $AmOH_4$ | | | | | |

Stable Oxidation Number Equals the Group Number

Lower Oxidation Numbers More Stable

Stable Oxidation Number is 2 less than the Group Number

SOURCE: Forms chosen based on data given in M. Pourbaix, *Atlas of Electrochemical Equilibria in Aqueous Solutions*, NACE, Houston, 1974, with preference being given to positive oxidation states when possible.

NOTE: Shaded areas represent insoluble compounds.

**Table 2.10**  Precipitation of Hydroxides of Cations in Natural Waters

| Cation category | pH of precipitation[a] | Form in | |
|---|---|---|---|
| | | normal waters | acid waters |
| Nonacidic | >14 | Hydrated ion | Hydrated ion |
| Feebly acidic | 11.5–14 | Hydrated ion | Hydrated ion |
| Weakly acidic | 6–11.5 | Hydrated ion | Hydrated ion |
| Moderately acidic | 1–6 | Hydroxide | Hydrated ion |
| Strongly acidic | −4–1 | Hydroxide | Hydroxide |
| Very strongly acidic | < −4 | Hydroxide or oxo anion | |

[a] The pH at which precipitation of the hydroxide is expected, taken as equal to $pK_a$.

it is possible for some of these elements to go into solution. The adverse consequences of acid rain for fish seem to involve just this process. Lakes that are poorly buffered can undergo quite substantial changes in pH when acid rainfall runs off into them. The pH of such a lake can drop to the point that normally insoluble aluminum is converted to $Al^{3+}$ (aq). The toxic effects on fish seem to be due not to the enhanced concentration of hydrogen ion but to the presence of aluminum ion, which is normally not encountered in natural waters.

A rise in the pH of an acidic body of water by dilution or neutralization can cause the weakly and moderately acidic ions to precipitate as hydroxides. This could be a way of removing aluminum ion from a lake "killed" by acid rain, but caution is called for! This difference in pH between the acid water and the more neutral environment of the gills of fish apparently causes gelatinous aluminum hydroxide to precipitate there, coating the gills. The fish seem to sneeze themselves to death in an attempt to get rid of this gelatinous precipitate. Similar chemical reactions are involved in the *acid mine drainage* problem of old coal mines, in which the mineral pyrite, $FeS_2$, is slowly oxidized by air to iron(III) sulfate, $Fe_2(SO_4)_3$. This compound dissolves in the water draining from the mine and, of course, undergoes extensive hydrolysis to give sulfuric acid. As the pH of the acid water is raised again by dilution with uncontaminated streams of water, the hydroxy cation of iron converts to yellow insoluble iron(III) hydroxide, which precipitates as unsightly *yellow boy* along the stream banks.

Some elements that are desired nutrients for plants and animals are also weakly or moderately acidic cations and so may be unavailable in neutral or slightly basic solutions (unless some other type of chemistry can be used to keep them in solution). Many essential micronutrients are weakly acidic +2-charged cations: $Zn^{2+}$, $Cu^{2+}$, $Co^{2+}$, $Fe^{2+}$, $Mn^{2+}$. These precipitate from dilute solutions at pH's above 5.3 to 8.5 [13], so liming (adding CaO to) an acidic soil or lake may remove essential micronutrients. Iron is more commonly found in the form of $Fe^{3+}$, a moderately acidic cation that precipitates from dilute solution at pH's above about 2.0. Thus iron-deficiency anemia cannot be cured by ingesting iron(III) salts or even by eating nails, since in the intestine (where absorption of the iron must occur) the pH is far higher than 2.0.

## 2.13

## Study Objectives

1. Using its charge, radius, and electronegativity, classify a metal ion as nonacidic, feebly acidic, weakly acidic, moderately acidic, strongly acidic, or very strongly acidic.

2. From its charge and number of oxo groups, calculate the $pK_{b1}$ of an oxo anion; then classify it as nonbasic, feebly basic, moderately basic, or very strongly basic.

3. Using its number of oxo groups, classify an oxo acid as weakly acidic, moderately acidic, strongly acidic, or very strongly acidic.

4. Predict the extent of hydrolysis of a metal ion or metal salt in water and tell whether it is possible to reverse this hydrolysis and how.

5. Explain the relationship of hydration energy and acidity of a cation to its charge, radius, and electronegativity.

6. Predict which hydrated metal ions will be colored.

7. Give the formulas of and name oxo anions and acids.

8. Name metal salts of oxo anions or write the formulas of salts given their names.

9. Explain *why* some oxo anions are more basic than others.

10. Explain why, for some elements, hydroxo anions are more commonly found than oxo anions.

11. Describe the effects of changing maximum coordination numbers on the hydrolysis of halides; on the formulas of oxo anions.

12. Using the most likely oxidation state of a given element, predict the form in which it would likely be found (hydrated cation, insoluble oxide or hydroxide, or oxo anion) in nonpolluted or (acid-base) polluted natural waters.

## 2.14

Exercises

1. Give two examples of metal (or nonmetal) cations for each of the following categories: **a.** Gives a neutral solution in water; **b.** gives a faintly acidic solution in water, but the acidity is masked by that due to dissolved carbon dioxide; **c.** gives a weakly acidic solution (comparable in acidity to vinegar); **d.** hydrolyzes reversibly to give a strongly acidic solution; **e.** hydrolyzes irreversibly in water.

*2. Consider the following cations: **a.** $U^{3+}$;    **b.** $Ag^+$;    **c.** $Pa^{5+}$;    **d.** $C^{4+}$; **e.** $As^{3+}$;    **f.** $Tl^+$;    **g.** $Th^{4+}$.

    **2.1** Classify the acidity of each of these cations and describe the reactions of their chlorides with water.

    **2.2** Which of these would give cloudiness or precipitation upon dissolving in water? What could you do to rectify this if it occurred?

    **2.3** If the solutions of these were adjusted to final pH's of 5.5 to 7, in what chemical form would each element be present?

*3. Use equations (2.10) and (2.11) to calculate $pK_a$ values for each of the cations in the previous question and determine whether any of the ions needs to be shifted to another category of acidity.

4. For each category of metal ion listed, give **a.** the typical range of $Z^2/r$ ratios found if the Pauling electronegativity of the metal is less than 1.8; **b.** the typical range of $Z^2/r$ ratios found if the Pauling electronegativity is greater than 1.8; **c.** the approximate pH range at or above which the hydroxide of the metal will precipitate; **d.** two real metal ions from the $d$-block of metals that fall in that category. Categories:
   **4.1** Feebly acidic
   **4.2** Moderately acidic
   **4.3** Weakly acidic
   **4.4** Strongly acidic

5. Describe briefly what will happen when you try to dissolve each of the following compounds in water: **a.** KCl; **b.** $NbCl_5$; **c.** $AlBr_3$; **d.** $CBr_4$; **e.** $BaI_2$; **f.** $IF_7$. Which of these would be likely to fume in air and why?

6. Referring to Table 2.3, compute the smallest and the largest radius that will allow a $+1$ ion to fit into each category of cation acidity; do the same for $+2$, $+3$, and $+4$ ions. Then referring to Table C (and the electronegativity table), list all ions from Table C that fit into each of the categories. Locate the ions of each category on a periodic table to see what sort of periodic pattern results.

†7. Name each of the following oxo anions, calculate its approximate $pK_{b1}$, and tell whether its solution in water will be neutral, feebly basic, moderately basic, or so strongly basic that it will react with the water to form a hydroxo anion: **a.** $CO_3^{2-}$; **b.** $BrO_4^-$; **c.** $IO_6^{5-}$; **d.** $XeO_6^{4-}$; **e.** $AsO_3^{3-}$; **f.** $IO^-$.

†8. Write the formulas of each of the following oxo anions and classify each as nonbasic, feebly basic, moderately basic, or very strongly basic: **a.** silicate; **b.** tellurate; **c.** perbromate; **d.** sulfite; **e.** hypochlorite; **f.** perneptunate; **g.** nitrite; **h.** ferrate.

9. Write the formula (including charge) of the oxo anion of **a.** boron with oxidation number $+3$, and **b.** bromine with oxidation number $+3$.

10. Without using equation (2.14), cancel oxo groups and negative charges to select the appropriate category of basicity for each of the following hypothetical oxo anions: **a.** $MO_5^{2-}$; **b.** $MO_5^{5-}$; **c.** $MO_7^{6-}$. List them in order of increasing basicity.

†11. Name each of the following salts: **a.** $UO_2SO_4$; **b.** $TiCl_3$; **c.** $SO_2Cl_2$; **d.** $FePO_4$; **e.** $Ag_5IO_6$; **f.** $Hg_3TeO_6$.

†12. Give the formulas of the following salts: **a.** bismuthyl nitrate; **b.** strontium perchlorate; **c.** europium(II) sulfite;

    **d.** iron(II) phosphate;     **e.** chromium(II) carbonate;
    **f.** cesium phosphate;     **g.** zinc(II) perbromate;
    **h.** potassium perbromate;     **i.** calcium bromite;     **j.** calcium borate.

13. Explain how and why the basicity of an oxo anion depends on its charge, the number of oxo groups present, and the electronegativity. Use your reasoning to try to predict the approximate basicity of the following more complicated oxo anions: $(O_3P—O—PO_3)^{4-}$ (pyrophosphate ion); $(O_3P—O—PO_2—O—PO_3)^{5-}$ (tripolyphosphate ion).

14. When the use of tripolyphosphates in detergents was under attack, sodium carbonate and sodium silicate were tried as replacements for use in high concentration in detergents. But concern was expressed that such detergents would be very dangerous if infants swallowed them, and might also be quite corrosive to the washing machines. Explain why.

15. Calculate the expected $pK_{b1}$'s of the following oxo anions, which are hypothetical or quite unstable. **a.** $BrO_2^-$;     **b.** $MnO_4^{3-}$;     **c.** $UO_4^{2-}$.

†16. Write the formula of each of the following oxo acids and tell whether its solution in water will be very strongly acidic, strongly acidic, moderately acidic, or weakly acidic: **a.** permanganic acid;     **b.** selenic acid;     **c.** arsenious acid;     **d.** selenous acid;     **e.** telluric acid;     **f.** molybdic acid.

†17. For each of the following elements, give its expected oxidation number, and give the formula of the form (species) in which you would expect to find it in water of pH 5.5 to 7. **a.** Li; **b.** Al; **c.** W. Referring to your answers to questions 13.3 and 13.4 in Chapter 1, do the same for the following: **d.** element number 121; **e.** Rn; and the elements directly below **f.** Np; **g.** No; **h.** Os; **i.** Pb.

18. A lake near the Oak Ridge National Laboratory in Tennessee has become contaminated with plutonium from the reprocessing of spent fuel from nuclear reactors. Predict whether most of the plutonium in this lake is likely to be dissolved in the water or will be found in the sediments at the bottom of the lake. Also predict whether this situation might be altered if the lake were strongly subjected to the effects of acid rain.

19. Two radioactive elements, technetium (Tc, no. 43) and promethium (Pm, no. 61) do not occur naturally on earth but are found in the fallout from atomic bomb explosions. Predict the formulas of the forms (cation, anion, or oxide) in which each would likely be found in a lake of pH 5.5 to 7.

20. Which of the following hydrated metal ions would you expect to be colored? $Bi^{3+}$, $Rb^+$, $Tl^+$, $Cr^{2+}$, $Dy^{3+}$, $Ce^{4+}$.

21. If you did not do Experiments 1 and 2, go back and try to apply the principles of this chapter to predict what would have happened in each experiment. Also answer the questions included in each experiment.

## Notes

[1]  Latimer, W. M., K. S. Pitzer, and C. M. Slansky, *J. Chem. Phys.*, 7, 108 (1939). The original equation given by Latimer has been updated by Roger W. Todd to fit the more modern thermodynamic data and radii.

[2]  Burgess, J., *Metal Ions in Solution*, Wiley, New York, 1979.

[3]  Baes, C. F., Jr., and R. E. Mesmer, *The Hydrolysis of Cations*, Wiley-Interscience, New York, 1976, p. 417.

[4]  The pH of precipitation of a metal hydroxide can be determined from its solubility product expression:

$$K_{sp} = [M^{z+}][OH^-]^z \tag{A}$$

If we multiply this by $[H^+]^z/[H^+]^z$ we obtain:

$$K_{sp} = \frac{[M^{z+}]K_w^z}{[H^+]^z} \tag{B}$$

This equation can be rewritten as an expression for the pH of precipitation [J. E. Fergusson, *Inorganic Chemistry and the Earth: Chemical Resources, Their Extraction, Use, and Environmental Impact*, Pergamon, Oxford, 1982, p. 262.]:

$$pH = \left(\frac{1}{z}\right)\log K_{sp} - \left(\frac{1}{z}\right)\log[M^{z+}] + 14 \tag{C}$$

We can develop an approximate expression for $K_{sp}$ that relates it to $pK_a$ for the metal ion involved. To do this we take the $K_{sp}$ expression (A) and multiply it $z$ times by the equilibrium expression corresponding to $pK_a$:

$$K_a = \frac{[M(OH)^{(z-1)+}][H^+]}{[M^{z+}]} \tag{D}$$

This gives us

$$K_{sp} \times (K_a)^z = \frac{[M(OH)^{(z-1)+}]^z[H^+]^z[OH^-]^z}{[M^{z+}]^{z-1}} \tag{E}$$

This formidable expression is just the product of $K_w^z$ times the equilibrium expression for equation (2.5); hence

$$K_{sp} = \frac{10^{-14z} \times 10^{-5.6}}{(K_a)^z} \tag{F}$$

Taking the logarithm of this expression and substituting it into (C), we obtain our final equation (2.7):

$$pH = pK_a - \left(\frac{1}{z}\right)\log[M^{z+}] - \frac{5.6}{z} \tag{2.7}$$

This expression shows the relationship of $pK_a$ to precipitation, but it is less precise than equation (C) because the number 5.6 in equation (2.7) has a standard deviation of $\pm 3.0$.

[5]  Baes and Mesmer, op.cit., Chapter 18.

[6]  Barnum, D. W., *Inorg. Chem.*, 22, 2297 (1982).

[7]  Dewar, M. J. S., *Organometallics*, 1, 1705 (1982).

[8]  In dot structures of oxo anions these oxo groups will often appear with double bonds to the metal atom but may alternatively show only a single coordinate covalent bond.

[9]  This and the following conclusions concerning $pK_b$ are derived from generalizations given for $pK_a$'s of oxo acids in F. A. Cotton and G. Wilkinson, *Advanced Inorganic Chemistry: A Comprehensive Text*, 4th ed., Wiley-Interscience, New York, 1980, p. 235.

[10]  Jolly, W. L., *Modern Inorganic Chemistry*, McGraw-Hill, New York, 1984, Ch. 3.

[11]  Useful compilations of $pK_a$'s can be found in many places, such as in F. A. Cotton and G. Wilkinson, *Advanced Inorganic Chemistry: A Comprehensive Text*, 4th ed., Wiley-Interscience, New York, 1980, p. 235.

[12]  Baes and Mesmer, op.cit., p. 410.

[13]  Moore, J. W., and E. A. Moore, *Environmental Chemistry*, Academic Press, New York, 1976, p. 360.

# Ionic Solids and Precipitation Reactions of Hydrated Ions

## 3.1

### Solubility Rules for Ionic Solids

In Chapter 2 we examined the acid-base properties of cations in aqueous solution as well as the acid-base properties of oxo anions. We did not consider possible interactions between cations and anions present in the same solution (except for the reaction of cations with hydroxide ions to give insoluble metal hydroxides or oxides). But if a solution contains cations it must also contain anions to preserve its electroneutrality. In Experiment 1 the solutions contained an anion, $Cl^-$, that did not generally perturb the chemistry of the cations being studied; in Experiment 2, the cation present, $Na^+$, did not appreciably alter the reactions of the anions. But, as you may see in Experiment 3, many combinations of cations and anions react with each other to give insoluble precipitates. (In Chapter 5 we will examine combinations that give rise to oxidation-reduction reactions.)

It is important to be able to anticipate when a precipitation reaction will occur in a solution containing several cations and anions. This happens, for example, in the body when the products of the concentrations of certain cations and anions come to exceed the solubility products of their salts, as when calcium ions and oxalate ions precipitate as calcium oxalate, one form of kidney stone. A laboratory worker preparing a buffered solution of a certain metal ion for biological or medical studies may be upset to find the metal ion giving a precipitate with the anion of the buffering agent. Conversely, a worker preparing a solution of a metal ion or of an oxo anion nearly always does this by dissolving an ionic salt in water. In preparing a standard solution of the barium ion for use in analytical chemistry, he or she may go to the stockroom to find a barium salt, and find not one but many different salts; he or she certainly does not want to select a salt that is insoluble in water! If a table containing the relevant solubility products ($K_{sp}$'s) is at hand, there is no problem; but often this is not the case. Qualitative knowledge of the solubility (precipitation) properties of inorganic salts is then quite useful.

Unfortunately, the solubility properties of ionic salts in water present certain complexities not found in the solubilities of other classes of materials. There is a general saying, "Like dissolves like," which is often encountered in general chemistry. Nonpolar covalent materials quite generally dissolve in nonpolar covalent solvents; polar hydrogen-bonding molecules generally dissolve in hydrogen-bonding solvents. Ionic salts are usually not used as solvents, however, because they are not generally liquids below inconveniently high temperatures. (There are a few exceptions, and chemistry in molten ionic salts as solvents is an important area of current research.) Unless some specific bonding interaction is present, ionic salts will not dissolve in nonpolar covalent solvents. But with respect to the most important polar covalent solvent, water, no simple blanket statement can be made—many ionic salts are soluble in water, and many are not.

The need to be able to anticipate the solubility properties of salts leads many general chemistry textbooks to present a series of solubility rules, arranged by anion: "All sulfates are soluble except those of $Ca^{2+}$, $Sr^{2+}$, $Ba^{2+}$, $Hg^{2+}$, $Pb^{2+}$, and $Ag^+$," and so on. Due to the limited appeal of memorizing rules such as these, the rules do not cover all the cations, and cover only a few anions, so even if the student remembers these rules after the exam, they may not cover the situation at hand. A more reliable source of information is a chemical handbook such as the *Handbook of Chemistry and Physics* [1], which has an extensive tabulation of "Physical Constants of Inorganic Compounds." Even this tabulation can cover only a fraction of all the possible or known inorganic salts, however. Thus there is a need for generalized solubility rules that can easily be applied to predict the solubility of a wide variety of inorganic salts.

At this point your instructor may want you to carry out (or observe and discuss) Experiment 3, which gives observations from which you may be able to devise generalized solubility rules.

## 3.2

## Thermodynamics and Generalized Solubility Rules for Salts

Several approaches to the problem of solubilities of inorganic salts have been taken by various authors. All are based on the thermodynamics of the precipitation reaction (3.1) (or of its reverse, the process of dissolution of an ionic salt):

$$y M^{m+}(aq) + m X^{y-}(aq) = M_y X_m(s) + p H_2O \qquad (3.1)$$

Thermodynamically, whether reaction (3.1) gives rise to a precipitate (with unit concentrations or activities of the ions) depends on the Gibbs free energy change, $\Delta G°$, for the reaction, which of course equals $\Delta H° - T\Delta S°$. For many reactions at room temperature the entropy term $-T\Delta S°$ is small compared with the enthalpy term $\Delta H°$ and can be neglected. But it is usually found (as in Experiment 3) that there is little heat evolution or absorption involved in the precipitation (or dissolution) of an ionic salt; hence in analyzing the thermodynamics of this reaction, it is necessary to consider both the enthalpy change and the entropy change. Thermodynamic data for the precipitation of some salts are shown in Table 3.1.

**Table 3.1**  Thermodynamic Data on Precipitation

| Salt | $\Delta G°$ | $\Delta H°$ | $-T\Delta S°$, 298 K | Solubility (mol/kg $H_2O$) |
|------|------|------|------|------|
| **I. Acidic Cations + Basic Anions** | | | | |
| $Be(OH)_2$ | $-121$ | $-31$ | $-90$ | 0.000008 |
| $Mg(OH)_2$ | $-63$ | $-3$ | $-61$ | 0.0002 |
| $Ca(OH)_2$ | $-28$ | 16 | $-44$ | 0.025 |
| $Li_2CO_3$ | $-17$ | 18 | $-34$ | 0.18 |
| $MgCO_3$ | $-45$ | 28 | $-74$ | 0.0093 |
| $CaCO_3$ | $-48$ | 10 | $-57$ | 0.0002 |
| $SrCO_3$ | $-52$ | 3 | $-56$ | 0.00007 |
| $BaCO_3$ | $-47$ | $-4$ | $-43$ | 0.00011 |
| $FePO_4$ | $-102$ | 78 | $-180$ | slight |
| **II. Nonacidic Cations + Nonbasic Anions** | | | | |
| $KClO_4$ | $-12$ | $-51$ | 39 | 0.054 |
| $RbClO_4$ | $-14$ | $-57$ | 43 | 0.027 |
| $CsClO_4$ | $-12$ | $-55$ | 44 | 0.034 |
| $NaNO_3$ | 6 | $-21$ | 27 | 8.59 |
| **IIIa. Acidic Cations + Nonbasic Anions** | | | | |
| $Mg(NO_3)_2$ | 89 | 85 | 4 | 1.65 |
| $Ca(NO_3)_2$ | 32 | 20 | 13 | 2.08 |
| $Sr(NO_3)_2$ | $-3$ | $-18$ | 14 | 1.89 |
| $Ba(NO_3)_2$ | $-13$ | $-40$ | 27 | 0.33 |
| $Mg(ClO_4)_2$ | 144 | 141 | 4 | 2.24 |
| $Ba(ClO_4)_2$ | 46 | 12 | 34 | 5.91 |
| **IIIb. Nonacidic Cations + Basic Anions** | | | | |
| $KOH$ | 62 | 55 | 7 | 19.1 |
| $RbOH$ | 74 | 63 | 11 | 17.6 |
| $CsOH$ | 83 | 71 | 12 | 26.4 |
| $K_2CO_3$ | 36 | 35 | 1 | 8.12 |
| $Rb_2CO_3$ | 50 | 41 | 9 | 19.5 |
| $Cs_2CO_3$ | 73 | 62 | 10 | 8.0 |
| **IV. Sulfates (Feebly Basic Anions)** | | | | |
| $Al_2(SO_4)_3$ | 96 | 338 | $-241$ | 0.92 |
| $BeSO_4$ | 59 | 123 | $-64$ | 2.40 |
| $MgSO_4$ | 30 | 91 | $-61$ | 2.88 |
| $CaSO_4$ | $-27$ | 18 | $-45$ | 0.014 |
| $SrSO_4$ | $-34$ | 9 | $-43$ | 0.0006 |
| $BaSO_4$ | $-50$ | $-19$ | $-31$ | 0.00001 |

SOURCES: C. S. G. Phillips and R. J. P. Williams, *Inorganic Chemistry*, Oxford University Press, Oxford, 1965, p. 254; D. A. Johnson, *Some Thermodynamic Aspects of Inorganic Chemistry*, Cambridge University Press, Cambridge, 1968, p. 107; B. G. Cox and A. J. Parker, *J. Amer. Chem. Soc.*, 95, 6879 (1973); *Handbook of Chemistry and Physics*, 36th ed., Chemical Rubber Publishing Co., Cleveland, 1954, p. 1682.

NOTES: Numbers given in the table are the standard free energies, enthalpies, and entropies of precipitation—i.e., for reaction (3.1) of the salts listed, in kJ/mol. Solubilities listed are often for the hydrated salts, for temperatures between 0 °C and 30 °C, and are calculated from data in the *Handbook of Chemistry and Physics*.

The data in Table 3.1 suggest the variety of causes that must exist for the solubility or insolubility of given salts. Solubilities of less than about 1 mol/L (neglecting activity effects), which show up in the table as negative free energies of precipitation, may be due to negative entropy terms ($-T\Delta S°$, i.e., positive entropies) for precipitation (as in part I of the table) or to negative enthalpies of precipitation (as in part II). Solubilities of greater than 1 mol/L seem generally to be due to positive enthalpies of precipitation (part III). To develop generalized solubility rules, however, one must go beyond such a tabulation and determine which cations and which anions will give combinations falling in parts I, II, and III of the table. This has been done mathematically by Johnson [2] in terms of the radii and of the charges of the cations and anions involved, and by Rich [3] in terms of the basicity of the anions involved. (Obviously, the approaches are related since acidity and basicity of cations and anions are functions of their charges and radii.) For ease of use and compatibility with the concepts of the previous chapter, we present the following generalized solubility rules based on acidity and basicity as related to Table 3.1. In the following sections we will return to thermodynamics to seek the physical justification for these rules.

I. ACIDIC CATIONS + BASIC ANIONS   Since most metal cations found in aqueous solution are at least weakly acidic, and most oxo anions are at least moderately basic, the reaction of an acidic cation and a basic anion is especially important. Part I of Table 3.1 shows that these ions do indeed generally react to give precipitates: $\Delta G°$ for the precipitation reaction (3.1) is negative. Hence, in general, *acidic cations and basic anions give insoluble salts*. By *acidic cations* we mean any weakly, moderately, strongly, or very strongly acidic cations. The term *basic anions* refers to any anions that are weakly, moderately, strongly, or very strongly basic. (Although not all these categories are found among simple oxo anions, they will be found later among other types of anions.)

This generalized solubility rule approximately sums up two commonly given "solubility rules": (*a*) All carbonates are insoluble except those of the Group 1 elements and $NH_4^+$ (which acts like $K^+$ in many respects); and (*b*) all hydroxides are insoluble except those of the Group 1 elements, $Sr^{2+}$, and $Ba^{2+}$. The generalized solubility rule gives up some exactness in terms of the slightly different behavior of hydroxides and carbonates, but it is clearly applicable to the salts of a number of other oxo anions, such as phosphate, arsenate, silicate, and borate, and it also suggests the behavior of some cations not considered in rules (a) and (b): Thus TlOH, $Eu(OH)_2$, and $Tl_2CO_3$ are also all soluble.

II. NONACIDIC CATIONS + NONBASIC ANIONS   Although there are relatively few nonacidic cations and nonbasic anions, this combination usually also gives rise to moderately insoluble salts, as shown by the small negative free energies in part II of Table 3.1. Hence, the generalized solubility rule here is that *nonacidic cations and nonbasic anions give insoluble salts*.

IIIA. ACIDIC CATIONS + NONBASIC ANIONS/IIIB. NONACIDIC CATIONS + BASIC ANIONS   The positive free energy changes shown for the precipitation of combinations of acidic cations plus nonbasic anions (part IIIa of the table) and for combinations of nonacidic cations and basic anions (part IIIb) suggest the generalized solubility rule: *Cross-combinations give soluble salts.* By *cross-combinations* we mean nonacidic cations plus basic anions and acidic cations plus nonbasic anions. The most common solubility rule related to this rule of thumb is that for nitrate, "All nitrates are soluble." Since most cations are acidic, their soluble oxo salts will be of nonbasic anions such as nitrate and perchlorate. (The nitrates of some of the most nonacidic cations such as $Ba^{2+}$ and $Cs^+$ are not highly soluble, however.)

IV. SULFATES (FEEBLY BASIC ANIONS)   Some anions cannot clearly be called basic or nonbasic in terms of these trends; there are *feebly basic* anions such as $SO_4^{2-}$, $SeO_4^{2-}$, and $MoO_4^{2-}$. The solubility of salts of these ions is particularly complex and difficult to predict; as can be seen from Table 3.1, their tendency to precipitate or not to precipitate varies with the relative magnitude of the opposing entropy and enthalpy terms. We will need to carry out a more detailed thermodynamic analysis before we come up with a satisfactory rule for predicting the solubilities of salts of feebly basic anions. For now we may note that *many feebly acidic cations and feebly basic anions give insoluble salts.*

V. It is perhaps useful to point out here that *the "solubilities" of insoluble salts of basic anions are enhanced in solutions of strong acids.* This is not a true case of solubility, perhaps, but of reaction: A basic anion tends to react more completely with the strong acid $H_3O^+$ than with a less acidic metal ion, thus generating a new salt of that metal ion and a weak acid in solution. (Of course this may not occur appreciably if the metal ion is also strongly acidic.) The equilibrium that results from the competition between $H_3O^+$ and the metal ion will be especially enhanced if the weak oxo acid decomposes; thus carbonates and sulfites readily "dissolve" in strong acids, since the carbonic and sulfurous acids produced decompose to carbon dioxide and sulfur dioxide, gases that escape from the solution and allow the equilibrium to shift further toward dissolution.

EXAMPLE   Without reference to a table of solubility products, classify each of the following salts as soluble or insoluble in water and in strongly acidic solutions: $Ag_2SeO_4$, $K_3PO_4$, $ZnSO_4$, and $Th_3(PO_4)_4$.

Solution   ■ The charges of the cations and anions in each salt must be identified so that their acidity and basicity can be classified. The cations are $Ag^+$, $K^+$, $Zn^{2+}$, and $Th^{4+}$. Referring to Table C, we see that $Z^2/r$ ratios for these four cations are computed as 0.008, 0.007, 0.046, and 0.145, respectively, which allows the first two to be categorized as nonacidic, the third as weakly acidic, and the fourth as moderately acidic. The table of electronegativities, however, shows that Ag has an electronegativity exceeding 1.8, so it should be reclassified as feebly acidic.

■ The anions are $SeO_4^{2-}$, $PO_4^{3-}$, and $SO_4^{2-}$. Any oxo anion with a $-3$ or higher charge (such as phosphate) is automatically at least moderately basic. A calculation using equation (2.14) (or reference to Table 2.6) classifies sulfate and selenate as feebly basic; hence the salt $Ag_2SeO_4$ is a combination of a feebly acidic cation and a feebly basic anion and should be insoluble; $K_3PO_4$ is a combination of a nonacidic cation and a basic anion and should be soluble; $ZnSO_4$ is a combination of an (weakly) acidic cation and a feebly basic anion and should be soluble; $Th_3(PO_4)_4$ is a combination of an (moderately) acidic cation and a (moderately) basic anion and should be insoluble. A check of the *Handbook of Chemistry and Physics* confirms all these predictions.

■ Solubility in acid is of interest for compounds that are insoluble in water; we may then want to dissolve them in a strong acid such as HCl. If the anion of the insoluble salt is basic, and the cation is less strongly acidic than the (very strongly acidic) hydrogen ion, we would anticipate a chemical reaction to give the metal ion in solution and the oxo acid. The feebly basic $SeO_4^{2-}$ ion will react somewhat with $H^+$, so we expect (and find) somewhat improved solubility in strong acids. The moderately basic anion $PO_4^{3-}$ will react more with the very strong acid $H^+$ than with the moderately acidic $Th^{4+}$ ion, so we also find it to be soluble in 30% HCl.

□

## 3.3

## Entropy and Precipitation: The Structures of Hydrated Ions and Liquid Water

It may seem surprising at first that precipitation reactions of acidic cations and basic anions should be due to a negative entropy term, $-T\Delta S$, since this term corresponds to increasing randomness or disorder in the system, which is not what we expect in a reaction that produces a crystalline solid precipitate. But the reacting cations and anions exist as hydrated ions, and upon formation of the precipitate, water molecules are released. If enough water molecules are released (and if these molecules are sufficiently disordered as part of the water solvent), then the resulting positive entropy change (negative $-T\Delta S$) may exceed the ordering effect of producing a crystalline precipitate.

From Table 3.1 we can see that large negative $-T\Delta S$ terms occur characteristically only upon the reaction of acidic cations with basic anions. It has been suggested by Latimer [4] that the entropy change in precipitation reactions is independent of the type of crystalline product formed; this has made it possible, with some further assumptions, to assign separate entropies of precipitation for the hydrated cations and for the hydrated anions. (The actual practice has been to consider the reverse of equation (3.1)—the dissolution of an ionic solid—and to tabulate entropies of solution for the cations and anions [5].) Theoretically, these entropies of solution of cations or anions should also be related to the familiar ratio $Z^2/r$ [6], as suggested by Figure 3.1.

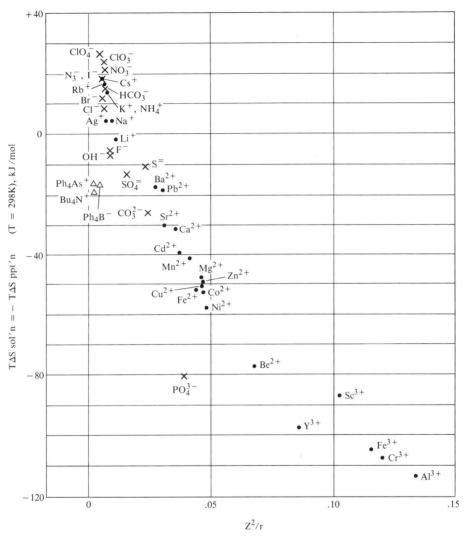

**Figure 3.1** 298 °K entropy terms ($T\Delta S_{solution} = -T\Delta S_{precipitation}$) for cations and anions as a function of $Z^2/r$. Crosses equal anions; dots equal metal cations; triangles equal organic ions. Thermodynamic data from [5]; thermochemical radii of oxo anions from J. E. Huheey, *Inorganic Chemistry*, 3d ed., Harper and Row, New York, 1983, p. 78. Points in parentheses are based on estimated radii.

It is readily apparent from Figure 3.1 that *all but the nonacidic cations* and *all but the nonbasic anions* make negative contributions to the entropy term for reaction (3.1): Most ions release enough water molecules to give a net disordering effect upon precipitation or a net ordering effect upon dissolution of a salt. Such ions are referred to as **electrostatic structure makers**.

We also note that the more acidic or basic the ion (the higher its charge or

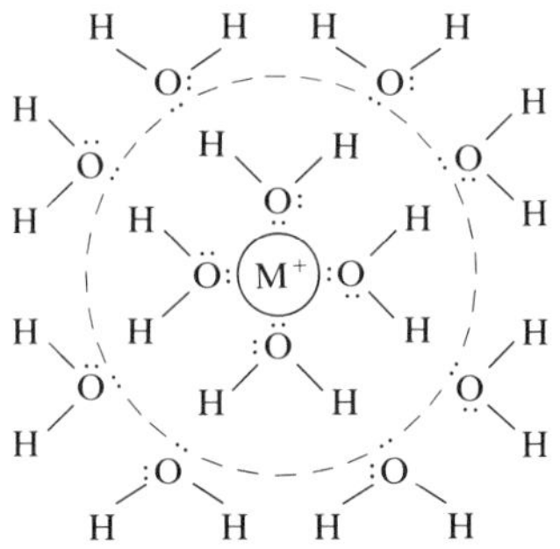

**Figure 3.2**   Primary and secondary hydration spheres of a hydrated cation.

the smaller its radius), the greater is the magnitude of the ion's structure-making property. The inverse relationship to size may seem surprising: Since larger cations have greater maximum coordination numbers, we might expect them to attach more water molecules. But evidently the smaller ion attaches more water molecules or at least orders them more effectively. Similarly, although increasing the charge of a cation does not increase its maximum coordination number, it does increase the ordering of water molecules in the hydrated ion.

These seeming contradictions can be explained by postulating that hydrated cations and anions have more complex structures than we have assumed up to now. Certainly all hydrated metal ions and oxo anions have a layer of water molecules surrounding the ion, as shown in Figure 2.1. This layer is referred to as the **primary hydration sphere** of the hydrated ion, and the larger the "bare" cation or anion, the more water molecules that can be accommodated in this primary hydration sphere. But ions that are at all acidic or basic exert such a strong attraction for the water molecules in their primary hydration sphere, pulling electron pairs toward a cation or hydrogens toward an anion, that these water molecules exert an enhanced hydrogen-bonding attraction for other water molecules and organize them into a **secondary hydration sphere** around the first layer (Figure 3.2). Note that this phenomenon has the same cause as does the phenomenon of hydrolysis that we discussed in Chapter 2. (In this case, however, the attraction does not have to be strong enough to pull the water molecule of the second layer apart—it only needs to attach it by strengthened hydrogen bonds.) Thus we can see why this phenomenon should be directly related to the acidity or basicity of the cation or anion.

This secondary hydration sphere may consist of more than one layer of water molecules—the stronger the attraction of the bare ion for water molecules (i.e., the greater its acidity or basicity), the more water molecules that may be attached in additional layers. Each subsequent layer would be expected to involve weaker attractions, however, so water molecules would stay immobilized by hydrogen bonds for shorter and shorter periods of time. Many methods have been used to study this phenomenon of multiple hydration layers, but these diverse methods of measuring the number of waters attached or the radius of the hydrated ion give

**Table 3.2**  Hydration Numbers and Hydrated Radii of Some Hydrated Ions

| Ion | $Z^2/r$ | Hydration Number | Hydrated Radius (pm) |
|---|---|---|---|
| $Cs^+$ | 0.0055 | 6 | 228 |
| $K^+$ | 0.0066 | 7 | 232 |
| $Na^+$ | 0.0088 | 13 | 276 |
| $Li^+$ | 0.0111 | 22 | 340 |
| $Ba^{2+}$ | 0.0268 | 28 | |
| $Sr^{2+}$ | 0.0303 | 29 | |
| $Ca^{2+}$ | 0.0351 | 29 | |
| $Mg^{2+}$ | 0.0465 | 36 | |
| $Cd^{2+}$ | 0.0549 | 39 | |
| $Zn^{2+}$ | 0.0599 | 44 | |

SOURCES: Hydration numbers from A. T. Rutgers and Y. Hendrikx, *Trans. Faraday Soc.*, 58, 2184 (1962). Hydrated radii from R. P. Hanzlik, *Inorganic Aspects of Biological and Organic Chemistry*, Academic Press, New York, 1976, p. 31.

NOTE: $Z^2/r$ ratios corrected for electronegativity using equation (2.11).

inconsistent answers, since each method requires a different amount of time for the water molecules to remain attached in order to be counted. But nearly all measurements give consistent *trends* in hydration numbers among series of ions, such as the results for the ions in Table 3.2, obtained from transference measurements by way of the *Stokes radii* or *hydrated radii* of the hydrated ions, also in Table 3.2.

In summary, the reason that most acidic cations and basic anions react to give precipitates is due to the disorder resulting from the release of numerous water molecules from the multiple hydration spheres of the cations and of the anions. This is apparent when we write the seemingly simple precipitation reaction of, say, magnesium and carbonate ions in full, including a reasonable number of water molecules:

$$Mg(H_2O)_{36}{}^{2+} + CO_3(H_2O)_{28}{}^{2-} \rightleftharpoons MgCO_3(s) + 64\,H_2O \tag{3.2}$$

Since the *hydrated* ions of smaller bare ions may be larger than those of larger bare ions, we sometimes see unexpected chemical results. For example, an important question in biochemistry has to do with how metal ions can migrate across cell membranes, which consist largely of lipids in which metal ions are quite insoluble. One model of this process, called the *pore model* [7], envisions the cell membrane as containing pores through which hydrated ions can pass without having to dissolve in the lipids. The antibiotic gramicidin A in fact contains a hollow center of about 400-pm diameter that, when in biological membranes, causes cells to pass ions readily. The relative rate of passage of ions through this pore is $Cs^+ > Rb^+ > K^+ > Na^+ > Li^+$, which makes sense if the hydrated cesium ion is indeed the smallest hydrated ion in this series.

3.4

## Electrostatic Structure-Breaking Ions and the Structure of Liquid Water

Since all metal ions form hydrated ions in which water molecules are presumably more ordered than they are in liquid water, it may seem surprising that not all metal ions are classified as electrostatic structure makers. As can be seen from Figure 3.1, the nonacidic cations ($Na^+$ through $Cs^+$) and the nonbasic anions have positive entropies of solution and thus have been termed **electrostatic structure breakers**. To understand how this can be, we must realize that liquid water itself has a structure that can be broken by large ions dissolved in it.

Due to the high polarity of the O—H bond, each water molecule (like many other kinds of molecule) can form hydrogen bonds to other molecules of the same kind. Unlike any other small molecule, however, the $H_2O$ molecule can form *two* hydrogen bonds using its two hydrogen atoms, and it can complete *two other* hydrogen bonds using its two unshared electron pairs. Thus in ice, a very extensive network of hydrogen bonds allows each water molecule to be attached to four other water molecules (Figure 3.3). This arrangement of hydrogen bonds between water molecules results in large cavities within the network of water molecules, which causes ice to be of rather low density. Molecules are normally more tightly packed in the solid state than they are in the liquid state. Thus the solid form of a substance is usually more dense than its liquid form. But such is not true of ice—ice floats in water!

When ice melts to form liquid water, enough hydrogen bonds must be broken to allow molecules to flow around each other: Thus ice has an unusually high heat of fusion. But there is good evidence that not all the hydrogen bonds are broken,

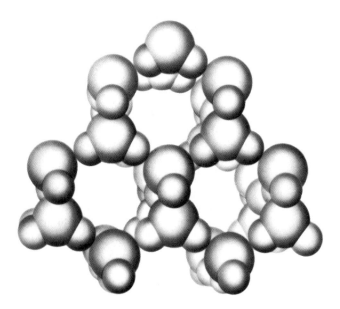

**Figure 3.3** The structure of ice. The large spheres represent oxygen atoms, and the small ones hydrogen. Reproduced from *Chemical Principles*, by W. L. Masterton, E. J. Slowinski, and C. L. Stanitski. Copyright © 1981 by Saunders College Publishing. Reprinted by permission of CBS College Publishing.

and not all the extensive organization and structure of ice is lost. The structure of liquid water is a controversial topic, but one school of scientists [8] describes it as containing short-lived "icebergs" of structured water amid other non-hydrogen-bonded water molecules (about 80% of the water molecules are supposed to be present in icebergs). Some of the free water molecules then can fill up the cavities present in the icebergs, giving liquid water a density higher than that of ice. (X-ray studies indicate that each water molecule has an average of 4.4 other water molecules near to it, rather than 4.0 as in ice.)

The large size of the nonacidic cations and anions is thought to cause disruption of the iceberg structure in the nearby liquid water, hence giving these ions their positive entropies of solution and earning them the title *electrostatic structure breakers*. Interestingly, nonacidic cations and anions having organic chains attached (such as the tetrabutylammonium cation, $(C_4H_9)_4N^+$, or the tetrabutylborate anion, $(C_4H_9)_4B^-$) have the opposite effect. Their long hydrocarbon chains (containing nonpolar C—H and C—C bonds) do not hydrogen bond to water, but instead slip inside the cavities of the icebergs, forcing out the loose water, which results in the production of more iceberg structure. Thus such ions are known as **hydrophobic structure makers**.

## 3.5

## Enthalpy and Precipitation: The Importance of Lattice Energies

Regardless of whether they are inorganic electrostatic structure breakers or organic hydrophobic structure makers, nonacidic cations and nonbasic anions behave similarly with regard to solubility: They form insoluble salts with each other. Clearly, their structure-making or structure-breaking properties must not be responsible, since they differ. Examination of the data in Table 3.1 shows that the insolubility of these salts is due, not to the entropy term, but to the *enthalpy* term. Likewise, the solubility of cross-combination salts (part III of Table 3.1) is due to the enthalpy term; the salts listed include one type of ion that is structure making and one that is structure breaking, so the entropy term is very small for these soluble salts. Clearly, we must also understand the nature of the enthalpy term in order to understand the solubility or insolubility of salts.

The enthalpy of precipitation of an ionic salt (i.e., $\Delta H$ for reaction (3.1)) can be related to the properties of the cation and anion involved if we break down the relatively complex reaction (3.1) into three simpler steps: (1) The hydrated cation is dehydrated and converted to a gaseous bare cation; (2) the same is done to the anion; and (3) the gaseous cations and anions are then allowed to come together to form the ionic solid. (This process is shown schematically in Figure 3.4.) By Hess's law, the enthalpy change for the overall reaction (3.1) must equal the sum of the enthalpy changes for each of the component steps.

Steps 1 and 2 are simply the reverse of the processes of formation of the hydrated cations and hydrated anions, so their enthalpy changes are readily obtained by reversing the signs of the hydration enthalpies given in Tables 2.1 and

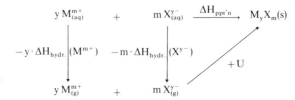

**Figure 3.4** Thermodynamic cycle for analysis of the precipitation of a solid.

2.5. The enthalpy change for step 3, called the **lattice energy**, is due to the attraction of oppositely charged ions for each other, which results in their condensing to form a crystalline solid. These three energy terms add up to give the overall enthalpy of precipitation, given in Table 3.1. Thus we can evaluate the lattice energies of several salts. For example, for the reaction of nonacidic $K^+$ with nonbasic $ClO_4^-$, we have the overall relationship

$$\Delta H_{pptn}(KClO_4) = -\Delta H_{hyd}(K^+) - \Delta H_{hyd}(ClO_4^-) + U(KClO_4) \tag{3.3}$$

where $U$ represents the lattice energy. Taking data from Tables 2.1, 2.5, and 3.1, we find numerically:

$$-51 \text{ kJ} = -(-321) \text{ kJ} - (-227)\text{kJ} + U(KClO_4) \tag{3.4}$$

from which we find $U(KClO_4)$ to be $-599$ kJ, a very large energy term indeed.

EXAMPLE    Calculate the lattice energies of $Ba(ClO_4)_2$ and of $Mg(ClO_4)_2$.

Solution    In doing this summation, it is necessary to multiply each energy (in kJ/mol) from the tables by the number of moles of that species involved. The hydration enthalpies are given in kilojoules per mole of ion; for each of these salts we are using 2 mol of perchlorate ion. The enthalpies of precipitation are per mole of product; the question implies that 1 mol of each salt is produced. Thus for $Ba(ClO_4)_2$ we have $(+12 \text{ kJ/mol}) \times (1 \text{ mol}) = -(-1304 \text{ kJ/mol}) \times (1 \text{ mol}) - (-227 \text{ kJ/mol}) \times (2 \text{ mol}) + U$; solving for $U$ gives us $-1746$ kJ/mol. Similarly the data for $Mg(ClO_4)_2$ give us a lattice energy of $-2235$ kJ/mol.    □

Clearly, lattice energies are very large (negative) energy terms. Hydration enthalpies are also very large (negative) energies; when we take the difference of these, (usually) only a very small enthalpy of precipitation remains. Nonetheless, this small enthalpy of precipitation is what causes $KClO_4$ to be insoluble and $Ba(ClO_4)_2$ and $Mg(ClO_4)_2$ to be soluble. But the greatest lattice energy, that of magnesium perchlorate, is not enough to cause that compound to precipitate, while the smallest lattice energy, that of potassium perchlorate, is sufficiently large to cause that compound to be insoluble! What matters is not how large the lattice energy is in an absolute sense, but whether the lattice energy is larger than the combined hydration enthalpies of the component ions. (Magnesium perchlorate also has the largest hydration enthalpies.) In the next sections we will examine the nature of the lattice energy in much greater depth, and we will see that it has much in common with the hydration enthalpies. We will try to see why, in the case of nonacidic cations and nonbasic anions, the lattice energies are larger (more negative) than the sums of the hydration energies, thus producing insoluble salts, while in the case of acidic cations plus nonbasic anions (or nonacidic cations plus basic anions) the reverse is true, producing soluble salts.

## 3.6

## Ionic Solids: Coulombic Attractions and Lattice Energies

The large lattice energies of ionic solids result from the strong attractions of the oppositely charged cations and anions. The energy of such an attraction can be derived from Coulomb's law in physics and can be expressed as

$$E = \frac{Z_+ Z_- e^2 N}{4\pi\varepsilon_0 r} \tag{3.5}$$

In this equation, the variables are the charge on the cation, $Z_+$, the charge on the anion, $Z_-$, and the distance between the two, $r$. Thus it has dimensions of charge squared over distance and is another example of this familiar type of ratio in chemistry. This equation includes certain constants needed to give the answers in SI units: $e$ is the charge on the electron, $1.602 \times 10^{-19}$ C (coulombs); N is Avogadro's number (so that the result will be expressed per mole of ionic compound); and $\varepsilon_0$ is the dielectric constant (permittivity) of a vacuum, $8.854 \times 10^{-12}$ $C^2 m^{-1} J^{-1}$. We prefer, however, to use pm instead of SI units of length (m), and to have our answer in kJ/mol not in J/mol. Incorporating the appropriate conversion factors and collecting all constant terms into one term, we simplify this equation to

$$E = \frac{138,900 Z_+ Z_-}{r} \tag{3.6}$$

This is an exothermic interaction ($E$ is negative), since the charge of the anion is negative.

In developing the relationship between this attractive energy and the lattice energy, we find it desirable to use spherical cations and anions—i.e., not oxo anions. Hence, we will consider the attraction of 1 mol of sodium ions and 1 mol of chloride ions; experimentally the lattice energy of NaCl is found to be $-774$ kJ/mol. Let us start with cations and anions at opposite ends of the universe ($r$ = infinity), so that the energy of attraction is zero. We now allow the cations and anions to approach until each cation just touches one other anion, but no other interactions occur; now we have a mole of gaseous **ion pairs** of NaCl. Experimentally the NaCl distance in such an ion pair is 236 pm. Substituting this distance in equation (3.6), we find an attractive energy of $-589$ kJ/mol.

Although bringing the ions closer would bring the answer closer to $-774$

**Table 3.3**  Born Exponents and Electron Configurations of Ions

| Born Exponent | Principal Quantum Number of Outermost Electrons of Ion |
|---|---|
| 5 | 1 ($1s^2$: H$^-$, Li$^+$) |
| 7 | 2 ($2s^2 2p^6$: F$^-$, Na$^+$) |
| 9 | 3 ($3s^2 3p^6$, perhaps $3d^{10}$: Cl$^-$, K$^+$, Zn$^{2+}$, Ga$^{3+}$) |
| 10 | 4 ($4s^2 4p^6$, perhaps $4d^{10}$: Br$^-$, Rb$^+$, Cd$^{2+}$, In$^{3+}$) |
| 12 | 5 ($5s^2 5p^6$, perhaps $5d^{10}$: I$^-$, Cs$^+$, Au$^+$, Tl$^{3+}$) |

kJ/mol, this cannot happen because the like-charged electrons of the oppositely charged ions cause repulsion between the ions. The energy of the repulsion is given by

$$E_{\text{repulsion}} = \frac{NB}{r^n} \tag{3.7}$$

where $B$ is a constant and $n$ is known as the Born exponent; its value depends on the principal quantum numbers of the electrons involved in the repulsion (Table 3.3) and can be determined by measuring the compressibility of an ionic compound.

The total energy of the ion pairs will be the sum of the attraction represented by equation (3.6) and the repulsion represented by equation (3.7) and is a minimum (at its most negative) when the ions are at their equilibrium separation $r_0$ (in the case of NaCl, 236 pm apart). In the general case, the minimum in energy is found (using calculus) by taking the derivative of the total energy with respect to distance and setting that derivative equal to zero:

$$0 = \frac{dE_{\text{total}}}{dr} = \frac{-138{,}900 Z_+ Z_-}{r_0^{\,2}} - \frac{nNB}{r_0^{\,n+1}} \tag{3.8}$$

This equation can be solved for the constant $B$, and the new value for $B$ can be substituted back into the total energy (3.6 plus 3.7) to give the final form for the total energy of 1 mol of ion pairs:

$$E_{\text{total}} = \frac{138{,}900 Z_+ Z_-}{r_0}\left(1 - \frac{1}{n}\right) \tag{3.9}$$

To evaluate this expression for the NaCl ion pair, we note that the Born exponent for Na$^+$ is 7 and that for Cl$^-$ is 9; hence, with an average Born exponent of 8, the last term of (3.9) becomes $(1 - \frac{1}{8})$, and the *total* energy of formation of the NaCl ion pair from the infinitely separated ions is $\frac{7}{8}$ of $-589$ kJ/mol, or $-515$ kJ/mol.

This is still quite a bit short of the experimental lattice energy, but of course we know that solid NaCl does not consist of isolated ion pairs. The maximum coordination numbers of Na$^+$ and of Cl$^-$ are by no means met in the ion pair; additional chloride ions from other ion pairs can still be attracted to the sodium

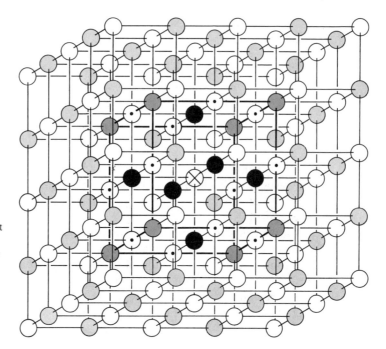

**Figure 3.5** Crystal structure of NaCl. Starting with the $Na^+$ ion marked $\otimes$, there are 6 nearest neighbors ($\bullet$), 12 next nearest neighbors ($\odot$), 8 next, next nearest neighbors (darkly shaded), etc. From Figure 3.3 (p. 56). *Inorganic Chemistry: Principles of Structure and Reactivity*, 2d ed. by James E. Huheey. Copyright © 1978 by James E. Huheey. Reprinted by permission of Harper & Row Publishers, Inc.

ion in one ion pair, and vice versa. This attraction leads ultimately to the structure of solid NaCl (Figure 3.5), in which each $Na^+$ surrounds itself with as many $Cl^-$ ions as can fit (six), and each $Cl^-$ likewise is surrounded by six $Na^+$ ions. The equilibrium Na—Cl distance $r_0$ in this structure is 283 pm.

This crystal lattice has many more Coulombic attractions between oppositely charged ions than does the ion pair. In addition there are many Coulombic repulsions between like-charged ions. Each one can be evaluated using equation (3.9); they must then be counted up, the distance involved in each interaction noted, and all the energy terms added over the entire crystal!

To begin, we note from Figure 3.5 that each $Na^+$ ion has 6 nearest-neighbor $Cl^-$ ions at a distance of $r_0$; these contribute 6 times the energy term of equation (3.9) to the total lattice energy, $U$. However, the next-nearest neighbors to each $Na^+$ ion are 12 other $Na^+$ ions, each across the diagonal of a square from the original $Na^+$ and thus at a distance from it of $\sqrt{2}$ times $r_0$; these *subtract* a total of $12/\sqrt{2}$ times the energy term of equation (3.9) from the lattice energy $U$. Likewise the next, next nearest neighbors to each $Na^+$ ion are the eight $Cl^-$ ions across the body diagonal of a cube (at a distance of $\sqrt{3}r_0$) from the $Na^+$; this adds $8/\sqrt{3}$ energy terms to the lattice energy. And so the process continues, giving rise to the infinite series shown as the last term in equation (3.10).

$$U = \frac{138,900 Z_+ Z_-}{r_0}\left(1 - \frac{1}{n}\right)\left(\frac{6}{1} - \frac{12}{\sqrt{2}} + \frac{8}{\sqrt{3}} - \frac{6}{\sqrt{4}} + \frac{24}{\sqrt{5}} - \cdots\right) \qquad (3.10)$$

The infinite series in equation (3.10) converges rather slowly to a limiting number that is characteristic of the NaCl crystal lattice geometry but that is independent of the charges on the ions and the distance $r_0$; this number is called the **Madelung constant**, $M$; for the NaCl lattice type, it has the value 1.74756, and applies to any of the dozens of ionic compounds in which the ions are arranged in this manner. Thus equation (3.10) can be rewritten as

$$U = \frac{138,900MZ_+Z_-}{r_0}\left(1 - \frac{1}{n}\right) \qquad (3.11)$$

Finally, we can substitute into equation (3.11) the values of the Madelung constant, ionic charges, and interionic distance to obtain the theoretical lattice energy of NaCl, $U = -751$ kJ/mol. This is quite close to the experimental lattice energy of $-774$ kJ/mol.

Small corrections for other minor energy terms can be added, but this treatment shows us the major factor responsible for the stability of an ionic lattice: the strong Coulombic attraction of the oppositely charged ions, which extends in all directions from a given ion and throughout space. This attraction is strong enough to overcome the Coulombic repulsions that must exist between like-charged ions and the repulsions that exist between outer electrons of adjacent ions.

## 3.7

## Radius Ratios and Lattice Types

The NaCl lattice type shown in Figure 3.5 is only one of many known arrangements of ions (lattice types) in ionic solids. Some other lattice types that are important for binary ionic substances are shown in Figure 3.6.

Three major reasons why most ionic solids must adopt a lattice type other than the NaCl type are

**1.** The NaCl lattice incorporates an equal number of cations and anions. If the *stoichiometry* of the compound is other than 1 : 1, another lattice type is required.

**2.** If there is significant covalent bonding between the cation and the anion, the orbitals of the two touching atoms must overlap well; consequently, certain approximate bond angles are needed in the lattice type. The NaCl (or some other) lattice type may not incorporate these angles. Thus for a lattice involving saturated carbon, bond angles somewhere near 109° are needed, whereas the "bond" angles Cl—Na—Cl and Na—Cl—Na in the NaCl structure are all 90° or 180°. Some unusual lattices can result, as will be seen, for example, in Chapter 9 for some partially covalent halides. We will leave this consideration for that chapter, however, and (after checking to see that our electronegativity differences between touching atoms are at least close to 2.0) consider only highly ionic compounds in this chapter.

**3.** The NaCl lattice incorporates coordination numbers of 6 for the cation and 6 for the anion. This coordination number is not optimal for many combinations of

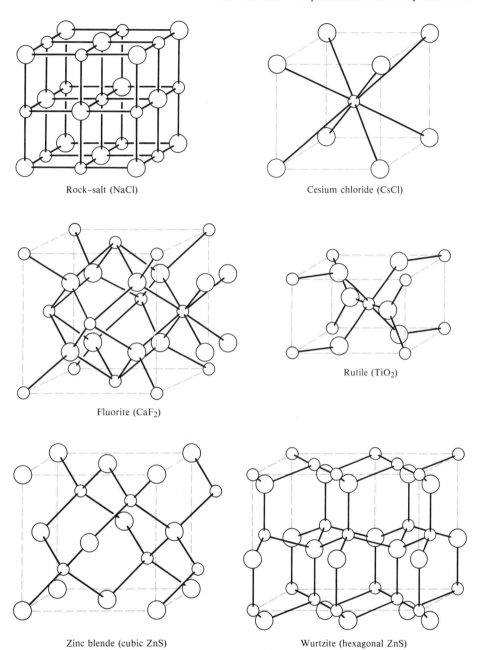

Rock–salt (NaCl)

Cesium chloride (CsCl)

Fluorite (CaF$_2$)

Rutile (TiO$_2$)

Zinc blende (cubic ZnS)

Wurtzite (hexagonal ZnS)

**Figure 3.6**   Important lattice types for binary ionic compounds. Small circles denote metal cations; large circles denote anions. Reproduced with permission from F. A. Cotton and G. Wilkinson, *Advanced Inorganic Chemistry*, 4th ed., Wiley-Interscience, New York. Copyright © 1980 by John Wiley & Sons.

cations and anions. (a) If the cation is too small relative to the anion, it may not touch the anions, while the large anions still touch each other. This reduces the attractive forces while keeping the repulsive forces strong—a situation to be avoided, if possible, by going to a lattice in which the coordination number of the cation is lower so that the smaller number of anions can touch the cation without touching each other. (b) If the cation is too large relative to the anion, it may be possible to fit additional anions around the cation without introducing excessive anion-anion repulsion; hence the compound will be expected to adopt a different lattice, one in which the cation has a higher coordination number.

Thus the relative size of the cation and anion (the **radius ratio** of the cation to the anion) is an important element in determining the type of crystal lattice that an ionic compound will adopt. We can use solid geometry to work out the ideal radius ratio for "perfect packing" (cations and anions just touching each other; anions and anions just touching each other) of many lattices. For example, for the NaCl lattice type (Figure 3.5 or 3.6) we can consider a square of two $Na^+$ ions at opposite corners and two $Cl^-$ ions at the other corners. The $Na^+$ and $Cl^-$ ions touch along the edges, and the larger $Cl^-$ ions touch each other across the diagonal of the square. Consequently we have the following two equations:

$$r_{cation} + r_{anion} = a \qquad \text{(the edge length)} \qquad (3.12)$$

$$2r_{anion} = \sqrt{2}a \qquad \text{(the diagonal length)} \qquad (3.13)$$

By eliminating $a$ between these two equations and rearranging, we can solve for the "ideal" radius ratio $r_{cation}/r_{anion} = \sqrt{2} - 1 = 0.414$. Likewise, starting with the cube of the CsCl lattice (Figure 3.6) and using the fact that the cube diagonal is equal to $\sqrt{3}$ times the edge, we can calculate the ideal radius ratio for this lattice to be 0.732. If a cation and an anion have a radius ratio between 0.414 and 0.732, the NaCl lattice will be expected to be chosen, since enlarging the cation in a NaCl lattice spreads the touching anions apart, reducing repulsions, while shrinking the cation in the CsCl structure reduces cation-anion contact, reducing attractions; hence the range of radius ratios between 0.414 and 0.732 should be ideal for the NaCl crystal structure to be adopted. Similar ranges for the other common binary lattice types for different stoichiometries are shown in Table 3.4.

It is possible to predict coordination numbers for salts not corresponding to any of the types specified in Table 3.4:

**1.** The radius ratio should predict the coordination number for the less abundant ion in any lattice type and for any stoichiometry. (The less abundant ions have more neighbors of the opposite charge and hence the most critical problems of crowding.) Except in some compounds such as $K_2O$, the less abundant ion is the cation; hence from the radius ratios given at the bottom of Table 3.4, the expected coordination number of the cation (more properly, the less abundant ion) can be predicted. (Intermediate coordination numbers such as 5, 7, and 9 do occur with some stoichiometries, but it is difficult to predict exactly when.)

**Table 3.4** Lattice Types and Madelung Constants for Different Stoichiometries and Radius Ratios of Cations and Anions

| Radius Ratio (Cation/Anion) | Lattice Type | Coordination Number of Cation | Anion | Madelung Constant | Reduced Madelung Constant |
|---|---|---|---|---|---|
| **A. 1:1 Stoichiometry of Salt (MX)** | | | | | |
| 0.225–0.414 | Wurtzite (ZnS) | 4 | 4 | 1.63805 | 1.63805 |
| | Zinc blende (ZnS) | 4 | 4 | 1.64132 | 1.64132 |
| 0.414–0.732 | Rock salt (NaCl) | 6 | 6 | 1.74756 | 1.74756 |
| 0.732–1.000 | CsCl | 8 | 8 | 1.76267 | 1.76267 |
| **B. 1:2 Stoichiometry of Salt (MX$_2$)** | | | | | |
| 0.225–0.414 | Beta-quartz (SiO$_2$) | 4 | 2 | 2.201 | 1.467 |
| 0.414–0.732 | Rutile (TiO$_2$) | 6 | 3 | 2.408[a] | 1.605 |
| 0.732–1.000 | Fluorite (CaF$_2$) | 8 | 4 | 2.51939 | 1.6796 |
| **C. 2:3 Stoichiometry of Salt (M$_2$X$_3$)** | | | | | |
| 0.414–0.732 | Corundum (Al$_2$O$_3$) | 6 | 4 | 4.1719[a] | 1.6688 |
| **D. Other Stoichiometries and Lattice Types** | | | | | |
| Never favored | Ion pair | 1 | 1 | 1.00000 | 1.0000 |
| 0.000–0.155 | | 2 | | | |
| 0.155–0.225 | | 3 | | | |
| 0.225–0.414 | | 4 | | | |
| 0.414–0.732 | | 6 | | | |
| 0.732–1.000 | | 8 | | | |
| 1.000 | | 12 | | | |

NOTE: Reduced Madelung constant = Madelung constant × $2/p$, where $p$ = number of ions in the simplest formula of the salt.
[a] Exact value dependent on details of the structure.

**2.** The average coordination number of the anion (the more abundant ion) can then be calculated from the stoichiometry of the salt of M and X:

$$(\text{Coord. no. of M}) \times (\text{no. of M in formula}) = \quad\quad (3.14)$$
$$(\text{coord. no. of X}) \times (\text{no. of X in formula})$$

Normally, anions are larger than cations, so the radius ratio is less than 1.00; but occasionally the cation is the larger. In this case we often find that one of the known lattice types is adopted, but the positions of the cations and the anions are reversed—the cation, which is now the larger ion, takes the position held by the usually larger anion. Thus if the radius ratio is greater than 1.00, we can calculate the inverse radius ratio $r_{anion}/r_{cation}$ to pick the lattice type in which the cations and anions will reverse roles. The lattice type is then named with the prefix *anti-*.

EXAMPLE    Predict the lattice types that are adopted by the following oxides: $Tl_2O$, BeO, and $ThO_2$.

Solution    ■ First, we obtain the appropriate radii from Tables C and 2.5 and then calculate the radius ratios $r_{cation}/r_{anion}$: $164/126 = 1.30$, $59/126 = 0.47$, and $108/126 = 0.86$, respectively. (Since the first radius ratio is greater than 1.00, we calculate the inverse radius ratio $126/164 = 0.77$).

■ Second, we refer to part D of Table 3.4 to obtain the expected coordination number of the cation (6 for $Be^{2+}$ and 8 for $Th^{4+}$) or the anion if it is less abundant in the chemical formula. This is the case for $O^{2-}$ in $Tl_2O$, for which we predict the coordination number 8.

■ Third, we calculate the coordination number of the other ion using equation (3.14). We compute that each $Tl^+$ should be surrounded by four $O^{2-}$ ions, each $O^{2-}$ in BeO ought to be surrounded by six $Be^{2+}$ ions, and each $O^{2-}$ in $ThO_2$ should be surrounded by four $Th^{4+}$ ions.

■ Finally, if these stoichiometries and sets of coordination numbers match any of the entries in parts A through C of Table 3.4, we can read the lattice type from the table. We thus predict that BeO ought to adopt the NaCl structure, $ThO_2$ ought to adopt the fluorite structure, and $Tl_2O$ should adopt the antifluorite structure, in which each $Tl^+$ ion is surrounded by four $O^{2-}$ ions, and each $O^{2-}$ ion is surrounded by eight $Tl^+$ ions. According to Wells [9], $ThO_2$ does indeed adopt the fluorite structure, but BeO actually adopts the wurtzite structure, and $Tl_2O$ adopts a structure with a coordination number of 6 for oxygen.    □

In general, the success of the radius ratio rule in predicting lattice types is limited. The trends are generally right—for a given anion, larger cations adopt lattices in which the cations have larger coordination numbers—but the radius ratio at which the transition from one lattice type to another occurs is often substantially different than predicted. Certainly, one reason for this in many cases is the second one given above—covalency in the bonding. Another source of error may arise from the fact that the measured radii, particularly of cations, change when their coordination numbers change: As more anions crowd around the cation, they repel each other more and are less able to compress the cation. Thus, for example, the radius of $Na^+$ varies with its coordination number, as shown in Table 3.5. Despite the middling success of quantitative predictions with the radius ratio rule,

Table 3.5    Effect of Coordination Number on the Radius of the Na⁺ Ion

| Coordination Number of $Na^+$: | 4 | 5 | 6 | 7 | 8 | 12 |
|---|---|---|---|---|---|---|
| Radius of $Na^+$ (pm): | 113 | 114 | 116 | 126 | 132 | 138 |

SOURCE: Data from R. D. Shannon, *Acta Crystallogr. Sect. A*, 32, 751 (1976).

though, it gives us further insight into the consequences of Coulombic attractions and repulsions in ionic compounds.

EXAMPLE    Compute the lattice energy of $ThO_2$.

Solution    Equation (3.11) requires us to provide the cation and anion charges ($+4$ and $-2$), the Madelung constant, the interionic distance, and the average Born exponent for the two ions. In the previous example we detemined that $ThO_2$ should and does adopt the fluorite lattice, which from Table 3.4 has a Madelung constant of 2.51939. The interionic distance is presumed to be the sum of the ionic radii of $Th^{4+}$ and $O^{2-}$, $108 + 126 = 234$ pm. From Table 3.3 we find the Born exponent for $O^{2-}$ (electron configuration $2s^2 2p^6$) to be 7. $Th^{4+}$ has the valence electron configuration $7s^0$ or, considering its outermost electrons, $6s^2 6p^6$. The Born exponent for this configuration is not listed, so let us extrapolate a value of 14. Averaging the two Born exponents 7 and 14 gives us $n = 10.5$ to use in equation (3.11). Substituting these numbers gives us

$$\frac{138,900 \cdot (2.51939) \cdot (4) \cdot (-2)}{234} \left(1 - \frac{1}{10.5}\right)$$

which is a lattice energy of $-10,824$ kJ/mol.    □

## 3.8

## Stability of Lattices and the Solubility Rules

Suppose that we wish to compare the lattice energies of three salts of the same $1:1$ stoichiometry, $A^+X^-, B^+Y^-$, and $C^+Z^-$, chosen such that (1) the sum of the cationic and anionic radii, $r_0$, is the same in all three cases; (2) the average Born exponent is the same in each case; but (3) the radius ratio differs such that the first salt takes one of the ZnS structures, the second salt takes the NaCl structure, and the third salt takes the CsCl structure. From equation (3.11) we can see that the lattice energies of the three salts will differ due to the different Madelung constants, and from Table 3.4 we can see that the higher the radius ratio, the greater the Madelung constant. Thus $C^+Z^-$ should have the most negative lattice energy. We conclude that *more-stable lattices are formed by cations and anions that are relatively close in size*.

Now let us compare the total hydration energies of the cations and anions in each of these three salts. In the manner of the Latimer equation (2.1), we expect the sum of these energies to be proportional to $Z^2/r$ for the cations plus $Z^2/r$ for the anions (ignoring the correction for the radius of water). By assuming some arbitrary values for these radii, we may verify that the sum of hydration energies should be most negative for the salt with the most disparate radii, $A^+X^-$; hence according to the thermochemical cycle for solubility (Figure 3.4), the enthalpy of precipitation should be most positive for $A^+X^-$ and most negative for $C^+Z^-$. As noted by Morris [10], the most insoluble salts are expected to be those in which the hydration energies of the cations and anions are most nearly matched. These are also expected to be the cations and anions that are most nearly matched in the strength of their acidity and basicity, respectively.

Looking back at our solubility rules (Section 3.2), we can now explain more fully our rules for those classes of salts for which solubilities are determined principally by the enthalpy (rather than the entropy) of precipitation. The ultimate reason that nonacidic cations and nonbasic anions give rise to insoluble salts is that such cations and anions are both large (i.e., similar in size), so they form an especially stable crystal lattice but do not give especially good hydration energies (in comparison to the lattice energy).

Synthetic inorganic and organic chemists often take advantage of this fact when they are synthesizing new organic or inorganic ions. A new ion is likely to be large; just about all the small ones have already been made! In order to isolate such a new ion from solution (in water *or* other solvent), the chemist adds a solution of a large ion of the opposite charge; the ionic product often then crystallizes from solution.

Synthetic chemists have also noticed that (other things being equal) matching the charges of the two ions also seems to help in precipitating or crystallizing a given ion. Comparing Madelung constants of lattices involving different charges on one of the ions cannot be done directly, since different stoichiometries must result. Kaputstinskii [11] found it useful to convert the Madelung constants for stoichiometries other than 1 : 1 to a 1 : 1 basis, e.g., for 2 mol of ions, by dividing by one-half the number of ions in the formula. Such *reduced Madelung constants* (Table 3.4) are close to 1.6 for most lattice types, but are the largest for 1 : 1 salts, suggesting that the best packing of ions occurs with equal numbers of cations and anions. Hence to precipitate a large $-1$-charged (i.e., nonbasic) anion we are likely to do best if we add a large $+1$ (nonacidic) cation.

The favorable lattice energies resulting from good matching of cation and anion charges and radii not only render such products more insoluble, aiding their isolation, but also may stabilize the salts against thermal decomposition. For example, many oxo anions such as carbonate and sulfate decompose upon heating to give gases such as carbon dioxide and sulfur trioxide, plus the oxide ion, which is much smaller than the original oxo anion and thus forms more stable lattices with small (more acidic) metal ions. Since the larger carbonate and sulfate ions form better lattices with large cations, carbonates and sulfates of large (nonacidic) cations are stable to higher temperatures than are the carbonates and sulfates of small, acidic cations. (The carbonate ion is more susceptible to this effect, and in general, carbonates of even moderately acidic cations are not stable at all at room temperature.)

Likewise, the reason that acidic cations and nonbasic anions give rise to soluble salts is that such ions are quite different in size and give poorer lattice energies than hydration energies. In extreme cases the cation may be so much smaller than the anion that even in one of the ZnS structures there may not be good contact between cations and anions. In solution, however, good contact may be achieved between cations and small water molecules, and there is "bonus" energy resulting from the formation of multiple spheres of hydration; hence not only are such salts quite soluble in water, but if the solution is evaporated until the salt does indeed crystallize, it normally crystallizes as a **hydrated salt**, in which the primary hydration sphere of the cation (usually) is retained. (Of course, the radius ratio of the *hydrated* cation to the anion is much larger than that of the anhydrous cation to the anion.) For example, most $d$-block metals form hydrated sulfates, nitrates, chlorides, etc., which are often written $CuSO_4 \cdot 5 H_2O$, and so on, but which could more accurately be written as including hydrated cations, such as $[Cu(H_2O)_4]^{2+}$. When such salts are obtained anhydrous, they often are very good **desiccants** or drying agents, capable of removing water from solvents or from the air; $Mg(ClO_4)_2$, $CaCl_2$, and $CaSO_4$ are used in this manner.

We are now better able to consider the case of solubility rules for feebly basic anions such as sulfate. The difficulty in prediction arises from the fact that these anions are just basic enough to have $-T\Delta S$ terms moderately favoring precipitation with acidic cations; but these anions are so much larger than most acidic cations that the $\Delta H$ term is usually unfavorable (Table 3.1, part IV). Only with large, feebly acidic cations (especially with charges matched to that of the anion) is the lattice sufficiently satisfactory to allow insolubility. Consequently we can slightly modify our generalized solubility rule in Section 3.2 to read *large, $-2$-charged (feebly basic) anions such as sulfate are best precipitated with large, $+2$-charged (feebly acidic) cations such as barium ion*. Since the somewhat smaller feebly basic anions such as $NO_2^-$ and $ClO_2^-$ do not have both $-T\Delta S$ and $\Delta H$ working for insolubility, we find that all the salts of these anions are apparently soluble, although many of them are unstable to decomposition via oxidation-reduction reactions.

## 3.9

## Close Packing of Anions

An alternative way of looking at crystal lattices that is useful for many ionic compounds is based on the fact that anions are normally larger than cations and are often in contact with each other in one of two common arrangements that are the closest possible ways of packing together spheres of equal size. In this view of lattices, the much smaller cations then fill some of the holes or interstices between the anions. Thus if we were packing basketballs and baseballs for shipment in the same large box, we would figure out first how to pack the larger basketballs in the most efficient way possible, with confidence that the baseballs would fit in the spaces between the larger spheres.

In a single layer, spheres of equal size can be packed closest in a pattern in which each sphere is surrounded by six other spheres in a hexagonal pattern (Figure 3.7a). In packing another layer below the given layer, we do not superimpose the spheres directly atop each other, but we shift the layers so that each sphere nestles into the cavity formed by three of the spheres in the first layer (Figure 3.7b). Between the three spheres in the original layer and the sphere in the new layer there is still a small open space, called a **tetrahedral hole**, since it is surrounded by four large spheres; a small sphere put in here would have a coordination number of four.

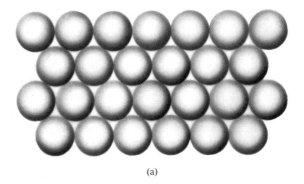

(a)

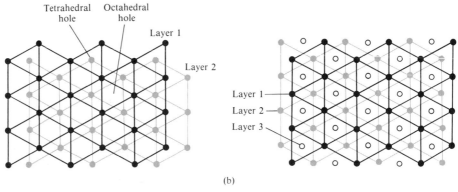

(b)

**Figure 3.7** Close packing of spheres in (a) a single layer; (b) two layers, or if the layer of light-colored spheres is repeated not only below but also above the layer of dark spheres, the three-layer pattern of hexagonal close packing; (c) the three-layer pattern of cubic close packing. Reproduced with permission from F. A. Cotton and G. Wilkinson, *Basic Inorganic Chemistry*, Wiley-Interscience, New York. Copyright © 1980 by John Wiley & Sons, Inc.

As can also be seen from Figure 3.7b, this arrangement of the lower layer of spheres fills in only half of the three-sided cavities of the upper layer; the other half of the three-sided cavities of the upper layer match up with three-sided cavities in the lower layer to produce larger **octahedral holes** surrounded by six larger spheres. It can be shown that for each sphere in a given layer, there are two tetrahedral holes but only one octahedral hole. The stoichiometry of the salt does not normally allow all the holes to be filled with cations, but (if close packing is utilized) the structure can be described in terms of the type and fraction of holes occupied.

The two possible patterns of close packing differ in the way a third layer (of anions) is placed over the first two. With one possibility, called **hexagonal close packing** (hcp), the third (top) layer is matched up exactly with the bottom layer so that the light-colored pattern shown in Figure 3.7b is repeated. This pattern is then normally continued throughout the whole crystal, with layers of spheres alternately in the A position (light) and B position (dark).

In the alternative pattern of close packing, the third (top) layer coincides

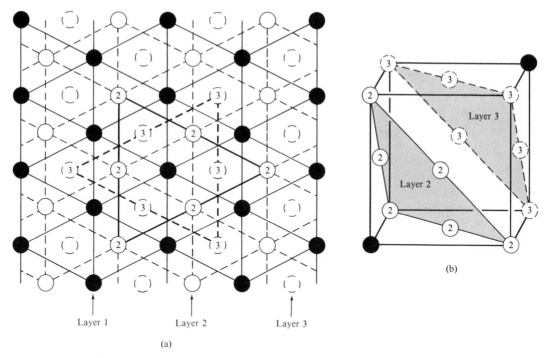

Layer 1          Layer 2          Layer 3

(a)

(b)

**Figure 3.8** Fourteen of the close-packed spheres of Figure 3.7c shown from another angle to demonstrate their cubic pattern. The numbers (1, 2, 3, 4) on each sphere correspond to the number of the layer in Figure 3.7c in which each is located; layer 4 is superimposed directly over layer 1. Adapted with permission from F. A. Cotton and G. Wilkinson, *Basic Inorganic Chemistry*, Wiley-Interscience, New York. Copyright © 1980 by John Wiley & Sons, Inc.

directly neither with the bottom nor with the middle layer. Instead the spheres of the third layer fall directly over the three-sided cavity of both the lower and the middle layer (Figure 3.7c). This pattern is called **cubic closest packing**, because, as shown in Figure 3.8, from another perspective this give rise to a cubic lattice of large spheres.

Although these close-packing descriptions are used more frequently with other types of lattices than those found in Table 3.4 (and illustrated in Figure 3.6), it is useful to show how they may be applied to these simple lattice types. Thus the two ZnS lattice types differ in the type of close packing: The zinc blende structure (Figure 3.6) has cubic close packing of its sulfide ions, with zinc ions being located in one-half of the tetrahedral holes. The wurtzite structure has an obvious hexagonal pattern (hcp sulfide ions), again with zinc ions in half the tetrahedral holes. NaCl has cubic close packing of chloride ions, with sodium ions in all the octahedral holes. The fluorite structure has cubic close packing of the *calcium* ions, with the more abundant fluoride ions filling all the tetrahedral holes. The rutile structure can only indirectly be derived by distorting a hcp structure; the CsCl structure is not close packed.

## 3.10

---

# Writing and Interpreting Net Ionic Equations

### 3.10.1   Writing Net Ionic Equations

Equations for chemical equilibria are made easier to write if the dissolved ionic compounds involved in them are written as separate ions (e.g., as $Na^+(aq)$ and $NO_3^-(aq)$) rather than as complete formula units ($NaNO_3(aq)$). Such equations are known as **ionic equations**. It is often found that in equilibria involving such ionic compounds, one of the ions appears both as reactant and as product—i.e., does not react at all. (We say that such an ion is a **spectator ion**.) Deleting spectator ions from an ionic equation simplifies the equation and gives a **net ionic equation**. It is easier to write an equilibrium constant expression starting from a net ionic equation. Such an equation is also more general—we can see that it applies as well for other ionic salts with different spectator ions.

In writing ionic equations, (1) *soluble* ionic salts are written as separate ions, while (2) *insoluble* ionic salts are written as complete formula units (often with (s) or an arrow after them). Similarly, (3) *very strong* and *strong* oxo acids exist in solution predominantly as separate ions, and are written as such, while (4) *moderate* and *weak* oxo acids exist in solution mostly as molecules and are written as complete formula units.

For example, one of the precipitation reactions in Experiment 3 involves combining solutions of $Mg(NO_3)_2$ and $Na_3PO_4$; a precipitate of $Mg_3(PO_4)_2$ results. Writing full formula units for this reaction gives us the long "molecular" form of the reaction:

$$3Mg(NO_3)_2 + 2Na_3PO_4 \rightleftharpoons Mg_3(PO_4)_2 + 6NaNO_3 \qquad (3.15)$$

We obtain the ionic form of this reaction by writing the three soluble salts as separate ions:

$$3Mg^{2+}(aq) + 6NO_3^-(aq) + 6Na^+(aq) + 2PO_4^{3-}(aq) \rightarrow \qquad (3.16)$$
$$Mg_3(PO_4)_2(s) + 6Na^+(aq) + 6NO_3^-(aq)$$

Although reaction (3.16) does not yet save any space, we note that $6Na^+$ and $6NO_3^-$ occur on both sides of the equation. These spectator ions can be deleted to give the net ionic equation:

$$3Mg^{2+}(aq) + 2PO_4^{3-}(aq) \rightarrow Mg_3(PO_4)_2(s) \qquad (3.17)$$

Not only is this equation more compact, but it also suggests that *any* soluble $Mg^{2+}$ salt and *any* soluble $NO_3^-$ salt would give the same reaction. Hence dozens of reactions are summarized by this one equation.

### 3.10.2   Interpreting Net Ionic Equations; Using Inorganic Handbooks

Due to the efficiency of net ionic equations in presenting information, they are commonly encountered in textbooks. To carry out these reactions, however, some interpretation is needed, since we must use salts that also contain the (unspecified) spectator ions.

Let us suppose that we want to prepare 0.100 mol of cerium(III) phosphate, $CePO_4$. Knowing that this salt is the combination of an acidic cation and a basic anion, we suspect that it can be prepared by precipitation:

$$Ce^{3+}(aq) + PO_4{}^{3-}(aq) \rightarrow CePO_4(s) \tag{3.18}$$

To carry out this reaction, we must decide what cerium salt and what phosphate salt to use. To get satisfactory reaction rates, we normally choose *soluble* salts, which we can choose with the aid of our solubility rules. For $Ce^{3+}$, we may select a salt of a nonbasic anion such as $ClO_4{}^-$ or $NO_3{}^-$ (or $Cl^-$, as we will see in a later chapter), or of a feebly basic ion such as $SO_4{}^{2-}$. For $PO_4{}^{3-}$ we select the salt of a nonacidic cation such as $Na^+$ or $K^+$.

But we must also guard against the possibility that the spectator ions we choose will combine to give a second precipitate, which would contaminate our product. In theory, the nonacidic cation we have chosen should react with the nonbasic anion we have chosen to give a second precipitate! Fortunately, this will not always be the case. As we will see later, although nonbasic, the chloride ion is small, so that most chlorides are soluble. The nitrate ion is smaller and less nonbasic than the perchlorate ion, so that it gives precipitates only with nonacidic cations from fairly concentrated solutions. Similarly, the sodium ion is the least nonacidic of all nonacidic cations, so that it usually does not give precipitates with nonbasic anions from dilute solutions. Of course the sulfate ion is acceptable, so we actually have several choices available to us, depending on what is available in the stockroom. However, we would not want to take the perchlorate of one ion and the potassium salt of the other!

When we go to weigh out the reactants, we must include the weight of any water of hydration that the salt includes. Sometimes the label on the bottle mentions the presence of waters of hydration only in the fine print, if at all! To learn the likely number of waters of hydration in the reactants or products and to check that our predictions of solubility are correct, it is useful to look up the reactants and products in the inorganic section of a chemistry handbook such as the *Handbook of Chemistry and Physics*. Let us suppose that we have decided to use cerium(III) nitrate and sodium phosphate to synthesize the cerium(III) phosphate. Under "cerium" in the handbook we find the nitrate hexahydrate, $Ce(NO_3)_3 \cdot 6\,H_2O$. Looking under the columns labeled "Solubility ... in cold water" we find either abbreviations such as *i.*, *d.*, *sl.s.*, *s.*, or *v.s.* or numerical data such as $0.053^0$. The abbreviations mean, respectively, *insoluble*, *decomposed*, *slightly soluble*, *soluble*, or *very soluble*; the number means that 0.053 g of salt dissolves in 100 ml of water at the temperature of the superscript. (An infinite sign indicates complete solubility.) If we find i. or d. for a reactant, we had better choose again; if we find sl.s. or a number under (say) 5, we will need a lot of water to dissolve the reactant. Beside $Ce(NO_3)_3 \cdot 6\,H_2O$ we find v.s., which makes this a good choice. To make 0.100 mol of $CePO_4$ we need 0.100 mol of the hydrated nitrate, which (from its listed formula weight) is 43.4 g.

To find the data for the phosphate, we must look under the counterion we have chosen. Under "sodium" we find many "phosphate" entries. Some of these are partially protonated forms such as $NaH_2PO_4 \cdot H_2O$. These might be suitable, but it is better to choose a hydrate of $Na_3PO_4$. As expected, these are soluble. A 10-hydrate and a 12-hydrate are listed. We should see which is in the stockroom and use its formula weight to calculate the quantity to be used in the reaction. After checking to be sure that $CePO_4$ is insoluble and that $NaNO_3$ is soluble (they are), we are ready to go to work in the lab [12].

## 3.11

## Study Objectives

1. Predict whether a given salt of an oxo anion is soluble or insoluble in water.
2. Calculate energy terms from thermochemical cycles for the precipitation of ionic salts from water.
3. Describe the structures of ice, liquid water, and hydrated ions.
4. Compare different ions in terms of their primary and secondary hydration spheres, their hydration numbers, and their classifications as electrostatic structure makers, electrostatic structure breakers, or hydrophobic structure makers.
5. In terms of entropy and enthalpy changes and the above classifications, explain *why* a given class of oxo salt is soluble or insoluble.
6. Calculate the Madelung constant for a simple geometry.
7. Calculate the lattice energy of a salt from its ionic radii, given a table of Madelung constants and Born exponents.
8. Know how the radius of an ion changes as its coordination number changes.
9. Use the stoichiometry of a salt and the ratio of its ionic radii to predict the coordination numbers of its anions and cations and (if possible) the lattice type it adopts.
10. Pick out examples of salts that are likely to be good desiccants and that are likely to crystallize from water as hydrates.
11. Tell how to isolate a new ionic product you have synthesized.
12. Describe the difference between cubic close packing and hexagonal close packing. Know how many tetrahedral and how many octahedral holes occur per anion and how these relate to the stoichiometry of the salt.
13. Convert the molecular form of an equation to the net ionic form. Convert the net ionic form to a useful molecular form, and check in a handbook to determine whether the salts involved likely are hydrates.

## 3.12

## Exercises

*1. Which of the following salts will be insoluble in water? $Co(NO_3)_2$, $CsBrO_4$, $CePO_4$, $Cs_3AsO_4$, $BaSeO_4$, $Hg_5(IO_6)_2$, $TiO_2$.

2. Which of the following salts will be insoluble in water? $CuMoO_4$, $Cs_2MoO_4$, $Cr_2TeO_6$, $Cr(NO_3)_3$, $HgO$, $NaOH$, $RbTcO_4$, $KIO_4$, $K_5IO_6$.

†**3.** Write general chemistry–type specific solubility rules for the salts of the following oxo anions: **a.** chromate; **b.** ferrate; **c.** pertechnetate; **d.** silicate; **e.** tellurate. These rules should state whether most salts of this anion are soluble or insoluble; then list some ions that are exceptions.

**4.** Describe the structures of ice and of liquid water, and explain why ice floats in water.

*5. Use one of the following ions $(Al^{3+}, Li^+, Rb^+, (C_4H_9)_4N^+)$ to complete each of the following: **a.** the ion that attaches the fewest water molecules when it dissolves in water; **b.** the ion that forms the largest hydrated ion; **c.** a hydrophobic structure maker; **d.** an electrostatic structure maker; **e.** an electrostatic structure breaker.

**6.** Consider the following set of ions: $K^+$, $Zn^{2+}$, $Al^{3+}$, $SiO_4^{4-}$, $ClO_4^-$, $SeO_4^{2-}$. **a.** Which cation will form the largest hydrated ion? **b.** Which cation(s) will be electrostatic structure makers? **c.** Which cation(s) will disrupt the icebergs in liquid water? **d.** Which anion is most likely to be an electrostatic structure breaker? **e.** Which cation(s) will give insoluble salt(s) with the $ClO_4^-$ ion? **f.** Which cation(s) will give insoluble salt(s) with the $SiO_4^{4-}$ ion? **g.** Write the formula of any salt containing just these ions that will precipitate for reasons connected with an entropy change. **h.** Write the formula of any soluble salt containing just these ions and explain (in terms of lattices, etc.) why it is soluble.

**7.** Give an example of a salt composed of a cation and an anion, each of which is an electrostatic structure maker. Is such a salt likely to be soluble or insoluble? Give the physical reason for this (in terms of the structure of the hydrated ions, the suitability of its lattice, etc.). Is this solubility/insolubility associated with entropy or enthalpy effects?

†**8.** Enthalpies of precipitation are known for the following halides: LiF, $-5$ kJ/mol; NaCl, $-4$ kJ/mol; AgCl, $-65$ kJ/mol; $CaCl_2$, $+83$ kJ/mol.
  **8.1** Use data on hydration enthalpies (Tables 2.1 and 2.5) to calculate the lattice energies of these compounds.
  **8.2** All but one of these halides (which one?) have the NaCl structure. Calculate the lattice energies of these compounds from their ionic radii, Madelung constants, etc. (Use the radius ratio to choose the lattice type for the halide with the non-NaCl structure.) For which halide is there the greatest discrepancy between your answers to the previous part and to this part? How does the cation in this halide differ from the other three cations, and how might this explain the discrepancy?

†**9.** The hydration energy of the sulfate ion is not given in Table 2.5.
  **9.1** Use the known lattice energies for $CaSO_4$ ($-2653$ kJ/mol), $SrSO_4$ ($-2603$ kJ/mol), and $BaSO_4$ ($-2423$ kJ/mol), along with data from Tables 2.1 and 3.1, to compute the hydration energy of the sulfate ion. How well do your results agree for each of these three salts?

**9.2** Compare your result with the hydration energy of the perchlorate ion, which has a similar size, and explain any differences between the two.

10. Usually, $\Delta H°$ for the precipitation of an ionic salt is about equal to zero. Use this approximate value and the hydration energies for $Al^{3+}$ and for $ClO_4^-$ (Tables 2.1 and 2.5) to estimate the lattice energy of aluminum perchlorate.

*11. Using Tables C, 2.5, and 3.4: **a.** calculate the radius ratio for $CeO_2$ and predict the coordination numbers of the cerium and oxide ions and the type of lattice that $CeO_2$ would adopt; **b.** calculate the lattice energy of $CeO_2$.

12. Cesium and gold form an ionic compound, $Cs^+Au^-$, with a cesium-gold distance of 369 pm. **a.** What type of lattice will CsAu adopt? **b.** Estimating a suitable Born exponent for the $Au^-$ ion, calculate the lattice energy of CsAu. **c.** Assuming that the enthalpy of precipitation of CsAu is zero, calculate the enthalpy of hydration of the $Au^-$ ion.

13. Calculate the Madelung constant for: **a.** a simple cube consisting of four cations alternating corners with four anions; **b.** an infinite linear arrangement of alternating cations and anions. This case involves an infinite series. Do you know the sum of this series from your calculus course? If not, add up the first ten terms.

14. Aluminum phosphate is quite insoluble. Which effect is responsible for its insolubility? **a.** An enthalpy effect resulting from the very stable lattice formed; **b.** an entropy effect resulting from the very stable lattice formed; **c.** an enthalpy effect resulting from the release of numerous waters of hydration when forming aluminum phosphate from its ions; **d.** an entropy effect resulting from the release of numerous waters of hydration when forming aluminum phosphate from its ions; **e.** an enthalpy effect resulting from the disruption of the structure of liquid water by the large ions; **f.** an entropy effect resulting from the large ions disrupting the structure of liquid water.

15. Explain, with reference to the structure of water, of crystal lattices, and/ or of hydrated ions, why: **a.** iron(III) perchlorate is soluble in water; **b.** iron(III) phosphate is insoluble in water; **c.** rubidium permanganate is fairly insoluble in water; **d.** tetrabutylammonium tetrabutylborate is insoluble in water.

16. Will $BeSO_4$ or $BaSO_4$ be more likely to be **a.** a good desiccant, or **b.** crystallize as a hydrate?

17. Liming a lake to neutralize acid rain in it could seriously affect the availability of nutrients such as phosphate and molybdate, but not nitrate. Explain.

18. Assume that the rare, radioactive elements astatine and polonium form anions $At^-$ and $Po^{2-}$, each of which has a radius of about 226 pm. Which of the following lattice types ($ZnS, NaCl, CsCl, SiO_2, TiO_2, CaF_2$) would be expected to be adopted by: **a.** the astatide of each Group 1 metal (Na through

Cs); **b.** the astatide of each Group 2 metal (Mg through Ba); **c.** the polonide of each Group 2 metal (Mg through Ba).

19. Using your results from the previous question and Table 3.4, calculate the lattice energy of barium astatide. If we assume that the enthalpy of solution of barium astatide in water in zero, calculate the hydration energy of the astatide ion. Draw the hydrated astatide ion, showing how the water molecules would orient themselves.

20. In organic lab you have just synthesized the tropylium ion, $C_7H_7^+$, in solution. Tell how you would isolate it from solution to get a stable crystalline solid.

21. Generalized solubility rule IV in section 3.2 says that many feebly acidic cations and feebly basic anions give insoluble salts. Both $Li^+$ and $Ba^{2+}$ are feebly acidic cations, and $NO_2^-$, $ClO_2^-$, $SO_4^{2-}$, and $SeO_4^{2-}$ are all feebly basic anions. Although $BaSO_4$ and $BaSeO_4$ are indeed insoluble, $Ba(NO_2)_2$, $Ba(ClO_2)_2$, $Li_2SO_4$, and $Li_2SeO_4$ are all soluble. Can you explain this?

22. Which of the following structural descriptions is/are inconsistent with the stoichiometry of the salt being described?
    **22.1** $CdCl_2$ adopts a hcp lattice of chloride ions in which all the octahedral holes are occupied by cadmium ions.
    **22.2** $CdCl_2$ adopts a ccp lattice of chloride ions in which half of the octahedral holes are occupied by cadmium ions.
    **22.3** $Li_2SO_4$ adopts a hcp lattice of sulfate ions in which all the tetrahedral holes are occupied by lithium ions.
    **22.4** $(CH_3)_4NF$ adopts a ccp lattice of tetramethylammonium ions in which all the octahedral holes are occupied by fluoride ions.

†23. Rewrite each of the following molecular equations as a net ionic equation:
    **23.1** $Ba(ClO_4)_2(aq) + H_2SO_4(aq) \rightarrow BaSO_4(s) + 2\,HClO_4(aq)$
    **23.2** $Na_2SeO_3(aq) + H_2SO_4(aq) \rightarrow H_2SeO_3(aq) + Na_2SO_4(aq)$
    **23.3** $2\,H_2SO_4(aq) + Na_4SiO_4(aq) \rightarrow Si(OH)_4(s) + 2\,Na_2SO_4(aq)$
    **23.4** $3\,Th(NO_3)_4(aq) + 4\,K_2NaPO_4(aq) \rightarrow Th_3(PO_4)_4(s) + 8\,KNO_3(aq) + 4\,NaNO_3(aq)$
    **23.5** $2\,NaHSO_3(aq) + H_2SO_4(aq) \rightarrow Na_2SO_4(aq) + 2\,H_2O(l) + 2\,SO_2(g)$

†24. Using a chemistry handbook, write specific instructions (compounds used, weight of each compound needed) to carry out the following net ionic equations and produce 0.20 mol of product:
    **24.1** $3\,Co^{2+}(aq) + 2\,VO_4^{3-}(aq) \rightarrow Co_3(VO_4)_2(s)$
    **24.2** $2\,Ag^+(aq) + SO_4^{2-}(aq) \rightarrow Ag_2SO_4(s)$
    **24.3** $3\,Mg^{2+}(aq) + 2\,AsO_4^{3-}(aq) \rightarrow Mg_3(AsO_4)_2(s)$
    **24.4** $3\,UO_2^{2+}(aq) + 2\,VO_4^{3-}(aq) \rightarrow (UO_2)_3(VO_4)_2(s)$

25. If you did not do the experiment at the beginning of this chapter, go back and try to apply the principles of this chapter to predict what would have happened at each step. Also answer the questions included.

**Notes**

[1]  Annual Editions, Chemical Rubber Publishing Co., Cleveland.

[2]  Johnson, D. A., *Some Thermodynamic Aspects of Inorganic Chemistry*, Cambridge University Press, Cambridge, 1968, Ch. 5.

[3]  Rich, R., *Periodic Correlations*, Benjamin, New York, 1965, Ch. 8.

[4]  Latimer, W. M., *The Oxidation States of the Elements and Their Potentials in Aqueous Solution*, 2d ed., Prentice-Hall, Englewood Cliffs, N. J., 1952.

[5]  Cox, B. G., and A. J. Parker, *J. Amer. Chem. Soc.*, 95, 6879 (1973).

[6]  Phillips, C. S. G., and R. J. P. Williams, *Inorganic Chemistry*, Oxford University Press, New York, 1965, pp. 258–261.

[7]  Hanzlik, R. P., *Inorganic Aspects of Biological and Organic Chemistry*, Academic Press, New York, 1976, p. 61.

[8]  Burgess, J., *Metal Ions in Solution*, Ellis Horwood, Chichester, England, 1978, pp. 16–17; H. S. Frank and W.-Y. Wen, *Faraday Discuss. Chem. Soc.*, 24, 133 (1957).

[9]  Wells, A. F., *Structural Inorganic Chemistry*, 4th ed., Oxford University Press, Oxford, 1975.

[10]  Morris, D. F. C., *Struct. Bonding*, 6, 157 (1969).

[11]  Kapustinskii, A. F., *Quart. Rev. Chem. Soc.*, 10, 283 (1956); see also T. C. Waddington, *Adv. Inorg. Chem. Radiochem.*, 1, 157 (1959).

[12]  If the cation is strongly acidic and/or the anion is strongly basic, hydrogen ions may be more abundant than the desired cation, or hydroxide ions may be more abundant than the desired anion. The hydroxide of the cation, or the oxo acid form of the anion, may then precipitate along with or instead of the desired salt. Sometimes this may be avoided by the proper order of addition or by techniques such as homogeneous precipitation—the literature should be consulted for techniques appropriate for the desired salt.

# Oxides and Polynuclear Oxo Anions of the Elements: Their Physical, Chemical, and Environmental Properties

In this chapter we will examine some of the chemical and physical properties of two classes of compounds, the oxides of the elements and the salts of polynuclear oxo anions of the elements. The chemical properties we will focus on are the acid-base properties of these compounds, applying the theories developed in the last chapter. Physical properties (such as the ease of melting and boiling and thus whether given compounds are solids, liquids, or gases at room temperature and atmospheric pressure) are of great practical importance: Before preparing and handling a new compound, a scientist needs to know whether that compound is a solid, liquid, or gas so that he or she will know how to handle and measure it out. Many of the oxides and the salts of the polynuclear oxo anions are very important in industry, geology, and soil science, as well as in pollution chemistry. We will see how the applied properties of these substances relate to their physical and chemical properties. Since some of the oxides are ionic and some are covalent (and polynuclear oxo anions include both types of bonds), we will see how these physical and chemical properties relate to the type of bonding.

## 4.1

## Periodic Trends in Acid-Base and Solubility Properties of Oxides

Due to its very small size and high charge, the oxide ion ($O^{2-}$) is a very strongly basic anion that cannot exist in water but instead reacts completely to generate the hydroxide ion (still a very strong base):

$$O^{2-} + H_2O \rightarrow 2\,OH^- \tag{4.1}$$

This reaction is an illustration of the **leveling** property of very strong bases: Any base that is stronger than the characteristic base of the solvent will react with the solvent (in this case, water) to generate its characteristic base (i.e., hydroxide ion). We have already seen this happen to the very strong oxo anions such as $SbO_6{}^{7-}$. Similarly, the very strongly acidic hypothetical cations such as $S^{6+}$ are leveled by reaction with water to generate the characteristic acid of water solutions, $H_3O^+$. Thus no acid stronger than $H_3O^+$ and no base stronger than $OH^-$ can persist in water.

Since the oxide ion is a very strong base, we expect and find that the solubility principles of the previous chapter apply to it. Among metal oxides, only the oxides of the nonacidic and feebly acidic cations dissolve in water, in which they react very exothermically to give the hydroxides of these cations. (By extension, only the hydroxides of the nonacidic and feebly acidic cations dissolve in water.) The solutions of these oxides or hydroxides, of course, have high pH's and are very basic; thus these oxides are logically called **basic oxides** even though they do not contain hydroxide ions.

The oxides of the nonacidic cations react so strongly with water that they are seldom seen; they cannot be prepared by heating the hydroxides to drive off water. Even the hydroxides are deliquescent and release heat on dissolving. By far the most important strongly basic compound of the nonacidic cations is NaOH, which is manufactured on a very large scale by electrolysis of NaCl solutions (an oxidation-reduction process that will be discussed later).

The oxides of the feebly acidic cations are more commonly seen than those of the nonacidic cations. CaO, commonly called lime, is the cheapest basic oxide and is easily manufactured by heating the abundant $CaCO_3$ (limestone):

$$CaCO_3(s) + \text{heat} \rightarrow CaO(s) + CO_2(g) \qquad (4.2)$$

Oxides of the feebly acidic cations react exothermically with water to give the hydroxide (with CaO this reaction is called *slaking*):

$$CaO + H_2O \rightleftharpoons Ca(OH)_2 \qquad (4.3)$$

This reaction can be reversed by heating the hydroxide to a high temperature (853 K) to drive off the water. The hydroxides of these feebly acidic cations are not notably deliquescent.

The oxides and hydroxides of the weakly and moderately acidic cations are insoluble in water, and so alter its pH little or not at all. Nonetheless, both show a characteristic of other bases: They react with (i.e., neutralize) strong acids:

$$FeO + 2\,H_3O^+(aq) \rightleftharpoons Fe^{2+}(aq) + 3\,H_2O \qquad (4.4)$$

This large body of metal oxides is also called "basic oxides."

Correspondingly, acidic properties persist in many of the covalent oxides of the very acidic (hypothetical) cations. Many of these oxides dissolve in water to give solutions of the oxo acid in which the element has the same oxidation number:

$$P_4O_{10} + 6\,H_2O \rightarrow 4\,H_3PO_4 \qquad (4.5)$$

These oxo acids ionize (to a greater or lesser extent, depending on their strength) to give hydronium (hydrogen) ions and a pH below 7, so it seems logical to call

Oxides and Polynuclear Oxo Anions of the Elements

**Table 4.1**  Major Acidic and Amphoteric Oxides
of the *p*- and *d*-Block Elements

| | | | | |
|---|---|---|---|---|
| $B_2O_3$ | $CO_2$ | $N_2O_5$ | | |
| | | $N_2O_3$ | | |
| $Al_2O_3$ | $SiO_2$ | $P_4O_{10}$ | $SO_3$ | $Cl_2O_7$ |
| | | $P_4O_6$ | $SO_2$ | $Cl_2O$ |
| $Ga_2O_3$ | $GeO_2$ | $As_2O_5$ | $SeO_3$ | |
| | | $As_4O_6$ | $SeO_2$ | $Br_2O$ |
| $In_2O_3$ | $SnO_2$ | $Sb_2O_5$ | $TeO_3$ | $XeO_4$ |
| | $SnO$ | $Sb_2O_3$ | $TeO_2$  $I_2O_5$ | $XeO_3$ |
| | $PbO_2$ | | | |
| | $PbO$ | | | |

| | | | |
|---|---|---|---|
| $V_2O_5$ | $CrO_3$ | $Mn_2O_7$ | |
| $Nb_2O_5$ | $MoO_3$ | $Tc_2O_7$ | $RuO_4$ |
| $Ta_2O_5$ | $WO_3$ | $Re_2O_7$ | $OsO_4$ |

NOTE: Oxides above and to the right of the dashed line dissolve in water (at least to some extent) and act as acids. Oxides between the dashed and solid lines are not soluble in water but do dissolve in (react with) strong bases. Oxides below and to the left of the solid line are amphoteric. (These are not included for the *d*-block elements.) The oxides of Tl and $Bi_2O_3$ are exclusively basic oxides.

these oxides **acidic oxides** or **acidic anhydrides**. (The latter term suggests that they are acids that are missing the elements of water).

Soluble acidic oxides of the *p*- and *d*-block elements are shown in Table 4.1. The oxide will be soluble if its reaction product with water is a *strong* or *very strong* acid. These acids are completely ionized into highly hydrated hydrogen ions and oxo anions, which helps shift the solubility equilibrium toward high solubility. If the oxo acid produced is *moderately acidic*, the oxide may or may not be soluble. (Some ions are produced, and the un-ionized acid may hydrogen-bond to the water, which may or may not suffice to ensure solubility.) If the oxo acid produced is *weakly acidic*, the oxide is usually (but not always) insoluble in water. Some of these acidic oxides (especially those of sulfur and nitrogen) are air pollutants; these oxides react with the moisture of the air and produce "acid rain."

The oxides between the dashed and solid lines in Table 4.1, although not soluble enough in water to alter its pH appreciably, do react with solutions of strong bases to generate oxo or hydroxo anions and thus also can be justifiably termed "acidic oxides."

As mentioned in Section 2.7, hydroxo anions can be formed in strongly basic solution even by many metal ions that are only moderately acidic; hence there are oxides of metals that are insoluble in water but that dissolve in both strong acids and strong bases; such oxides are termed **amphoteric oxides**. In Table 4.1 these are shown (for *p*-block elements only) to the left of and below the solid line.

In addition, there are a few oxides, such as $NO_2$ and $ClO_2$, that do not correspond in oxidation state to a stable or known oxo acid; these can give rise to a mixture of oxo acids or anions by **disproportionation**:

$$2\,NO_2 + 2\,OH^- \rightarrow NO_2^- + NO_3^- + H_2O \qquad (4.6)$$

Only three nonmetal oxides from the upper-right portion of the $p$-block have such low oxidation numbers for the nonmetal atom that their aqueous solutions are neutral, namely, $CO$, $N_2O$, and $NO$.

Because the acidity of a cation rises rapidly with its charge, there are several $d$-block elements possessing several oxidation numbers (such as chromium) that have one or more oxides that show only basic properties (e.g., chromium(II) oxide, $CrO$), one or more oxides that are amphoteric (e.g., chromium(III) oxide, $Cr_2O_3$), and one or more oxides that possess only acidic properties (chromium(VI) oxide, $CrO_3$). Clearly, the higher the oxidation number of a given element, the more acidic the corresponding oxide will be.

Since basic oxides can react with the hydronium ion, a strong aqueous acid, and acidic oxides can react with the hydroxide ion, a strong aqueous base, it is not too surprising that basic oxides and acidic oxides can react with each other:

$$2\,MgO \text{ (basic oxide)} + SiO_2 \text{ (acidic oxide)} \rightarrow Mg_2SiO_4 \qquad (4.7)$$

The products of these reactions are salts of oxo acids and, since water is not involved in the reaction, can be salts of oxo anions such as $SiO_4^{4-}$ that are too basic to persist in aqueous solution. As we will discuss later in this chapter, direct reactions of acidic and basic oxides are of enormous practical importance in such areas as control of pollution by gaseous acidic oxides and in the production of materials such as concrete, glass, and ceramics.

**EXAMPLE**  Complete and balance the following chemical equations for reactions of oxides:
(a) $N_2O_3 + H_2O \rightarrow ?$
(b) $BaO + OsO_4 \rightarrow ?$

**Solution**  ■ a. $N_2O_3$ is the oxide of a very acidic nonmetal "cation," $N^{3+}$, from the upper right of the periodic table; hence it is expected to be an acidic oxide and will likely dissolve in water to give the oxo acid containing nitrogen in the $+3$ oxidation state, $HNO_2$. The complete balanced equation is

$$N_2O_3 + H_2O \rightarrow 2\,HNO_2$$

■ b. $BaO$ is the oxide of a feebly acidic cation, $Ba^{2+}$, so the strong basicity of the oxide ion will be virtually undiminished: This is a basic oxide. The hypothetical $Os^{8+}$ cation in $OsO_4$ will be so acidic that the oxide itself might also be expected to be acidic. The product of the reaction of a basic oxide with an acidic oxide will be the salt of an oxo anion. For the relatively large Os, an oxo anion with six oxygens is expected: $OsO_6^{4-}$ (Table 2.7). Balancing the $-4$ charge of this anion with two $Ba^{2+}$ cations, we complete and balance the equation:

$$2\,BaO + OsO_4 \rightarrow Ba_2OsO_6 \qquad \square$$

## 4.2

# Interunit Forces and Physical Properties of Compounds

Before we attempt to predict whether a given oxide will be a gas, liquid, or solid at room temperature, it is useful to review and compare the kinetic-molecular descriptions of the gaseous, liquid, and solid states. As postulated in the **kinetic-molecular theory**, units (molecules or ion pairs, etc.) in the gaseous state exert no attractive forces on each other, move independently of each other, and remain relatively far apart. In practice, however, there are attractive forces between units; the weakest of these is the van der Waals force, which occurs between units of all substances. The van der Waals force results when a momentary unsymmetrical distribution of electrons in one molecule (a temporary dipole) induces an opposed momentary unsymmetrical distribution in a neighboring molecule; the momentarily oppositely charged ends of the two units then attract each other. In small molecules the van der Waals force is very weak as compared with covalent bond energies or Coulombic attractions. But in very large molecules with numerous highly polarizable (relatively loosely bound) electrons the van der Waals forces may be so substantial that the kinetic energy available to molecules at room temperature is not enough to overcome such an attraction; hence substances with large molecules cannot be gaseous at room temperature and atmospheric pressure. Of course, they may become gases at higher temperatures or lower pressures. In practice, essentially [1] only monoatomic noble gases or small covalent molecules are gases at room temperature and atmospheric pressure.

The kinetic-molecular picture of the liquid state postulates units that are held close by forces that are strong compared with the thermal energy at that temperature. But the units are free to flow around each other, which implies that their interunit attractive forces can be stretched. The units must also be either relatively small or at least not locked together in rigid two- or three-dimensional lattices by excessively strong, unbendable interunit forces. Medium-sized covalent molecules, or smaller molecules linked by somewhat stronger forces such as hydrogen bonding (i.e., water and alcohols), are most likely to be liquids at room temperature.

In the solid state, units are packed closely together and exert attractive forces on one another that are strong enough to hold the units in position in a lattice despite their vibrational energies. Large covalent molecules often have strong enough van der Waals forces to be solid at room temperature, but their melting points are not nearly so high as those of compounds that have units linked by the much stronger Coulombic forces (ionic salts) or covalent bonds. The latter are called **macromolecules** or **polymers**. Examples shown in Figure 4.1 include silicon dioxide, in which $SiO_2$ units are linked in three dimensions, and arsenic(III) oxide, with a two-dimensional or layer linkage of units; the unshared pairs of $p$ electrons on As atoms block linkage in the third dimension. Polymers linked in only one dimension (such as chromium(VI) oxide, shown in Figure 4.1, or polyethylene, $(CH_2CH_2)_x$), have fewer linkages to break or may be flexible enough to allow fluid motion of units among each other; their melting points may be substantially lower than those

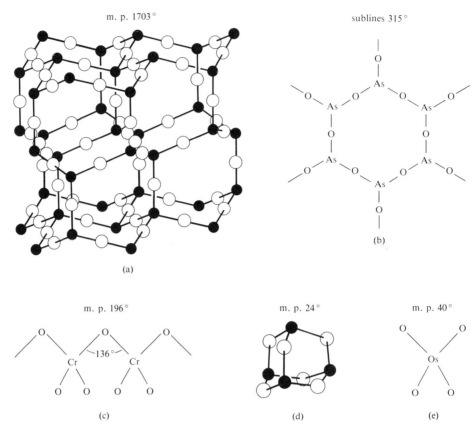

m. p. 1703°

sublines 315°

(a)

(b)

m. p. 196°

~136°~

(c)

m. p. 24°

(d)

m. p. 40°

(e)

**Figure 4.1** Melting points and the cross-linking of structural units; open circles represent oxygen atoms. (a) $SiO_2$ (beta-tridymite), with a three-dimensional structure and a coordination number of 2 for oxygen; (b) $As_2O_3$, with a two-dimensional (layer) structure and a coordination number of 2 for oxygen; (c) $CrO_3$, with a one-dimensional (chain) structure and an average coordination number of 1.33 for oxygen; (d) $M_4O_6$ (M = P, As, or Sb), an oligomer with a finite molecular size and a coordination number of 2 for oxygen; (e) $OsO_4$, a monomeric covalent molecule with coordination number of 1 for oxygen. Structures are from *Structural Inorganic Chemistry*, 3d ed., by A. F. Wells. Copyright © 1962 by Oxford University Press. Reprinted by permission.

of the more fully cross-linked two- or three-dimensional polymers but will normally be higher than those of comparable monomers.

## 4.3

## The Physical States and Structures of the Oxides of the Elements

The room-temperature physical states (solid, liquid, or gaseous) and the relative melting and boiling points of compounds are strongly influenced by whether the simplest-formula units of that compound are discrete or whether they link together into oligomers (medium-sized molecules containing just a few times the number of atoms found in the simplest formula), one-dimensional polymers, or higher-dimensional polymers. If all the atoms or ions of one kind have a coordination number of 1, linkage is impossible. If only a few of the atoms or ions of one kind exceed a coordination number of 1, the linkage will not be extensive, and oligomers or one-dimensional polymers are likely to result. To produce two- or three-dimensional polymers or ionic lattices, there must be enough atoms (or ions) of each kind that have coordination numbers of at least 2 to complete infinite linkage in two or three dimensions.

Periodic trends in physical states and melting and boiling points should, therefore, be related to the coordination numbers of the most abundant atoms. These coordination numbers, in turn, should be related to the ratio of the cationic and anionic radii. Radius-ratio calculations, of course, are not highly reliable. The coordination number will also be reduced if there is double bonding between the atoms, and we are not yet ready to predict when that will occur. Nonetheless, let us try to predict some periodic trends using the following calculations.

**1.**  Calculate the radius ratio $r_{\text{"cation"}}/r_{\text{"anion"}}$. Referring to Table 3.4, use this ratio to predict the coordination number of the least abundant (normally the metal) atom or ion. If the radius ratio is close to the cutoff points, there is a good deal of uncertainty in the result; either coordination number may prove correct. Some compounds are **polymorphic**—that is, they may have one form with one coordination number and one with the other. Note that if the cation or metal is a *p*-block element with less than the maximum oxidation number, some space must be reserved for the unshared pair of electrons, so a lower coordination number will be predicted. This will also be the case for some of the heavier *d*-block elements, which routinely take unusually low coordination numbers.

**2.**  Calculate the coordination number of the most abundant atom or ion, using equation (3.14). This is the atom or ion that will have the lowest coordination number and that may not be able to link units into oligomers or polymers.

**3.**  If the coordination number of the most abundant atom or ion is 1, the species is predicted to occur as monomeric molecules, with relatively low melting or boiling points. To predict whether it is a gas, liquid, or low-melting solid at room temperature, we need to look at the size of the molecule and whether or not it can engage in hydrogen bonding.

**4.**  If the coordination number of the most abundant atom or ion is greater than 2, this will usually suffice to link the units into polymers, which will have high melting and boiling points. Indeed, in practice such species may decompose before they can melt or boil; if they decompose into small gaseous molecules, they may **sublime**—pass directly from the solid to the gaseous state.

**5.**  If the average coordination number of the most abundant atom or ion is between 1 and 2, there may be enough linkage possibilities to make a one-dimensional polymer, but probably not a two- or three-dimensional polymer. The lower this number is, the more likely it is that a small oligomer will be formed. The most likely prediction in this case is that the melting and boiling points will be somewhat higher than those found for monomeric molecules, but not nearly so high as those found for polymers or ionic compounds. Exact structural predictions are unfortunately not easy here; thus the 2-coordinate oxygen atoms in $As_2O_3$ (Figure 4.1b) are used to produce a two-dimensional polymer, but in $As_4O_6$ the same oxygen atoms, with the same coordination number, link intramolecularly to give an oligomer (Figure 4.1d).

Although exact predictions of structures are not to be expected, it should be possible to predict periodic *trends* in melting and boiling points and to understand why dramatic changes occur when they do in a period or a group.

EXAMPLE Predict the coordination numbers, the physical states, and the relative melting and boiling points of the oxides of the second-period elements in their maximum oxidation state. State whether each should be monomeric, oligomeric, or polymeric or in an ionic lattice.

Solution ■ From Table B we see that these oxides will be $Li_2O$, $BeO$, $B_2O_3$, $CO_2$, and $N_2O_5$. The ionic radius of oxide ion is 126 pm (Table 2.5); we obtain the ionic radii of the "cations" from Table C. The radius ratios are 0.714, 0.464, 0.325, 0.238, and 0.214, respectively. The least abundant atom in $Li_2O$ is oxygen; in the others it can be taken as the nonoxygen atom. Hence the coordination numbers predicted are: either 6 or 8 for O in $Li_2O$; either 6 or 4 for Be; 4 for B; 4 or 3 for C; and 3 for N.

■ We now compute the (lower) coordination numbers of the most abundant atom in each compound: For lithium, this is either 3 or 4; for oxygen either 6 or 4 in BeO, $2\frac{2}{3}$ in $B_2O_3$, either 2 or $1\frac{1}{2}$ in $CO_2$, and $1\frac{1}{5}$ in $N_2O_5$. These are average coordination numbers; fractional numbers mean that some atoms will have the higher (nearest) whole number coordination number and some the lower.

■ For the first three compounds, all coordination numbers are predicted to be greater than 2, so we predict that these will be ionic or polymeric covalent compounds, with high melting and boiling points, and we predict that these will all be solids at room temperature. Our predictions are verified: $Li_2O$ has a melting point of 1427 °C, BeO of 2530 °C, and $B_2O_3$ of 450 °C (but with a boiling point of 1860 °C) [3].

■ For carbon dioxide, the prediction is uncertain but wrong in either case, since our calculations cannot predict the special stability of oxygen double-bonded to carbon. $CO_2$ is actually a monomeric molecule and a gas at room temperature, which sublimes at −79 °C at atmospheric pressure.

■ For dinitrogen pentoxide the prediction is correct: One of the five oxygens bridges or links the two nitrogens, while each of the other oxygen atoms is bonded to only one nitrogen atom; each nitrogen has a coordination number of three. The solid form of this compound consists of $NO_2^+$ and $NO_3^-$ ions, with lower coordination numbers than expected. The compound is volatile and explodes readily, so that no melting or boiling point is recorded for it. □

We may also predict lattice types for the ionic compounds, using Table 3.4, and compare them with results given by Wells [2a]: For $Li_2O$ we predict either an antirutile or antifluorite lattice and find the latter; and for BeO we predict either the NaCl or ZnS lattice and find the latter (wurtzite). The data for $B_2O_3$ does *not* suggest the corundum lattice type; two forms of this compound are known, in which the coordination numbers of boron are 3 and 4, respectively.

Finally we note that the coordination numbers of the second-period elements in each of these compounds did not exceed our expected maximum coordination number for the second period of 4. Our predictions indicated that the largest ions, $Li^+$ and $Be^{2+}$, could have exceeded this number; $Li^+$ is known to do so in its halides.

For purposes of examining periodic trends, in Table 4.2 we summarize similar calculations for the third-period and the early sixth-period elements in their group oxidation states. The following periodic trends are illustrated by such calculations:

**1.** At the left of a period, solids of high melting points prevail. As we go to the right the coordination number of the more abundant (oxygen) atoms declines. Approximately at the point at which this number drops below 2, the melting points become sharply lower, as one-dimensional polymers, oligomers, and monomers prevail.

**Table 4.2** Predictions of Physical Properties and Structures of Some Oxides

| Period 3 | $Na_2O$ | MgO | $Al_2O_3$ | $SiO_2$ | $P_2O_5$ | $SO_3$ | $Cl_2O_7$ | |
|---|---|---|---|---|---|---|---|---|
| Radius ratio | 0.921 | 0.683 | 0.532 | 0.429 | 0.413 | 0.341 | 0.325 | |
| *Predicted* | | | | | | | | |
| C.N. of metal | 4 | 6 | 6 | 6 or 4 | 4 | 4 | 4 | |
| C.N. of oxide | 8 | 6 | 4 | 3 or 2 | 1.6 | 1.33 | 1.14 | |
| *Observed* | | | | | | | | |
| C.N. of metal | 4 | 6 | 6 | 4 | 4 | 4 | 4 | |
| C.N. of oxide | 8 | 6 | 4 | 2 | 1.6 | 1.33 | 1.14 | |
| Lattice type | antiflourite | NaCl | corundum | beta-silica | oligomer | oligomer | molecular | |
| Melting point | (1275) | 2800 | 2050 | 1723 | (300) | 17 | −91 | |
| **Period 6** | $Cs_2O$ | BaO | $Lu_2O_3$ | $HfO_2$ | $Ta_2O_5$ | $WO_3$ | $Re_2O_7$ | $OsO_4$ |
| Radius ratio | 1.437 | 1.183 | 0.793 | 0.675 | 0.619 | 0.587 | 0.537 | 0.476 |
| (Inverse " ") | (0.696) | (0.846) | | | | | | |
| *Predicted* | | | | | | | | |
| C.N. of metal | 3 | 8 | 8 | 6 | 6 | 6 | 6 | 6 |
| C.N. of oxide | 6 | 8 | 5.33 | 3 | 2.4 | 2 | 1.71 | 1.5 |
| *Observed* | | | | | | | | |
| C.N. of metal | 3 | 6 | 7 | 8 | 6 | 6 | 5 | 4 |
| C.N. of oxide | 6 | 6 | 4.67 | 4 | 2.4 | 2 | 1.42 | 1 |
| Lattice type | * | NaCl | * | Rutile | * | * | 2-dim. polymer | molecular |
| Melting point | 490 | 1920 | 2487 | 2900 | 1870 | 1473 | 296 | 40 |

NOTES: For sources of data see [2] and [3]. Melting points in degrees Celsius; if in parentheses, these are sublimation temperatures. C.N. = average coordination number. *These compounds have three-dimensional ionic or polymeric covalent lattice types not discussed in this text; see [2].

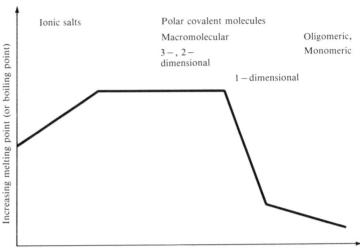

Ionic salts     Polar covalent molecules

Macromolecular                        Oligomeric,
3 −, 2 −                               Monomeric
dimensional

1 − dimensional

Increasing melting point (or boiling point)

**Figure 4.2** Typical trends in melting (or boiling) points of oxides of the elements in their group oxidation states, as the oxidation and group numbers increase.

Increasing oxidation number of the central atom

**2.** Small atoms with high oxidation numbers will form oxides with the lowest melting points (those that are gases or liquids at room temperature, or that are low-melting solids). Such atoms must be surrounded by many of their "own" oxygen atoms and thus have little or no room for "bridging" oxygens from other units. Thus the calculations in Table 4.2 successfully predict sharply lower melting points for the compounds $P_2O_5$, $SO_3$, $Cl_2O_7$, $Re_2O_7$, and $OsO_4$.

**3.** Among metal oxides forming ionic lattices, melting and boiling points tend to rise as the charge on the metal ion rises, up to a point. This is a consequence of the increase in Coulombic forces as the ionic charge increases. Thus melting points increase substantially from $Na_2O$ to $MgO$, and from $Cs_2O$ to $HfO_2$. The trend does not continue indefinitely, however. Evidently this tendency is counteracted after a while by the poorer linkage that results when the coordination number of oxygen falls, or in many cases by the fact that the elements with higher oxidation numbers also have higher Pauling electronegativities, so that their polymeric oxides can more readily "decompose" to gaseous small-molecule covalent oxides. Figure 4.2 shows schematically the horizontal periodic trends typically found in melting or boiling points of oxides of the elements in their group oxidation states.

## 4.4

## Covalent Oxides: Nomenclature and Periodic Trends in Structure and Physical State

Most metal oxides are high-melting solids, the melting points of which follow the trends given above reasonably well. More nonmetal oxides are molecular or oligomeric; since many of these are also quite important, we will devote more individual attention to them in this section. Figures 4.3, 4.4, and Table 4.3 sum-

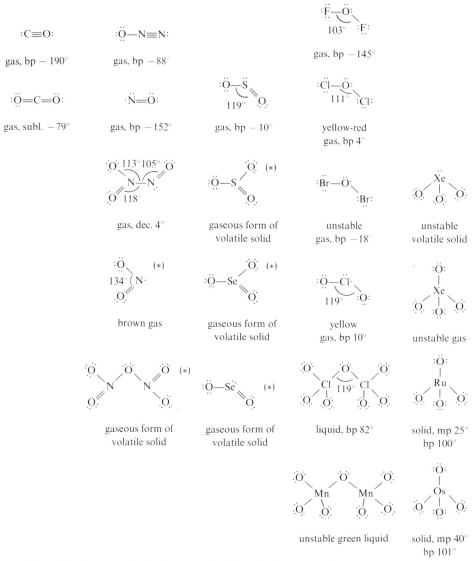

**Figure 4.3** Structures, physical states, and melting or boiling points of the monomeric molecular (covalent) oxides. Structures with an asterisk (*) beside them also have alternate structural forms that are oligomeric or polymeric; these are shown in Figure 4.4 or Table 4.3. Not shown are $Tc_2O_7$, which has a structure similar to that of $Mn_2O_7$ and $Cl_2O_7$, and $Re_2O_7$, which has a similar structure in the gaseous state but normally is a volatile solid with a polymeric structure.

marize the Lewis dot structures, physical states at room temperature, melting points, and structural forms of most of the important covalent oxides. For convenience, the oxides that are small covalent molecules (i.e., whose molecular formulas and simplest formulas are the same) are tabulated and discussed together, followed by the oligomeric oxides, then the polymeric or ionic oxides.

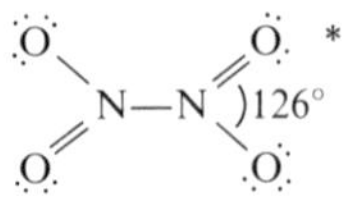

gas, bp 21°

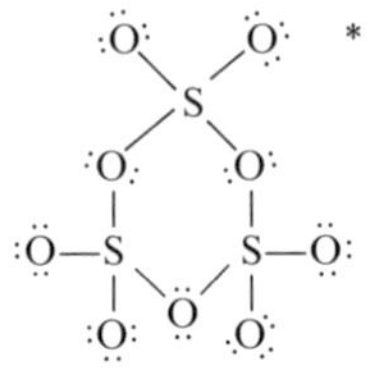

liquid, mp 17°, bp 45°

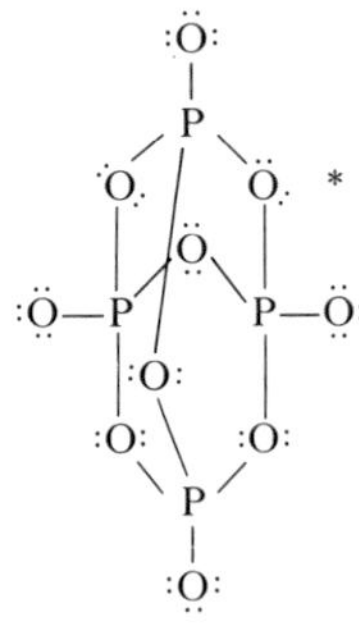

$Se_3O_9$, structure like $S_3O_9$*

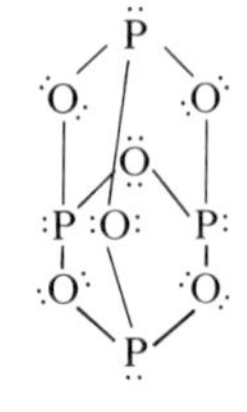

solid, sublimes 360°

$As_4O_6$, structure like $P_4O_6$*
solid, mp 309°

$Sb_4O_6$, structure like $P_4O_6$*
solid, mp 656°

solid, mp 25°

**Figure 4.4** Structures, physical states, and melting or boiling points of the oligomeric covalent oxides. Structures with an asterisk (*) beside them also have structural forms that are monomeric or polymeric.

### 4.4.1 Nomenclature

Before discussing the structures and physical states of these compounds, we find it worthwhile to mention an alternative system of nomenclature that is often used for covalent substances (this system also applies to halides, sulfides, and other binary compounds). Thus although $Cl_2O_7$ may be perfectly acceptably named chlorine(VII) oxide, it is also commonly named **di**chlorine **hept**oxide. This alternative system of nomenclature is also often used with the more covalent oxides, etc., of the d-block elements (i.e., those of the d-block elements in their higher oxidation states).

As we see from the example, prefixes are added to the name of the element and to the name of the "anion" (i.e., oxide in this case) to indicate the number of such atoms present in the molecule or in the simplest formula. The prefixes used are *di-* if two atoms are present, *tri-* for three, *tetra-* for four, *penta-* for five, *hexa-* for six, *hepta-* for seven, *octa-* for eight, *nona-* for nine, *deca-* for ten, *undeca-* for eleven, etc. If the name of the anion begins with a vowel (e.g., oxide), the last vowel is dropped for smoothness in pronunciation. The prefix *mono-* is normally omitted and hence is assumed if the name contains no prefix. But it may be used to emphasize that the

**Table 4.3**  Macromolecular Oxides of the Nonmetals

| $(SiO_2)_x$ | $(P_2O_5)_x^*$ | $(SO_3)_x^*$ | |
|---|---|---|---|
| CN(Si) = 4 | CN(P) = 4 | CN(S) = 4 | |
| CN(O) = 2 | CN(O) = 1.6 | CN(O) = 1.33 | |
| 3-dim | 2-dim | 1-dim | |
| solid, mp 1723 | solid | solid, mp 17 | |
| $(GeO_2)_x$ | $(As_2O_5)_x$ | $(SeO_3)_x^*$ | |
| CN(Ge) = 4, 6 | CN(As) = 5 | CN(Se) = 4 | |
| CN(O) = 2, 3 | CN(O) = 2 | CN(O) = 1.33 | |
| 3-dim | 3-dim | 1-dim | |
| solid, mp 1116 | solid, dec. 315 | solid, mp 120 | |
| $(SnO_2)_x$ | $(Sb_2O_5)_x$ | $(TeO_3)_x$ | |
| CN(Sn) = 6 | CN(Sb) = 6 | CN(Te) = 6 | |
| CN(O) = 3 | CN(O) = 2.4 | CN(O) = 2 | |
| 3-dim | 3-dim | 3-dim | |
| solid, mp 1127 | solid, dec. 380 | solid, dec. 395 | |
| $(PbO_2)_x$ | | $(SeO_2)_x^*$ | |
| CN(Pb) = 6 | | CN(Se) = 3 + : | |
| CN(O) = 3 | | CN(O) = 1.5 | |
| 3-dim | | 1-dim | |
| solid, dec. 290 | | solid, mp 315 | |
| $(SnO)_x$ | $(Sb_2O_3)_x^*$ | $(TeO_2)_x$ | $(I_2O_5)_x$ |
| CN(Sn) = 4 + : | CN(Sb) = 3 + : | CN(Te) = 6 + : | |
| CN(O) = 4 | CN(O) = 2 | CN(O) = 3 | |
| 2-dim | 2-dim | 3-dim | 3-dim |
| solid, dec. 1080 | solid | solid, mp 733 | solid, dec. 300 |
| $(PbO)_x$ | $(Bi_2O_3)_x$ | $(PoO_2)_x$ | |
| CN(Pb) = 4 + : | CN(Bi) = 6 + : | CN(Po) = 6 + : | |
| CN(O) = 4 | CN(O) = 4 | CN(O) = 3 | |
| 2-dim | 3-dim | 3-dim | |
| solid, mp 888 | solid, mp 874 | solid, dec. 500 | |

NOTES: CN(Si) = average coordination number of silicon. 2-dim = forms a two-dimensional or layer polymer. mp = melting point. dec. = temperature at which the solid decomposes. * = also has monomeric or oligomeric forms; see Figures 4.3 and 4.4. An unshared pair of electrons for the atom in parentheses is indicated by :.

particular oxide being discussed has just one oxygen and is not another common oxide with some other number of oxygens; thus we say "carbon monoxide" to make it clear that we are not talking about any oxide of carbon in general.

In the case of *p*-block nonmetals, the nonmetal with the lower electronegativity is generally named first. Thus the compound $F_2O$ shown in Figure 4.3 is not called difluorine oxide, but rather oxygen difluoride. It is customarily considered a fluoride rather than an oxide and will not be discussed further in this chapter.

### 4.4.2    Structures and Physical Properties

MONOMERIC MOLECULES, FIGURE 4.3    As previously discussed, these small covalent molecules, found in the upper-right part of the periodic table or among $d$-block elements with oxidation states of $+7$ or $+8$, have relatively weak forces between the molecular units. The smaller, lighter molecules in this group have few polarizable electrons; hence they have quite weak van der Waals forces—these molecules are gases. The molecules with quite a few atoms (the di-element heptoxides) or those with atoms with many electrons (xenon, ruthenium, etc.) have van der Waals forces strong enough to enable them to be liquids or quite volatile solids at room temperature.

Several of the structures shown in Figure 4.3 represent the structures found in the vapors above volatile (usually one-dimensional) polymeric solids. The solid-state structure of $N_2O_5$ is different: In the solid state $N_2O_5$ is ionic, consisting of $NO_2^+$ cations that are **isoelectronic** with (have the same electron-dot structure as) carbon dioxide, along with the stable, nonbasic nitrate anion. Unlike most ionic compounds, this one is volatile at room temperature, since it can easily convert to a stable covalent structural form in the gas phase. The solid-state structure of $N_2O_3$ has an unusually long N—N bond and resembles an ionic structure, too. (The cation $NO^+$, isoelectronic with CO, is rather stable; the structure of solid $N_2O_3$ resembles a $NO^+$ ion paired with a $NO_2^-$ ion.)

Note that NO, $NO_2$, and $ClO_2$ each have an odd number of electrons. These are examples of a class of molecules known as **free radicals**, which are usually exceedingly reactive, since they tend to pair up with each other (or with other radicals) to share the odd electrons and form additional covalent bonds, releasing the covalent bond energy. $NO_2$ does this fairly readily, forming the oligomer $N_2O_4$ (Figure 4.4), but little energy is released; the other two show almost no tendency to do this. Clearly it is impossible to draw satisfactory Lewis dot structures for these three substances or to use the octet rule to explain why these three free radicals are so much more stable than most free radicals. Another bonding theory, the **molecular orbital theory**, is necessary to account for this phenomenon, but it is impractical to try to introduce this theory and apply it to these substances at this time.

We have seen before (Section 1.5) that the octet rule sometimes must be violated; according to rule 6 in Section 1.5, this occurs fairly frequently among elements of the third period and below, since these atoms are large enough to form more than four covalent bonds. Aside from the three oxides mentioned in the previous paragraph, all the Lewis dot structures shown in Figure 4.3 obey the octet rule; but alternative structures can be drawn that legitimately violate this "rule." For example, we could draw a structure for $SO_3$ in which *all three* sulfur-oxygen bonds are double bonds (Figure 4.5). This would seem to accommodate the physical evidence better than the structure in Figure 4.3, since it is found experimentally that (a) all sulfur-oxygen bonds are equal in length, and (b) all the bonds are much shorter

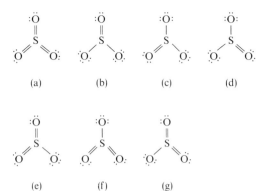

(a)       (b)       (c)       (d)

(e)       (f)       (g)

**Figure 4.5**  Possible resonance forms for the $SO_3$ molecule.

than expected for a sulfur-oxygen single covalent bond. (Recall that double bonds are typically shorter than single bonds.) We may also, however, accommodate this evidence by drawing two other Lewis dot structures analogous to the one in Figure 4.3, but with the double bonds to the *other* oxygen atoms (Figure 4.5b, c, d). If we assume that the true structure is an average of these three **resonance** forms, we can account for the observations. (The molecular orbital theory, about which you will learn later, handles this awkward problem much more neatly.) But there seems to be no reason not to include the other resonance forms shown in Figure 4.5. You will sometimes see Lewis dot structures for oxides and oxo anions of third-period elements and below drawn with these extra double bonds. Such structures are not wrong, although we are avoiding them for now for simplicity's sake.

OLIGOMERIC MOLECULES    Figure 4.4 shows the structures of *p*-block oxides that have molecular formulas that are double or triple their simplest formulas (i.e., are oligomers). With the exception of $N_2O_4$, these molecules are large enough to have van der Waals forces strong enough to keep them in the liquid or solid state at room temperature. The majority of these also have allotropes in which the bridging (2-coordinate) oxygen atoms are used to link units together into polymeric units.

In general, the oxides that are monomeric or oligomeric are also those that are most soluble in water (Section 4.1). This is intuitively reasonable, since these oxides will not have strong lattice energies that must be overcome to form hydrated oxo anions.

POLYMERIC OXIDES    Table 4.3 presents data on the major oxides of the *p*-block elements that form polymeric covalent networks. (The details of the structures of these networks are not sufficiently important for us to present them here; see [2]. Neither have we listed the numerous *d*-block oxides that might be considered to have polymeric covalent structures.) As expected, in general the melting points of these oxides are much higher than those of the monomeric or oligomeric oxides; many of these oxides decompose before they can be heated to a high enough temperature to melt. The one-dimensional polymers at the upper right ($SO_3$, $SeO_3$, and, from the *d*-block, $CrO_3$) are, as expected, easier to melt than the higher-dimensional polymers. Some of the oxides at the lower left of Table 4.3, $SnO_2$ and $PbO_2$, adopt a structure that we think of as typically ionic, namely, the rutile lattice. The melting points of these compounds show no discontinuity with the melting points of their neighbors—Coulombic attractions and interunit covalent bonds are both strong interunit forces, about equally capable of resisting melting.

## 4.5

## Practical Uses and Environmental Chemistry of Volatile Oxides and Oxo Acids

A number of the volatile (and usually acidic) oxides, and some of the oxo acids formed from them, are of immense significance in our technological civilization. These include some of the chemicals produced in the greatest total tonnage in the world. Any chemicals used in such high tonnage are bound to escape into the environment to some degree. Since these are mainly acidic oxides, we can anticipate that they will not be innocuous components of the environment; since these are volatile oxides, we can anticipate that there will be air-pollution problems connected with them. In this section we will deal with the uses and environmental chemistry of the most important of these oxides; there are many useful sources of this information which should be consulted for more details [4, 5, 6, 7, 8, 9]. Given the periodic trends we have seen in the previous sections and chapters, we should also be able to predict some uses and environmental chemistry for the other oxides, should they come into widespread use.

### 4.5.1    Carbon Oxides

CARBON MONOXIDE    CO is produced industrially on a huge scale as a mixture with $H_2$, by the reaction of steam with hot coal:

$$C(s) + H_2O(g) \rightarrow CO(g) + H_2(g); \qquad \Delta H^\circ = +131 \text{ kJ/mol} \tag{4.8}$$

(Since this reaction is quite endothermic, air must periodically be blown through the coal to provide energy via oxidation of the coal.) This mixture of carbon monoxide and hydrogen is known as **water gas**; after adjustment of its hydrogen content it is known as **synthesis gas**, since it is used in the industrial production of a number of important organic chemicals. Intensive research is currently focused on the chemistry of carbon monoxide, since this mixture could be a source of organic chemicals from coal when our present source of many organic chemicals, petroleum, is exhausted. For example, synthesis gas can be converted to hydrocarbons via the **Fischer-Tropsch process**:

$$n\,CO + (2n + 1)H_2 \rightarrow C_nH_{2n+2} + n\,H_2O \tag{4.9}$$

Environmentally, the main anthropogenic (human-created) source of carbon monoxide is the incomplete combustion of fuel in automobile engines (197 million tons per year). In urban environments this is a problem due to the high toxicity of carbon monoxide (of which we have no warning, since it is odorless). In terms of the global atmospheric environment, however, carbon monoxide is not considered a serious pollutant, since (a) natural production of CO (i.e., by the oxidation of the $CH_4$ produced by anaerobic decomposition of organic material in swamps and in the tropics) far outweighs human production, and (b) natural processes continuously remove CO from the atmosphere. These processes include microbial degradation in the soil and reactions in the atmosphere with reactive free radicals such as hydroxyl ($\cdot OH$) and hydroperoxyl ($HO_2\cdot$). (Free radicals can persist in the atmosphere— especially in the thin upper atmosphere—longer than in solution, since the concentration of any species with which they may react is much lower there than in solution.)

CARBON DIOXIDE   $CO_2$ is produced industrially by a number of reactions, such as the combustion of carbonaceous fuels, and by the **water gas shift reaction**:

$$CO + H_2O \rightleftharpoons CO_2 + H_2 \qquad (4.10)$$

Because of its physical properties, it is widely used as a refrigerant: the solid form, Dry Ice (tm), sublimes at $-78\,°C$, thus cooling its environment without generating any messy liquids or toxic gases. It is also used on a large scale in fire extinguishers and in carbonated beverages.

Although of very low toxicity, carbon dioxide *is* considered to be a potentially serious global atmospheric pollutant. It is, of course, removed from the atmosphere by plant photosynthesis and put into the atmosphere by plant and animal respiration; presumably these processes have reached a balance over time. However, the atmospheric concentration of $CO_2$, as measured in such remote locations as Antarctica and the top of Mauna Loa, Hawaii, has been increasing 0.2% a year since about 1870. This is of concern because of the vital role of $CO_2$ in the atmosphere: It absorbs infrared light (heat) emitted by the earth and returns some of it to the surface of the earth, thereby warming it; this process is popularly known as the "greenhouse effect." An increase of only $2\,°C$ or $3\,°C$ would have profound effects on the climate of the earth; between 1880 and 1940 the mean temperature of the earth rose $0.4\,°C$. This might have been due to the evident changes in the atmospheric $CO_2$ concentration, which is thought to be the result of the combustion of fossil fuels and (more importantly) the cutting and burning of many of the tropical forests of the world. (A catastrophe in climate is not assured, however; there is an opposing scattering of incoming solar energy by suspended atmospheric particulates, which may be responsible for the cooling of the earth by $0.1\,°C$ since 1940.)

## 4.5.2   Nitrogen Oxides and Nitric Acid

Of the nitrogen oxides shown in Figure 4.3, only three are stable enough to be of practical use or environmental importance. One of these, $N_2O$ (commonly called nitrous oxide), is used as an anesthetic and as a propellant (to provide pressure to expel ingredients) in aerosol cans; it is involved in reactions in the upper atmosphere that could deplete the **ozone layer**.

There are two common sources of NO: catalytic oxidation of ammonia (which comes ultimately from petroleum and air) and direct combination of nitrogen and oxygen of the air in an electrical discharge (lightning) or at around 2000 °C (e.g., in a power plant during the burning of coal or in an automobile engine):

$$4\,NH_3 + 5\,O_2 \rightarrow 4\,NO + 6\,H_2O \tag{4.11}$$

$$N_2 + O_2 \rightleftharpoons 2\,NO \tag{4.12}$$

After a few days in the atmosphere, NO is oxidized by oxygen to $NO_2$; hence in air-pollution work these two are often collectively referred to as $NO_x$. Nitrogen dioxide and its dimer, dinitrogen tetroxide, are readily interconverted in an equilibrium that is visible due to the brown color of nitrogen dioxide:

$$N_2O_4 \rightleftharpoons 2\,NO_2; \qquad \Delta H^\circ = +57 \text{ kJ/mol} \tag{4.13}$$

In the solid state this system is colorless, since it is completely in the form of $N_2O_4$. At its boiling point (21 °C) the liquid is deep brown due to a 0.1% content of $NO_2$. The vapor becomes steadily darker with increasing temperature due to the increasing dissociation of the dimer, which is nearly complete at 140 °C.

$NO_2$ is an acidic oxide that reacts with water to produce nitric acid:

$$3\,NO_2 + H_2O \rightarrow 2\,HNO_3 + NO \tag{4.14}$$

This reaction occurs in the atmosphere and is one of the sources of acid rain and of nitrate as a plant nutrient; it also is carried out in industry in the manufacture of nitric acid, which is used on a large scale in the manufacture of ammonium nitrate fertilizer, nylon, steel, and in rockets (as the oxidizer of the rocket fuel).

$NO_x$ and nitric acid are involved in several environmental problems. The acid rain problem, at least as far as damage to lakes is concerned, apparently results more from the solubilization of toxic metal ions at low pH's, as discussed in Section 2.12, than from the toxicity of nitrate ion or even directly of hydrogen ion.

There has been concern about the injection of $NO_x$ into the upper atmosphere (stratosphere) due to high-temperature combustion in the engines of supersonic transport aircraft (SST's). Ozone ($O_3$) is an important component of the upper atmosphere, since it absorbs high-energy ultraviolet radiation from the sun, preventing it from reaching the surface of the earth, where it would cause extensive skin cancer and genetic mutations. Nitric oxide is known to catalyze the destructive reaction of ozone with **atomic oxygen** (also present in the upper atmosphere):

$$NO + O_3 \rightarrow NO_2 + O_2 \tag{4.15}$$

$$NO_2 + O \rightarrow NO + O_2 \tag{4.16}$$

The NO produced in the second of these reactions is then able to reinitiate the first, thus functioning as a catalyst.

In the lower atmosphere, $NO_2$ is involved in a complex series of **photochemical** reactions in air that is also contaminated with unburned hydrocarbons (from automobile exhaust) and in the presence of bright sunlight (e.g., in Los Angeles). These reactions produce ozone, aldehydes, and organic nitrates such as peroxyacetyl nitrate (PAN) and peroxybenzoyl nitrate (PBN), which are powerful eye irritants and are quite damaging to vegetation.

### 4.5.3 Physical Properties of Concentrated Oxo Acids

We have not yet had a chance to apply the concepts of this chapter concerning physical properties to the oxo acids discussed in Chapter 2. While we are on the subject of nitric acid, this may be a good time to do so. Unlike oxides, oxo acids have hydrogen atoms directly bonded to oxygen and can thus engage in hydrogen bonding. This interunit attraction (previously seen in water itself) is often substantially stronger than van der Waals attractions, but is not nearly so strong as Coulombic forces or covalent bonds between units. Hence pure oxo acids, although smallish molecules, are not gases, but are liquids or solids with rather high boiling points. Pure $HNO_3$, for example, has a boiling point of 83 °C.

Generally speaking, pure oxo acids are rather difficult to make. Weak oxo acids often decompose to oxides that are polymeric and insoluble in water (Table 4.3) or to gaseous oxides that are readily lost from solution (even in solution, most dissolved $CO_2$ is present not as carbonic acid, but as hydrated $CO_2$ molecules). Strong oxo acids such as nitric acid exothermically form hydrated oxo anions and $H_3O^+$ ions in solution; thus they are soluble in water. When one tries to isolate the oxo acids in anhydrous form, enough stabilization is lost that in many cases the pure oxo acid cannot exist or has greatly reduced stability. The more stable oxo acids are commonly handled as concentrated solutions in water. In many cases these are prepared by distilling off water until a mixture of water and acid of constant composition finally distills over (sometimes under reduced pressure to reduce their fairly high boiling points to a temperature low enough that the oxo acid will not decompose). Nitric acid is commonly seen as concentrated $HNO_3$ (68% $HNO_3$, 15$M$). It is corrosive to the skin, reacting with skin protein to produce a yellow material called xanthoprotein.

### 4.5.4 Sulfur Oxides and Sulfuric Acid

$SO_2$ is made commercially by the combustion of sulfur, $H_2S$, or sulfide ores such as $FeS_2$. It is a colorless, poisonous gas with a choking odor and a relatively high boiling point ($-10$ °C), and is useful as a solvent, refrigerant, food preservative, and (mainly) in the manufacture of sulfuric acid. In this process the $SO_2$ must first be oxidized by air to $SO_3$, which is a kinetically slow process, so a catalyst of platinum sponge, $V_2O_5$, or NO is required. The $SO_3$ resulting from this oxidation reacts exothermically with water to give $H_2SO_4$, but this reaction is impractical in industry, since a mist of $H_2SO_4$ would be produced in the air and would pass out into the atmosphere; hence the gas stream is bubbled through concentrated sulfuric acid, with which it reacts to generate a polynuclear sulfuric acid (such as will be discussed later in this chapter):

$$SO_3 + H_2SO_4 \rightarrow H_2S_2O_7 \tag{4.17}$$

This acid is then *carefully* reacted with the proper amount of water to give concentrated sulfuric acid.

Sulfuric acid is the leading industrial chemical in terms of the number of tons produced per year. Concentrated sulfuric acid is 98% $H_2SO_4$ by weight, is about 18 M, and boils at 338 °C. It has a very strong affinity for water and releases a great deal of heat on absorbing water. Contact with the skin causes dehydration and burns; should a spill occur, the acid should immediately be flushed away with large quantities of water for *at least 15 minutes*. The dilution of concentrated sulfuric acid should be carried out *cautiously*—the acid should be poured slowly into water with good stirring to dissipate the heat. Adding water to the acid can cause dangerous spattering of concentrated acid. Sulfuric acid will even remove the elements of water from many organic molecules, converting carbohydrates to carbon, for example.

The uses of sulfuric acid are so many and varied that the figures for production of sulfuric acid in a given country are considered a reliable indicator of that country's industrial capacity. The largest usage is in the production of fertilizer (Section 4.5.5); other major uses are in the refining of petroleum, the manufacture of chemicals, and in metallurgy.

ENVIRONMENTAL CHEMISTRY    Natural sources produce large amounts of $SO_2$ via decay of organic matter to $H_2S$, which is rapidly oxidized to $SO_2$ in the atmosphere. Anthropogenic $SO_2$ is produced in comparable quantities during the roasting of sulfide ores and the burning of oil and coal (which often contains substantial amounts of $FeS_2$). In the atmosphere, $SO_2$ is also oxidized to $H_2SO_4$; this process is also speeded by catalysts such as water droplets and *d*-block metal ions found in atmospheric particles of soot. These conditions, which once prevailed in London fog (as opposed to Los Angeles or photochemical smog), have been found in many cities of the world in which homes were heated by burning soft coal. It may well be imagined that breathing in droplets of sulfuric acid does not have a desirable effect on the lungs and body; the resulting strain on the lungs and heart has shortened the lives of many people during such episodes of smog.

In the vicinity of smelters in which sulfide ores are roasted, the concentration of sulfuric acid and sulfur oxides has been so great that artificial deserts have been created. This problem has been alleviated by the construction of very high smokestacks ("the solution to pollution is dilution"), which unfortunately has resulted in the spread of the sulfur oxides and sulfuric acid over whole continents and has helped cause the current problem of acid rain, which may have pH's as low as 2.1. (The contributions of coal-fired power plants, though never concentrated enough to give rise to local deserts, are collectively larger than those of smelters.) This acid rain is corroding away many historic monuments and statues made of susceptible salts of basic oxo anions (marble and limestone are largely $CaCO_3$). Lakes that are in contact with limestone deposits are protected by the same reaction, which neutralizes the acid rain, but those which are not so fortunate become quite acidic, with harmful consequences already discussed. Effects on trees and vegetation are now also being discovered.

Due to these problems, a considerable amount of research has been done on methods of control of sulfur dioxide emissions from smelters and power plants. This would seem to be a simple matter, since acidic oxides such as $SO_2$ and $SO_3$ would be expected to react readily with inexpensive basic oxides or hydroxides. Thus a solution of $Ca(OH)_2$ can be sprayed down the smokestack of the plant in a **scrubber** to react with the sulfur oxides according to the reaction

$$Ca(OH)_2 + SO_2 + \tfrac{1}{2}O_2 \rightarrow CaSO_4 + H_2O \tag{4.18}$$

(Much of the calcium sulfite expected from this reaction is oxidized by the air to calcium sulfate).

Finding a suitable chemical reaction is only the first step. A coal-fired power plant emits much more $CO_2$ than $SO_2$; $CO_2$ is also a weakly acidic oxide that would use up much of the $Ca(OH)_2$. As a consequence, for each ton of coal burned, up to 0.2 ton of limestone would be required, and an enormous quantity of wet $CaSO_4$ would be generated. We have now converted an air-pollution problem to a water-pollution problem or (if we dry out the wet $CaSO_4$) a solid-waste problem, since there are not enough uses for $CaSO_4$ to be able to market such quantities. (Furthermore, the lime solution would cool the exhaust gases so much that they would no longer rise up and go out of the smokestack, so the stack gases would have to be reheated!)

A number of other alternative processes have been studied [6, 7]. For example, using $Mg(OH)_2$ instead of $Ca(OH)_2$ has advantages despite the fact that $Mg(OH)_2$, being the hydroxide of a weakly acidic cation, is insoluble in water. After the $SO_2$ reacted with the slurry of $Mg(OH)_2$, the resulting $MgSO_3$ could be heated (in another location) to regenerate the $SO_2$:

$$MgSO_3 \rightleftharpoons MgO + SO_2 \tag{4.19}$$

This reaction is much more feasible than the analogous reaction with $CaSO_3$, because there is a greater mismatch of cation and anion radius with $MgSO_3$, which favors the formation of $MgO$ at a lower temperature than $CaO$ can be formed. The $MgO$ could be recycled to form $Mg(OH)_2$; thus there would be no solid-waste problem and no great investment in $Mg(OH)_2$. The $SO_2$ is formed at a concentration great enough to allow the manufacture of sulfuric acid, which could be sold. However, sulfuric acid is the cheapest acid, and the sale of this acid would not pay for the heat used to decompose the $MgSO_3$.

Although many other alternatives exist, each has its drawbacks too. It is estimated that equipping the power plants of the United States with devices to remove most of the $SO_2$ will cost about \$32 billion by 1990. Unless subsidized by the government, this amount will be added to consumers' electric power bills.

## 4.5.5 Phosphorus Pentoxide and Phosphoric Acid

The acidic $P_4O_{10}$ (commonly misnamed phosphorus pentoxide from its simplest formula, without the di-!) reacts very completely with water and hence is used as a drying agent. But it is dangerous to use it for drying organic liquids, because it releases so much heat that it can set the organic liquid on fire! It is involved indirectly also in the production of high-purity "syrupy" (85% or 15 $M$ concentrated) phosphoric acid, made by oxidizing elemental phosphorus in the presence of water:

$$P_4 + 5O_2(\rightarrow P_4O_{10}) + 6H_2O \rightarrow 4H_3PO_4 \tag{4.20}$$

This high-purity acid is used in making detergents, toothpaste, and in foods such as colas (about 0.05% $H_3PO_4$, pH = 2.3!), but it is too expensive for the main use of phosphoric acid, making fertilizers. For this purpose phosphoric acid, being a moderately weak acid, is produced by the reaction of a strong acid with the moderately basic phosphate ion:

$$Ca_5(PO_4)_3F + 5H_2SO_4 \rightarrow 3H_3PO_4 + 5CaSO_4 + HF \tag{4.21}$$

This reaction mixture is diluted with water and the insoluble calcium sulfate is filtered off; the solution may then be concentrated. This process costs only one-third as much as the preceding process, but as the phosphate rock used (fluoroapatite) contains many impurities, so does the resulting phosphoric acid. This reaction also produces a serious air pollutant, gaseous HF. For use in fertilizers, it is not necessary to protonate the $PO_4^{3-}$ ion completely. Fluoroapatite is too insoluble to be utilized by plants as a nutrient, but partial protonation of the phosphate ion reduces its basicity, so that the salt $Ca(H_2PO_4)_2$ becomes soluble enough to be used as a nutrient ("superphosphate" fertilizer).

The three main nutrient elements provided in fertilizers are K, P, and N; the potassium and phosphorus are provided in the form of $K^+$ and $PO_4^{3-}$ and the nitrogen either as the $NH_4^+$ ion or the $NO_3^-$ ion. Interestingly, the compositions of fertilizers are often expressed in terms of acidic and basic oxide contents: Potassium ion is expressed as the weight percent of $K_2O$, and phosphorus ion content is expressed as the weight percent of $P_4O_{10}$ or $P_2O_5$ needed to give the actual content of K and P via acidic oxide–basic oxide reactions. (Since the nitrogen may be present as $NH_4^+$, which cannot arise from an acidic or basic oxide, its content is given more simply as %N.) A fertilizer labeled 16–48–0 contains 16% N, 48% $P_2O_5$, and 0% $K_2O$.

## 4.5.6   Other Volatile Oxides

Many of the remaining oxides shown in Figure 4.3 involve elements with oxidation numbers greatly exceeding their most stable oxidation numbers. Many of these oxides ($ClO_2$, $Cl_2O_7$, $XeO_3$, $XeO_4$, and $Mn_2O_7$) are treacherously explosive and find no use (except $ClO_2$, which is not explosive if diluted enough with $CO_2$ and which is used as a bleach). $RuO_4$ and $OsO_4$ are not explosive but are still strong oxidizing agents. $OsO_4$ is widely used in biology to stain tissues—it oxidizes the organic material in them and is reduced to brown $OsO_2$—but it is quite hazardous because of its high volatility: It readily oxidizes the organic material in the eye, too!

Although $Cl_2O_7$ is too unstable to use, the same is not true of its corresponding (very strong) acid, $HClO_4$. Perchloric acid is available commercially as a 72% solution, with a boiling point of 203 °C. Contact of concentrated $HClO_4$ with organic materials and other easily oxidized materials should be avoided, however, since the products may be treacherous explosives.

<br>

4.6
<hr>

# Uses of Metal Oxides; Mixed Metal and Nonstoichiometric Metal Oxides

The physical properties of the ionic metal oxides contrast strikingly with those of the more volatile oxides we have been discussing [2a, 5, 10]. Their very high melting points and low volatilities make some of them (such as MgO) useful as **refractories** for providing surfaces capable of withstanding very high temperatures. MgO is used not only to line furnaces but also to cover the heating elements of electric ranges, since it conducts heat much more readily than it conducts electricity. A related use is that of *thoria* (actually 99% $ThO_2$ + 1% $CeO_2$) to provide luminosity to gas flames for lighting purposes: The oxides become white-hot without melting.

In contrast, titanium dioxide is used extensively for its intense whiteness when cold, e.g., in white paints, where it has replaced very toxic lead compounds. Naturally occurring $TiO_2$ (the minerals anatase and rutile) is normally darkened with impurities, however, so that $TiO_2$ must be manufactured chemically. This is not easy, since the $Ti^{4+}$ ion is strongly acidic, and consequently $TiO_2$ is quite insoluble. The titanium ore ilmenite, $FeTiO_3$, will dissolve in hot concentrated sulfuric acid, however; after the solution is diluted and the $FeSO_4 \cdot 7H_2O$ is crystallized out, the remaining solution of titanyl sulfate ($TiOSO_4$) is hydrolyzed to give pure $TiO_2$.

There are a number of technologically important oxides, the simplest formulas of which look like oxo salts (e.g., $BaTiO_3$), but which are not formed from acidic oxides and the structures of which do not involve identifiable oxo anions. Such compounds are called **mixed metal oxides**. They may be produced by reacting the component oxides, but these oxides are generally insoluble in water and also melt at impractically high temperatures, so they often must be made by heating finely divided and well-mixed powders to very high temperatures (up to 2500 °C), where-upon the ions gradually diffuse into each others' lattices (this process is known as **sintering**). The mixed-metal oxides are best regarded as consisting of lattices of oxide ions together with two (or more) different types of metal ions. In many of these the oxide ions are close packed (Chapter 3), and one kind of metal ion may occupy tetrahedral holes, and the other kind, octahedral holes in the close-packed structure.

An important example is the class of mixed-metal oxides known as the **spinels**, $AB_2O_4$. (This class is named after the mineral spinel, $MgAl_2O_4$.) In spinels the oxide ions are cubic close packed. Normally the A metal ions are +2-charged ions of radius between 80 and 110 pm, which occupy one-eighth of the tetrahedral holes in the oxide-ion lattice. Spinels are known in which the A metal ions are the +2 ions of Mg, Cr, Mn, Fe, Co, Ni, Cu, Zn, Cd, and Sn. Normally the B metal ions are +3-charged ions of radius between 75 and 90 pm, which occupy one-half of the octahedral holes; these include the +3 ions of Ti, V, Cr, Mn, Fe, Co, Ni, Rh, Al, Ga, and In.

Of particular interest are spinels in which both A and B are the same element: They form compounds of stoichiometry $M^{2+}(M^{3+})_2(O^{2-})_4$ or $M_3O_4$, which seem to have fractional oxidation numbers because of the presence of two different oxidation states in the same compound. These compounds have unusual colors and electrical and magnetic properties: $Mn_3O_4$, $Fe_3O_4$, and $Co_3O_4$ are all black, and much darker than the simple oxides of these metals. $Fe_3O_4$ has one million times the electric conductivity of $Fe_2O_3$ and is most noted for its magnetic properties (it is the mineral *magnetite* or *lodestone*). An electric current consists of moving charged particles; ionic *solids* have charged particles that cannot move, so they are normally nonconductors. (In the liquid state or in solution, motion becomes possible, so that ionic compounds can then conduct.) Both the dark color and the improved conductivity result from the relatively easy transport of an electron from a $M^{2+}$ ion to a $M^{3+}$ ion to produce a $M^{3+}$ and a $M^{2+}$ ion; this oxidation-reduction reaction is particularly easy when the two metals are the same. Spinels are consequently very important in the solid-state electronics industry for their electric and magnetic properties.

Metals that have more than one cation are also involved in another kind of "mixed oxide" known as **nonstoichiometric oxides** (or compounds, since these phenomena are not restricted to oxides). Thus iron(II) oxide as normally prepared gives an actual elemental analysis corresponding approximately to $Fe_{0.95}O$; there is a **defect** consisting of missing iron atoms. The compound, however, must still be electrically neutral; this is accomplished by the replacement of three $Fe^{2+}$ ions by two $Fe^{3+}$ ions, leaving a hole or vacancy but keeping the overall electroneutrality. Again, unusual colors and enhanced electrical conductivity can result from this sort of process.

Another important class of mixed-metal oxides are the **perovskites**, $ABO_3$, of which the prototype is $CaTiO_3$ (perovskite). It has an unusual cubic close-packed lattice of oxide *and calcium* ions (Figure 4.6), in the octahedral holes of which the much smaller $Ti^{4+}$ ions can "rattle around." This structure type requires that the sum of the charges of the A and B metal ions be $+6$ and that they be quite different in size: Thus, other perovskites such as $Li^+Nb^{5+}O_3$ are also known. If the temperature is not too high, the $Ti^{4+}$ ions tend to be off the center of the lattice unit cell, giving rise to an electric charge separation or dipole; such materials are known as **ferroelectrics**. Application of mechanical pressure on one side of a perovskite crystal causes the $Ti^{4+}$ ions to migrate, generating an electric current; application of an electric current causes mechanical motion of the ions. The pressure effect, known as the **piezoelectric effect**, makes perovskites useful in converting mechanical energy to electric energy or vice versa (as in microphones). The electrical effect makes these materials useful as capacitors and for deflecting laser beams in electronics.

There exist a number of nonstoichiometric mixed-metal oxides. For example, J. T. Kummer and N. Wever of Ford Motor Company in 1967 discovered the material commonly called *sodium beta-alumina* [5], which has some $Na_2O$ added to the spinel-like structure of $Al_2O_3$ (its approximate composition is $Na_{2.6}Al_{21.8}O_{34}$). The structure consists of layers of the $Al_2O_3$ structure bridged by 2-coordinate oxide ions. Between the layers, among the bridging oxide ions, are large open spaces that

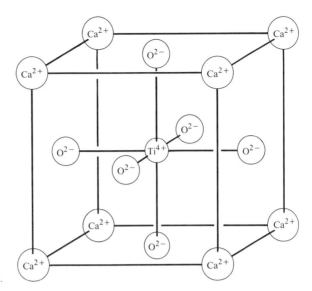

**Figure 4.6** The structure of perovskite.

are only partially occupied by $Na^+$ ions. These $Na^+$ ions move very rapidly upon application of an electric field; thus the compound is a useful solid-state electrolyte that can serve, in effect, as the salt bridge in batteries. Since $Na_2O$ is not readily available, sodium beta-alumina is made by heating sodium carbonate with $Al_2O_3$ at 1500 °C ($CO_2$ being lost).

Another solid-state reaction, that of tungsten trioxide with sodium metal, leads to the *tungsten bronzes*, $Na_xWO_3$, in which $x$ varies between 0.3 and 0.9, and the color varies between bluish-black and gold. These appear to be perovskite structures that are deficient in sodium ions; their conductivity arises from the apparent failure of the sodium to reduce the $W^{6+}$ ions to lower tungsten ions—the electrons interact with each other much as in a pure metal, giving rise to high electrical conductivity and a metallic luster to the compound.

4.7

# Polysilicates: Basic Structural Types, Uses, and Chemistry

Of overwhelming importance in the geochemistry of the earth's crusts are the compounds formed by the reaction of the acidic oxide silica ($SiO_2$) with various basic metal oxides. These are not really mixed-metal oxides such as were discussed in the last section, since definite silicon oxo anions, with covalent Si—O bonds, are present. But most of them do not have the simple silicate ion $SiO_4^{4-}$ that we discussed in Chapter 2, but rather have many 2-coordinate oxygen atoms linking different silicon atoms into oligomeric or one-, two-, or three-dimensional polysilicate ions. These more complex silicates are the most common of the **polynuclear** oxo anions (the nonmetal atoms are treated as "nuclei" of the oxo anions). After examining the polysilicates in some detail, at the end of the chapter we will look briefly at the polynuclear oxo anions of other elements.

The simple silicate ion $SiO_4^{4-}$ (also called *orthosilicate* ion, to distinguish it from the polysilicate ions) is not found in a wide variety of minerals: It is a very strong base that cannot persist in aqueous solution, but it does occur in nature as insoluble salts of acidic cations. Some of the mineral forms containing the orthosilicate ion are phenacite, $Be_2SiO_4$; willemite, $Zn_2SiO_4$; zircon, $ZrSiO_4$; the garnets, $(M^{2+})_3(M^{3+})_2(SiO_4)_3$ ($M^{2+}$ = Ca, Mg, Fe; $M^{3+}$ = Al, Cr, Fe); and olivine, $(M^{2+})_2SiO_4$ ($M^{2+}$ = Mg, Fe). Although with these relatively acidic cations there is doubtless some covalent character to the M—O bonds, we will treat these as salts.

LINEAR POLYSILICATES    In virtually all silicates, silicon has a coordination number of 4. Polymeric structures require **bridging** (2-coordinate) oxygens; to make room for a bridging oxygen an oxide ion must be removed from the "receiving" silicon nucleus:

$$2\,SiO_4^{4-} + 2\,H^+ \rightarrow [O_3Si\!-\!O\!-\!SiO_3]^{6-} + H_2O \tag{4.22}$$

The resulting disilicate ion then has a somewhat lower charge density of $-3$ per silicon nucleus and is consequently less basic. (Note that this polysilicate ion, as well as the others to be discussed, has one negative charge for each nonbridging or **terminal** oxygen atom). The disilicate ion is uncommon in nature, being found in the rare mineral thortveitite, $Sc_2Si_2O_7$. Longer chain structures (trisilicates and tetrasilicates) are even rarer.

CYCLIC POLYSILICATES    More commonly, the ends of these long chains of silicate groups come together to eliminate oxide ions and form *cyclic silicates* (Figure 4.7):

$$Si_3O_{10}^{8-} + 2\,H^+ \rightarrow Si_3O_9^{6-} + H_2O \tag{4.23}$$

These cyclic *metasilicates* can be regarded as oligomers of the unknown $SiO_3^{2-}$ ion; they have two bridging and two terminal oxygen atoms around each silicon atom and thus achieve a lower charge density of $-2$ per silicon nucleus. Cyclic trimers, $[SiO_3]_3^{6-}$, and hexamers, $[SiO_3]_6^{12-}$, are most common and are found in such minerals as benitoite, $BaTi(Si_3O_9)$, and beryl, $Be_3Al_2(Si_6O_{18})$.

CHAIN POLYSILICATES    Just as we previously saw among the polymorphs of oxides such as $SO_3$, bridging oxygens may alternately be used to form chain (one-dimensional or linear) polymers $[SiO_3]_n^{2n-}$ rather than oligomers. Linear polymers are actually more common among polysilicates and result in the important class of minerals known as the pyroxenes, which includes enstatite, $MgSiO_3$; diopside, $CaMgSi_2O_6$; spodumene, $LiAlSi_2O_6$; and pollucite, $CsAlSi_2O_6$. The negative charge still remains at $-2$ per silicon nucleus.

Chains, of course, may be linked together side to side by replacing an oxide ion with another bridging oxygen atom; if this is done on alternate $SiO_3$ groups in each chain, the result is the double-chain structure $[Si_4O_{11}]_n^{6n-}$. This operation

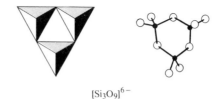

$[Si_3O_9]^{6-}$

**Figure 4.7** Schematic and ball-and-stick models of two cyclic polysilicate ions: (a) $Si_3O_9{}^{6-}$; (b) $Si_6O_{18}{}^{12-}$. In the ball-and-stick model, the small closed circles represent Si and the large open circles represent O. In the schematic diagram, each tetrahedron represents the four oxygen atoms coordinated to a silicon atom at the center; bridging oxygens are located at the intersections of neighboring tetrahedra. Reproduced with permission from N. N. Greenwood and A. Earnshaw, *Chemistry of the Elements*, copyright 1984 by Pergamon Press.

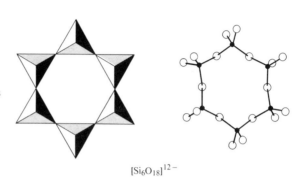

$[Si_6O_{18}]^{12-}$

results in a further reduction in both the charge and the number of oxygen atoms per silicon nucleus, as can be seen if we write the simplest formula of this ion as $[SiO_{2.75n}]^{-1.5n}$. The double-chain amphibole asbestos minerals such as crocidolite, $Na_2Fe_5(OH)_2[Si_4O_{11}]_2$, have long been prized for their fire- and heat-resistance and for their fibrous nature (undoubtedly rooted in the long chain structure of the anion), which allows the weaving of insulating, nonflammable garments, as well as the fabrication of more than 3000 other products. Now these minerals are feared, since it has been realized that the inhalation of the tiny fibers of asbestos often leads, after 20 or 30 years, to asbestosis (nonmalignant scarring of the lungs) or rare cancers such as mesothelioma. The Environmental Protection Agency has proposed a complete ban on asbestos products. Finding materials to replace asbestos in its 3000 uses poses quite a challenge to the industrial inorganic chemist!

SHEET POLYMERIC SILICATES  If the side-to-side linking of chains is continued indefinitely, still more oxide ions are eliminated and a two-dimensional polymeric or sheet silicate $[Si_4O_{10}]_n{}^{4n-}$ is produced. The sheet polymeric silicates have been extremely important to humans for millennia, due in part to their ready cleavage into thin sheets and other properties that can be related to their layer structures. These minerals include the micas (muscovite, biotite), the clay minerals (montmorillonite, kaolinite or china clay, and vermiculite), talc, soapstone, and chrysotile asbestos. The layer structures of clays provide surfaces on which cations are adsorbed in soils; clay surfaces might even have served as templates upon which the first life-molecules formed [11]!

If this process is continued and sheets are linked into three-dimensional polymers, *all* the oxide ions are ultimately eliminated, *all* the remaining oxygens are converted into bridging oxygens, and the uncharged oxide silica $[SiO_2]_n$ is produced. Having no negative charge, it is no longer basic at all; it is, of course, an acidic oxide. Silica is a very important compound that has more than 22 phases (some of them necessarily impure, but at least 12 of which can be pure $SiO_2$). Many of its forms are familiar to collectors: alpha-quartz (a major component of granite and sandstone), rose quartz, smoky quartz, flint, heliotrope, jasper, onyx, amethyst, citrine, agate, chalcedony, and others. Industrially useful forms include kieselguhr and diatomaceous earth. Less common polymorphs include tridymite, cristobalite, coesite, and the remarkable stishovite, found in Meteor Crater, in which silicon is 6-coordinate in a rutile lattice. It is not surprising that the high pressure that is presumably produced by the impact of a meteorite is what is needed to force six oxygen atoms around a small silicon atom.

As summarized in Figure 4.8, the successive steps of polymerization of the simple silicate ion have resulted in (1) a successive reduction of the overall ratio of oxygen atoms to silicon atoms, from 4 : 1 in the orthosilicate ion to 2 : 1 in silica; (2) a decrease in the number of terminal oxygen atoms per silicon nucleus; and (3) a decrease in the charge per silicon nucleus in the anion. The chemical formulas of minerals are often written to set off the polysilicate ions from other anions, such as hydroxide, that may also be present; if this is done it is relatively easy to interpret the formula to tell what kind of polysilicate ion is present.

EXAMPLE    Identify the degree of polymerization of the polysilicate ions found in the following minerals: (a) pyrophyllite, $Al_2Si_4O_{10}(OH)_2$; (b) grunerite, $Fe_7Si_8O_{22}(OH)_2$; (c) spessartite, $Mn_3Al_2Si_3O_{12}$; and (d) bustamite, $CaMn(SiO_3)_2$.

Solution    If the polysilicate ion can be identified, calculation of the ratio of oxygen to silicon will allow a choice among the structures shown in Figure 4.8, with the exception of the choice between chain and cyclic structures. (Even without reference to the figure or the structures of the ions, we can state that the lower the ratio O : Si, the more polymerized the ion is.) The ratio O : Si in these minerals is 10 : 4 or 2.5 in pyrophillite, 22 : 8 or 2.75 in grunerite, 4.0 in spessartite, and 3.0 in bustamite. Thus the simple orthosilicate ion is present in spessartite (a type of garnet), and the degree of polymerization increases in the sequence bustamite < grunerite < pyrophillite. Reference to the table (or consideration of the structures) identifies grunerite as a double-chain polysilicate and pyrophyllite as a sheet silicate. Bustamite could be either a ring or a chain silicate; it happens to be the latter.

GLASS    The completely polymerized, acidic silica is often reacted at very high temperatures (1700 °C) with basic oxides to generate polysilicates that are cooled too rapidly to allow any of the orderly polysilicate ions mentioned above to form. The resulting material, which has a random mixture of all sorts of polysilicate ions (Figure 4.9), has no definite freezing point, but instead just thickens on cooling to form an **amorphous** solid or **glass**. Glass is made by fusing (melting) sand and recycled glass with $Na_2CO_3$ and limestone (which act as sources of the basic oxides $Na_2O$ and $CaO$).

| Structure | Formula name | | Total O/Si Ratio | O/Si ratio at each nucleus | | Charges per Si |
|---|---|---|---|---|---|---|
| | | | | Bridging O | Terminal O | |
| $= SiO_4$ | $SiO_4^{4-}$ | Orthosilicate | 4 | 0 | 4 | −4 |
|  | $Si_2O_7^{6-}$ | Disilicate | 3.5 | 1 | 3 | −3 |
|  | $Si_3O_9^{6-}$ | Cyclic silicate | 3 | 2 | 2 | −2 |
|  | $Si_6O_{18}^{12-}$ | Cyclic silicate | 3 | 2 | 2 | −2 |
|  | $(SiO_3^{2-})_n$ | Pyroxene (chain silicate) | 3 | 2 | 2 | −2 |
| 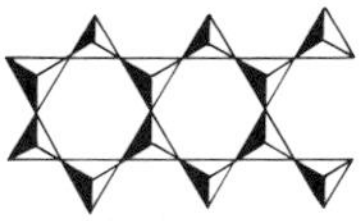 | $(Si_4O_{11}^{6-})_n$ | Amphibole (double chain) | 2.75 | 2.5 | 1.5 | −1.5 |
|  | $(Si_4O_{10}^{4-})_n$ | Infinite sheet silicate | 2.5 | 3 | 1 | −1 |
| | $(SiO_2)_n$ | Silica | 2 | 4 | 0 | 0 |

**Figure 4.8** Structures of some polysilicates and resulting ratios of oxygen atoms and charge to the number of silicon atoms present. Reproduced with permission from J. E. Ferguson, *Inorganic Chemistry and the Earth: Chemical Resources: Their Extraction, Use, and Environmental Impact*, Pergamon Press, Oxford, 1982.

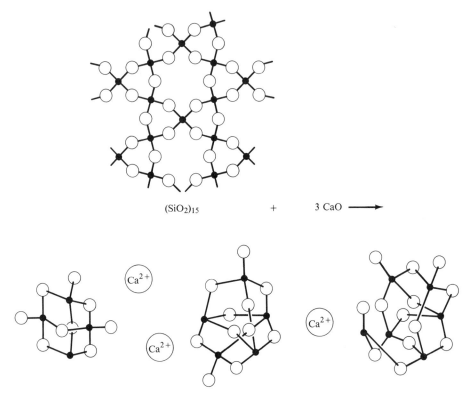

$(SiO_2)_{15}$          +          $3\ CaO$ ⟶

**Figure 4.9**   Schematic diagram of the breaking of silicon-oxygen bridges in the formation of glass.

Specialty glasses are made by altering the composition of acidic and basic oxides in the glass. For example, Pyrex (tm) glass, which is unusually resistant to thermal shock, incorporates 10 to 25% of the acidic boron oxide $B_2O_3$. Glass can be colored by incorporating *d*-block metal oxides as part of the basic oxide component. Incorporation of 20 to 60% of the basic PbO gives dense, highly refractive *flint glass*, used in crystal and in optical equipment. Addition of SrO gives a glass that absorbs the X-rays emitted by color television sets; addition of $La_2O_3$ gives glass with fine optical qualities suitable for camera lenses.

SOIL CHEMISTRY   As noted above, increasingly polymerized polysilicate ions have decreasing charges per silicon nucleus, so these ions are less basic. This fact has important consequences in soil chemistry: The more basic the polysilicate anion of a mineral, the more readily it will react with weak acids and undergo **weathering** [6, 8]. Rainwater is acidic even in the absence of sulfur and nitrogen oxides due to dissolved carbon dioxide. This solution of carbonic acid is weakly acidic, and over the ages it will react with the less polymerized silicate anions to remove oxide ions (as water), thus inducing replacement of the oxide ion with a bridging oxygen to produce a more highly polymerized silicate:

$$Mg_2SiO_4 + 2\,H^+ \rightarrow Mg^{2+}(aq) + H_2O + MgSiO_3 \qquad (4.24)$$

$$MgSiO_3 + 2\,H^+ \rightarrow Mg^{2+}(aq) + H_2O + SiO_2 \qquad (4.25)$$

Thus soils containing large amounts of orthosilicates such as olivine are characterized as "youthful" soils: They may have crystallized from a melt (i.e., magma) recently, or they may be present in a desert region, in which the water necessary to form the acidic rainfall and the hydrated ions is absent. (The "dry valleys" of Antarctica, which seldom see liquid water, also have youthful soils.) Somewhat more slowly the cyclic and chain polysilicates will also weather, as suggested in equation (4.25); these will be followed by the double-chain silicates.

At the intermediate stage of weathering, such as is found in the temperate regions under a cover of grass or trees, layer silicates such as clays tend to predominate, along with some quartz. As can be seen from equations (4.24) and (4.25), this weathering process is accompanied by a loss of cations, which is especially prominent for nonacidic and feebly acidic cations. Thus the soil is less fertile than it was, due to the loss of the nonacidic plant nutrient potassium ion. (Desert soils, when first irrigated, are often very fertile.) The layer silicates present in the intermediate soils can still hold cations on their negatively charged surfaces, however; these can readily be exchanged for other ions (such as $H^+$) and are thus released as plants need them; such soils are found in the still-quite-fertile corn and wheat belts of the world.

In the tropics, however, when the trees are cut down and frequent rain and heat speed up the weathering process, the aging process becomes quite advanced. Such soils have high levels of oxides of the most acidic cations present in silicate minerals, such as anatase and rutile ($TiO_2$), zircon ($ZrO_2$), hematite ($Fe_2O_3$), and gibbsite ($Al(OH)_3$). These soils can no longer hold the less acidic nutrient metal ions and are quite infertile. When tropical rain forests are removed in "slash and burn" agriculture, the soil can be used for agriculture for only a few years; after that it becomes infertile (and rock-hard).

## 4.8
## Isomorphous Substitution and Diagonal Relationships of Cations: Aluminosilicates

The various polysilicate ions have negative charges that must be counterbalanced by appropriate cations. The terminal oxygen atoms sticking out from the polysilicate rings, chains, and layers provide negatively charged surfaces that often amount to approximately close-packed surfaces of negative charge. In the layers between chains or silicate layers there are thus tetrahedral and octahedral (but sometimes other shapes of) holes that are then filled by the cations needed to neutralize the polysilicates' negative charge. In different types of polysilicates the total charge to be neutralized will differ, as will the size of the holes.

The type of cations found in a particular type of polysilicate will depend on (1) the size of those cations and (2) the charge of those cations. Examination of

Table C will show that quite a few sets of ions can be found that have the same charge and very similar ionic radii. Generally speaking, as a mineral is formed by the cooling of molten magma, there is little reason for one of these matched types of ions to be preferred over another: Silicate minerals often have a mixture of cations present, which vary depending on the composition of the melt from which the mineral grew. For example, $Mg^{2+}$ and $Fe^{2+}$ not only have identical charges but have very similar radii (86 and 92 pm, respectively). Thus the mineral olivine, with an ideal composition of $Mg_2SiO_4$, is often "impure" and can contain varying percentages of **isomorphous substitution** [6] of the $Fe^{2+}$ ion in place of an equal number of $Mg^{2+}$ ions. Thus the formula of olivine is often written $(Mg, Fe)_2SiO_4$ to indicate that there are two magnesium or iron(II) cations present per mole of orthosilicate ion, although there is no definite relationship between the number of magnesium and the number of iron(II) cations.

Thus the first principle of isomorphous substitution is that *one cation may substitute for another in a lattice* (i.e., form what is sometimes called a *solid solution*) *if they have identical charges and differ in radii by not more than 10 to 20%.*

This principle applies not only to silicates but also to other ionic salts. Thus to isolate a tiny amount of radium ion from a large amount of uranium ore, we could (using the solubility principles of Chapter 3) add sulfate ion to precipitate radium sulfate while leaving uranium and most other cations in solution. But this would pose two problems: (1) Only a tiny amount of precipitate would be formed, which would be difficult to handle without losing it; and (2) there might be so much solution present that even the low solubility product of radium sulfate might not be exceeded, so no precipitate would form. These problems could be overcome by adding not only sulfate ion but also barium ion. A larger amount of barium sulfate would then precipitate and the radium would substitute isomorphously for the barium. This technique is known as coprecipitation and is used extensively in **radiochemistry**. The barium ion is said to act as a *carrier* for the $Ra^{2+}$. (Of course, we would then be confronted with the formidable problem of separating the very similar barium and radium ions, but at least we would be working with a much smaller volume of material.)

Another illustration of this process occurs in a familiar experiment of growing crystals of ionic compounds. For example, large, beautiful octahedral crystals of *alum*, $KAl(SO_4)_2 \cdot 12H_2O$, are readily grown from solution. But there exists a whole series of similar compounds that also form large, beautiful octahedral crystals having the same lattice types (these compounds are said to be isomorphous). In the formula for alum, the $K^+$ ion can be replaced by other $+1$ cations of similar radius, such as $Rb^+$ and $NH_4^+$. The $Al^{3+}$ ion can be replaced by numerous other $+3$ ions of similar radius, such as $Cr^{3+}$ (giving purple crystals of *chrome alum*) or $Fe^{3+}$ (giving pale violet crystals of *ferric alum*). The sulfate ion can even be replaced by the selenate ion. If nearly any combination of these three ingredients is mixed and crystallized, large crystals of an alum are formed; if a mixture, say, including both $Al^{3+}$ and $Cr^{3+}$ is used, crystals can be grown containing both ions, having whatever shade of light purple you desire! (Such a crystal is sometimes said to be *doped* with a certain percentage of the less abundant ion.)

Perhaps the most extensive case of isomorphous substitution occurs in the minerals monazite, $MPO_4$, and bastnaesite, $MFCO_3$. In 1794 J. Gadolin investigated a mineral obtained from the village of Ytterby, Sweden; from this mineral he extracted a metal oxide that he named *yttria*. But other chemists, in working with this material, kept getting slightly different properties; eventually it was realized that this oxide was a mixture. So the mixture was separated; but the "pure" components also turned out to be mixtures. Eventually all the $f$-block elements of the fourth period (except Tc), plus La and Y, turned out to be present in these minerals, isomorphously substituted for each other. (Refer to Tables A through C to note the extreme similarities of these elements to each other.) As element after element was discovered, the chemists were harder and harder pressed to come up with new names for them. Thus it came to pass that the humble village of Ytterby has more elements named after it (four) than do any of the great cities of the world. (Which four are named after Ytterby?)

The separation of these elements involves quite a complex process of ion exchange. The most troublesome case of isomorphous substitution, however, is that of the elements Zr (Pauling electronegativity 1.33, ionic radius 86 pm) and Hf (Pauling electronegativity 1.3, ionic radius 85 pm). Hf occurs isomorphously substituted in all zirconium compounds to the same extent (about 2%), so there were no chemical discrepancies in the "pure" samples of zirconium prepared in 1825 and thereafter. As a result the presence of Hf went undetected for a whole century!

There are often very practical reasons for substituting one metal ion for another in an ionic lattice (or even in compounds such as metalloenzymes). The $s$- and $p$-block metal ions are all colorless. If they can be isomorphously substituted with $d$- and $f$-block ions, they will acquire colors that can be usefully studied, because the colors of $d$- (and to some extent $f$-) block ions are very sensitive to their environment in the lattice or enzyme. (The explanation of this sensitivity, and what can be learned by examining the **spectra** of these colored ions, uses the **crystal field theory**, which you will find in a more advanced inorganic chemistry text). Metal ions also have other properties that tell us about their environment: Some are fluorescent; some have unpaired electrons with magnetic properties that can be studied; others with appropriate nuclei can be studied by nuclear magnetic resonance (NMR) or Mössbauer spectroscopy. Some of the geologically and biochemically most important metal ions, such as $K^+$ and $Zn^{2+}$, however, are "silent metals" that lack most or all of these properties. Silent $K^+$ (radius = 152 pm) can often usefully be substituted with fluorescent and NMR-active $Tl^+$ (radius = 164 pm); colorless $Zn^{2+}$ (88 pm) is usefully replaced by colored $Co^{2+}$ (88 pm); silent $Ca^{2+}$ (114 pm) can be replaced by $Eu^{2+}$ (131 pm), with seven unpaired electrons.

There are only a limited number of ions that match the common cations in silicates both in size and in charge, so it might be thought that isomorphous substitution is relatively rare. *The second principle of isomorphous substitution* allows more versatility, however. Substituting ions must be about the same size as the ions replaced in order not to change the lattice type, although the charge requirement can be overcome. The basic principle is that the *total* charge of the replacing ions must equal the total charge of the replaced ions. This means that isomorphous substitution can occur even if the new ions C have charges one greater than the old ions A, *if there is simultaneous substitution by new ions D of the same size as but with charges one less than the old ions B*. This conserves the electroneutrality of the silicate, since the sum of charges of the new ions C and D equals the sum of charges of the old ions A and B.

This condition greatly increases the number of possible substitutions in silicates: Thus the common $K^+$ ion can be replaced not only by the rare $Rb^+$ and $Tl^+$ ions, but also by the common $Ba^{2+}$ ion. Likewise, the common $Ca^{2+}$ ion (114 pm) can be replaced not only by $Sr^{2+}$ (132 pm), but also by $Na^+$ (116 pm) and $Y^{3+}$, $La^{3+}$, and the sixth-period $f$-block ions (100 to 117 pm). $Mg^{2+}$ (86 pm) can be replaced by $Li^+$ (90 pm) or by $Fe^{3+}$ (78 pm). Of greatest importance, $Si^{4+}$ itself (54 pm) can be replaced by the very common $Al^{3+}$ ion (67 pm) to produce the important **aluminosilicates**.

For instance, the most abundant of all minerals (about 60% of the earth's crust) are the **feldspars**, which fall into two categories. One involves simultaneous isomorphous substitution of the 150-pm ions $K^+$ and $Ba^{2+}$ and of $Al^{3+}$ and $Si^{4+}$; compositions range from $KAlSi_3O_8$ (orthoclase) to $BaAl_2Si_2O_8$ (celsian). The other category, the plagioclase feldspars, involves the smaller 115-pm $Na^+$ and $Ca^{2+}$ ions; compositions range continuously from $NaAlSi_3O_8$ (albite) through $Na_{0.33}Ca_{0.67}Al_{1.67}Si_{2.33}O_8$ (labradorite) to $Na_0Ca_1Al_2Si_2O_8$ (anorthite).

As a consequence of this isomorphous substitution of sets of four ions, the cations in most silicate minerals are extensively substituted, and silicates do not usually make economical sources (**ores**) for most of the elements, since they generally have low concentrations of the desired element in hard-to-separate mixtures. But we can predict sources for some of the rarer elements if we are willing to pay the price: Lithium ion (used in treating manic-depressive patients) can generally be found in magnesium minerals; beryllium ion (59 pm) resembles in a good deal of its chemistry the aluminum ion (67 pm). These **diagonal relationships** of elements (especially in the second period) to the elements one group to the right and one period down extend even to the nonmetallic elements, since the similar size of such atoms will affect, for example, their maximum coordination numbers in similar ways. (Examination of Table A shows that there is also some similarity in the Pauling electronegativities of second-row elements and those of their diagonal partners in the third row.)

Isomorphous substitution can involve the silicon in the polysilicate ion itself, which results in a *heteropolysilicate* in which all nuclei are not the same type of atom. Normally, the very abundant $Al^{3+}$ replaces $Si^{4+}$, giving rise to the aluminosilicates. This substitution is best indicated by writing the formula of the aluminosilicate ion in brackets, enclosing the structural aluminum ions but not those aluminum ions that serve only to neutralize the negative charge. Thus the mica mineral muscovite may be written as $KAl_2[AlSi_3O_{10}](OH)_2$, showing that only one-third of the $Al^{3+}$ ions enter into the sheet polysilicate ions. (The other $Al^{3+}$ ions, being moderately acidic, serve to bind pairs of polyaluminosilicate sheets into three-layer structures.) To determine the structure of this material from its formula, one would calculate the ratio of oxygens to silicon *plus nuclear aluminum* atoms: $10/(3 + 1) = 2.5$, i.e., a sheet structure. However, the negative charge of this anion is increased by one unit per each framework aluminum atom, since there is one less proton in the nucleus of the aluminum atom than in the nucleus of the silicon atom that it replaced.

EXAMPLE     Which of the following minerals could arise by isomorphous substitution processes in leucite, $K(AlSi_2O_6)$? (a) $K(YSi_2O_6)$; (b) $Rb(AlSi_2O_6)$; (c) $Ba(BeSi_2O_6)$; (d) $Ba(AlSi_2O_6)$.

Solution     ■ In addition to the two principles of isomorphous substitution, two even more fundamental principles of ionic compounds must be observed. The first and most paramount is that *the total charge of all the cations must equal the total charge of all the anions.* This requirement is implicit in the two principles of isomorphous substitution, but it also gives a separate simple test in this example: If the total charge of all cations going into the structure is not the same as the total charge of all cations coming out, the new compound cannot exist, let alone be isomorphous. In (a) $Y^{3+}$ replaces $Al^{3+}$; in (b) $Rb^+$ replaces $K^+$; in (c) $Ba^{2+}$ and $Be^{2+}$ replace $K^+$ and $Al^{3+}$; in (d) $Ba^{2+}$ replaces $K^+$. Substitution product (d) cannot exist.

■ The second principle is that *in order for the substitution to be isomorphous, the total number of cations going in must be nearly equal to the total number coming out.* (If the two numbers are not quite equal, a defect structure [Section 4.6] results. This is uncommon in polysilicates.) All four possible substitution products given above obey this principle.

■ Once these two principles are satisfied, the two principles of isomorphous substitution will be satisfied if the cations going into the replacement structures are within 10 to 20% of the radii of the cations coming out. In (a) $Y^{3+}$ has a radius of 104 pm, which is too much bigger than $Al^{3+}$ (67 pm) for the substitution to be isomorphous. However, (b) and (c) are satisfactory, since in (b) $Rb^+$ (166 pm) is close in size to $K^+$ (152 pm), and in (c) $Ba^{2+}$ (149 pm) is close to $K^+$ (152 pm), and $Be^{2+}$ is close to $Al^{3+}$.     □

To date we have discussed only one three-dimensional silicate, $SiO_2$ itself. There are no possibilities for isomorphous substitution in $SiO_2$, other than by the very rare $Ge^{4+}$ ion, since there is no other ion to be replaced simultaneously. However, a number of structures of aluminosilicates can be imagined as arising from the replacement of $Si^{4+}$ by $Al^{3+}$ *and* a $+1$ ion. To make room for the extra ion, holes must be opened up in the structure of $SiO_2$. Due to the resulting open three-dimensional structures, these compounds are known as *framework* aluminosilicates. Since the substitution of two ions for one leads to a change in structure, this change is *not* an example of isomorphous substitution.

Perhaps the best known of the framework aluminosilicates are the **zeolites**, with anions of general formula $[Al_xSi_yO_{2x+2y}]^{x-}$ and an oxygen-to-(aluminum plus silicon) ratio of 2 but a negative charge. Figure 4.10 shows how the three-dimensional polyaluminosilicate ion structure opens up to leave room for the

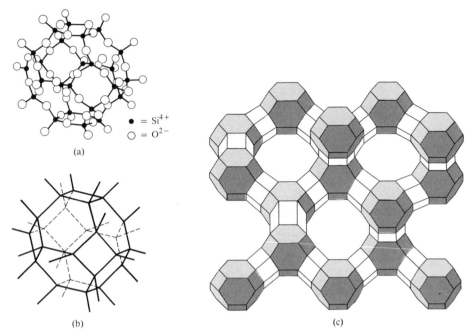

$\bullet$ = $Si^{4+}$
$\bigcirc$ = $O^{2-}$

(a)

(b)                                              (c)

**Figure 4.10** Structural representations of the zeolite structure. (a) Connection of aluminosilicate nuclei to form an open, basketlike framework; (b) line representation of the framework in (a); (c) connection (via bridging oxygens) of basketlike structures to give rise to a three-dimensional polymer with large open cavities, in which the cations reside. Reproduced with permission from *Environmental Chemistry*, by J. W. Moore and E. A. Moore. Copyright © 1976 by Academic Press.

cations needed to neutralize the charge of the anion. These cations are normally in the form of hydrated ions; the water of hydration can be driven off by heating to leave a structure with large vacancies. If the cation is $Na^+$, larger cavities are left than if the cation is $K^+$; still larger cavities are left if the cation is $Ca^{2+}$, since only half as many of these need be used. The resulting zeolites have cavities of tailored sizes that are used as Molecular Sieves (tm) to adsorb molecules of different sizes from liquids. Among the small molecules most readily adsorbed are water molecules, which re-form the hydrated metal ions: Thus these zeolites are very effective drying agents. As we will see in a later chapter, metal ions in the absence of their water molecules of hydration are very effective catalysts for many reactions; zeolites are used industrially in the **cracking** of petroleum to form gasoline.

Hydrated metal cations in such a structure, being large cations in a lattice with very large anions, are loosely bound and can readily be exchanged with other cations. Zeolites are used for *water softening*: the removal of ions of $+2$ charge found in tap water that would precipitate the anions used in detergents. A concentrated solution of NaCl is first percolated through the solid (insoluble) zeolite ion-exchanger, which replaces whatever hydrated ions are present with hydrated

$Na^+$ ions. Then the tap water is run through the zeolite; $Ca^{2+}$ and other $+2$ ions become associated with the anion in the solid phase, while $Na^+$ ions go into solution. When the $Na^+$ ions in the zeolite are depleted, the solid is again "recharged" by running concentrated NaCl solution through it.

This ion-exchange function is not an exclusive property of zeolites but can occur with any substance having a polymeric ion that is insoluble (since it cannot be broken up into small hydrated ions). Synthetic organic polyanions are also used for this purpose; two-dimensional sheet polysilicates, the clays, also perform this function naturally in the soil. (Clays, we recall, are characteristic of moderately weathered soils in the temperate zones of the earth.) Enormous quantities of hydrated cations can be held to the negatively charged (due to terminal oxygen atoms) surfaces of clays; these cations include many nutrient ions such as $Ca^{2+}$, $Mg^{2+}$, $K^+$, and $Na^+$. As slightly acidic rainwater percolates over these clays, these nutrient ions are slowly released by ion exchange with the hydronium ion in the water and made available for plants to use. The largest and the least charged of these ions are held most loosely and are released first (and become depleted first). Recalling that smaller ions have larger secondary hydration spheres, we find that the order of release is $Na^+$ before $K^+$ before $Mg^{2+}$ before $Ca^{2+}$. Since plants make little use of $Na^+$, the first deficiency that occurs among cationic nutrients is of $K^+$. Anionic nutrients such as $NO_3^-$ and $PO_4^{3-}$ cannot be bound to clays at all, so they are rapidly lost from soils. Thus we see the need for fertilizers containing (especially) potassium, nitrogen, and phosphorus.

## 4.9

## Polynuclear Oxo Anions of p- and d-block Elements: Selected Structures and Uses

Although they are much less important than the polysilicates, polynuclear oxo anions are formed by several other p- and d-block elements. As we might expect from the diagonal relationship of elements, the most extensive series of these is the polyborates, in which tetrahedral $BO_4$ and triangular $BO_3$ groups are linked together in many ways. The polyborate of the greatest importance is **borax**, $Na_2B_4O_7 \cdot 10H_2O$, mined in Death Valley, California; it actually contains the $H_4B_4O_9^{2-}$ ion shown in Figure 4.11. Although numerous other types of borates are known, in view of the overall scarcity of boron and the limited use of its oxo salts other than borax, we will not discuss these.

Aluminum forms some polyaluminates, of which the most important is the $Al_6O_{18}^{18-}$ ion found (as its calcium salt) in Portland cement (see Figure 4.11). In contrast to the less negatively charged polysilicates, this ion reacts rapidly with water to degrade the anion to various dialuminates. This reaction is responsible for the rapid setting of Portland cement.

The polynuclear phosphates, though few in number, are of industrial and especially biochemical importance. The diphosphate ion is also called pyrophosphate because it is formed by heating a dihydrogen phosphate:

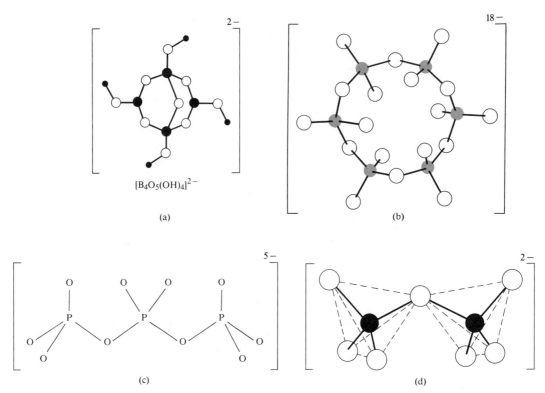

$$[B_4O_5(OH)_4]^{2-}$$

(a)

(b)

(c)

(d)

**Figure 4.11** Selected polynuclear oxo anions of the *p*-block elements: (a) the $H_4B_4O_9^{2-}$ ion found in borax; (b) the $Al_6O_{18}^{18-}$ ion found in Portland cement; (c) the $P_3O_{10}^{5-}$ ion found in the detergent builder sodium tri(poly)phosphate and (covalently bonded to organic groups) in ATP, DNA, and RNA; (d) the structure of the disulfate (and dichromate) ions. Reproduced by permission from N. N. Greenwood and A. Earnshaw, *Chemistry of the Elements*, copyright 1984 by Pergamon Press.

$$2\,Na_2HPO_4 + heat \rightarrow Na_4P_2O_7 + H_2O \tag{4.26}$$

This anion has one bridging oxygen atom linking two $PO_4$ tetrahedral nuclei and occurs (linked through one other oxygen covalently to adenosine) in the important biochemical adenosine diphosphate (ADP).

Of greater importance (Figure 4.11) is the triphosphate or tripolyphosphate ion, $P_3O_{10}^{5-}$, synthesized similarly to the diphosphate:

$$2\,Na_2HPO_4 + Na_2HPO_4 + heat \rightarrow Na_5P_3O_{10} + 2\,H_2O \tag{4.27}$$

This compound is used in large quantities as the "phosphate" found in detergent. (It is used as a chelating ligand to tie up the $+2$ ions in tapwater; see Chapter 7). The environmental problems with using so much of this material in detergents stem from its hydrolysis to give the simple phosphate ions (e.g., the reverse of reaction 4.27), which are plant nutrients in lakes. Many lakes that have high inputs of phosphates have been overgrown with algae.

The tripolyphosphate ion, covalently linked to sugar molecules through one of the end oxygen atoms, is found in all nucleic acids (DNA and RNA) and in adenosine triphosphate (ATP). Hydrolysis of one phosphate nucleus off of ATP produces ADP and releases a small, biochemically convenient amount of energy. This hydrolysis is readily reversed in the body, so the system of ADP and ATP functions in energy transfer in the body.

An intermediate in the manufacture of sulfuric acid is the disulfate ion, $S_2O_7^{2-}$ (Figure 4.11). More familiar is a $d$-block analogue, the dichromate ion, obtained when chromate ion is partially neutralized:

$$2\,HCrO_4^- \rightleftharpoons Cr_2O_7^{2-} + H_2O \tag{4.28}$$

A similar divanadate ion, $V_2O_7^{4-}$, is also formed by vanadium, but generally, vanadium and the heavier elements of Groups 5 and 6 form larger polynuclear ions based not on tetrahedra but (since the ions are larger) on octahedra linked together in tight structures by oxygen atoms, which may link many more than two metal nuclei (see Figure 4.12). Numerous such ions are known or suspected; we will highlight only a few, such as the decavanadate ion $V_{10}O_{28}^{6-}$. Niobium and tantalum do not seem to form monomeric oxo anions (perovskite-type mixed oxides are more common) but do give rise to polyniobates and polytantalates such as $M_6O_{19}^{8-}$.

The polymolybdates and polytungstates are especially numerous and elaborate in structure. Acidification of solutions of the molybdate ion ($MoO_4^{2-}$) results in the heptamolybdate ion $Mo_7O_{24}^{6-}$, the octamolybdate ion $Mo_8O_{26}^{4-}$ (Figure 4.12), and even the enormous ion $Mo_{36}O_{112}^{8-}$. The tungstate ion gives rise to a different series of polytungstates, of which the best known are the dodecatungstates $H_2W_{12}O_{42}^{10-}$ and $H_2W_{12}O_{40}^{6-}$.

HETEROPOLYMETALLATE IONS   As shown in Figure 4.12, there is a prominent tetrahedral hole inside the 12 tungstate units present in the $H_2W_{12}O_{40}^{6-}$ ion. This can be filled with any of at least 35 different cations to give a polytungstate ion also containing another type of nucleus and thus termed a **heteropolytungstate**; similar heteropolymolybdates also form. These ions are particularly important in determining the amount of phosphate contained in unknown solutions or samples; addition of molybdate ion to a solution containing phosphate ion, followed by acidification, produces the yellow phosphomolybdate ion $[PMo_{12}O_{40}]^{3-}$. The intensity of the yellow color can then give a measure of the phosphate ion concentration; greater sensitivity can be achieved by adding a reducing agent that reduces up to six $Mo^{6+}$ ions in the phosphomolybdate ion to $Mo^{5+}$ ions. The resulting *defect structure* is intensely blue due to the same kind of electron-transfer processes as occur in mixed oxides such as $Fe_3O_4$ (Section 4.6). Or the anion can be precipitated—it is an *extremely* large, *extremely* nonbasic anion (3 negative charges per 12 nuclei!), which gives crystalline precipitates with large nonacidic cations such as large organic ammonium ions (this type of anion is used to isolate such ions from natural products). These precipitates have enormous molecular weights of over 1000, so just a few milligrams of phosphate can become quite an easily weighable amount of a phosphomolybdate precipitate.

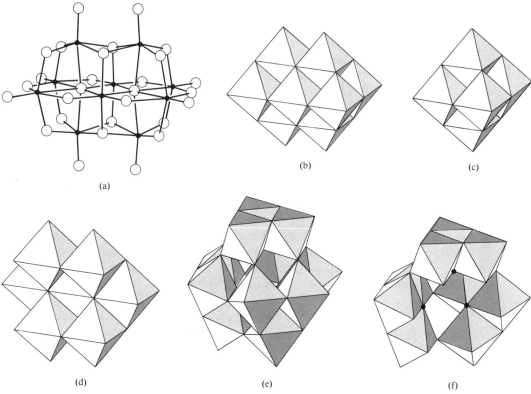

**Figure 4.12** Structures of some polynuclear anions of the *d*-block metals: (a) the decavanadate ion $V_{10}O_{28}^{6-}$; (b) the decavanadate ion represented by ten interlinked octahedra (two are obscured); (c) the hexaniobate or hexatantalate ions $M_6O_{19}^{8-}$ or the hexamolybdate ion $Mo_6O_{19}^{2-}$; (d) the heptamolybdate ion $Mo_7O_{24}^{6-}$; (e) the dodecatungstate ion $H_2W_{12}O_{40}^{6-}$; (f) the dodecatungstate ion with three octahedral nuclei removed to show the inner cavity; the dark dots represent three of the four oxygen atoms to which the phosphorus or other heteroatom attaches in the phosphotungstate, phosphomolybdate, and related heterpolyatomic anions. Reproduced with permission from N. N. Greenwood and A. Earnshaw, *Chemistry of the Elements*, copyright 1984 by Pergamon Press.

## 4.10

## Acid-Base Chemistry of Ions in Water: A Summary

Our world is bathed in water, which has one of the simplest molecular formulas and Lewis dot structures of any molecule. But we have seen in these last three chapters that water itself is no simple substance, and the acid-base chemistry of the elements dissolved in or in contact with water is no simple chemistry. In these chapters we have pretended that all elements in positive oxidation states begin their existences in water as hydrated cations, but we saw that there is quite an elaborate series of reactions between cations and water—especially as the pH of the water is raised—that gives rise to several series of types of compounds. In order of increasing degree of "neutralization" of the acidity of the cation we saw:

1. Hydrated cations, with high positive charges "per nucleus";
2. hydroxo and oxo cations, with reduced positive charges per nucleus;
3. neutral hydroxides, oxides, and oxo acids;
4. hydroxo anions or polymeric oxo anions, with low negative charges per nucleus; and
5. simple oxo anions, with higher negative charges per nucleus.

In addition, we saw that cations can neutralize their positive charges by combining with appropriate oxo anions to give insoluble salts.

In the next chapters, we will go beyond acid-base chemistry to investigate the oxidation-reduction chemistry of ions. Our study of reduction will also bring us to the elements themselves and to the anions of the nonmetals in which the nonmetals have negative oxidation numbers. Our study of oxidation will give us additional ways in which the oxo anions of the elements can be synthesized. We will see that the acid-base chemistry of water continues to play an important role in this separate class of chemical reactions.

## 4.11

## Study Objectives

1. Rank a series of oxides or hydroxides from the same group, period, or element (with different oxidation numbers for the element) in order of increasing basicity. Decide which are acidic, which amphoteric, and which basic.
2. Decide whether a given oxide is likely to be (a) soluble in water, (b) insoluble in water but soluble in strong acids, (c) insoluble in water but soluble in strong bases, or (d) insoluble in water but soluble both in strong acids and in strong bases.
3. Write balanced chemical equations including the products when (a) basic oxides react with water or with strong acids, (b) acidic oxides react with water or with strong bases, (c) amphoteric oxides react with strong acids or with strong bases, and (d) acidic and basic oxides react with each other in the absence of water.
4. Know and explain the relationship between the monomeric, oligomeric, or polymeric structure of a compound, its physical state (solid, liquid, gas), and its melting point.
5. Predict trends in melting points among the oxides (a) across a period of elements, (b) up or down a group of elements, and (c) of a given element in different oxidation states.
6. Know the common nonmetal oxides and tell whether each has a structure based on (a) small molecules, (b) oligomeric molecules, or (c) macromolecules or an ionic lattice. Tell whether each is (a) a solid with high melting and boiling points or (b) a gas, liquid, or solid with relatively low melting and boiling points.
7. Know the names of, main uses of, and environmental problems (if any) associated with the main oxides of the nonmetals.

8. Identify mixed-metal oxides that can adopt the spinel structure, the perovskite structure, and nonstoichiometric (defect) structures.

9. Given the formula of a silicate mineral, tell whether it contains simple silicate ions, linear oligomeric silicate ions, cyclic or chain polysilicate ions, linked chains, sheet polysilicate ions, or three-dimensional structures.

10. Given a table of ionic radii, select ions that are likely to be found in the same silicate (or other) minerals by the process of isomorphous substitution; identify some ions that are unlikely to substitute for any other common ions and explain why.

11. Given formulas of several silicate minerals, tell which will be weathered most rapidly.

12. Know which elements are prone to forming polymeric oxo anions; describe the structures and uses of some of these.

## 4.12

### Exercises

†1. Classify each of the following oxides as *acidic, basic, or amphoteric or neutral* and decide whether it will be soluble in water: **a.** $Na_2O$; **b.** $Cr_2O_3$; **c.** $CO_2$; **d.** $SiO_2$; **e.** $P_4O_{10}$; **f.** $BaO$; **g.** $Tl_2O$; **h.** $SO_2$; **i.** $Al_2O_3$.

*2. Which of the oxides in the previous question consist of small molecules? Which consist of oligomeric molecules? Which might reasonably be described as either ionic lattices or macromolecules?

†3. Arrange the following oxides in order of decreasing acidity/increasing basicity:
3.1 $Na_2O, Cr_2O_3, CO_2, SiO_2, P_4O_{10}$
3.2 $ZrO_2, CO_2, SrO, Rb_2O, Y_2O_3$
3.3 $MnO, MnO_2, Mn_2O_3, Mn_2O_7$
3.4 $TiO_2, TeO_2, SO_2, ThO_2$

†4. Put the following series of oxides in order of increasing melting points and decide which are gases at room temperature:
4.1 $Na_2O, Cr_2O_3, CO_2, SiO_2, P_4O_{10}$
4.2 $ZrO_2, CO_2, SrO, Rb_2O, Y_2O_3$

5. Assuming that all the following compounds are stable, predict where melting points will increase and where they will decrease in the following series of oxides:
5.1 $Cl_2O_5, Br_2O_5, I_2O_5$
5.2 $MnO, Mn_2O_3, MnO_2, Mn_2O_7$

*6. Complete and balance the following chemical equations (or tell if no reaction will occur):
6.1 $Tl_2O(s) + H_2O(l) \rightarrow$
6.2 $I_2O_5(s) + H_2O(l) \rightarrow$
6.3 $ClO_2(g) + OH^-(aq) \rightarrow$

**6.4** $La_2O_3(s) + H^+(aq) \rightarrow$

**6.5** $B_2O_3(s) + OH^-(aq) \rightarrow$

**6.6** $FeO(s) + P_4O_{10}(s) \rightarrow$

**6.7** $MnO(s) + H^+(aq) \rightarrow$

7. Complete and balance the following chemical equations (or tell if no reaction will occur):

    **7.1** $SrO(s) + MoO_3(s) \rightarrow$

    **7.2** $SrO(s) + ZrO_2(s) \rightarrow$

    **7.3** $N_2O_5(s) + H_2O(l) \rightarrow$

    **7.4** $CaO(s) + TeO_3(s) \rightarrow$

    **7.5** $CaO(s) + MnO(s) \rightarrow$

    **7.6** $Fe_3O_4(s) + H^+(aq) \rightarrow$

    **7.7** $Cl_2O_7(g) + H_2O(l) \rightarrow$

*8. Consider the following set of oxides: $SrO$, $ZrO_2$, $MoO_3$, and $RuO_4$. If needed, take the radius of $Ru^{8+}$ to be 52 pm. **a.** Which of these oxides will be soluble in water and give a basic solution? Write a chemical equation for this process. **b.** Which of these oxides will be soluble in water and give an acidic solution? Write an equation for this process. **c.** Which of these oxides (if any) will consist of small molecules? **d.** Which of these oxides (if any) will most easily become a gas?

9. Describe the air-pollution problems associated with nonmetal oxides specified by your instructor and devise some possible abatement procedures that might be tried.

10. **a.** Write three balanced chemical equations showing the three steps by which elemental sulfur in coal is converted to sulfuric acid in acid rain. **b.** Sulfur dioxide can be removed from smokestack gases by reaction with magnesium oxide (scrubbing). Write a chemical equation for this process. **c.** Calculate the number of grams of magnesium oxide that would be needed to clean the smokestack gases from burning 1,000,000 g of coal that is 3.2% S by weight.

11. With reference to Tables B and C, describe some metal oxides with low oxidation numbers that might be nonstoichiometric in a manner analogous to $Fe_{0.95}O$. Why would you not expect oxides with very high oxidation numbers, such as $Mn_2O_7$ and $OsO_4$, to be nonstoichiometric (e.g., $Mn_{2.2}O_7$ or $Os_{1.13}O_4$)?

*12. Which of the following formulas correspond to *possible* nonstoichiometric oxides? $Ca_{0.95}O$, $Fe_{0.95}O$, $Co_{0.95}O$, $C_{0.95}O$, $Cr_{0.95}O_3$, and $Eu_{0.95}O$.

*13. In the following list, circle the oxides that are likely to be spinels and underline the oxides that could be perovskites: $NiFe_2O_4$, $BaFe_2O_4$, $BaTiO_3$, $BeTiO_3$, $BaSO_3$, $TiZn_2O_4$, $Ni_3O_4$, $Pb_3O_4$, and $NaTaO_3$.

14. Consider the following series of oxides: $MnO$, $MnO_2$, $Mn_2O_3$, $Mn_2O_7$, and $Mn_3O_4$. **a.** Which of these will be most acidic? **b.** Which of these will be

most basic? **c.** Which is most likely to be soluble in water? **d.** Which of these is *least* likely to show a nonstoichiometric (defect) structure? **e.** Which (if any) of these will show a perovskite structure? **f.** Which (if any) of these will show a spinel structure? **g.** Which (if any) of these will consist of (nonpolymerized) covalent molecules? **h.** Which will have the lowest melting point? **i.** Which one is a liquid at room temperature? **j.** The rest will be which: gases or solids? **k.** Which is most likely to be amphoteric?

15. Give the symbols for six elements that readily form polymeric oxo anions.

*16. Use one of the following minerals—(1) wollastonite = $CaSiO_3$; (2) talc = $Mg_3(OH)_2(Si_4O_{10})$; (3) grunerite = $Fe_7(OH)_2(Si_4O_{11})_2$; (4) monticellite = $CaMgSiO_4$; (5) stishovite = $SiO_2$—to complete each of the following: **a.** Contains a monomeric silicate ion; **b.** contains a chain polysilicate ion; **c.** contains a double-chain polysilicate ion; **d.** contains a sheet polysilicate ion; **e.** the mineral that will weather the most rapidly; **f.** the mineral that will weather the most slowly.

17. Classify each of the following silicates as (1) a framework (three-dimensional) aluminosilicate; (2) a sheet polysilicate; (3) a chain or cyclic polysilicate; (4) a simple silicate; or (5) a double-chain polysilicate:
**a.** bustamite = $CaMn(SiO_3)_2$;     **b.** spodumene = $LiAl(SiO_3)_2$;
**c.** tremolite = $Ca_2Mg_5(OH)_2(Si_4O_{11})_2$;     **d.** coffinite = $U(SiO_4)$;
**e.** natrolite = $Na_2(Al_2Si_3O_{10})\cdot 2H_2O$.

18. Show that you understand the condensed drawings of the fragments of polysilicate structures shown in Figure 4.8 by redrawing them, using closed circles for Si atoms and open circles for O atoms, as in Figure 4.7. Redraw the following: **a.** $(SiO_3{}^{2-})_n$;     **b.** $(Si_4O_{11}{}^{6-})_n$;     **c.** $(Si_4O_{10}{}^{4-})_n$.

19. You are studying the mineral hornblende, $Ca_2Mg_5(OH)_2(Si_4O_{11})_2$, and find samples in which isomorphous substitution of the magnesium and the calcium has occurred. Which of the following are possible minerals that could result from isomorphous substitution processes in hornblende?
**a.** $Y_2Mg_5(OH)_2(Si_4O_{11})_2$;     **b.** $Na_2Mg_5(OH)_2(Si_4O_{11})_2$;
**c.** $Na_2Mg_3(Fe^{III})_2(OH)_2(Si_4O_{11})_2$;     **d.** $Y_2Mg_3Li_2(OH)_2(Si_4O_{11})_2$;
**e.** $Sr_2Mg_5(OH)_2(Si_4O_{11})_2$.

20. With reference to Table C, find other "cations," each of a different charge, that can substitute for $P^{5+}$ in the phosphomolybdate ion. Give the formula and charge of each heteromolybdate ion that results; rank these heteromolybdates in order of increasing basicity.

21. You are in charge of disposing of asbestos being removed from schools, where it was used as insulation. Devise a reasonable scheme for chemically destroying the asbestos.

22. In Figure 4.13 is shown the relative availability of nutrient elements in soil as a function of soil pH. Using the principles in the last three chapters, insofar

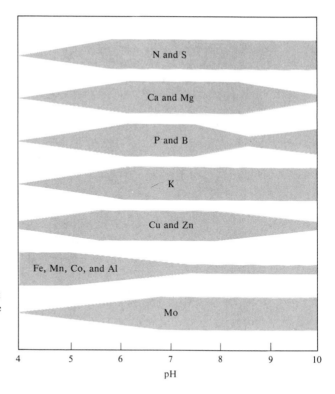

**Figure 4.13** Relative availability of nutrient elements in soil as a function of soil pH. (The broader the bar, the more available is the nutrient.) Reproduced with permission from *Environmental Chemistry*, by J. W. Moore and E. A. Moore. Copyright © 1976 by Academic Press. Reprinted by permission.

as you can, explain why each element is available and unavailable at the pH's given.

23. You have produced minute quantities of radioactive element number 109, the element below iridium, and are investigating its positive oxidation states.

   **23.1** Suppose that you have oxidized it very strongly in a hot acidic solution and find the radioactivity coming from the vapors above the solution; you conclude that you have a volatile oxide. What oxidation state do you probably have? Give arguments why you chose that oxidation state and not some other. Draw a likely structure of the oxide.

   **23.2** Suppose that you subsequently obtain the element in the $+6$ oxidation state in basic solution and find that it gives a precipitate not only with $Ba^{2+}$ but also with most acidic cations. What does this lead you to suspect about the formula of the $+6$ species?

   **23.3** You have only a trace of the element left, in the $+7$ oxidation state in basic solution. Outline a plan for precipitating it from solution.

Notes    [1]   In principle, small ion multiples (pairs, triplets, etc.) can also be gaseous if the number of oppositely charged ions surrounding the central ion equals the maximum coordination number of the central atom (as determined by the radius ratio). For example, the electronegativity difference between Si and F is over 2.0, so that the bonding between these two atoms surely has a high degree of ionic character. Assuming that the bonding is completely ionic, we find the radius ratio $Si^{4+}/F^-$ to be 0.454, which is almost small enough to require the coordination number for $Si^{4+}$ not to exceed 4. The $Si^{4+}$ ion of one $SiF_4$ "ion quintet" then may not be able to attract the fluoride ions of another $SiF_4$ closely enough to overcome the repulsions that would exist between that ion quintet's fluoride ions and the touching fluorides of the next ion quintet; hence we could predict this ionic compound to be a gas at room temperature. On the other hand, if we think of $SiF_4$ as a covalent substance, we expect a small molecule with tightly held electrons and weak van der Waals forces: i.e., a gas at room temperature. $SiF_4$ is indeed a gas at room temperature. This does not establish whether the bonding is predominantly ionic or covalent, but it does demonstrate that it consists of small units ($SiF_4$ molecules or ion quintets) with weak forces between the units. For practical purposes we usually treat the infrequent cases like $SiF_4$ as covalent molecules, thus ignoring any such category as "ion multiplets."

[2]  Structural data are obtained from **a.** A. F. Wells, *Structural Inorganic Chemistry*, 5th ed., Clarendon Press, Oxford, 1984, or from **b.** F. A. Cotton and G. Wilkinson, *Advanced Inorganic Chemistry: A Comprehensive Text*, 4th ed., Wiley-Interscience, New York, 1980.

[3]  Melting and boiling points are taken from M. C. Ball and A. H. Norbury, *Physical Data for Inorganic Chemists*, Longmans, London, 1974, or from the *Handbook of Chemistry and Physics*, 50th ed., Chemical Rubber Publishing Co., Cleveland, 1969. Values were also checked for consistency with data given in the *Alfa Catalog 1983–1984*, Morton-Thiokol Inc., Danvers, Mass., 1983.

[4]  *Kirk-Othmer Encyclopedia of Chemical Technology*, 3d ed., Wiley, New York, 1978 (24 volumes).

[5]  Greenwood, N. N., and A. Earnshaw, *Chemistry of the Elements*, Pergamon Press, Oxford, 1984.

[6]  Moore, J. W., and E. A. Moore, *Environmental Chemistry*, Academic Press, New York, 1976.

[7]  Manahan, S. E., *Environmental Chemistry*, 3d ed., Willard Grant Press, Boston, 1979.

[8]  Fergusson, J. E., *Inorganic Chemistry and the Earth: Chemical Resources, Their Extraction, Use, and Environmental Impact*, Pergamon Press, Oxford, 1982.

[9]  Rochow, E. G., *Modern Descriptive Chemistry*, Saunders, Philadelphia, 1977.

[10]  Huheey, J. E., *Inorganic Chemistry: Principles of Structure and Reactivity*, 3d ed., Harper and Row, Cambridge, 1983.

[11]  Ferris, J. P., "The Chemistry of Life's Origin," *Chem. Eng. News*, August 27, 1984, p. 22.

# Oxidation-Reduction Chemistry of the Elements

## Standard Reduction Potentials; Periodic Trends in Stability of High Oxidation States of Oxo Anions and Acids

In the previous chapters we examined the acid-base and precipitation reactions of compounds of the elements in fixed oxidation states. We saw, though, that most elements have more than one positive oxidation state; in addition, all can be prepared as the free elements with oxidation state 0, and some have negative oxidation numbers. Reactions in which elements change their oxidation numbers are known as **oxidation-reduction** or **redox** reactions and will be the focus of this chapter.

Redox reactions normally require the presence of two reactants: In one reactant the oxidation number of an element is reduced, and in the other an oxidation number is increased. For example, in the reaction

$$Pb(s) + NaNO_3(s) \rightarrow PbO(s) + NaNO_2 \tag{5.1}$$

the oxidation number of nitrogen is being reduced from $+5$ in the nitrate ion to $+3$ in the nitrite ion; we say that the nitrogen atom or the whole nitrate ion is being reduced. This reduction is accomplished by the lead, which we call the **reducing agent**. The oxidation number of lead is being increased from 0 in the lead metal to $+2$ in the lead(II) oxide; we say that the lead is being **oxidized**. Since this oxidation is accomplished by the action of the nitrate ion, we call the nitrate ion the **oxidizing agent**.

Thermodynamically, reaction (5.1) and other redox reactions will go if the free energy change, $\Delta G$, for the reaction is negative. Experimentally, instead of measuring $\Delta G$ we measure the cell voltage generated when the reaction occurs spontaneously in a voltaic (galvanic) cell. (Or we measure the voltage required to force it to go in an electrolytic cell.) This is possible because oxidation-reduction reactions generally involve transfer of electrons from the element or compound being oxidized to the element or compound being reduced. If these reactions occur in separate compartments of an electrochemical cell, the electrons can be channeled through electrodes, wires, and a potentiometer, and the voltage generated ($E$) can be measured. A spontaneous reaction occurring in a galvanic cell is considered to generate a positive voltage, $E$. If the reaction occurs under reversible conditions the voltage can be related to the free energy change by the equation

$$\Delta G = -nFE \qquad (5.2)$$

where $n$ is the number of electrons exchanged in the reaction as written, and $F$ is the conversion between electrochemical and thermodynamic units, 96.5 kJ/V.

Since in an electrochemical cell the oxidation and the reduction occur in separate compartments or half-cells, each part of the reaction (called a **half-reaction**) can be treated separately. The overall cell voltage is the sum of the two half-cell voltages. We can measure only this *sum*, so we do not know the actual voltage generated by any half-cell. By convention we assign a voltage of 0.000 V to the half-cell in which hydrogen ion is being reduced to hydrogen gas under standard conditions:

$$2\,H^+(aq, \text{activity} = 1) + 2\,e^- \rightleftharpoons H_2(\text{pressure} = 1\text{ atm}); \qquad E^\circ = 0.000\text{ V} \quad (5.3)$$

The voltage generated by this and other half-cells depends on the conditions of the reactants and products, so we define conditions such that each element is in its standard state. For gases, the standard state means a pressure of 1.000 atm for an ideal gas (or its equivalent if the gas is not ideal); for pure liquids, solvents, and solids, this is the pure liquid or solid at 1.000 atm pressure. The standard state for a solute is a concentration that gives an activity or thermodynamic concentration of 1.000, in units of either molalities or molarities. In dilute aqueous solutions molalities and molarities converge to the same value, as do activities and concentrations. Unfortunately a concentration of 1.000 mol/L is not very dilute, but we will not concern ourselves here with the distinction between activities and concentrations, which results from disruptions of the primary and secondary hydration spheres of the ions in nondilute solutions. If all reactants and products are in their standard states, we place a superscript "o" on the $E$ (and the $\Delta G$) to designate this fact.

We may now set up electrochemical cells in which all reactants and products are present in their standard states. In one half-cell, hydrogen gas is being oxidized to hydrogen ion, and in the other half-cell a chemical species is being reduced (reversibly) to a product in which an element is in a lower oxidation state. The voltage generated in this cell (or required as an input if the reaction is not spontaneous) is called the **standard reduction potential**, $E^\circ$, of that species. The more positive the $E^\circ$ of a species, the more easily it is reduced (and the more reactive it is as an oxidizing agent). If all elements in the test species are in thermodynamically stable oxidation states, the species usually has a negative standard reduction potential. The species then cannot be reduced by hydrogen (under standard conditions), and it is thus a poor oxidizing agent.

Extensive tabulations have been made of standard reduction potentials of various chemical species. We will examine only selected sets of these potentials in order to note periodic trends in thermodynamic stability (with regard to redox reactions) of similar types of chemical species under standard conditions. We will begin with the oxo anions and oxo acids of the *p*-block elements in their highest (group) oxidation state and examine the standard reduction potentials that result when these species are reduced by two electrons. These reduction potentials are tabulated in Table 5.1. Note that these reactions involve the hydrogen ion (e.g., equation (5.4)). Since standard conditions are employed, the hydrogen-ion concentration (more properly, activity) must be 1 M; hence the species listed are those most stable at a pH of 0 (i.e., only nonbasic and feebly basic oxo anions are listed as oxo anions; the others are more properly listed as oxo acids).

$$ClO_4^-(aq) + 2e^- + 2H^+ \rightleftharpoons ClO_3^-(aq) + H_2O \qquad (5.4)$$

We first may note the horizontal trends in $E°$'s among *p*-block elements in their highest oxidation states. As we go to the right and the oxidation number increases, the species become less stable and better oxidizing agents. Thus in the fifth period, $Cd^{2+}$ is not readily reduced to Cd metal, whereas $H_4XeO_6$ is easily reduced; its very positive $E°$ signifies that it is a powerful oxidizing agent indeed. This is not surprising: An oxidation number of $+8$ for xenon indicates that a lot of oxygen atoms are pulling a lot of electron density from it—some resistance to this on the part of the xenon atom has to be expected. (In the second period this trend is so strong that oxygen cannot be made to adopt an oxidation number of $+6$; nor F, $+7$; nor Ne, $+8$.)

Much more surprising are the vertical trends among the *p*-block elements, which can be completely seen only in Group 15/VA. As reflected in its negative $E°$, phosphoric acid is by far the most nearly devoid of oxidizing properties of the acids, anions, and oxides of the Group 15/VA elements. Likewise, in Group 16/VIA, sulfuric acid is the poorest oxidizing agent and the most stable species. In general, *in the p-block, the group oxidation number is most stable in the third period.*

It may also be noted that *in the p-block, the group oxidation number is least stable at the top (second period) and at the bottom (sixth period).* Thus nitrate ion in acid solution (i.e., nitric acid) is a useful oxidizing agent; acid solutions of "bismuthate" ion are very powerful oxidizing agents. Various reasons can be given for this peculiar periodic tendency; the more fundamental ones will have to await the final chapter of the book. But for now we may note that an oxidation number over 5 in an oxo anion requires a coordination number of at least 4; this is not possible in the oxo anions of the second period, since the atoms are too small. We also noted previously that some of the elements of the last period also often adopted unexpectedly low coordination numbers. (In Chapter 8 we will also see that the bonds to oxygen are relatively weak in these periods.)

A more subtle vertical periodicity occurs in Groups 16/VIA and 17/VIIA (and almost in Group 15/VA) on going from the third through the fourth to the fifth periods; the stability of the high oxidation state follows the sequence Period 3 element $\gg$ Period 4 element $<$ Period 5 element. Although small, this reversal is enough to have some interesting chemical consequences: For example, the perbromate ion was not synthesized until 1968, long after perchlorates and periodates were made and came into routine use.

In Table 5.2 similar $E°$ data are gathered for oxo anions and oxides of the

**Table 5.1**  Standard Reduction Potentials of Elements in Group Oxidation Numbers Being Reduced by Two Electrons

| | | $CO_2 \rightarrow CO$ | $NO_3^- \rightarrow HNO_2$ | | $ClO_4^- \rightarrow ClO_3^-$ |
|---|---|---|---|---|---|
| | | $-0.12$ V | $+0.94$ V | | $+1.19$ V |
| $Zn^{2+} \rightarrow Zn$ | $Ga^{3+} \rightarrow Ga^+$ | $GeO_2 \rightarrow Ge^{2+}$ | $H_3PO_4 \rightarrow H_3PO_3$ | $SO_4^{2-} \rightarrow H_2SO_3$ | $BrO_4^- \rightarrow BrO_3^-$ |
| $-0.76$ V | $-0.40$ V | $-0.50$ V | $-0.28$ V | $+0.17$ V | $+1.74$ V |
| $Cd^{2+} \rightarrow Cd$ | $In^{3+} \rightarrow In^+$ | $Sn^{4+} \rightarrow Sn^{2+}$ | $H_3AsO_4 \rightarrow H_3AsO_3$ | $SeO_4^{2-} \rightarrow H_2SeO_3$ | $H_5IO_6 \rightarrow IO_3^-$ |
| $-0.40$ V | $-0.43$ V | $+0.15$ V | $+0.56$ V | $+1.15$ V | $+1.60$ V |
| $Hg^{2+} \rightarrow Hg$ | $Tl^{3+} \rightarrow Tl^+$ | $PbO_2 \rightarrow Pb^{2+}$ | $Sb_2O_5 \rightarrow SbO^+$ | $H_6TeO_6 \rightarrow TeO_2$ | $H_5AtO_6 \rightarrow AtO_3^-$ |
| $+0.85$ V | $+1.25$ V | $+1.46$ V | $+0.58$ V | $+1.02$ V | $+1.6$ V |
| | | | $Bi_2O_5 \rightarrow BiO^+$ | $PoO_3 \rightarrow PoO_2$ | $H_4XeO_6 \rightarrow XeO_3$ |
| | | | $+1.60$ V | $+1.52$ V | $+2.36$ V |

SOURCES: Standard reduction potentials are taken from J. E. Huheey, *Inorganic Chemistry: Principles of Structure and Reactivity*, 2d ed., Harper and Row, New York, 1978, and from B. Douglas, D. H. McDaniel, and J. J. Alexander, *Concepts and Models of Inorganic Chemistry*, 2d ed., John Wiley, New York, 1983.

**Table 5.2**  Standard Reduction Potentials of $d$-Block Oxo Anions and Acids

| | | | |
|---|---|---|---|
| $HCrO_4^- \rightarrow Cr^{3+}$ <br> $+1.20$ V | $MnO_4^- \rightarrow MnO_2$ <br> $+1.68$ V | | $FeO_4^{2-} \rightarrow Fe^{3+}$ <br> $+2.20$ V |
| $H_2MoO_4 \rightarrow Mo^{3+}$ <br> $+0.34$ V | $TcO_4^- \rightarrow TcO_2$ <br> $+0.74$ V | $RuO_4 \rightarrow RuO_2$ <br> $+1.40$ V | $RuO_4^{2-} \rightarrow Ru^{3+}$ <br> $+0.90$ V |
| $WO_3 \rightarrow W^{3+}$ <br> $-0.07$ V | $ReO_4^- \rightarrow ReO_2$ <br> $+0.51$ V | $OsO_4 \rightarrow OsO_2$ <br> $+0.96$ V | |

SOURCES: Standard reduction potentials are calculated or taken from J. E. Huheey, *Inorganic Chemistry: Principles of Structure and Reactivity*, 2d ed., Harper and Row, New York, 1978, and from B. Douglas, D. H. McDaniel, and J. J. Alexander, *Concepts and Models of Inorganic Chemistry*, 2d ed., John Wiley, New York, 1983.

$d$-block elements in very high oxidation states. The horizontal periodic trends are the same as for the $p$-block elements: To the right, the increasingly high oxidation numbers lose stability. But the vertical periodicity is different. This time the fourth-period elements show the greatest reluctance to adopt the group oxidation number, which is *more* stable in the fifth and especially the sixth periods. Indeed, it may be noted that there is more stability for oxo anions and acids having the group oxidation numbers for $d$-block elements in the fifth and especially in the sixth periods than for $p$-block elements in any period!

### 5.2

## pH and the Stability of High Oxidation States; Syntheses of Oxo Anions and Their Use as Oxidizing Agents

Virtually all redox reactions carried out in the real world involve reactants or products present under other than standard conditions. For example, as a reaction proceeds, the concentrations of reactants and products change, which changes the driving force for the reaction (LeChatelier's principle). We know, for example, that as the chemical reaction in a dry cell proceeds to completion, its voltage drops until the cell is "dead." The effect of nonstandard concentrations and other activities on potential at room temperature is given by the well-known Nernst equation

$$E = E^\circ - \frac{0.059}{n} \log Q \tag{5.5}$$

where $n$ represents the number of electrons transferred and $Q$ is the reaction quotient, which has the same form as the equilibrium expression but can include concentrations that apply when the system is not at equilibrium.

For example, in the reduction of a metal ion to the metal a typical half-reaction can be written

$$Hg^{2+}(aq) + 2e^- \rightleftharpoons Hg(l) \tag{5.6}$$

The metal, as long as it is pure, is always in the standard state, so its activity does not vary, although the concentration of the metal ion can vary a great deal. This will alter the half-cell potential from the standard value (0.85 V for $Hg^{2+}$):

$$E = 0.85 - \frac{0.059}{2} \log \frac{1}{[Hg^{2+}]} = 0.85 + \frac{0.059}{2} \log [Hg^{2+}] \tag{5.7}$$

Increasing the concentration of the mercury ion will tend to drive the reaction (5.6) to the right and will increase the potential.

For oxo anions, oxo acids, and oxides and hydroxides the concentration dependence of the potential is more complicated. For example, for the reduction of $FeO_4^{2-}$ to $Fe^{3+}$ the balanced half-reaction is

$$FeO_4^{2-} + 3\,e^- + 8\,H^+ \rightleftharpoons Fe^{3+} + 4\,H_2O \tag{5.8}$$

Consequently the half-cell potential depends not only on the concentrations of the $FeO_4^{2-}$ and $Fe^{3+}$ ions, but also *strongly* on the hydrogen-ion concentration (pH):

$$E = 2.20 - \frac{0.059}{3} \log \frac{[Fe^{3+}]}{[FeO_4^{2-}][H^+]^8} \tag{5.9}$$

$$= 2.20 - 0.157\,\text{pH} - 0.020 \log \frac{[Fe^{3+}]}{[FeO_4^{2-}]}$$

The very high half-cell potential for ferrate ion means that this ion is quite unstable under acidic conditions. Removing hydrogen ion from the equilibrium (5.8) will strongly shift the equilibrium to the left, favoring stability in strongly basic solution. At pH = 14 the potential is 0.00 V minus the last term of equation (5.9)—but as there can be very little $Fe^{3+}$ ion in solution at pH 14, this works out to 0.72 V for 1 $M$ ferrate ion.

This equilibrium condition has important practical consequences: *The synthesis of a salt of an oxo anion with a very highly oxidized central atom is normally carried out in basic solution.* On the other hand, such oxo-anion salts are often useful as oxidizing agents, but they are most powerful as oxidizing agents in acidic solutions. To carry out an oxidation in basic solution, an oxidizing agent that involves few or no hydrogen ions in the reduction (such as chlorine) is preferred. The choice of the oxidizing agent may vary depending on kinetic, economic, or other factors, but for species with standard reduction potentials greater than about +1.2 V (in acid solution), the use of strongly basic solutions and strong oxidizing agents (or electrolysis) is required [1]. Some of the strong oxidizing agents commonly chosen are listed in Table 5.3 along with their standard reduction potentials, which naturally exceed +1.2 V in acid solution. (In basic solution, where these oxidizing agents are used, their reduction potentials are less if hydrogen and oxygen are involved, but so are the reduction potentials of the oxo anions being synthesized.)

Thus a very unstable species such as the ferrate ion is normally prepared in very alkaline solution, using chlorine (which is converted in part to hypochlorite ion in such solutions):

$$2\,Fe(OH)_3 + 3\,OCl^- + 4\,OH^- \rightarrow 2\,FeO_4^{2-} + 5\,H_2O + 3\,Cl^- \tag{5.10}$$

**Table 5.3**  Standard Reduction Potentials of Some Useful Oxidizing Agents in Acidic and Basic Solution (pH = 0 and 14)

| Oxidizing Agent and Half-Reactions | $E°_{acid}$ | $E°_{base}$ |
|---|---|---|
| **Fluorine** | | |
| $F_2 + 2e^- = 2F^-$ | +2.87 V | +2.87 V |
| **Chlorine** | | |
| $Cl_2 + 2e^- = 2Cl^-$ | +1.36 V | +1.36 V |
| **Ozone** | | |
| $O_3 + 2e^- + 2H^+ = O_2 + H_2O$ | +2.076 V | |
| $O_3 + 2e^- + H_2O = O_2 + 2OH^-$ | | +1.24 V |
| **Hydrogen peroxide** | | |
| $H_2O_2 + 2e^- + 2H^+ = 2H_2O$ | +1.776 V | |
| $HO_2^- + 2e^- + H_2O = 3OH^-$ | | +0.88 V |
| **Oxygen** | | |
| $O_2 + 4e^- + 4H^+ = 2H_2O$ | +1.229 V | |
| $O_2 + 4e^- + 2H_2O = 4OH^-$ | | +0.40 V |
| **Nitric acid; nitrates** | | |
| $NO_3^- + e^- + 2H^+ = NO_2 + H_2O$ | +0.79 V | |
| $NO_3^- + 3e^- + 2H_2O = NO + 4OH^-$ | | +0.427 V |

SOURCES: Standard reduction potentials are taken or derived from data given in B. Douglas, D. H. McDaniel, and J. J. Alexander, *Concepts and Models of Inorganic Chemistry*, 2d ed., Wiley, New York, 1983.

To further assist in the synthesis, the ferrate ion can be precipitated by the use of an appropriate cation ($Ba^{2+}$ in this case, since ferrate is feebly basic). This reduces the concentration of ferrate ion in solution, which (from equation (5.9)) further improves the cell potential. Periodate is similarly prepared by alkaline oxidation of iodate ion, and tellurate from tellurite.

Another useful oxidizing agent in such syntheses is hydrogen peroxide, $H_2O_2$. Although the reduction of $H_2O_2$ does involve hydrogen ion, it involves less than most oxo anions do, so basic conditions are still useful. Thus chromate may be synthesized as follows:

$$2 Cr(OH)_3 + 3 H_2O_2 + 4 OH^- \rightarrow 2 CrO_4^{2-} + 8 H_2O \tag{5.11}$$

Another powerful oxidizing agent, ozone ($O_3$) in basic solution, is similarly used to oxidize xenon trioxide to perxenate; fluorine gas is used to oxidize bromate ion to perbromate ion.

Difficult oxidations can often be carried out by heating solid reactants (oxide of the element, basic oxide, and an oxidizing agent such as sodium peroxide) until they melt and react:

$$Na_2O + 2\,Na_2O_2 + Bi_2O_3 \rightarrow 2\,Na_3BiO_4 \qquad (5.12)$$

Often the oxygen of the air will suffice as the oxidizing agent:

$$10\,Li_2O + 4\,NpO_2 + 3\,O_2 \rightarrow 4\,Li_5NpO_6 \qquad (5.13)$$

As previously mentioned in Chapter 2, salts of some oxo anions decompose on heating to evolve gaseous oxides and leave behind a basic oxide. If a change in oxidation state is also involved, these reactions can be used for high-temperature solid-state syntheses of oxo anions:

$$2\,KNO_3 + Se \rightarrow K_2SeO_4 + 2\,NO \qquad (5.14)$$

This reaction is similar to that involved in the detonation of gunpowder, in which saltpeter ($KNO_3$) is used to oxidize carbon and sulfur to their oxides. Similar reactions are used to prepare potassium manganate ($K_2MnO_4$) from $MnO_2$, and potassium ruthenate ($K_2RuO_4$) from ruthenium metal.

Many oxo anions can also be prepared by **electrolysis** of aqueous solutions containing the elements in low oxidation states. Reduction of the water at the cathode evolves hydrogen and produces $OH^-$ to make the solution basic; the oxo anion is produced at the anode. Hypochlorite ion ($ClO^-$) can be produced by electrolysis of cold dilute solutions of NaCl; further reaction yields chlorate ion ($ClO_3^-$); electrolysis of solutions of the chlorate ion yields the perchlorate ion. Permanganate ion is similarly produced by electrolysis of the manganate ion ($MnO_4^{2-}$).

If the oxo acid or anion has a standard reduction potential of about $+0.3$ to about $+1.2$ V (i.e., the oxo anion does not involve too high an oxidation state), basic conditions may not be necessary. Strongly oxidizing acids such as nitric acid may then be used. In this manner iodine is oxidized to iodic acid, selenium to selenous acid, $As_2O_3$ to arsenic acid, and technetium and rhenium compounds to $HTcO_4$ and $HReO_4$. Or it may be possible to oxidize the element in air to give its acidic oxide. In this case the oxo acid may be prepared by reaction of the acidic oxide with water, or the oxo salt by reaction of the acidic oxide with a basic oxide. As you should recall from Chapter 4, nitric and sulfuric acids are prepared in this manner.

If the desired oxo acid or anion has a reduction potential below about $+0.2$ V, simpler syntheses may be possible, since the element may well occur in nature in that oxidation state. Treatment of a salt with a strong acid may liberate the oxo acid. (Recall from Chapter 4 that phosphoric acid is normally made in this manner.) In some cases hydrolysis of the halide (as in Experiment 1) may be a suitable synthesis:

$$PCl_3 + 3\,H_2O \rightarrow H_3PO_3 + 3\,HCl \qquad (5.15)$$

If the oxo anion involves the element in a low oxidation state, one may carry out an electrochemical *reduction* of a higher oxo anion or use a chemical reducing agent (e.g., the synthesis of nitrite ion, equation (5.1)). Dissolution of the element itself in a strongly basic solution may be appropriate, as in the synthesis of hypophosphorous acid:

$$P_4 + 4\,OH^- + 4\,H_2O \rightarrow 4\,H_2PO_2^- + 2\,H_2 \qquad (5.16)$$

Similar very vigorous reactions occur with the common metals zinc and aluminum (to produce their hydroxo anions); the heat and hydrogen gas generated by this reaction are used to clean stopped-up drains. For this reason, too, household lye (NaOH) should never be used in aluminum pans. A similar reaction is the preparation of sodium hypochlorite (household bleach):

$$Cl_2 + 2\,OH^- \rightleftharpoons ClO^- + Cl^- + H_2O \qquad (5.17)$$

In this case chloride ion is an inevitable byproduct; in acidic solution the above reaction is not favored, as the chloride and hypochlorite ions react to give chlorine gas. For this reason household bleach should never be combined with any acidic cleansing agents.

As we have seen, the synthetic inorganic or industrial chemist has a variety of methods at his or her disposal to prepare oxo acids and anions. Although many of these can be purchased, some cannot due to their instability; even some of those that can be purchased may deteriorate during their storage or transportation; hence synthesis may be necessary. Useful sources of synthetic details include the serial **Inorganic Syntheses**, which tends to focus on more exotic chemicals; for simple compounds such as are covered in this book, the work by Brauer [2] is probably most useful.

For purposes of classroom study, the particular choice of oxidizing agent made to synthesize a particular oxo acid or salt is not so important for us to know as the overall trends:

■ Synthesis of a strongly oxidizing oxo anion (one with a standard reduction potential over $+1.2$ V) requires electrochemical methods or strong oxidizing agents and strongly basic conditions. Such an oxo anion normally involves a central-atom oxidation number greater than 6, or somewhat lower if the element is in the second or sixth period of the $p$-block or the fourth period of the $d$-block elements.

■ If the oxo acid or anion is not strongly oxidizing (standard reduction potential between 0.3 and 1.2 V), gas-phase oxidation to the oxide followed by reaction with water, or oxidation in acidic solution with a cheap oxidizing agent like nitric acid, is generally preferred.

■ Nonoxidizing oxo acids can often be prepared by using acid-base reactions, rather than oxidation-reduction reactions, since the element with that same oxidation number may be prevalent in nature.

As we mentioned previously, if chemists want to *use* oxo anions or acids as oxidizing agents, they generally use acidic solutions in order to enhance their reduction potentials. Many of these oxo anions find use in organic chemistry in different sorts of oxidations, but these are beyond the scope of this book. In analytical chemistry dichromate and permanganate are often used to determine the concentration of various reducing agents by titration; other oxidizing agents such as bismuthate and perchlorate are used in preparing samples for titration. Overall, nitric acid is undoubtedly most frequently used (among the oxo acids) for oxidative purposes.

## 5.3

## Pourbaix (pE/pH) Diagrams and the Forms in Which the Elements Occur in Natural Waters under Different Conditions of Aeration

In Section 2.12 we summarized the effects of pH on the form that an element in a given oxidation state would take in natural waters. This dependence on pH was summarized graphically in Figure 2.7b, which shows the predominant phosphorus(V)-containing species in equilibrium with aqueous solutions as a function of pH. But the phosphorus may not remain in the $+5$ oxidation state if suitable reducing agents are present; hence to know the fate of an element in natural waters it is not sufficient to know only the pH of the aqueous environment. We must also know if it is oxidizing (i.e., well aerated) or reducing (i.e., polluted with organic wastes, etc.). This condition can be expressed in terms of the reducing potential, $E$, of the solution, and the graph of Figure 2.7b can be expanded to include $E$ as well as pH as a variable. Such a diagram is very useful and is known variously as a **Pourbaix diagram** [3], a predominance-area diagram, an $E°$–pH diagram, or a pE–pH diagram [4].

In such a diagram a low $E$ (or pE) means a reducing solution and a high $E$ means an oxidizing solution. Any given (well-mixed) solution can be characterized by a particular $E$ and a particular pH. Finding this point on the Pourbaix diagram of an element will give us the thermodynamically most stable (in principle, most abundant) form of that element at that $E$ and pH. For example, we can see from Figure 5.1 that at a reduction potential of $+0.8$ V and a pH of 14, the predominant form of iron in a solution with a total of 1 mol of iron present per liter is $FeO_4^{2-}$.

A second use of Pourbaix diagrams is to give a visual representation of the oxidizing and reducing abilities of major stable compounds of the elements. Terms such as "oxidizing agent" and "reducing agent" are often confused by students, especially since the terms "the element being oxidized" and "the element being reduced" are often used in the same settings. Conjuring up a mental image of the Pourbaix diagram can help with these terms. Strong oxidizing agents and oxidizing conditions are found at the *top* of Pourbaix diagrams and not elsewhere; strong oxidizing agents have *lower* boundaries that are also high on the diagram. (Of course, strong oxidizing agents also have high oxidation numbers for the element in question.)

Reducing agents and reducing conditions prevail at the *bottom* of a Pourbaix diagram and not elsewhere; strong reducing agents have *low* upper boundaries on the diagram. They also are generally the compounds of the element in which the oxidation number of the element has been *reduced* to its lowest value. (For metals the strongest common reducing agent is generally the metal itself.)

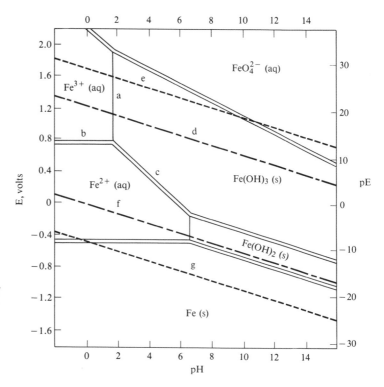

**Figure 5.1** A simplified Pourbaix diagram for 1 $M$ iron solutions. Solid single lines (such as $a$) separate species related by acid-base equilibria; solid double lines ($b$ and $c$) separate species related by redox equilibria. Longer dash lines ($d, f$) enclose the theoretical, and dotted lines ($e, g$) enclose the practical region of stability of the water solvent to oxidation or to reduction.

In contrast, a species that prevails from top to bottom of a Pourbaix diagram (at least at the pH in question) has no oxidizing or reducing properties at all within that range. Such a species would be considered a very poor oxidizing *and* a very poor reducing agent.

EXAMPLE    On the Pourbaix diagram for iron find (a) the chemical form of iron that is the strongest oxidizing agent; (b) the form of iron that is the strongest reducing agent; (c) the form of iron that would predominate in a neutral solution at a potential of 0.00 V; (d) the standard reduction potential for the reduction of $FeO_4^{2-}$ to $Fe^{3+}$; (e) the standard reduction potential for the reduction of $Fe^{2+}$ to Fe metal.

Solution    ■ (a) The most strongly oxidizing form of iron shown in Figure 5.1 is the form with the highest lower boundary to its region of predominance, the ferrate ion $FeO_4^{2-}$. This form contains iron with its highest oxidation number, $+6$.

■ (b) Iron metal has the lowest upper boundary on the Pourbaix diagram and is the most strongly reducing form of iron shown.

■ (c) A neutral solution has a pH of 7. If we draw a line straight up from the pH of 7, it will intersect a line drawn across from the potential of 0.00 V in the area labeled $Fe(OH)_3(s)$. This is the most abundant form of iron present in a solution

147

having a total of 1 mol of iron in any form present per liter. (Since we are close to the regions labeled $Fe^{2+}$(aq) and $Fe(OH)_2$(s), at equilibrium lesser amounts of these will probably also be present.)

■ (d), (e) Standard reduction potentials are found at pH = 0, since 1 $M$ is the standard concentration of $H^+$. The standard reduction potential for the reduction of ferrate ion to iron(III) ion is found at the intersection of the pH = 0 vertical line with the boundary separating $FeO_4^{2-}$ and $Fe^{3+}$, or approximately at 2.2 V on the diagram. Similarly, the standard reduction potential for the reduction of $Fe^{2+}$(aq) to metallic iron is found at about $-0.5$ V.     □

Pure acid-base equilibria are reflected in Pourbaix diagrams by vertical boundaries such as (a) in Figure 5.1, which shows the pH at which half of the 1 $M$ iron is in the form of $Fe^{3+}$ and half is precipitated as $Fe(OH)_3$. Thus the Pourbaix diagrams already incorporate the $Z^2/r$ calculations and acid-base equilibria of Chapter 2; these calculations need not be repeated. The position of such an acid-base equilibrium is, of course, dependent on the total concentration of iron; reducing the total concentration of iron (and therefore of $Fe^{3+}$) will, by LeChatelier's principle, reduce the driving force for precipitation. Thus if we reduce the total iron concentration from 1 $M$ to $10^{-6}$ $M$ (as is more realistic for geochemists and corrosion engineers), the boundary is shifted from a pH of about 1.7 to about 4.2. In general, in more dilute solutions the soluble species have larger predominance areas.

Redox equilibria of species not involving hydrogen or hydroxide ions are reflected in Pourbaix diagrams by horizontal boundaries such as (b). Redox equilibria that also involve hydrogen or hydroxide ions are also (in part) acid-base equilibria and so have diagonal boundaries such as (c). These diagonal boundaries generally slope from upper left to lower right, since basic solutions tend to favor the more oxidized species.

The upper dashed line (d) represents the potential of water saturated with dissolved $O_2$ at 1 atm pressure—i.e., of very well aerated water. It also represents the potential above which water is oxidized to oxygen according to equation (5.18):

$$2\,H_2O \rightleftharpoons 4\,H^+(aq) + O_2 + 4\,e^- \qquad E° = +1.229\ V \tag{5.18}$$

In principle this reaction can happen with any dissolved oxidizing agent of $E°$ greater than 1.229 V, but in practice it takes about another 0.5 V of potential—i.e., the potential of the upper dotted line (e)—to overcome the *overvoltage* of oxygen formation. (Even then the reaction may be kinetically slow.) Any species that lies completely above line (e) at the pH in question should be unstable in aqueous solution and should gradually oxidize the water to $O_2$ gas. For example, it may be seen that about 1 $M$ ferrate ion is not stable in water solution at a pH of less than about 10.

The lower dashed line (f) represents the potential of a solution saturated with dissolved $H_2$ at 1 atm pressure—i.e., one with a high level of reducing agents in solution. This is also the potential below which the hydrogen ion in water theoretically is reduced to hydrogen according to equation (5.3). A similar overvoltage effect prevents significant release of hydrogen until the lower dotted line (g) is reached. (Thus although in theory iron metal is unstable in water, the overvoltage holds down the rate of its reaction with water unless the pH is quite low.) Any species lying wholly below line (g) at the pH in question is likely to react with water, producing $H_2$ gas and a chemical species of that element that is higher on the Pourbaix diagram. Thus iron metal reacts with acidic solutions to give $Fe^{2+}$(aq) and $H_2$.

We note that the predominance area for iron(II) species ($Fe^{2+}$ and $Fe(OH)_2$) narrows considerably at higher pH's. In some cases (for example, chlorine(0), which appears as $Cl_2$ in Figure 5.2) the predominance area for a given oxidation state may disappear completely above or below a given pH. If the element is in an intermediate oxidation state, at appropriate pH's the element will undergo a type of redox reaction called **disproportionation**, in which part of the element in the unstable oxidation state will act as oxidizing agent and part as reducing agent. Thus above a pH of about 0, chlorine water will begin to disproportionate to the chlorine($-$I) and chlorine(I) oxidation states according to equation (5.17).

We note that in the Pourbaix diagram for chlorine (Figure 5.2; see also Figure 5.3) predominance areas are missing for the hypochlorite, chlorite, and chlorate ions. This may mean either that the electrochemical data for these ions are unknown and hence not included in the diagram or (in this case) that these species are thermodynamically unstable to disproportionation. They are well-known species, however: The rates of the disproportionation reactions are slow enough that these chlorine species can be observed and used. (This is much less true for the corresponding bromine and iodine species.) The disproportionation of chlorate in particular is slow enough that many chlorates are known and handled in commerce; but heating potassium chlorate above 400°C speeds up the disproportionation

$$4\,KClO_3 \rightarrow 3\,KClO_4 + KCl \tag{5.19}$$

In Chapter 3 we discovered solubility rules for predicting when cations and anions would react to give precipitates. At that time we were careful to deal with the most stable oxidation states of the elements, since cations and anions can undergo oxidation-reduction reactions as well, if the predominance areas of their oxidation states do not coincide. Thus ferrate ion is expected to be a feebly basic anion like sulfate and to give precipitates with feebly acidic cations. It does indeed with $Ba^{2+}$, but with another feebly acidic cation, $Eu^{2+}$, a redox reaction occurs instead, since $Eu^{2+}$ is a good reducing agent, with no part of its predominance area above a potential of $-0.429$ V (Figure 5.4). There is no overlap of this region with that of ferrate ion, so reaction occurs to give species that do have overlap (europium(III) and iron(III)).

Pourbaix diagrams with appropriate total concentrations allow us to make more realistic predictions of the forms that the different elements will take in natural waters (Sec. 2.12). The surface waters of a clean lake are well aerated and have dissolved oxygen concentrations that are high enough to make their potentials reasonably close to the oxygen standard reduction potential (Figure 5.5). In a lake that is highly polluted with organic reducing agents, in the bottom layer of a

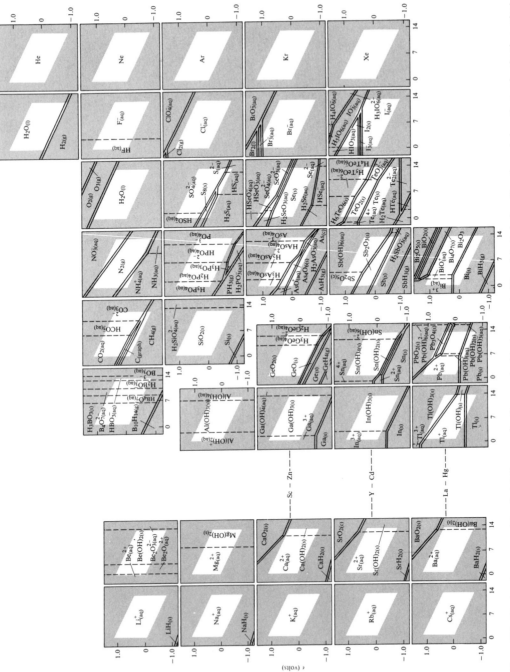

**Figure 5.2** Predominance area (Pourbaix) diagrams for the *s*- and *p*-block elements. The white "windows" are bounded on the top by the theoretical *E* for the oxidation of water, and on the bottom by the theoretical *E* for the reduction of water. From J. A. Campbell and R. A. Whiteker, "A Periodic Table Based on Potential-pH Diagrams." *J. Chem. Educ.*, 46, 92 (1969). Reprinted with permission.

150

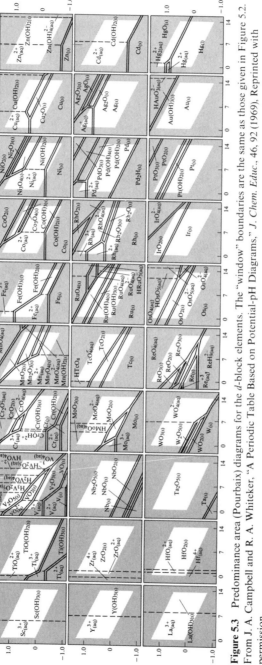

**Figure 5.3** Predominance area (Pourbaix) diagrams for the *d*-block elements. The "window" boundaries are the same as those given in Figure 5.2. From J. A. Campbell and R. A. Whiteker, "A Periodic Table Based on Potential-pH Diagrams," *J. Chem. Educ.*, *46*, 92 (1969), Reprinted with permission.

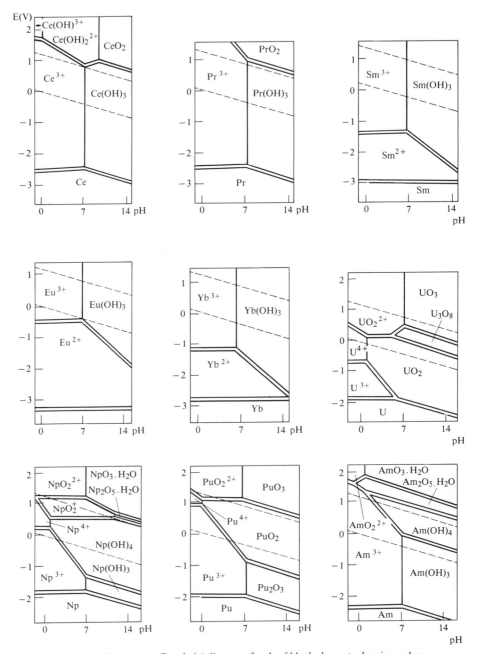

**Figure 5.4** Predominance area (Pourbaix) diagrams for the *f*-block elements showing redox chemistry. Adapted from M. Pourbaix, *Atlas of Electrochemical Equilibria in Aqueous Solutions*, National Association of Corrosion Engineers, Houston, 1974. Reprinted with permission.

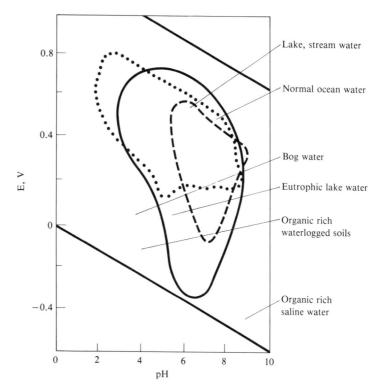

**Figure 5.5** The approximate potential and pH values for some environmental waters, indicated by the lines shown or by the regions enclosed as follows: ——— soil water, – – – – subsurface water, ..... acid mine-drainage water. Reprinted with permission from J. E. Fergusson, *Inorganic Chemistry and the Earth*, copyright 1982 by Pergamon Press.

thermally stratified lake, or in a swamp, conditions may be not only quite anaerobic, but also actively reducing, so that the lower boundary of reduction to $H_2$ may be approached. Under these conditions, many of the most common nutrient elements are converted to unfamiliar forms: From Figure 5.2 we see that carbon may be converted to methane, $CH_4$, sometimes referred to as *swamp gas*. Nitrogen is reduced to the ammonium ion, $NH_4^+$, or to ammonia, $NH_3$. The latter affects the odor of the water and is quite toxic. Sulfur is reduced to hydrogen sulfide, $H_2S$; emissions of this gas to the atmosphere are a major source of acid rain after the sulfur has been reoxidized (Sec. 4.5.4). Given overvoltages, it is even possible for phosphorus to be reduced to phosphine, $PH_3$, a toxic, foul-smelling gas that commonly ignites upon exposure to air, giving rise to eerie light emission over swamps at night.

## 5.4

## The Stability of Low Oxidation States: The Reduction of Metal Ions

At this point your instructor may ask you to perform or observe and discuss Experiment 4 in Chapter 12. This experiment involves interpretation of the **activity**

**series of metals**, which you may have investigated in general chemistry. In this case, your instructor may assign only Part F of Experiment 4.

In Section 5.2 we examined periodic trends in the stability of high oxidation states of elements in their oxo anions and acids; now we will examine periodic trends in the stability of the zero oxidation states of elements (the elemental forms). This tendency can also be measured electrochemically: The more positive the standard reduction potential of a metal cation to the metal, the more easily the metal itself is obtained. In a Pourbaix diagram, the predominance areas of easily obtained metals will be larger and will penetrate closer to (or even into) the stability region of water. Table 5.4 summarizes the standard reduction potentials (in acid solution) of the most common metal ion of each metallic element. For elements with more than one metal ion the values would naturally be at least somewhat different if a differently charged cation was being reduced to the metal.

For a qualitative judgment of the relative stability of metals with respect to oxidation by hydrogen ion or water, it is more convenient to examine, not the reduction of metal ions, but the reverse reaction, the oxidation of a metal to its hydrated cation (Experiment 4). The traditional *activity series of metals* is simply a listing of the metals in decreasing order of their reactivity with hydrogen-ion sources (water or acids). The more positive the standard reduction potential of a metal cation, the *harder* it is to oxidize a metal to its hydrated metal cation, so the later it falls in the activity series. Since activity is quantitatively measured electrochemically, the activity series is also called the *electromotive series*.

As you may have discovered in analyzing Experiment 4, the activity of a metal correlates fairly well with its Pauling electronegativity, as shown in Figure 5.6. Although this correlation is not exact enough for precise ordering of elements of very similar activity (recall that Pauling electronegativities have uncertainties of $\pm 0.05$), it does enable us to suggest groupings of elements, based on their Pauling electronegativities. These groupings will be useful later in grouping the tendencies of elements to undergo certain redox and other reactions.

**1.** The most active metals have low electronegativities. We may include in a group of *very electropositive metals* all metals with Pauling electronegativities below 1.4; their cations generally have standard reduction potentials of $-1.6$ V or below. This group includes the metals at the far left of the periodic table (Groups 1, 2, the *f*-groups, and Group 3). These metals are so reactive that they react with water to release hydrogen; they are very good reducing agents. (For some of the least active of these the water must be boiling or in the form of steam.) Their metal ions are not good oxidizing agents at all: They cannot be reduced to the metal in aqueous solution.

**2.** We may identify a group of *electropositive metals* with Pauling electronegativities between 1.4 and 1.9; their cations generally have standard reduction potentials between 0.0 and $-1.6$ V. This group includes the *d*-block elements of the fourth period and the *p*-block metals of the fourth and fifth periods. Although these metals do not react very readily with water, they do react with the hydrogen ion (acids).

**Table 5.4** Standard Reduction Potentials of Hydrated Metal Ions

| | 1 | 2 | 3 | 4 | 5 | 6 | 7 | 8 | 9 | 10 | 11 | 12 | 13/IIIA | 14/IVA | 15/VA | 16/VIA |
|---|---|---|---|---|---|---|---|---|---|---|---|---|---|---|---|---|
| 1 | H<br>+1<br>0.00 | | | | | | | | | | | | | | | |
| 2 | Li<br>+1<br>−3.04 | Be<br>+2<br>−1.85 | | | | | | | | | | | | | | |
| 3 | Na<br>+1<br>−2.71 | Mg<br>+2<br>−2.37 | | | | | | | | | | | Al<br>+3<br>−1.66 | | | |
| 4 | K<br>+1<br>−2.93 | Ca<br>+2<br>−2.87 | Sc<br>+3<br>−2.02 | Ti<br>+3<br>−1.21 | V<br>+3<br>−0.87 | Cr<br>+3<br>−0.74 | Mn<br>+2<br>−1.19 | Fe<br>+2<br>−0.47 | Co<br>+2<br>−0.28 | Ni<br>+2<br>−0.23 | Cu<br>+1<br>+0.52 | Zn<br>+2<br>−0.76 | Ga<br>+3<br>−0.56 | Ge<br>+2<br>0.00 | | |
| 5 | Rb<br>+1<br>−2.92 | Sr<br>+2<br>−2.89 | Y<br>+3<br>−2.37 | Zr<br>+4<br>−1.54 | Nb<br>+3<br>−1.10 | Mo<br>+3<br>−0.20 | Tc<br>+2<br>−0.50 | Ru<br>+3<br>+0.38 | Rh<br>+3<br>+0.80 | Pd<br>+2<br>+0.92 | Ag<br>+1<br>+0.80 | Cd<br>+2<br>−0.40 | In<br>+3<br>−0.34 | Sn<br>+2<br>−0.14 | | |
| 6 | Cs<br>+1<br>−3.08 | Ba<br>+2<br>−2.91 | Lu<br>+3<br>−2.30 | Hf<br>+4<br>−1.70 | Ta | W<br>+3<br>−0.11 | Re<br>+3<br>+0.3 | Os<br>+2<br>+0.85 | Ir<br>+3<br>+1.16 | Pt<br>+2<br>+1.2 | Au<br>+1<br>+1.68 | Hg<br>+2<br>+0.85 | Tl<br>+1<br>−0.34 | Pb<br>+2<br>−0.13 | Bi<br>+3<br>+0.29 | Po<br>+2<br>+0.65 |
| 7 | | Ra<br>+2<br>−2.92 | Lr<br>+3<br>−2.06 | | | | | | | | | | | | | |

| | 3f | 4f | 5f | 6f | 7f | 8f | 9f | 10f | 11f | 12f | 13f | 14f | 15f | 16f |
|---|---|---|---|---|---|---|---|---|---|---|---|---|---|---|
| 6 | La<br>+3<br>−2.36 | Ce<br>+3<br>−2.34 | Pr<br>+3<br>−2.35 | Nd<br>+3<br>−2.32 | Pm<br>+3<br>−2.29 | Sm<br>+3<br>−2.30 | Eu<br>+3<br>−1.99 | Gd<br>+3<br>−2.29 | Tb<br>+3<br>−2.30 | Dy<br>+3<br>−2.29 | Ho<br>+3<br>−2.33 | Er<br>+3<br>−2.31 | Tm<br>+3<br>−2.31 | Yb<br>+3<br>−2.22 |
| 7 | Ac<br>+3<br>−2.13 | Th<br>+4<br>−1.80 | Pa<br>+4<br>−1.62 | U<br>+4<br>−1.37 | Np<br>+4<br>−1.75 | Pu<br>+4<br>−1.25 | Am<br>+3<br>−2.07 | Cm<br>+3<br>−2.06 | Bk<br>+3<br>−1.97 | Cf<br>+3<br>−2.01 | Es<br>+3<br>−1.98 | Fm<br>+3<br>−1.95 | Md<br>+3<br>−1.66 | No<br>+3<br>−1.18 |

SOURCES: Data are taken from B. Douglas, D. H. McDaniel, and J. J. Alexander, *Concepts and Models of Inorganic Chemistry*, John Wiley, New York, 1983, and from M. C. Ball and A. H. Norbury, *Physical Data for Inorganic Chemists*, Longman, London, 1974.

NOTE: The first figure under the symbol of the element represents the charge of the cation being reduced to the element. (For ions with more than one cation, the more prevalent cation was chosen.) The second figure is the standard reduction potential for the reduction of that cation to the elemental form.

155

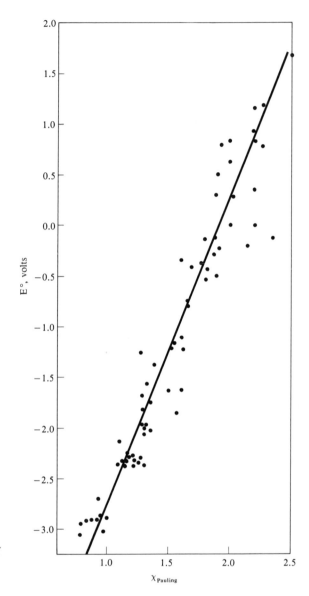

**Figure 5.6** The standard reduction potential ($E^0$) of the most common cation of each metallic element versus its Pauling electronegativity.

**3.** There is a group of *electronegative metals* with Pauling electronegativities between 1.9 and 2.54; their cations generally have positive standard reduction potentials. This group includes most of the *d*-block elements of the fifth and sixth periods and the *p*-block metals of the sixth period. This group of metals is not oxidized by the hydrogen ion; the metal ions of this group are good oxidizing agents that will oxidize (for instance) hydrogen gas, producing hydrogen ion and depositing the metal from aqueous solution. (As noted in the experiment, these cations of less active metals also oxidize more active metals to produce the cations of the more active metals and release the less active metals. Such a reaction is sometimes called a **displacement reaction**.)

This relative order of reactivity of metals applies not only to oxidation by water or the hydrogen ion, but (approximately) to oxidation by other oxidizing agents as well, such as oxygen itself or oxidizing oxo acids. Thus the very electropositive metals readily ignite in air, burning to produce oxides. (The heavier Group 1 metals produce peroxides (salts of the $O_2^{2-}$ ion) and superoxides (salts of the $O_2^{-}$ ion), since these larger and/or less charged anions give better lattice energies with these large cations.) Fires that result from the combustion of these metals cannot be fought with water, since water also reacts with these metals to produce $H_2$, which burns explosively. Neither can such fires be extinguished with $CCl_4$, which is actually a rather strong oxidizing agent, or with $CO_2$; such fires are generally best extinguished with sand ($SiO_2$).

The electropositive metals do not burn in air as readily. Their surfaces generally tarnish, and they become covered with a surface layer of oxide, which is often tough enough and insoluble enough to protect the surfaces against further oxidation, although thermodynamically the production of oxide is favored. (This is why the aluminum metal used in the experiment had to be cleaned before its activity could be determined.) If the surface is in contact with air and water the process of **corrosion** may occur, in which the metal acts as an anode and is oxidized:

$$Fe(s) = Fe^{2+} + 2e^- \tag{5.20}$$

Some other point on the surface of the metal, or the junction of the metal with a more electronegative metal, acts as a cathode, at which (typically) oxygen is reduced to hydroxide ion. The electrochemical circuit is completed by migration of hydroxide ions (or other ions—salt water assists corrosion strongly) back to the $Fe^{2+}$ ions to precipitate $Fe(OH)_2$, which further oxidizes and dehydrates to give hydrated iron(III) oxide, the familiar rust.

Since the mechanism of corrosion involves oxidation of iron to $Fe^{2+}$, corrosion of iron occurs principally at pH's and $E$'s within the predominance area of $Fe^{2+}$ (Figure 5.1). Corrosion does not occur in adequately basic solutions due to the coating of the iron with FeO (or $Fe(OH)_2$). If the $E$ is below the standard reduction potential of iron, the iron is stable and will not rust. This condition may be accomplished by coating the iron or by attaching a more active metal (such as Zn or Mg) to it; the more active metal will then corrode instead. Or the metal may actually be **passivated** by the use of a strong oxidizing agent (such as dichromate or pertechnetate), since that coats the iron with an impervious layer of $Fe_2O_3$. The prevention of corrosion is economically a very important activity; it has been estimated that as much as half of all the metal that has been produced by humans has corroded away.

Many of the electronegative metals are not corroded by oxygen. This property (plus their scarcity) has contributed to their being prized by humans for millennia as the *noble metals* and to their use in coinage and jewelry. Since oxygen is a stronger

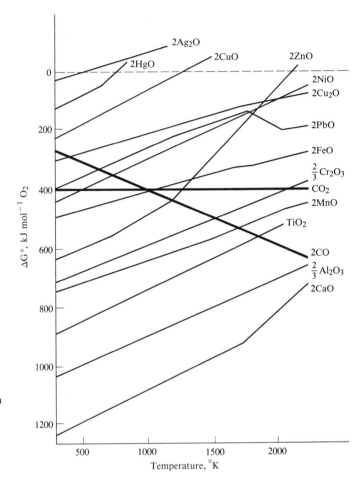

**Figure 5.7** An Ellingham diagram showing the free energies of decomposition of metal oxides as a function of temperature. (Discontinuities result from phase changes of either the metal or the oxide.) From D. J. G. Ives, *Monographs for Teachers, Number 3*, The Chemical Society, London, 1969. Reprinted with permission.

oxidizing agent than the hydrogen ion, some of the electronegative metals (which are not attacked by $H^+$) are oxidized slowly by oxygen, but the oxides formed are not very stable. They can generally be decomposed by heating:

$$2\,HgO(s) + heat \rightarrow 2\,Hg(l) + O_2(g) \tag{5.21}$$

This decomposition is favored at higher temperatures due to the increasing significance of the entropy change term $-T\Delta S$ at higher $T$; $\Delta S$ is positive due to the production of the gaseous oxygen. **Ellingham diagrams** (Figure 5.7) show the variation in the free energy of decomposition of oxides at higher temperature (electrochemical measurements are not practical at such high temperatures). The enthalpy changes required to decompose the oxides of the very electropositive or electropositive metals are too high to allow heat alone to be used to produce these metals from their oxides. But heat alone can be used to produce the electronegative metals from their oxides.

Although nonoxidizing acids cannot attack the electronegative metals, oxidizing acids such as concentrated nitric acid can attack and dissolve most of them:

$$Ag(s) + 2\,HNO_3 \rightarrow Ag^+(aq) + NO_3^-(aq) + NO_2(g) + H_2O \qquad (5.22)$$

But some of the electropositive metals are not attacked by $HNO_3$, because they are passivated.

*Qualitatively* we can extend the relationship of activity and electronegativity to the nonmetals: The more electronegative a nonmetal atom, the harder it is to oxidize. (One form of phosphorus inflames spontaneously in air when dry; at the other extreme, fluorine and the lighter Group 18/VIIIA elements cannot be oxidized.) But the standard reduction potentials of nonmetals do not fit the correlation of Figure 5.6, since they do not oxidize to form cations.

## 5.5

## Nonmetals and Their Monoatomic Anions

### 5.5.1 Redox Chemistry of Nonmetals and Their Anions

The nonmetallic elements can also act as oxidizing agents. Upon oxidizing an element, the nonmetals may be reduced to their monoatomic anions or the protonated forms of these (their hydrides, Chapter 9). As the electronegativity of a nonmetal increases, its activity as an *oxidizing* agent, as measured by its standard reduction potential, increases (Table 5.5). We may note that activity as an oxidizing agent increases to the right across a period and up a group.

Just as we divided the metals into three groups based on redox properties, we may also divide the nonmetals into two groups. The first of these we may call the *very electronegative nonmetals*—those with Pauling electronegativities above 2.8. These nonmetals, which are good oxidizing agents, include $F_2$, $Cl_2$, $Br_2$, $O_2$, and $N_2$.

Fluorine is such an active oxidizing agent that it will oxidize all metals and most other substances. (Even water, in the form of steam, will catch fire in an atmosphere of fluorine.) This poses a problem: In what container can one store fluorine gas? Fortunately, the metals nickel and copper are passivated by fluorine—impervious films of nickel or copper fluorides are formed on the surfaces of these metals.

The other halogens are successively less oxidizing. A significant reaction of the halogens is the displacement reaction: A more active halogen displaces a less active halogen from its anion. Bromine is made in this way from bromide ion in seawater:

$$2\,Br^- + Cl_2 \rightleftharpoons Br_2 + 2\,Cl^- \qquad (5.23)$$

Oxygen is actually a strong oxidizing agent, but its oxidizing reactions are generally quite slow; hence very flammable materials can last indefinitely in the presence of air, at least until a flame or spark begins the reaction. Nitrogen is also not really as inert as we commonly think, but its very stable triple bond means that high temperatures or strong reducing agents are required to reduce it.

**Table 5.5** Standard Reduction Potentials of the Nonmetallic Elements

**A. Standard Reduction Potentials in Acid Solution**

| | | | |
|---|---|---|---|
| $C \rightarrow CH_4$<br>+0.13 V | $N_2 \rightarrow NH_4^+$<br>+0.27 V | $O_2 \rightarrow H_2O$<br>+1.23 V | $F_2 \rightarrow F^-$<br>+2.87 V |
| $Si \rightarrow SiH_4$<br>+0.10 V | $P_4 \rightarrow PH_3$<br>+0.01 V | $S_8 \rightarrow H_2S$<br>+0.14 V | $Cl_2 \rightarrow Cl^-$<br>+1.36 V |
| $Ge \rightarrow GeH_4$<br>−0.86 V | $As \rightarrow AsH_3$<br>−0.60 V | $Se \rightarrow H_2Se$<br>−0.40 V | $Br_2 \rightarrow Br^-$<br>+1.07 V |
| | $Sb \rightarrow SbH_3$<br>−0.51 V | $Te \rightarrow H_2Te$<br>−0.72 V | $I_2 \rightarrow I^-$<br>+0.54 V |
| | $Bi \rightarrow BiH_3$<br>−0.80 V | $Po \rightarrow H_2Po$<br>−1.00 V | $At_2 \rightarrow At^-$<br>+0.3 V |

**B. Standard Reduction Potentials in Basic Solution**

| | | | |
|---|---|---|---|
| $C \rightarrow CH_4$<br>−0.70 V | $N_2 \rightarrow NH_3$<br>+0.1 V | $O_2 \rightarrow OH^-$<br>+0.40 V | $F_2 \rightarrow F^-$<br>+2.87 V |
| $Si \rightarrow SiH_4$<br>−0.93 V | $P_4 \rightarrow PH_3$<br>−0.89 V | $S_8 \rightarrow S^{2-}$<br>−0.48 V | $Cl_2 \rightarrow Cl^-$<br>+1.36 V |
| $Ge \rightarrow GeH_4$<br>< −1.1 V | $As \rightarrow AsH_3$<br>−1.21 V | $Se \rightarrow Se^{2-}$<br>−0.92 V | $Br_2 \rightarrow Br^-$<br>+1.07 V |
| | $Sb \rightarrow SbH_3$<br>−1.34 V | $Te \rightarrow Te^{2-}$<br>−1.14 V | $I_2 \rightarrow I^-$<br>+0.54 V |
| | $Bi \rightarrow BiH_3$<br>< −1.6 V | $Po \rightarrow Po^{2-}$<br>< −1.4 V | $At_2 \rightarrow At^-$<br>+0.3 V |

Data from B. Douglas, D. H. McDaniel, and J. J. Alexander, *Concepts and Models of Inorganic Chemistry*, 2d ed., John Wiley, New York, 1983.

The remaining nonmetals we may term the *electronegative nonmetals*, with Pauling electronegativities between 1.9 and 2.8. These find little laboratory use as oxidizing agents, but (for example) sulfur reacts slowly with some of the electronegative metals (such as Ag in silverware) to tarnish them. In the early history of the earth, reducing conditions prevailed, so many of these elements formed minerals in which they took negative oxidation numbers. The earth's atmosphere is now oxidizing, so many of these elements are now more often found oxidized (as oxo anions or oxides) than reduced.

The *monoatomic nonmetal anions* and (in most cases) their protonated forms exhibit the lowest possible oxidation number of their elements. As such they have the potential of transferring electrons back to other materials, i.e., of acting as *reducing agents*. The anions of the very electronegative nonmetals ($F^-$, $Cl^-$, $Br^-$, $O^{2-}$, and $N^{3-}$), of course, have very little tendency to do this; they are poor reducing agents. But the anions of the electronegative nonmetals show moderate to strong reducing properties.

**Table 5.6**  Aqueous Basicity of Monoatomic Anions and Their Partially Protonated Forms

**A. Monoatomic Anions**

| | | | |
|---|---|---|---|
| $C^{4-}$ very strong | $N^{3-}$ very strong | $O^{2-}$ very strong $pK_1 = -22$ | $F^-$ weak $pK_1 = 10.85$ |
| $Si^{4-}$ very strong | $P^{3-}$ very strong | $S^{2-}$ strong $pK_1 = 0$ | $Cl^-$ nonbasic $pK_1 = 20.3$ |
| $Ge^{4-}$ very strong | $As^{3-}$ very strong | $Se^{2-}$ moderate $pK_1 = 3.0$ | $Br^-$ nonbasic $pK_1 = 22.7$ |
| | | $Te^{2-}$ weak $pK_1 = 9.0$ | $I^-$ nonbasic $pK_1 = 23.3$ |

**B. Partially Protonated Anions**

| | | |
|---|---|---|
| $CH_3^-$ very strong $pK_4 = -30$ | $NH_2^-$ very strong $pK_3 = -25$ | $OH^-$ strong $pK_2 = -1.74$ |
| $SiH_3^-$ very strong $pK_4 = -21$ | $PH_2^-$ very strong $pK_3 = -13$ | $SH^-$ moderate $pK_2 = 7.11$ |
| $GeH_3^-$ very strong $pK_4 = -11$ | $AsH_2^-$ very strong $pK_3 = -9$ | $SeH^-$ weak $pK_2 = 10.3$ |
| | | $TeH^-$ feeble $pK_2 = 11.4$ |

SOURCES: $pK_b$ values are calculated from $pK_a$ values given in W. L. Jolly, *Modern Inorganic Chemistry*, McGraw-Hill, New York, 1984, p. 177; in R. V. Dilts, *Analytical Chemistry*, Van Nostrand, New York, 1974, p. 553; in F. A. Cotton and G. Wilkinson, *Basic Inorganic Chemistry*, John Wiley, New York, 1976; and in W. H. Nebergall, H. H. Holtzclaw, Jr., and W. R. Robinson, *General Chemistry*, 6th ed., Heath, Lexington, 1980.

## 5.5.2    Nonmetal Anions as Bases

The monoatomic anions of the nonmetallic elements and their partially protonated forms are shown in Table 5.6. These anions are generally named using the suffix -ide instead of -ate or -ite, as in the oxo anions. Thus $I^-$ is named iodide; $S^{2-}$, sulfide; $N^{3-}$, nitride; $C^{4-}$, carbide; etc. Collectively, the anions of Group 17/VIIA are often called the halide ions, those of Group 16/VIA, the chalcogenide ions, and those of Group 15/VA, the pnicogenide ions.

The basicities ($pK_b$'s) of these ions in aqueous solutions are summarized in Table 5.6. (Some values are estimated from values in other types of solutions.) As we would expect, the basicities of these ions increase strongly with increasing negative charge (i.e., from right to left in a period). Particularly if the anionic charge is $-3$ or $-4$, these anions are very strong bases that cannot persist in water. We also note, as expected, that the basicities of these anions decrease with increasing size (down a group).

*In aqueous solution* three of the halide ions—chloride, bromide, and iodide— are nonbasic, having no tendency to combine with hydrogen ions. This means that their basicity is sufficiently less than that of water molecules that they cannot

161

compete with water for hydrogen ions. Later we will see other circumstances in which basic properties of the halide ions are evident.

The *partially protonated forms* of the ions are also important ions that sometimes have special names. Although $SH^-$ and $SeH^-$ are just called the hydrogen sulfide ion and the hydrogen selenide ion, respectively, $OH^-$ is, of course, called the hydroxide ion, $NH_2^-$ is called the amide ion, and $CH_3^-$ may be called the methide ion or the methyl anion. The $pK_b$ values for the partially protonated $-1$ ions of the nonmetals are summarized in Table 5.6B. These, of course, are less basic than the parent monoatomic ions with multiple negative charges; even so these anions cannot persist in water for the elements of Groups 14/IVA and 15/VA. If we compare $-1$ ions in a given period, we find that basicity decreases strongly from left to right, as the electronegativities of the elements increase. (The effect of electronegativity is much more important here than among the oxo anions, since the negative charges are no longer concentrated on the same element, oxygen.)

The fully protonated conjugate acids of the monoatomic anions of the nonmetals are also called the hydrides of these elements and will be discussed in more detail in a later chapter. For now we note their nomenclature, which has some peculiarities. These hydrides are gases; the hydrides of the Group 16/VIA and 17/VIIA elements dissolve in water to give acidic solutions that are named differently than the pure compounds. Thus HF by itself is named hydrogen fluoride, but its aqueous solution is known as hydrofluoric acid; $H_2S$ by itself is named hydrogen sulfide, and its aqueous solution is sometimes referred to as hydrosulfuric acid. The same pattern applies to the other Group 16/VIA and 17/VIIA hydrides, except for $H_2O$, which is always just plain old water.

The hydrides of the Group 14/IVA and 15/VA elements have a different pattern of naming, in which the hydrogen is not directly referred to but is implied by a special suffix, -ane in Group 14/IVA and (usually) -ine in Group 15/VA. The formulas and names (which use Latin roots related to the symbol of the element when available) are summarized in Figure 5.8c.

A number of other important anions are related in structure to those shown in Table 5.6. The hydrogens in the partially protonated anions may be substituted with organic groups such as the methyl group, $CH_3$, or by other organic groups (collectively known as alkyl groups and symbolized by the letter R). Such substitution gives similar anions, such as $CH_3O^-$ (methoxide), $CH_3S^-$ (methyl sulfide), $(CH_3)_2N^-$ (dimethylamide), or collectively $RO^-$ (alkoxide), $RS^-$ (alkyl sulfide or mercaptide), and $R_2N^-$ (dialkylamide) ions.

### 5.5.3 Pseudohalide Ions

Several important polyatomic (nonoxo) anions with quite diverse structures bear some resemblance in their acid-base and oxidation-reduction chemistry to the halide ions and are known as **pseudohalide** ions; these are shown in Figure 5.9. One of their similarities to the halide ions lies in their relatively low basicity (feebly to moderately basic); other similarities in basic properties will show up in later

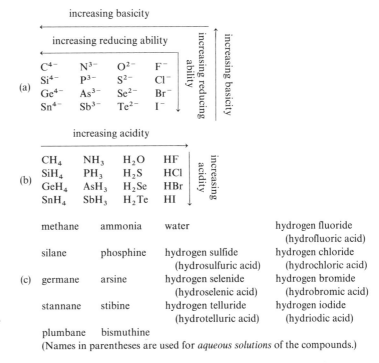

**Figure 5.8** Summary of periodic trends among the nonmetals in (a) basicity and reducing ability of the monoatomic anions; (b) acidity of the protonated anions (conjugate acids of the anions); and (c) nomenclature of the protonated anions (hydride derivatives).

| methane | ammonia | water | hydrogen fluoride (hydrofluoric acid) |
| silane | phosphine | hydrogen sulfide (hydrosulfuric acid) | hydrogen chloride (hydrochloric acid) |
| (c) germane | arsine | hydrogen selenide (hydroselenic acid) | hydrogen bromide (hydrobromic acid) |
| stannane | stibine | hydrogen telluride (hydrotelluric acid) | hydrogen iodide (hydriodic acid) |
| plumbane | bismuthine | | |

(Names in parentheses are used for *aqueous solutions* of the compounds.)

chapters. In redox chemistry the **cyanide**, **thiocyanate**, and **thiosulfate** ions resemble the halogens in that oxidation produces dimeric species:

$$2\,Cl^- \rightleftharpoons Cl_2(g) + 2\,e^- \tag{5.24}$$

$$2\,CN^- \rightleftharpoons (CN)_2(g) + 2\,e^- \tag{5.25}$$

$$2\,SCN^- \rightleftharpoons (SCN)_2(l) + 2\,e^- \tag{5.26}$$

$$2\,S_2O_3{}^{2-} \rightleftharpoons S_4O_6{}^{2-}(aq) + 2\,e^- \tag{5.27}$$

This reaction of the thiosulfate ion is used extensively in analytical chemistry in the quantitative analysis of oxidizing agents, since many oxidizing agents react with iodide ion to give (intensely colored) iodine, which reacts completely with easily prepared solutions of the thiosulfate ion to give iodide ion and the tetrathionate ion, $S_4O_6{}^{2-}$.

The **azide** ion is a very powerful reducing agent that reacts to give, not a dimer, but 3 mol of (very stable) nitrogen molecules:

$$2\,N_3{}^- \rightarrow 3\,N_2 + 2\,e^- \tag{5.28}$$

The azide ion will even reduce the sodium ion: Heating solid sodium azide produces sodium metal. The resemblance of the **cyanate** ion, $NCO^-$, to the halide ions lies more in the area of acid-base than of redox chemistry. Although they are not considered pseudohalide ions, two other nonmetallic anions are worthy of mention

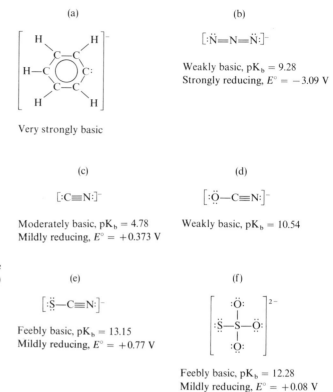

(a)

Very strongly basic

(b)

$[:\ddot{N}=N=\ddot{N}:]^-$

Weakly basic, $pK_b = 9.28$
Strongly reducing, $E° = -3.09$ V

(c)

$[:C\equiv N:]^-$

Moderately basic, $pK_b = 4.78$
Mildly reducing, $E° = +0.373$ V

(d)

$[:\ddot{O}{-}C\equiv N:]^-$

Weakly basic, $pK_b = 10.54$

(e)

$[:\ddot{S}{-}C\equiv N:]^-$

Feebly basic, $pK_b = 13.15$
Mildly reducing, $E° = +0.77$ V

(f)

Feebly basic, $pK_b = 12.28$
Mildly reducing, $E° = +0.08$ V

**Figure 5.9** Lewis dot structures and basic and reducing properties of some important polyatomic ions other than oxo anions. (a) The phenyl anion, $C_6H_5^-$; (b) the azide ion, $N_3^-$; (c) the cyanide ion, $CN^-$; (d) the cyanate ion, $CNO^-$; (e) the thiocyanate ion, $CNS^-$; (f) the thiosulfate ion, $S_2O_3^{2-}$. $pK_b$ values are taken from R. V. Dilts, *Analytical Chemistry*, Van Nostrand, New York, 1974, p. 533. Standard reduction potentials shown are in acid solution, taken from J. E. Huheey, *Inorganic Chemistry: Principles of Structure and Reactivity*, 3d ed., Harper & Row, Cambridge, 1983.

now, since they will be used in later chapters: the **phenyl** anion (Figure 5.9) and the **hydride** ion, $H^-$, both of which are powerfully basic and powerfully reducing.

The conjugate acids of the pseudohalide ions are named like oxo acids if the pseudohalide uses an -ate ending: HOCN, cyanic acid; HSCN, thiocyanic acid; $H_2S_2O_3$, thiosulfuric acid. They are named like the conjugate acids of the halides if the pseudohalide uses an -ide ending: HCN is hydrogen cyanide when pure or hydrocyanic acid in solution; $HN_3$ is hydrogen azide when pure or hydrazoic acid in solution.

## 5.6

## Explosives, Flammable Compounds, and Strong Oxidizing Agents

So far in this chapter we have examined compounds in unusually high oxidation states (principally oxo anions and oxo acids) and in unusually low oxidation states (the elements themselves and the anions and hydrides of the nonmetallic elements). These two contrasting classes, of course, seldom have overlapping predominance areas in Pourbaix diagrams, so in general they react with each other. Such oxidation-reduction reactions can be extremely exothermic, giving off much more energy than is usually observed for acid-base reactions. In the case of the combustion of the hydrides of carbon (hydrocarbons) as fuels, we carry out such an oxidation-reduction reaction for the express purpose of generating energy.

Many reactions of species with nonoverlapping predominance areas, although they are quite exothermic, may not occur immediately upon contact of the reactants due to the lack of an **activation energy** that is needed to start the reaction, or of a **catalyst** that will allow it to proceed by another route that does not require the activation energy. Thus strong oxidizing and strong reducing agents may often be mixed or placed in contact with no immediate result (e.g., gasoline in contact with air). But once a catalyst or a source of the activation energy (e.g., a spark or flame) is provided, the exothermic reaction may occur so rapidly that a fire or *explosion* may occur.

EXAMPLE    Which of the following appear the most likely to show explosive properties? $Eu(N_3)_2$, $AgN_3$, $Eu(ClO_4)_2$, $AgClO_4$.

Solution    ■ These salts contain the following cations and anions, with the standard reduction potentials listed: $Eu^{2+}$, a reducing agent with $E°$ about $-0.5$ V (from Figure 5.4); $Ag^+$, an oxidizing agent with $E°$ of $+0.80$ V (from Table 5.4); $ClO_4^-$, an oxidizing agent with $E°$ of $+1.19$ V (from Table 5.1); $N_3^-$, a reducing agent with $E°$ of $-3.09$ V (Figure 5.9). The predominance area for $ClO_4^-$ (Figure 5.2) does not extend below $+1.19$ V at pH $= 0$, while that for $Eu^{2+}$ does not extend above $-0.5$ V; there is no overlap. Hence these two ions should react with each other. Of course, their reaction may be kinetically slow enough to allow the salt to be isolated. But the electrochemical potential for reaction of these two under standard conditions, given by the difference $E°_{oxidizing\ agent} - E°_{reducing\ agent} = +1.7$ V, is quite large. (A difference of greater than or near to 1 V *may* be a sufficient potential for explosive properties.)

■ Similarly, the combination of $Ag^+$ and $N_3^-$ shows a very positive overall potential of $+0.80 - (-3.09) = +3.89$ V. The salt $AgN_3$ is indeed explosive. The combination $AgClO_4$ is one of two oxidizing agents and hence should be stable—at least in the absence of dust or other external reducing agents. The salt $Eu(N_3)_2$ similarly is a combination of two reducing agents and is thus not likely to explode.

□

One of the most important reasons for seeing periodic trends in redox chemistry is to become aware of species that when mixed have the potential to inflame or explode. If you do well in learning the skills involved in this chapter, your life expectancy may increase. As a first general lesson in **chemical safety**, always take precautions when you are carrying out a reaction of a strong oxidizing and a strong reducing agent. Unless you are doing this reaction in dilute aqueous solution (which slows reactions by dilution and which has a high heat capacity), you should take such precautions as providing external cooling, allowing a vent for any gases produced, and recalling the locations of the appropriate types of fire extinguishers.

If an explosion should occur, the glass of the reaction vessel may be shattered into shards that could be hurled into your skin and eyes, so **wear safety goggles** and **do the reaction behind a safety shield**, such as the window of a hood or fume cupboard. (The hood also serves to confine flames and toxic gases that might be given off.) Be especially careful if the expected exothermic reaction does *not* start upon initial cautious addition of the first small amount of the second reagent. This means that an activation energy or catalyst is needed but is (temporarily) absent. Under these conditions heating or disturbing the system may suddenly provide the activation energy for part of the molecules; the energy released may provide it for the rest, and the result may be a violent explosion.

Unfortunately it is not possible to provide a complete list of chemicals or combinations of chemicals that give rise to explosions—chemists are discovering new ones every month, usually unintentionally. (Such new discoveries are usually promptly reported in the Letters to the Editor section of *Chemical and Engineering News*.) Of course, many combinations of oxidizing and reducing agents react smoothly at room temperature, since only a small activation energy is needed, or the reactions are only mildly exothermic. But an incomplete list of potentially hazardous combinations can be suggested here.

**1.** A salt that is the combination of a strongly reducing cation and a strongly oxidizing anion may persist for kinetic reasons but is thermodynamically unstable and an explosion hazard. From the predominance diagram for nitrogen (Figure 5.2) it can be seen that the **ammonium** ion, $NH_4^+$, has a small predominance area; its salts with strongly oxidizing oxo anions are often explosive. Thus ammonium nitrate, a common fertilizer, when ignited with the help of fuel oil is a powerful explosive. There are also many organic ammonium ions containing one to four organic R groups in place of the ammonium hydrogen ions; these are large cations that form nicely insoluble perchlorates, but these perchlorates are treacherous explosives.

**2.** The same considerations apply to the combination of a strongly oxidizing cation and a strongly reducing anion. The cations of the electronegative metals are strongly oxidizing; the azide ion and the **acetylide** ion, $C_2^{2-}$, are strongly reducing. Thus lead azide is very shock-sensitive and is used to provide the activation energy needed to detonate more stable and powerful explosives. Sodium azide is proposed for use in the automobile air-bag; the large volume of $N_2$ released by decomposing azides, of course, adds to the force of such explosions. Similarly, acetylides such as $Ag_2C_2$ and $Cu_2C_2$ are very explosive. (As we will see in Chapter 8, however, many combinations of this sort are more stable than we would predict at this point.)

**3.** Naturally, mixtures of strongly oxidizing and reducing substances present the possibility of a fire or an explosion. Such substances should be stored in separate areas of the stockroom or laboratory. Mixing is particularly easy if both substances are gases; the physical chemistry that determines whether a fire or an explosion will result has been well studied [5]. Likely gaseous reducing agents include hydrogen and many of the hydrides of the nonmetals (such as the hydrocarbons); likely gaseous oxidizing agents include $F_2$, $Cl_2$, $Br_2$, $O_2$, and the volatile oxides of the elements (Chapter 4). $F_2$ is so vigorous an oxidizing agent that most substances will catch fire or explode in it; its usefulness is correspondingly limited. Although $O_2$ is a powerful oxidizing agent, it usually requires a significant activation energy—but as classroom demonstrations of ignition of a mixture of $H_2$ and $O_2$ demonstrate, this is readily available. Explosions or fires may also occur, however, even if only one or neither reactant is a gas. Active metals can be ignited in air; finely powdered combustible solids (such as dust in grain elevators) can form explosive mixtures with air; mixtures of solids, such as gunpowder (sulfur, carbon, potassium nitrate), also can be explosive.

**4.** It is possible to combine an oxidizing group of atoms (functional group) and a reducing group in the same molecule; such substances are often explosive. Familiar explosives such as nitroglycerine, $C_3H_5(NO_3)_3$, TNT (trinitrotoluene, $C_7H_5(NO_2)_3$), and picric acid, $C_6H_3O(NO_2)_3$, contain oxidizing nitrogen groups and reducing hydrocarbon groups, and upon reaction generate not only much energy but also large volumes of gaseous oxides of carbon and hydrogen, and $N_2$. Similarly hazardous organic perchlorates can be produced by combining perchloric acid and alcohols—reactions that should never be attempted.

**5.** Finally, for reasons that will be seen in the next two chapters, compounds containing covalent bonds between different highly electronegative nonmetals are not only good oxidizing agents, but the bonds are also often intrinsically unstable. Compounds containing these bonds should be treated as potentially explosive: O—O (e.g., hydrogen peroxide, $H_2O_2$, or organic peroxides, $R_2O_2$), N—N (e.g., azides, diazonium salts), O—Cl (perchloric acid, chlorates, chlorites, and the oxides of chlorine), O—N (the nitro organics mentioned in the previous paragraph and compounds such as silver fulminate, AgONC), and even N—Cl, N—Br, and N—I ($NI_3 \cdot NH_3$ is detonated by the force of a housefly stomping its foot).

## 5.7

# Industrial Processes for Extracting Elements from Their Ores

Although humans have been mining and using metals for thousands of years and continue to discover new deposits of ores, by and large we have used up many of the ores that have the highest percent composition of the element being sought. Consequently we now have to use many ores containing very low percentages of the elements, which require increasingly elaborate chemical processes to recover the metals and nonmetals [6, 7]. Many of these processes can be divided into a few general steps, however.

Often the first step is one called **beneficiation** of the ore. The particular mineral containing the desired element may be associated with large amounts of other materials such as rock and gravel. Beneficiation concentrates the ore so that energy and chemicals do not have to be wasted treating rock and gravel. To achieve beneficiation, chemical or physical differences between the ore and the other material are exploited. For example, the principal aluminum ore, bauxite, contains $Al_2O_3$ in association with $Fe_2O_3$ and $TiO_2$. In the Bayer process the different acid-base properties of these three oxides are utilized. The bauxite is heated with 30% NaOH solution at 190°C and 8 atm pressure to cause the amphoteric oxide $Al_2O_3$ to dissolve by forming $Al(OH)_4^-$. The basic oxides $Fe_2O_3$ and $TiO_2$ do not dissolve and are filtered off. Then the solution is cooled, which reduces the tendency of the aluminum to form the $Al(OH)_4^-$ ion, so purified $Al(OH)_3$ crystallizes out and is filtered off.

Copper is now being extracted from ores containing less than 1% Cu, so beneficiation is quite important. Copper ores are often sulfides, which have different physical properties than silicates; thus detergents made from organic acids containing sulfur atoms in place of oxo groups are used to wet the copper ore but not the rock. Vigorous agitation produces a copper sulfide–containing froth that floats to the surface of the water and can be skimmed away from the water and rock.

Once the element has been concentrated, it may need to be **chemically converted** to a different salt of the metal for further processing. For example, sulfide ores are much more difficult to process than oxide ores, so they are *roasted* in air to convert the sulfide to sulfur dioxide and the oxide of the metal:

$$2\,MS(s) + 3\,O_2(g) \rightarrow 2\,MO(s) + 2\,SO_2(g) \tag{5.29}$$

The sulfur dioxide released at smelters producing such metals as copper and nickel has been a serious air-pollution problem. In the case of titanium, the chloride is more easily processed than is the oxide, so rutile is heated with chlorine and carbon:

$$TiO_2(s) + 2\,C(s) + 2\,Cl_2(g) \rightarrow TiCl_4(l) + 2\,CO(g) \tag{5.30}$$

In some modern processes of *hydrometallurgy*, the beneficiation and the conversion are carried out in one step. Thus action of air and acid on copper sulfide ore can both separate copper from rock and convert it to a more usable form:

$$2\,Cu_2S(s) + 5\,O_2(g) + 4\,H^+(aq) \rightarrow 4\,Cu^{2+}(aq) + 2\,SO_4^{2-}(aq) + 2\,H_2O \tag{5.31}$$

The third step in the production of metals is **reduction**, in which the preferred salt of the metal ion is reduced to the elemental form. This process is closely related to the activity series of the metals: A reducing agent is needed that is more active than the metal being produced. The reducing agent chosen is the cheapest one that is active enough to do the job. Correspondingly, for nonmetals occurring in negative oxidation states in nature, the cheapest effective oxidizing agent is chosen.

In the case of the very inactive metals (the electronegative metals), no reducing agent at all may be needed; these metals often occur **native** (uncombined; in the elemental form) in nature. When these metals do occur combined (often as sulfides), the heat generated by the roasting process is often sufficient to decompose the oxide to the metal (see the Ellingham diagram, Figure 5.7):

$$Cu_2S(s) + O_2(g) \rightarrow 2\,Cu(l) + SO_2(g) \tag{5.32}$$

Likewise, the very inactive nonmetals (the noble gases of Group 18/VIIIA and $N_2$)

occur uncombined in the atmosphere and do not need to be oxidized; they (and $O_2$) are obtained by fractional distillation of liquid air. Oxygen, sulfur, and carbon also occur uncombined in nature, even though they are not particularly inactive; these deposits are products of past biological processes. Since neither oxidation nor reduction is necessary, carbon and sulfur are very cheap nonmetallic elements.

Most of the moderately active metals and nonmetals (the electropositive metals and the electronegative nonmetals) are reduced from their oxides by the inexpensive reducing agent carbon (coke from coal) and heat. As we can see from Figure 5.7, the decomposition of metal oxides is favored by heating, but the decomposition of carbon dioxide (and especially carbon monoxide) is not. (CO is a gaseous oxide with an increasingly favorable entropy for *formation* at higher temperatures.) Consequently at high enough temperatures carbon can reduce a more "active" metal:

$$FeO(s) + C(s) \rightarrow Fe(l) + CO(g) \tag{5.33}$$

This reaction can occur at any temperature above which the heavy line for CO (or $CO_2$) in Figure 5.7 crosses the line for the decomposition of the metal oxide in question. In practice, furnace temperatures in excess of $2000°K$ are impractical due to energy costs and/or probable damage to the furnace lining. The metals and nonmetals that can be produced by reduction with carbon are shown in Figure 5.10. Note that they fall in a relatively narrow range of electronegativities.

Some of the electropositive metals (such as Mn and Ti), although they can in principle be produced by carbon reduction, are active enough to react with carbon itself to produce metallic carbides that contaminate the product; hence, unless the carbide is a desirable component of an alloy, another, more expensive reducing agent must be chosen (Figure 5.10). Hydrogen may be used, but many of these metals also combine with hydrogen to give metal hydrides (Chapter 10); hence a more active metal such as aluminum may be used to displace these metals from their oxides in the spectacular *thermite reaction*:

$$Cr_2O_3(s) + 2\,Al(s) \rightarrow 2\,Cr(l) + Al_2O_3(s) \tag{5.34}$$

This reaction generates so much heat that the metal is produced as a red-hot liquid.

Many of the very electropositive metals can be produced using calcium or magnesium with the halide of the element in an inert atmosphere. Some of these metals are so active that, at the temperature of the reaction, they will also react with $N_2$ to give nitrides. The production of titanium metal is carried out at $900°K$ in an argon atmosphere by the Kroll process:

$$TiCl_4(g) + 2\,Mg(l) \rightarrow Ti(s) + 2\,MgCl_2 \tag{5.35}$$

Titanium is an abundant, very desirable metal that is as strong as steel but considerably lighter, so it would no doubt be very widely used were it not for the very high expense of this process. (Titanium must also be welded in an argon atmosphere, which is difficult.) In the time of Napoleon III, aluminum was made by a similar, very expensive process. It is said that Napoleon III once entertained at a state dinner

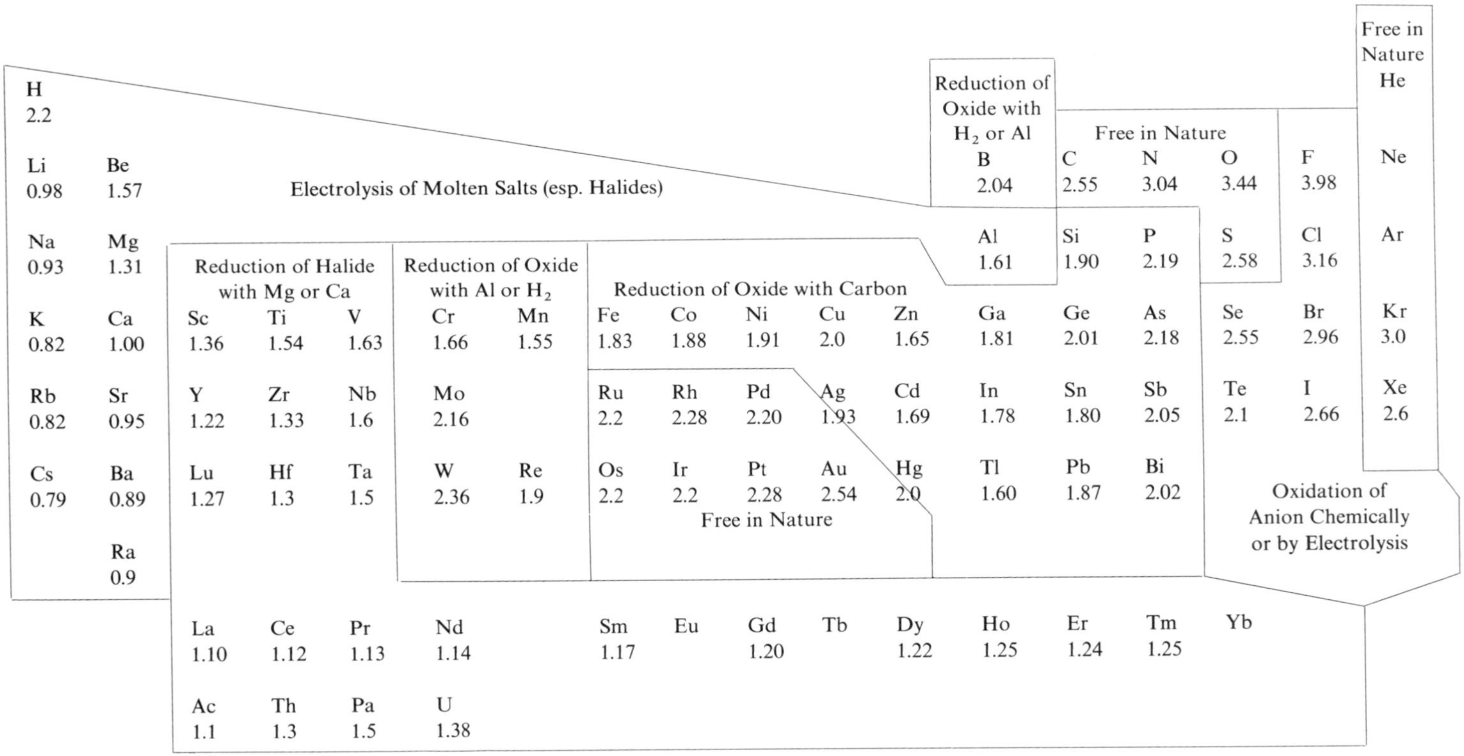

**Figure 5.10** The method of extraction (oxidation or reduction) of the elements as a function of periodic table position and electronegativity.

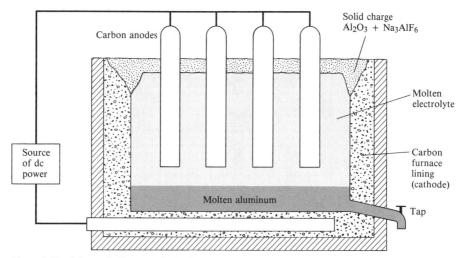

**Figure 5.11** Schematic diagram of a Hall-Heroult aluminum reduction cell. From *Environmental Chemistry*, by J. W. Moore and E. A. Moore. Copyright © 1976 by Academic Press. Reprinted with permission.

at which the lesser guests had to content themselves with eating from gold or silver plates while Napoleon III and the most important guests ate from aluminum plates!

For the most active metals, of course, no more active metal exists that can be used to displace them from their salts. These metals are produced by **electrolysis** of their fused (molten) halides (these metals are too active to be liberated by electrolysis of an aqueous solution). The most important of these metals are aluminum and sodium.

Aluminum is made by electrolysis of $Al_2O_3$, which is obtained by beneficiation and dehydration of bauxite. An oxide such as $Al_2O_3$ has too high a melting point to be used in fused form, however, so a nonaqueous solvent (not itself decomposed by electrolysis) had to be found for the $Al_2O_3$. This was no easy problem, but in 1886 Hall in the United States and Heroult in France found a suitable mineral, cryolite ($Na_3AlF_6$, found only in Ivigtuk, Greenland, but now made synthetically). The electrolysis (Figure 5.11) operates at a temperature of nearly 1000°C, a voltage of 4.5 V, and a current of 100,000 A. The furnace lining of graphite serves as the cathode, at which molten aluminum forms and is drawn off. Oxygen forms at carbon anodes, where it reacts with them to generate CO. This helps reduce the demand on electrical energy; nonetheless, the production of aluminum requires enormous inputs of electricity. Consequently the recycling of scrap aluminum (as in cans) is profitable (even in the absence of any environmental considerations), since this step is avoided.

For the most active (very electronegative) nonmetals electrolytic *oxidation* must also be used. Fluorine gas is produced by electrolysis of a molten adduct of KF and HF, using a stainless steel cathode and a carbon anode (the byproduct $H_2$ must be kept rigorously separate from the $F_2$ or violent explosions will result).

171

Chlorine gas is produced on a very large scale by electrolysis of NaCl, either fused (in which case sodium metal is also produced) or in solution (in which case $H_2$ and NaOH are the byproducts).

The less active nonmetals $Br_2$ and $I_2$ can be produced by chemical oxidation of their aqueous anions by the relatively inexpensive $Cl_2$. Bromine is obtained directly from $Br^-$ ion in seawater, despite its low concentration there, by passing $Cl_2$ through the seawater and blowing the volatile liquid $Br_2$ out with air. Iodine is similarly obtained from certain brines having a high concentration of $I^-$.

The final step in the production of most elements is **purification** of the metal (or nonmetal). One of the main contaminants for metals such as iron are residues of $SiO_2$ or polysilicates. These materials have high melting points and are difficult to remove from blast furnaces, so advantage is taken of their acidic-oxide character. The charge for a blast furnace (Figure 5.12) includes not only iron ore and coke but also limestone ($CaCO_3$), which at the temperature of the blast furnace breaks down into CaO, the cheapest basic oxide. This then combines with $SiO_2$ and the highly polymerized polysilicates to form less polymerized silicates or **slag**, which has a lower melting point as a result of the loss of polymerization. The slag (having a simplest composition of approximately $CaSiO_3$) and the iron both melt and fall to the bottom of the blast furnace, but the slag is less dense and floats on the molten iron; they can then be drawn off separately. (The CaO and coke together also help remove sulfur, a harmful impurity occurring as FeS, by the reaction

$$FeS + CaO + C \rightarrow Fe + CaS + CO \tag{5.36}$$

The CaS is also a component of the slag that is removed.)

The iron that is produced in the blast furnace, called *pig iron*, still contains more than 2% each of several impurities that make it brittle: excess carbon (as iron carbide), phosphorus (as iron phosphide), silicon, and manganese. In the production of *steel* these impurities are removed by several processes, the most modern of which is known as the *basic-oxygen* process. In this process $O_2$ is blown over or through the molten iron; the more active impurities (carbide, phosphide, Si, Mn) have a greater tendency to be oxidized to their oxides than iron, although some iron oxides are also produced. Since most of these oxides are acidic, limestone is again added as a basic-oxide source to form more slag, which melts and is poured off the molten steel.

The purification or *refining* of copper is also important, since it must be quite pure for electrical purposes and because its impurities include metals such as silver and gold, which are even more valuable than the copper. Impure copper is cast into plates that are used as the anode in an electrolytic cell. A controlled voltage is applied, which is sufficient to oxidize copper to $Cu^{2+}$ and other more active metals to their cations but which does not oxidize the less active metals such as Ag, Au, and the platinum metals. These fall to the bottom of the electrolytic cell as *anode mud* or *slimes* and are recovered. The cathode is a sheet of highly purified copper; at the controlled voltage only $Cu^{2+}$ is active enough to be reduced back to the metal. (The $Fe^{2+}$, $Ni^{2+}$, etc., remain in the solution as cations.)

Perhaps we ought to add another stage to the process of extraction of elements

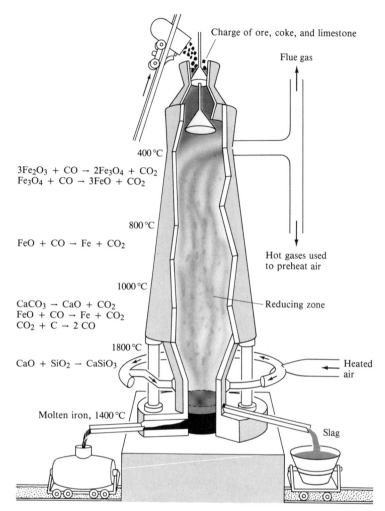

Charge of ore, coke, and limestone

Flue gas

400 °C

$$3Fe_2O_3 + CO \rightarrow 2Fe_3O_4 + CO_2$$
$$Fe_3O_4 + CO \rightarrow 3FeO + CO_2$$

800 °C

$$FeO + CO \rightarrow Fe + CO_2$$

Hot gases used to preheat air

1000 °C

$$CaCO_3 \rightarrow CaO + CO_2$$
$$FeO + CO \rightarrow Fe + CO_2$$
$$CO_2 + C \rightarrow 2 CO$$

Reducing zone

1800 °C

$$CaO + SiO_2 \rightarrow CaSiO_3$$

Heated air

Molten iron, 1400 °C

Slag

**Figure 5.12** Schematic diagram of a blast furnace, showing temperatures and chemical reactions occurring in different parts of the furnace. Adapted from *Modern Descriptive Chemistry* by E. G. Rochow. Copyright © 1977 by W. B. Saunders Company. Reprinted by permission of CBS College Publishing.

from their ores, that of **pollution control**. As you can appreciate on rereading the descriptions of the processes of beneficiation, conversion, reduction, and purification, many kinds of impurities form that can be solid, water, and air pollutants. The design of emission controls and waste disposal processes [6, 7] plays an increasingly important part in the economics of the production of the elements.

### 5.8

## Study Objectives

1. Balance oxidation-reduction equations and half-reaction equations (review from general chemistry).

2. Know the general periodic trends in the stability of high oxidation states of elements; classify high-oxidation-state oxo acids and anions as to relative strength as oxidizing agents.

3. Suggest methods of synthesizing given oxo acids or oxo anions.

4. Interpret Pourbaix diagrams; use them to suggest the forms in which given elements will occur in natural waters.

5. Use Pourbaix diagrams to identify forms of an element that will (at appropriate pH's) (a) oxidize water, (b) reduce water, or (c) disproportionate.

6. By comparing Pourbaix diagrams of different elements, predict whether a given combination of the cation of one element and the anion of another element will give rise to a redox reaction instead of a precipitate or a soluble salt; give possible products of such a reaction.

7. Predict the results of the reaction of any metal with cold water, cold HCl, hot HCl, or the solution of the ion of another metal.

8. Classify nonmetallic elements as to relative strength as oxidizing agents.

9. Name and draw the structures of the pseudohalide ions.

10. Name the monoatomic anions of the nonmetals and their protonated forms.

11. Classify the monoatomic anions of the nonmetals as to their relative strength as (a) bases and (b) reducing agents.

12. Recognize compounds likely to be flammable, oxidizing, or explosive; know appropriate safety precautions to take in handling these compounds.

13. Give a feasible and economically preferable method of preparing any element from the form in which it occurs in nature.

## 5.9

**Exercises**

*1. For which Group 15/VA element is the $+5$ oxidation state most stable?

2. For which Group 17/VIIA element is the $+7$ oxidation state least stable?

*3. For which of the following fourth period elements is the group oxidation state least stable? Zn, Ga, Ge, As, Se, Br.

4. Choose the best oxidizing agent in each set of ions:
a. $MnO_4^-$, $TcO_4^-$, $ReO_4^-$;    b. $GeO_4^{4-}$, $AsO_4^{3-}$, $SeO_4^{2-}$, $BrO_4^-$;
c. $SO_4^{2-}$, $SeO_4^{2-}$, $TeO_6^{6-}$, $PoO_6^{6-}$;
d. $NO_3^-$ in acid solution, $NO_3^-$ in basic solution, $NO_3^-$ in neutral solution;
e. $CrO_4^{2-}$, $MoO_4^{2-}$, $WO_4^{2-}$.

*5. For which two Group 15/VA elements is the $+5$ oxidation state least stable?

6. For which Group 15/VA element X should the following reaction proceed most readily?

$$4X + 5O_2 + 12OH^- \rightarrow 4XO_4^{3-} + 6H_2O$$

*7. The most suitable pH at which to carry out the synthesis of the $FeO_4^{2-}$ ion is which? 0, 7, or 14.

8. The most suitable pH at which to use the $FeO_4^{2-}$ ion as a powerful oxidizing agent is which? 0, 7, or 14.

*9. Choose from each set of oxo anions the one that requires the most strongly oxidizing conditions for synthesis:
  9.1 $FeO_4^{2-}$, $RuO_4^{2-}$, $OsO_6^{6-}$
  9.2 $GeO_4^{4-}$, $AsO_4^{3-}$, $SeO_4^{2-}$, $BrO_4^{-}$
  9.3 $SeO_4^{2-}$, $TeO_6^{6-}$, $PoO_6^{6-}$

10. The standard reduction potentials for the oxo anions $FeO_4^{2-}$, $RuO_4^{2-}$, and $OsO_6^{6-}$ are $+0.96$ V, $+1.40$ V, and $+2.20$ V, but not necessarily in that order. **a.** Which standard reduction potential belongs to which ion? **b.** Which of these ions would be the most powerful oxidizing agent? **c.** Write balanced chemical equations showing good methods of synthesis of each of these oxo anions or acids from the oxide of each element in its most common positive oxidation step (or from the element itself).

†11. According to the Pourbaix diagrams: **a.** What is the most oxidizing chemical form of cobalt? **b.** What is the most strongly reducing form of selenium in acidic solutions? **c.** Which $+2$ ion of the $f$-block metals from the sixth period is least strongly reducing? **d.** What is the most strongly oxidizing form of iridium? **e.** Which trioxide of the $f$-block metals from the seventh period is most strongly oxidizing? **g.** Which elements from the periodic table have only one stable oxidation state in aqueous solution?

12. Figure 5.4 shows the $E°/pH$ (Pourbaix) diagram for uranium. You are attempting to study the fate of waste uranium (from nuclear fuel reprocessing) in a natural lake.
  **12.1** If the lake is well aerated and not polluted, in what form will the uranium be found? Will it be present in solution or found in the sludge at the bottom of the lake?
  **12.2** If the lake is well aerated but is highly contaminated with acid rain (pH = 3), in what form will the uranium be found? Will it be present in solution or found in the sludge at the bottom of the lake?
  **12.3** Suppose some metallic uranium is dropped into this lake (with pH = 3), but it falls to the bottom of the lake, where oxygen is absent and reducing impurities (from decaying organic matter) are present. Will it remain as elemental uranium? If not, write a balanced chemical reaction for what will happen and describe its vigor.

13. Figure 5.3 shows the Pourbaix diagram for manganese. People often find that clear water drawn from wells deposits a black manganese-containing solid on standing in their toilet bowls. Explain what that solid is and why it forms in the toilet bowl and not underground.

*14. Figure 5.4 shows the Pourbaix diagram for plutonium. You are attempting to study the fate of waste plutonium (from atomic bomb assembly) in a lake.
  **14.1** If the lake is well aerated and of normal pH, in what chemical form will

the plutonium be found? Will it be present in solution or found in the sludge at the bottom of the lake?

14.2 Will metallic plutonium dumped into the lake remain in the metallic form or will it react with the water? If the latter, write a balanced chemical equation showing the reaction with the water.

14.3 Will any corrosion of the metallic plutonium by the water be more severe in acidic or in basic solution?

14.4 If the lake is anaerobic and highly polluted with acid rain, in what chemical form will the plutonium be found? Will it be present in solution or found in the sludge at the bottom of the lake?

†15. Referring to Pourbaix diagrams, for each of the oxo anions listed below, choose the best of the following three synthetic approaches: (1) The anion must be prepared in strongly basic solution by electrolysis or by use of a powerful oxidizing agent such as $Cl_2$, $F_2$, $O_2$, $O_3$, or $H_2O_2$; (2) the anion can be prepared in acidic solution (as an oxo acid) by the action of an oxidizing agent such as $HNO_3$; (3) the anion is likely to be prepared by reduction of a higher oxo anion or by the action of base on the element itself; (4) the oxo anion does not exist. Oxo anions: **a.** $KrO_4^{2-}$; **b.** $MnO_4^{2-}$; **c.** $Cr(OH)_6^{3-}$; **d.** $CrO_4^{2-}$; **e.** $IO_3^-$.

16. Using the oxidizing agent of your choice, write a balanced chemical equation for the synthesis of each real oxo anion from the previous question from an oxide of the element in a common oxidation state, or from the element itself.

†17. Examine the Pourbaix diagrams for the elements Am, Cr, Mn, Bi, and I. If one exists, find a chemical form of each of these elements that, at some pH, would **a.** disproportionate, **b.** oxidize water, and **c.** reduce water.

18. Arrange the following metals in an activity series: K, Au, Cu, Mg, Mn, Zn, Fe, Ni.

*19. Arrange in an activity series all the metals whose names begin with the letter *T*.

†20. Describe the activity of each of the following elements with (1) cold water; (2) cold HCl solution; (3) hot HCl solution; (4) concentrated $HNO_3$:
**a.** La; **b.** Pt; **c.** Co; **d.** Sc; **e.** Os; **f.** Sr; **g.** Cr; **h.** In; **i.** Ir; **j.** Nd; **k.** Cu; **l.** Zn; **m.** Sn; **n.** Ba.

21. Consider the following metals: Au, Be, Bi, Ce, Fe, Ga, Ir, Pu. List all the metals that will: **a.** react with water; **b.** react with HCl but not with water; **c.** not react either with HCl or with water; **d.** react with $Ag^+$ to give Ag metal; **e.** react with $Na^+$ to give Na metal; **f.** react with $I^-$ to give $I_2$.

22. Consider the following metals: Ba, Lu, Os, Ti, Bi, Sn, Cu, Au, Pu. **a.** Which of these will react with water? **b.** Which of these will react with HCl but not with water? **c.** Which of these will dissolve only in $HNO_3$ or will not dissolve in any of the above?

23. Consider the following four anions: $CH_3^-$, $C^{4-}$, $F^-$, $I^-$. **a.** Which is the least basic? **b.** Which is the most strongly basic? **c.** Which is the poorest reducing agent? **d.** Give the names of each of these anions.

24. Name and draw the Lewis dot structures of the monoatomic and pseudohalide ions and their conjugate acids.

*25. Which of the following elements will react (1) with a solution of $I^-$ to generate $I_2$; (2) with a solution of $Ag^+$ to generate Ag; (3) with neither? **a.** Zn; **b.** $Cl_2$; **c.** $At_2$; **d.** Au.

26. The cations of silver show the following standard reduction potentials:

$$Ag^+ + e^- = Ag(s) \qquad E^\circ = +0.799 \text{ V}$$
$$Ag^{2+} + e^- = Ag^+ \qquad E^\circ = +1.98 \text{ V}$$

   26.1 Is the $Ag^+$ ion more likely to act as a good oxidizing agent or a good reducing agent?

   26.2 Which one of the following is most likely to be capable of converting $Ag^+$ to $Ag^{2+}$? $F_2$, $F^-$, $I_2$, or $I^-$.

   26.3 Given that the $BrO_4^-$ ion has a standard reduction potential of $+1.743$ V (when being reduced to $BrO_3^-$), is $AgBrO_4$, if it exists, likely to be explosive?

   26.4 Given that the standard reduction potential for the reaction $H_2 + 2e^- = 2H^-$ is $-2.25$ V, is AgH, if it exists, likely to be explosive?

†27. For each of the following salts, tell (1) if it will be thermodynamically stable in the sense that its cations and anions will not undergo redox reactions with each other; (2) if it will be so unstable that, if it can be made at all, it will be potentially explosive: **a.** $CrFeO_4$; **b.** $NH_4NO_3$; **c.** $CsMnO_4$; **d.** $NH_4MnO_4$; **e.** $EuMnO_4$; **f.** $NH_4ClO_4$; **g.** $PuTe_2$; **h.** YbTe; **i.** $Ca_2C_2$; **j.** $Ag_2C_2$; **k.** $AgClO_4$; **l.** $Cr(BrO_3)_2$; **m.** $Tl(N_3)_3$.

28. Give one example of an explosive (or potentially explosive) compound in each of the following classes: **a.** a compound containing both strongly oxidizing and strongly reducing functional groups; **b.** a salt containing a strongly oxidizing cation and a strongly reducing anion; **c.** a compound containing unstable covalent bonds; **d.** a salt containing a strongly reducing cation and a strongly oxidizing anion.

29. Which would most likely be an explosive compound?
   **a.** methyl cyanate, $CH_3OCN$; **b.** methyl isocyanate, $CH_3NCO$; or **c.** methyl fulminate, $CH_3ONC$.

30. Give the symbols of three elements for which each of the following is the preferred commercial method of preparation: **a.** occur native (in the elemental form); **b.** prepared by electrolytic or chemical oxidation of their anions; **c.** prepared by reduction of their oxides or oxo salts with carbon;

**d.** prepared by reduction of their halides with a more active metal; **e.** prepared by electrolytic reduction of its cation in a fused-salt medium.

31. Using balanced chemical equations, explain **a.** why the element carbon (and no other) is so much better a reducing agent at higher temperatures than it is at room temperature; **b.** how copper is separated both from more active and from less active metals by electrolytic refining.

32. Explain **a.** why bauxite is found in Guyana rather than, say, Mauritania or Spitzbergen; and **b.** why aluminum metal is produced in Norway and Canada rather than Guyana.

33. Explain the function of each of the following processes in the production of a metal, and give an example for each: **a.** beneficiation of the ore; **b.** chemical conversion of the ore; **c.** reduction of the ore; **d.** purification of the metal.

†34. What is the purpose (beneficiation, chemical conversion, reduction, or purification) of each of the following chemical reactions in producing the metal specified?

34.1 Titanium: $TiO_2 + 2\,Cl_2 + 2\,C \rightarrow TiCl_4 + 2\,CO$

34.2 Aluminum: $2\,Al_2O_3 + 6\,C + \text{electrical energy} \rightarrow 4\,Al + 6\,CO$

34.3 Aluminum: $OH^- + (Fe_2O_3, Al_2O_3, TiO_2) \rightarrow$
$Fe_2O_3(s) + TiO_2(s) + Al(OH)_4^-(aq)$

34.4 Nickel (Mond process): $(Fe, Co, Ni) + CO \rightarrow$
$Fe(s) + Co(s) + Ni(CO)_4(g)$

34.5 Copper: $(Fe, Cu, Ag, Au) + \text{electrolytic oxidation} \rightarrow Fe^{2+}(aq) +$
$Cu^{2+}(aq) + Ag(s) + Au(s)$; solution + electrolytic reduction $\rightarrow$
$Cu(s) + Fe^{2+}(aq)$

35. With the help of other sources such as [6] and [7], describe and write chemical equations for as many of the four processes in the production of a metal as apply to the production of the following: **a.** steel;     **b.** aluminum; **c.** copper;     **d.** titanium.

36. Gold will not dissolve in sulfuric acid but will dissolve in concentrated selenic acid. Explain why.

37. If you did not perform Experiment 4, go back to it now. Describe what would have happened during the experiment, and answer the questions in part E.

---

Notes

[1] Standard reduction potentials reflect not only the stability of the higher oxidation state but also of the reduced state, so their values vary depending on the reduced state chosen. Thus a boundary such as +1.2 V can only be a rough guideline.

[2] Brauer, G., ed., *Handbook of Preparative Inorganic Chemistry*, 2 vols., 2d ed., Academic Press, New York, 1963.

[3]   Pourbaix, M., *Atlas of Electrochemical Equilibria in Aqueous Solutions*, Houston, National Association of Corrosion Engineers, 1974.

[4]   The pE scale is intended to represent the concentration of the standard reducing agent, the electron, as the pH scale represents the concentration of the standard acid, the hydrogen ion. pE values are obtained from reduction potentials by dividing by 0.059.

[5]   Jolly, W. L., *Modern Inorganic Chemistry*, McGraw-Hill, New York, 1984, p. 155.

[6]   Fergusson, J. E., *Inorganic Chemistry and the Earth: Chemical Resources, Their Extraction, Use, and Environmental Impact*, Pergamon, Oxford, 1982, pp. 82–106.

[7]   Moore, J. W., and E. A. Moore, *Environmental Chemistry*, Academic Press, New York, 1976, Ch. 11.

# Properties of the Elements Themselves

Some textbooks of descriptive inorganic chemistry begin with a study of the elements themselves. We have postponed this until now because the elements have such diverse properties. We will begin with some physical properties and, just as in Chapter 4 with oxides, explain these in terms of the degree of polymerization of atoms present in the structures of the molecules of the elements. A wide variety of structures will be apparent, which we will then explain in terms of the covalent bonding and other properties of the atoms in the periodic table. Finally, we will use the atomic properties of the elements introduced in this chapter to explain some of the periodic trends in the redox chemistry of the elements that we studied in the last chapter.

## 6.1

## Physical States and Heats of Atomization of the Elements

The elements show an enormous diversity in their simplest physical properties, such as their physical states and the ease with which they become gases or individual atoms. At one extreme, helium is a gas (at 1 atm pressure) at any temperature in excess of 4 °K. At the other extreme, tungsten metal does not become a gas until a temperature of 5936 °K is reached. Tungsten is so involatile at ordinary temperatures that, according to computations, if the entire universe consisted of tungsten at room temperature, *one atom* of tungsten would be in the vapor phase! These differences are rooted in the structures of the elements. Molecules of helium consist of individual atoms, while tungsten, other metals, and many nonmetals are polymerized at all temperatures below their boiling points.

Before we look at the physical properties of the elements individually, it is useful to look at the overall periodic trends. Two useful properties are presented here. One is the enthalpy change for the process by which the element in its usual form at room temperature (and 1 atm pressure) is converted to a gas. This is known as the **heat of vaporization** of the element. The other is the enthalpy change for the process, by which the element in its usual form at room temperature (and 1 atm pressure) is converted to *individual gaseous atoms*. This latter is known as the **heat of atomization** of the element and is the property we present in Table 6.1. The two enthalpies are identical if the element in the gaseous state consists only of single atoms and not of small molecules. This is definitely not the case for $H_2$, $N_2$, $O_2$, $F_2$, $Cl_2$, $Br_2$, and $I_2$, of course. For most of the other elements this is nearly the case, so for most elements there is only a minor difference in the two quantities.

In Table 6.1 we note some periodic trends that are quite different from any we have seen before. In crossing the *s*- and *p*-blocks in the second and third periods, we find the highest atomization energies in the *middle* of the period: Carbon and silicon are the hardest elements to atomize or vaporize in their periods. Likewise, in crossing the *s*- and *d*-blocks in the fifth and sixth periods, the highest atomization energies are also in the middle, at or near Group 6.

In a pattern that we have seen before, the vertical trends are opposite in the *p*- and in the *d*-blocks. Atomization energies are highest for elements at the *top* of the *p*-block (carbon in Group 14/IVA), while they are highest at the *bottom* of the *d*-block (tungsten in Group 6).

We will examine the reasons for these trends at several levels. First we will look at the structures of the elements. We already know from Chapter 4 that ease of vaporization depends on the degree to which a molecule is polymerized, so we may expect and do find quite a variety of structures among the elements. Indeed some of the elements exist in more than one structural form: These structural forms are known as **allotropes**. There are only a few structural forms for all metals, but the heats of atomization of metals still vary enormously. Thus we will need to examine the periodic trends in the strengths and types of the covalent and metallic bonds that hold the atoms of the elements together as larger molecules.

## 6.2

## Physical Properties, Allotropes, and Uses of the Elements

In this section we will examine the structures and some physical properties of the nonmetallic elements themselves, including their allotropes, and the uses of these elements [1]. Figure 6.1 summarizes briefly the degree to which the nonmetallic elements (in their various allotropes) are polymeric or consist of small molecules or individual atoms.

GROUP 18/VIIIA    The elements of Group 18/VIIIA are monoatomic. The individual atoms are attracted to each other only by the very weak van der Waals forces; hence these elements are gases at room temperature, with boiling points that increase as van der Waals forces increase down the periodic table (He, b.p. $-269\,°C$; Ne, $-246\,°C$; Ar, $-186\,°C$; Kr, $-152\,°C$; Xe, $-107\,°C$; Rn, $-62\,°C$). Although He and Ne are abundant in the universe, their low molecular weights allow them readily to achieve the velocity necessary to escape the earth's atmosphere; only Ar is abundant in our atmosphere (1%). It is often used when a very inert atmosphere is

**Table 6.1  Heats of Atomization of the Elements (kJ/mol)**

| | 1 | 2 | 3 | 4 | 5 | 6 | 7 | 8 | 9 | 10 | 11 | 12 | 13/IIIA | 14/IVA | 15/VA | 16/VIA | 17/VIIA | 18/VIIIA |
|---|---|---|---|---|---|---|---|---|---|---|---|---|---|---|---|---|---|---|
| 1 | H 218 | | | | | | | | | | | | | | | | | He 0 |
| 2 | Li 159 | Be 324 | | | | | | | | | | | B 563 | C 717 | N 473 | O 249 | F 79 | Ne 0 |
| 3 | Na 107 | Mg 146 | | | | | | | | | | | Al 326 | Si 456 | P 315 | S 279 | Cl 122 | Ar 0 |
| 4 | K 89 | Ca 178 | Sc 378 | Ti 471 | V 515 | Cr 397 | Mn 283 | Fe 415 | Co 426 | Ni 431 | Cu 338 | Zn 131 | Ga 277 | Ge 377 | As 303 | Se 227 | Br 112 | Kr 0 |
| 5 | Rb 81 | Sr 165 | Y 423 | Zr 605 | Nb 733 | Mo 659 | Tc 661 | Ru 652 | Rh 556 | Pd 377 | Ag 285 | Cd 112 | In 244 | Sn 302 | Sb 262 | Te 197 | I 107 | Xe 0 |
| 6 | Cs 76 | Ba 182 | Lu a414 | Hf 621 | Ta 782 | W 860 | Re 776 | Os 789 | Ir 671 | Pt 564 | Au 368 | Hg 64 | Tl 182 | Pb 195 | Bi 207 | Po 142 | At | Rn 0 |

| | 3f | 4f | 5f | 6f | 7f | 8f | 9f | 10f | 11f | 12f | 13f | 14f | 15f | 16f |
|---|---|---|---|---|---|---|---|---|---|---|---|---|---|---|
| 6 | La 423 | Ce 419 | Pr 356 | Nd 328 | Pm 301 | Sm 207 | Eu 178 | Gd 398 | Tb 389 | Dy 291 | Ho 301 | Er 317 | Tm 232 | Yb 152 |
| 7 | Ac a293 | Th 575 | Pa a481 | U 482 | Np a337 | Pu 352 | Am a239 | Cm | Bk | Cf | Es | Fm | Md | No |

SOURCES: Heats (enthalpies) of atomization of the s- and d-block elements were taken from W. L. Jolly, *Modern Inorganic Chemistry*, McGraw-Hill, New York, 1984, p. 292; those of the d-block elements were taken from W. W. Porterfield, *Inorganic Chemistry: A Unified Approach*, Addison-Wesley, Reading, Mass., 1984, p. 84; those of the f-block elements are from N. N. Greenwood and A. Earnshaw, *Chemistry of the Elements*, Pergamon, Oxford, 1984.

NOTE: Values preceded by [a] are enthalpies of *vaporization*, which are normally slightly less than true heats of atomization, since the metals vaporize in part as diatomic or polyatomic molecules.

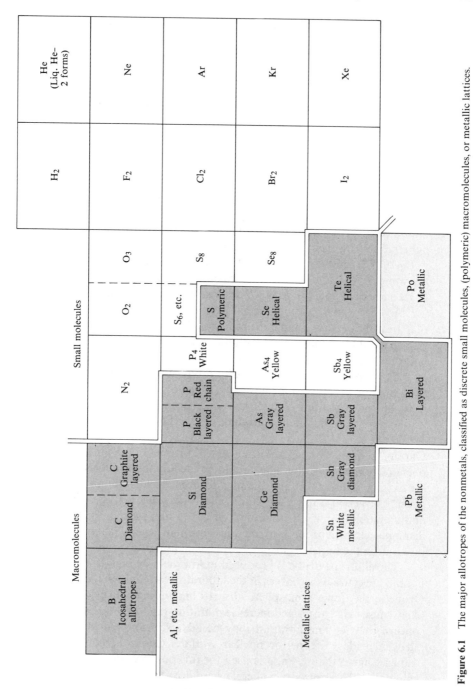

**Figure 6.1** The major allotropes of the nonmetals, classified as discrete small molecules, (polymeric) macromolecules, or metallic lattices.

needed (as for welding early *d*-block metals). Neon is used in neon lights, in which, upon electrical excitation, colored light is emitted.

Helium, despite its rarity on earth (it is principally obtained from certain natural gas deposits), is a very valuable material for low-temperature (cryogenic) research due to its low boiling point. Helium as a liquid is also unique in that it has two liquid phases, helium I and helium II. The latter is called a *superfluid*: It has zero viscosity and wets all surfaces, so that it has the ability to flow spontaneously out of a beaker into which it has been placed. It has many other strange properties that result from the especially strong manifestation of quantum-mechanical properties at low temperatures in light particles such as He atoms.

HYDROGEN AND GROUP 17/VIIA   The halogens consist of diatomic molecules; hydrogen also shares this property and is conveniently discussed here, too. These diatomic molecules are held together principally by van der Waals forces, so the lighter ones are gases ($H_2$, colorless, with a b.p. of $-253\,°C$; $F_2$, pale yellow, $-188\,°C$; $Cl_2$, greenish yellow, $-34\,°C$). As the molecular weights increase, however, the van der Waals and other weak covalent bonding interactions build up, so that $Br_2$ is a dark red liquid (b.p. $59\,°C$), whereas $I_2$ is a purple solid that sublimes readily.

As we saw in Chapter 5, the halogens (not $H_2$) are very active oxidizing agents that are quite corrosive and must be handled with some care. Bromine, for example, gives painful burns on contact with the skin; these burns take months to heal. The halogens are used commercially as oxidizing agents to prepare halogenated compounds. Fluorine is principally used to produce $UF_6$ for nuclear power production (by oxidation of $UF_4$). Chlorine is used on a very large scale to produce chlorocarbons by the oxidation of hydrocarbons. Bromine is similarly used to produce bromocarbons, and iodine to produce iodocarbons.

GROUP 16/VIA   Among the elements of Group 16/VIA we first encounter the phenomenon of allotropy. The double-bonded diatomic allotrope $X_2$ is stable only for oxygen, although $S_2$ and the like do occur at high temperatures in the vapor phase. Oxygen also has a less stable allotrope, **ozone** ($O_3$), which involves a single and a double bond. $O_2$ is now often called **dioxygen** by inorganic chemists to distinguish it from ozone and from the element in general; atoms of oxygen O, which do occur in the upper atmosphere, are called **atomic oxygen**.

The allotrope ozone is produced in low yield by high-energy processes acting upon $O_2$ (electrical discharges in the laboratory; sunlight in the upper atmosphere). Ozone in the upper atmosphere absorbs high-energy ultraviolet light that would likely cause skin cancer and increase mutation rates at the earth's surface. There is concern that its concentration may be adversely affected by pollutants such as nitrogen oxides, Freons, or nuclear warfare. Ozone is a much more powerful oxidizing agent than $O_2$ and is used for this purpose. But it is an undesirable air pollutant in the lower atmosphere, where it is irritating, attacks materials such as rubber, and reacts with other air pollutants such as unburned hydrocarbons to generate irritating pollutants such as PAN (Section 4.5).

The stable allotropes of the heavier Group 16/VIA elements involve two single bonds to each atom. Hence (as noted in Chapter 4 in our discussion of oxides) they may occur as cyclic oligomers or as linear (one-dimensional) polymers.

The stable allotrope of sulfur is a yellow solid of m.p. $119\,°C$ that contains the cyclic $S_8$ molecule (cyclooctasulfur, Figure 6.2). There are a number of additional cyclic allotropes such as $S_6$, $S_7$, $S_9$, $S_{10}$, $S_{11}$, $S_{12}$, $S_{18}$, and $S_{20}$. Above $160\,°C$, liquid $S_8$ shows some remarkable changes. S—S bonds in the rings begin to break, giving rise to open chains with unpaired electrons at each end, which join together to give longer and longer one-dimensional polymers. The chains may exceed 200,000 sulfur atoms in length at $180\,°C$. As a result, the color darkens and the liquid becomes quite viscous, since the long chains cannot flow effectively. Above $195\,°C$, entropy effects favor shorter chains, and the liquid decreases in viscosity up to its boiling point of $444\,°C$. If the hot liquid is rapidly cooled, various polymeric allotropes can be frozen out; these involve helical chains of sulfur atoms (Figure 6.2). If gaseous sulfur is heated sufficiently, entropy effects prevail and the small-molecule analogues of the allotropes of oxygen finally arise: cherry red $S_3$ and blue-violet $S_2$.

The allotropy of selenium is less complex: Red selenium contains $Se_8$ rings like those of $S_8$; the stable gray "metallic" form of selenium contains one-dimensional polymeric helical chains. Black (commercial) Se contains various sizes of large rings up to about $Se_{1000}$. Tellurium is still simpler, having only one allotrope, analogous to gray selenium. Polonium adopts a metallic lattice; thus the Group 16/VIA elements show almost the whole range of allotropic forms to be found among all the nonmetals (Figure 6.1).

Oxygen and sulfur are used industrially far more than the other Group 16/VIA elements. The largest use of pure oxygen is in the basic-oxygen process for purifying iron in producing steel; ozone is used in some organic syntheses and in one process for the purification of drinking water. Most sulfur is used to make sulfuric acid. Selenium is used to decolorize glass; higher concentrations give pink or brilliant red glasses. In accord with its more nearly metallic nature, tellurium is principally used in alloys.

GROUP 15/VA    The triple-bonded diatomic molecule $N_2$ constitutes 79% of the earth's atmosphere and is readily obtained from it by liquefaction of air followed by fractional distillation (b.p. of $N_2 = -195\,°C$). In contrast, the diatomic form $X_2$ for the heavier Group 15/VA elements is found only at high temperatures and low pressures in the vapor phase; allotropes having larger single-bonded molecules are favored under normal conditions.

The simplest phosphorus allotrope is known as white phosphorus, and consists of tetrahedral $P_4$ molecules (Figure 6.2). The bond angles in this tetrahedron are abnormally small ($60°$) and strained, which in part accounts for the fact that this allotrope is thermodynamically unstable and very much more reactive than the others (e.g., it ignites spontaneously in air). The two other major allotropes are known as red phosphorus and black phosphorus; the former is a one-dimensional polymer in which one of the tetrahedral P—P bonds has opened up and is replaced by a P—P bond that is external to the tetrahedron (Figure 6.2). This relieves strain and imparts greater stability and resistance to oxidation than is found in white

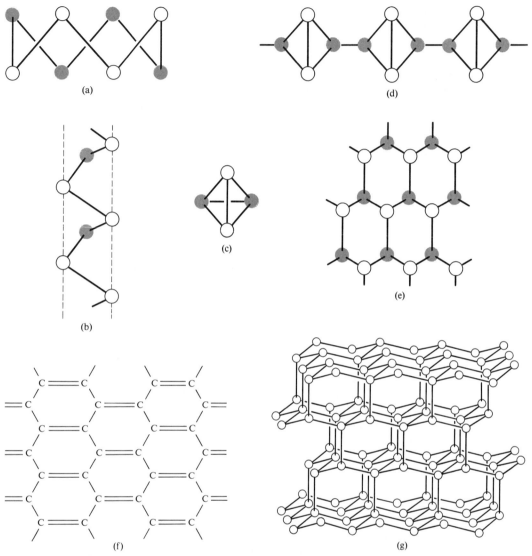

**Figure 6.2** The structures of some of the allotropes of the nonmetals. Shaded circles represent atoms below the plane of the paper, whereas open circles represent those above the plane of the paper. (a) $S_8$; (b) the helical structure of gray Se; (c) $P_4$ (white phosphorus); (d) red phosphorus (in one possible conformation); (e) black phosphorus (rhombohedral form); (f) graphite (one resonance structure); (g) diamond. (f) and (g) adapted from F. A. Cotton and G. Wilkinson, *Basic Inorganic Chemistry*. Copyright © 1976 by John Wiley & Sons, Inc. Reprinted by permission.

phosphorus. As we may expect, it also increases the melting point (44 °C for white; about 600 °C for red) and lowers its solubility in nonpolar solvents. It also greatly decreases its toxicity: Red phosphorus is essentially nontoxic, while white phosphorus is dangerously toxic, even by absorption through the skin. The two main allotropes called *black phosphorus* contain puckered sheets of phosphorus atoms (Figure 6.2); consistent with such structures, black phosphorus is flaky, with a graphitelike appearance.

Allotropy is less extensive in As, Sb, and Bi than in P. Tetrahedral $As_4$ and $Sb_4$ exist but are not thermodynamically stable at room temperature; the stable modifications of these three elements involve two-dimensional layer structures related to that of black phosphorus. In accord with the polymeric structures, these allotropes have rather high melting points (816 °C for As, which sublimes at 615 °C; 631 °C for Sb; but only 271 °C for Bi).

The rather inert $N_2$ is used principally as an inert atmosphere to prevent oxidation during metallurgical and chemical processes. $P_4$ finds its principal use in the production of high-purity phosphoric acid (Section 4.5.5). The more metallic As, Sb, and Bi are used in alloys.

GROUP 14/IVA    The stable form of carbon is graphite, in which the atoms participate in both single and double bonds, and are linked into a two-dimensional polymeric sheet structure based on six-membered rings (Figure 6.2). This structure gives graphite its soft, flaky nature, and makes it useful as a lubricant due to the absence of bonding between the layers. Only slightly less stable thermodynamically is the other major allotrope, diamond, which is a three-dimensional polymer involving only single bonds. Since diamond is bonded in all dimensions, it is the more dense allotrope (3.514 g/cm³ versus 2.266 g/cm³ for graphite); hence, although graphite is the more stable allotrope at normal pressures, under very high pressures the more dense form, diamond, is favored and formed. The "melting point" of both forms is very high (above 4000 °C) due to the necessity of rupturing so many C—C bonds; at high temperatures depolymerization to small gaseous carbon molecules actually occurs.

Silicon and germanium essentially have only one allotrope each, with the diamond structure. These have lower melting points than diamond (1420 °C for Si, 945 °C for Ge). Tin, on the other hand, has two allotropes: Its *gray* form, which has the diamond structure, is stable below 13 °C, while its *white* form, stable above that temperature, has a metallic lattice. The much less dense gray allotrope forms and crumbles on prolonged exposure of very pure tin to temperatures below 13 °C; this "tin disease" has been a very troublesome phenomenon in the tin pipes of old organs in European cathedrals. Continuing the type of periodic trend we have seen before (Figure 6.1), lead shows only a metallic form.

Carbon is used industrially in many forms. Graphite is used in making steel; diamond is used for its hardness in cutting and polishing. For many uses the less pure amorphous forms are more cost effective. Coke, made by heating coal, is used in steel manufacture. Carbon black or soot, made by incomplete burning of hydrocarbons, is used to strengthen rubber in tires. Activated carbon is used to decolorize

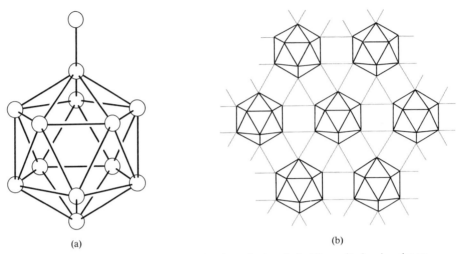

**Figure 6.3** (a) The structural unit of boron chemistry, the icosahedral $B_{12}$ unit, showing also an additional atom bonded to one of the icosahedral boron atoms. (b) The connections (in two dimensions only) of boron icosahedra in the alpha-rhombohedral allotrope of boron. Reprinted with permission from N. N. Greenwood and A. Earnshaw, *Chemistry of the Elements*, copyright © 1984 by Pergamon Press.

substances such as sugar. Silicon and germanium are used in the manufacture of transistors. Tin and lead have the applications of metals: tin in tin plate and alloys (such as brass); lead in automobile batteries.

BORON   Elemental boron forms numerous very complex allotropes, which are all exceptionally difficult to prepare pure. All these allotropes feature a structural unit very characteristic of boron chemistry: the icosahedron of 12 boron atoms (Figure 6.3a). In the different allotropes of boron these icosahedra are linked to each other in different ways to form three-dimensional polymers that have high melting points (e.g., 2300 °C) and high atomization energies, and which are very hard and quite resistant to chemical attack. Boron is used in fiber and filament form for certain uses in which high strength and light weight are very important (military aircraft, space shuttles, bicycle frames). One of the boron allotropes is shown (in two dimensions only) in Figure 6.3b.

## 6.3

# Electron Configurations, Covalent Bonding, and the Physical Properties of the Elements

Let us again ask ourselves why the elements are most effectively polymerized (and thus require the greatest input of energy to be atomized) in the middle of a block of elements. We know, of course, that the bonding between atoms in the nonmetallic

**Table 6.2** Horizontal Periodic Trends in Numbers of Unpaired Electrons Available for Bonding

| Lewis structure of atom: | Na | Mg | $\cdot\ddot{\text{Al}}$ | $\cdot\dot{\text{Si}}$ | $\cdot\ddot{\text{P}}\cdot$ | $:\ddot{\text{S}}\cdot$ | $:\ddot{\text{Cl}}\cdot$ |
|---|---|---|---|---|---|---|---|
| Valence-electron configuration: | $s^1$ | $s^2$ | $s^2p^1$ | $s^2p^2$ | $s^2p^3$ | $s^2p^4$ | $s^2p^5$ |
| Prepared-for-bonding configuration: | $s^1$ | $s^1p^1$ | $s^1p^2$ | $s^1p^3$ | $s^2p^3$ | $s^2p^4$ | $s^2p^5$ |
| Number of unpaired electrons available: | 1 | 2 | 3 | 4 | 3 | 2 | 1 |
| Lewis structure of atom: | Cs | Ba | $\cdot$Lu | $\cdot\dot{\text{Hf}}$ | $\cdot\dot{\text{Ta}}\cdot$ | $\cdot\ddot{\text{W}}\cdot$ | $:\text{Re}:$ |
| Valence-electron configuration: | $s^1$ | $s^2$ | $s^2d^1$ | $s^2d^2$ | $s^2d^3$ | $s^2d^4$ | $s^2d^5$ |
| Prepared-for-bonding configuration: | $s^1$ | $s^1d^1$ | $s^1d^2$ | $s^1d^3$ | $s^1d^4$ | $s^1d^5$ | $s^1d^6$ |
| Number of unpaired electrons available: | 1 | 2 | 3 | 4 | 5 | 6 | 5 |

elements is **pure covalent** bonding. A covalent bond is formed by the overlap of an appropriate orbital on one atom with an orbital on a neighboring atom, with each atom contributing one of the two electrons in the resulting bond. Since each $p$-block atom has four valence orbitals (one $s$ and three $p$), the maximum number of bonds can be achieved when each atom has four valence electrons, one per orbital (Table 6.2). As indicated in this table, the electrons may be shifted among the orbitals so as to achieve one electron per orbital; the resulting electron configuration is called the *prepared-for-bonding* electron configuration.

Adding electrons beyond four per atom reduces the number of covalent bonds that can be formed, since filled orbitals cannot effectively overlap half-filled or filled orbitals on other atoms. Thus Group 15/VA atoms can participate in only three covalent bonds. Group 16/VIA atoms are the last ones that can form the needed two bonds per atom to give a nonvolatile polymeric structure.

To a lesser extent, the same horizontal trend may be noted in the $d$-block elements: The maximum atomization energies seem to correspond to (at most) half filling of the one valence $s$ and five valence $d$ orbitals. (Since these elements also make some use of the following $p$ orbitals, an exact correspondence cannot be made.) This is a consequence of the first rule of Engel and Brewer [2]: "*The bonding energy* (i.e., minus the heat of atomization) *of a metal or alloy depends on the average number of unpaired electrons per atom available for bonding.*" This similarity to the principle observed with nonmetals seems to suggest that the bonding in metals, too, is pure covalent bonding—even though metals have quite different properties than non-metals (such as electrical conductivity) that lead us to call the bonding by a different name, **metallic bonding**.

As indicated, the first rule of Engel and Brewer also applies to **alloys** of the metals. Many alloys can be thought of as solid solutions of one metal in another, but many seem to correspond to distinct intermetallic compounds. Alloys between metals from the left side of the $d$-block with those from the right side, such as $ZrPt_3$, may have quite substantial heats of formation (up to about 330 kJ/mol). Metals such as Zr from the left side of the $d$-block have an insufficient number of valence $d$ electrons for an optimal bonding energy. Those such as Pt from the right side have paired too many of their $d$ electrons for optimal bonding. But when the two elements are alloyed, electron transfer from Pt to Zr can occur, giving rise to better electron configurations and total bonding energies for both.

We do note, however, that in the fourth period of the *d*-block and especially in both periods of the *f*-block, the maximum in bonding occurs well before the valence orbitals can be half-filled. It is interesting to note that these anomalies match fairly well the places in Table B that correspond to the first elements that commonly fail to achieve the group oxidation number: Cr in Period 4 of the *d*-block, and Ce in Period 6 and Np in Period 7 of the *f*-block. This correlation has led to the statement that *"electrons and orbitals which are of value in binding atoms to atoms of a different kind are also generally important in binding atoms to their own kind"* [3]. This statement in turn leads to the following conclusions: (a) 3*d* orbitals are less readily available for bonding than are 4*d* or 5*d* orbitals; (b) 4*f* orbitals are less readily available for bonding than 5*f* orbitals; and (c) *f* orbitals are generally less available for bonding than *d* or other orbitals.

## 6.4
## Types of Atomic Radii

The vertical periodic variations in atomization energies of the elements can be related to the varying radii of the atoms in the elements themselves or in other covalent compounds of the elements. The radii we will use are not the same as the ionic radii we have used to date.

We commonly distinguish five different types of atomic radii: covalent radii, metallic radii, ionic radii of cations, ionic radii of anions, and van der Waals radii. (There is no practical way of measuring the radii of gaseous atoms.) Table 6.3 compares the values of these different types of radii for selected atoms; you may want to refer to this table during the following discussion.

To illustrate the differences in these types of radii, we have drawn a diatomic species in Figure 6.4a. The inner circle for each atom represents the outermost extent of its *core* electrons; the outer circle for each atom represents the outer limit of its *valence* electrons. (Although we cannot measure such radii directly, it is useful to relate observed radii to the radii we anticipate for core and valence electrons.)

Let us first suppose that the species of Figure 6.4a are two *covalent molecules* of an element, such as $Cl_2$. In a covalent molecule the valence orbitals overlap so that a valence electron of each atom can be shared with the other atom. Overlap is improved by bringing the two atoms closer, but ultimately we would expect the overlapping valence electrons of one atom to "bump into" and be repelled by the core electrons of the other atom. In a molecule such as $Cl_2$ in which both atoms are identical, we define the **covalent radius** to be *one-half the internuclear distance*— i.e., the distance from either nucleus to the straight line in Figure 6.4. Clearly this distance should be intermediate between the radius of the core electrons and the radius of the valence electrons of either atom.

The two $Cl_2$ molecules of Figure 6.4 may be supposed to be in the solid state,

**Table 6.3**  Different Types of Atomic Radii (in pm) for Selected Atoms

| Element | Cationic | Covalent | Metallic | Anionic | van der Waals | Cov-Cat | An-Cov |
|---------|----------|----------|----------|---------|---------------|---------|--------|
| Li | 90 | 134 | 155 | | 180 | 44 | — |
| Na | 116 | 154 | 190 | | 230 | 38 | — |
| K | 152 | 196 | 235 | | 280 | 44 | — |
| Be | 59 | 125 | 112 | | | 66 | — |
| Mg | 86 | 145 | 160 | | 170 | 59 | — |
| B | 41 | 90 | 98 | | | 49 | — |
| Al | 67 | 130 | 143 | | | 63 | — |
| Ga | 76 | 120 | 140 | | 190 | 44 | — |
| In | 94 | 144 | 158 | | 190 | 50 | — |
| Tl | 102 | 147 | 159 | | | 45 | — |
| N | | 70 | | 132 | 150 | — | 62 |
| P | | 110 | 128 | | 185 | — | — |
| As | 60 | 122 | 148 | | 200 | 62 | — |
| Sb | 74 | 143 | 166 | | 220 | 69 | — |
| O | | 73 | | 126 | 150 | — | 53 |
| S | 43 | 102 | 127 | 170 | 180 | 59 | 68 |
| Se | 56 | 117 | 140 | 184 | 190 | 61 | 67 |
| Te | 70 | 135 | 160 | 207 | 210 | 65 | 72 |
| F | | 71 | | 119 | 155 | — | 48 |
| Cl | 41 | 99 | | 167 | 180 | 58 | 68 |
| Br | 53 | 114 | | 182 | 190 | 61 | 68 |
| I | 67 | 133 | | 206 | 204 | 66 | 73 |

SOURCES: Ionic radii are from Tables C and 2.5 of this book; covalent and van der Waals radii are from J. E. Huheey, *Inorganic Chemistry: Principles of Structure and Reactivity*, 3d ed., Harper and Row, New York, 1983, pp. 258–259; metallic and some covalent radii are from M. C. Ball and A. H. Norbury, *Physical Data for Inorganic Chemists*, Longman, London, 1974. The metallic radii are Pauling's metallic radii adjusted to correspond to a coordination number of 12 for the metal atoms.

NOTE: The column headed "Cov-Cat" gives the difference of the covalent and cationic radii of the element; the column headed "An-Cov" gives the difference of the anionic and covalent radii.

in a crystal in which each just touches other molecules above it and below it. We take *one-half the internuclear distance* to the nearest chlorine nucleus in the next molecule to be the **nonbonded** or **van der Waals radius** of the chlorine atom. Since this distance should more or less correspond to the radius of the valence electrons, we expect the van der Waals radius of an atom to exceed its covalent radius. Unfortunately it is difficult, for most elements, to find cases of molecules in which two atoms of that element touch but in which we are certain that they do not bond; hence quoted van der Waals radii differ considerably and are not known accurately.

In Figure 6.4b we represent four atoms of a solid metallic element. In solid metals each atom characteristically has 8 or 12 nearest-neighbor atoms, all at the same distance. We pushed the two molecules of Figure 6.4a together until they merged, with all distances becoming equal. Since this process has greatly increased the coordination number of each atom, this merger has resulted in a **metallic radius** for each atom that somewhat exceeds the covalent radius of the "unmerged" atoms

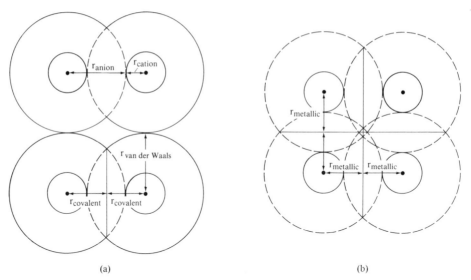

**Figure 6.4**  (a) Drawing of two adjacent molecules of a diatomic species; (b) drawing of four adjacent atoms in a metallic solid. The inner circles represent the outer extent of the core electrons of each atom; the outer circles represent the extent of the valence electrons. For explanations of the different lines and radii see the text.

of Figure 6.4a. Thus the metallic radius of an element falls between its covalent and van der Waals radius. Pauling [4] has carried out calculations designed to compensate for the high coordination number found in metallic solids; the resulting *single-bond metallic radii* are quite comparable to covalent radii found in covalent compounds of the metals (Table 6.4). Since pure covalent compounds of the metals are often hard to find, these single-bond metallic radii are often useful substitutes in looking at periodic trends in covalent radii.

Lastly, let us suppose that the diatomic species of Figure 6.4a is an *ion pair*: a cation touching an anion. For many elements, the characteristic cation has lost all its valence electrons; hence we may suppose that the cationic radius equals the radii of the core electrons, and should be smaller than any of our other radii. Monoatomic anions characteristically fill their valence orbitals, so we may suppose that the anionic radius equals the radius of the valence electrons.

In looking at periodic trends in atomic radii (such as in the next section), we must be sure to compare radii of the same type. But since a given element does not generally engage in all of these types of interactions, we often do not know a given type of radius for a particular atom, or we may not know it with much certainty. In such cases we can deduce trends approximately by noting the *usual* relationships among the radii, as illustrated in Table 6.3: (1) The smallest radius is the *cationic radius*; (2) roughly 45 to 65 pm larger than this are the *covalent radius* and *metallic radius*; (3) roughly 55 to 75 pm larger than the latter two radii are the *anionic radius* and the *van der Waals* radius.

**Table 6.4**  Covalent and Single-Bonded Metallic Radii of the Elements

| | 1 | 2 | 3 | 4 | 5 | 6 | 7 | 8 | 9 | 10 | 11 | 12 | 13/IIIA | 14/IVA | 15/VA | 16/VIA | 17/VIIA | 18/VIIIA |
|---|---|---|---|---|---|---|---|---|---|---|---|---|---|---|---|---|---|---|
| 1 | H 37 | | | | | | | | | | | | | | | | | He 32 |
| 2 | Li 134 122 | Be 125 89 | | | | | | | | | | | B 90 80 | C 77 | N 75 | O 73 | F 71 | Ne 69 |
| 3 | Na 154 157 | Mg 145 136 | | | | | | | | | | | Al 130 125 | Si 118 117 | P 110 110 | S 102 104 | Cl 99 | Ar 97 |
| 4 | K 196 202 | Ca 174 | Sc 144 | Ti 136 132 | V 122 | Cr 119 | Mn 139 118 | Fe 125 117 | Co 126 116 | Ni 121 115 | Cu 135 118 | Zn 120 121 | Ga 120 125 | Ge 122 124 | As 122 121 | Se 117 117 | Br 114 | Kr 110 |
| 5 | Rb 216 | Sr 191 | Y 162 | Zr 148 145 | Nb 134 | Mo 130 | Tc 127 | Ru 133 125 | Rh 132 125 | Pd 131 128 | Ag 152 134 | Cd 148 138 | In 144 142 | Sn 140 142 | Sb 143 139 | Te 135 137 | I 133 | Xe 130 |
| 6 | Cs 235 | Ba 198 | Lu 156 | Hf 144 | Ta 134 | W 130 | Re 128 | Os 133 126 | Ir 132 126 | Pt 131 129 | Au 140 134 | Hg 148 139 | Tl 147 144 | Pb 146 150 | Bi 146 151 | Po | At | Rn 145 |

| | 3f | 4f | 5f | 6f | 7f | 8f | 9f | 10f | 11f | 12f | 13f | 14f | 15f | 16f |
|---|---|---|---|---|---|---|---|---|---|---|---|---|---|---|
| 6 | La 169 | Ce 165 | Pr 164 | Nd 164 | Pm 163 | Sm 162 | Eu 185 | Gd 162 | Tb 161 | Dy 160 | Ho 158 | Er 158 | Tm 158 | Yb 170 |
| 7 | Ac | Th | Pa | U | Np | Pu | Am | Cm | Bk | Cf | Es | Fm | Md | No |
| | 165 | | | 143 | | | | | | | | | | |

Sources: Covalent radii are listed in the first row under the symbol for each element; these are taken from J. E. Huheey, *Inorganic Chemistry: Principles of Structure and Reactivity*, 3d ed., Harper and Row, New York, 1983, pp. 258–259, or, if not available there, from M. C. Ball and A. H. Norbury, *Physical Data for Inorganic Chemists*, Longman, London, 1974, pp. 139–144. Single-bonded metallic radii are listed in the second row under the symbol for each element; these are taken from Ball and Norbury, op.cit.

193

6.5

# Periodicity of Covalent Radii and Bond Energies

TRENDS IN COVALENT RADII   We note from Table 6.4 that neutral atoms (as found in the elements, metals, or in covalent compounds) tend to become smaller as we go from left to right in the periodic table. This trend is greatly diminished, however, as we cross the *d*- and *f*-blocks; indeed it is reversed somewhat at the end of the *d*-block. (The metallic radii of Eu and Yb are also anomalously large. As these are the *f*-block elements most prone to giving the $+2$ oxidation state, we assume that they are contributing fewer electrons to the formation of metallic bonds than the other *f*-block elements contribute.)

The general vertical trends strongly resemble the vertical trends in the radii of ions of the same charge (Table C): Down a group, atoms get larger. Exceptions are found in the same places as among the ions; thus the *d*-block atoms of the sixth

---

**Table 6.5**   Element-Element Covalent Bond Energies (kJ/mol)

**(a) Single (Sigma) Bond Energies**

| H—H<br>432 | | | | | | |
|---|---|---|---|---|---|---|
| Li—Li<br>105 | Be—Be<br>(208) | B—B<br>293 | C—C<br>346 | N—N<br>167 | O—O<br>142 | F—F<br>155 |
| Na—Na<br>72 | Mg—Mg<br>(129) | Al—Al | Si—Si<br>222 | P—P<br>201 | S—S<br>226 | Cl—Cl<br>240 |
| K—K<br>49 | Ca—Ca<br>(105) | Ga—Ga<br>113 | Ge—Ge<br>188 | As—As<br>146 | Se—Se<br>172 | Br—Br<br>190 |
| Rb—Rb<br>45 | Sr—Sr<br>(84) | In—In<br>100 | Sn—Sn<br>146 | Sb—Sb<br>121 | Te—Te<br>126 | I—I<br>149 |
| Cs—Cs<br>43 | | | | | | At—At<br>116 |

**(b) Pi-Bond Energies**

| | | | C—C<br>256 | N—N<br>387 | O—O<br>352 |
|---|---|---|---|---|---|
| | | | | P—P<br>140 | S—S<br>199 |
| | | | Ge—Ge<br>84 | As—As<br>117 | Se—Se<br>100 |
| | | | | Sb—Sb<br>87 | Te—Te<br>92 |

SOURCE: Data from J. E. Huheey, *Inorganic Chemistry: Principles of Structure and Reactivity*, 3d ed., Harper and Row, New York, 1983, Table E-1.
NOTE: Pi-bond energies are calculated as double-bond energies minus single-bond energies for the same element, or (triple-bond energies minus single-bond energies) times one-half.

period are not larger than the atoms of the fifth period. We also find the familiar anomalies in Group 13/IIIA: Ga is not larger than Al; Tl is scarcely larger than In.

TRENDS IN BOND ENERGIES   Our ability to analyze periodic trends in atomization energies is somewhat limited by our uncertainty as to how many electrons some of the metals contribute to the bonding. Trends show up more clearly among the related **bond dissociation energies** for homoatomic single bonds of the elements (Table 6.5). These are the energies required to break single covalent bonds; the atomization energies involve breaking variable numbers of bonds. We see that *single bond energies tend to increase as we proceed to the right in a given period.* We may also note the general vertical trend manifested in the *s-* and *p*-blocks—*bond dissociation energies increase as we go up a group.*

These trends may be connected with the preceding trends in covalent and metallic radii: *Small atoms generally form stronger covalent bonds.* This trend is illustrated in Figure 6.5. We will examine the reason for this trend in more detail in Chapter 11. For now we note that in bonds between smaller atoms, the shared

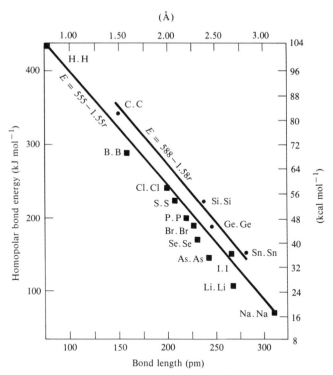

**Figure 6.5**   Relation between bond length and bond energy for bonds between like atoms. The upper line is for Group 14/IVA elements only; the lower line is for other *s-* and *p*-block elements. Reprinted with permission from J. E. Huheey and R. S. Evans, *J. Inorg. Nucl. Chem.*, 32, 383 (1970). Copyright 1970 by Pergamon Press.

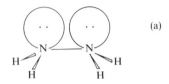

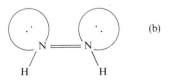

**Figure 6.6** (a) Electron pair–electron pair repulsion between two small second-period atoms, illustrated for one conformation of $H_2N$—$NH_2$; (b) reduction of the repulsion when a double bond is present, illustrated for HN=NH; (c) absence of the repulsion when a triple bond is present, illustrated for N≡N.

electrons are closer to (and therefore more strongly attracted to) both atomic nuclei.

An anomaly occurs in the upper right of the *p*-block. The N—N, O—O, and F—F single bonds are substantially weaker than P—P, S—S, and Cl—Cl bonds, even though the second-period atoms are much smaller. This is another case in which too small a size leads to diminishing returns. Molecules containing these single bonds (such as $H_2N$—$NH_2$, HO—OH, and $F_2$) also have unshared pairs of electrons. These electrons are brought so close to each other that substantial electron pair–electron pair repulsion results, weakening the net bonding (Figure 6.6).

Finally, we know that covalent bonds occur not only as single **sigma** bonds, but also as **pi** bonds, which are formed by side-to-side overlap of *p* (or *d* or *f*) orbitals with each other (Figure 6.7). Double bonds normally consist of one sigma and one pi bond, and triple bonds of one sigma and two pi bonds. As double and triple covalent bond energies are available for many of the *p*-block elements, we can subtract sigma- (single-) bond energies from these to obtain estimates for *pi-bond dissociation energies*, shown at the bottom of Table 6.5. (In the case of triple-bond energies, after subtraction the result is divided by two, since a triple bond contains two pi bonds.) The data show that *pi bonds are generally weaker than sigma bonds between the same elements* and that *pi bond energies drop off with increasing distance even faster than sigma bonds*. Although Figure 6.7 does not give a computationally precise representation of the overlap of atomic orbitals, it is designed to suggest the relative weakness of pi-type overlap and its sensitivity to distance. As Figure 6.6 also suggests, the repulsion of electron pairs on neighboring atoms is greatly diminished or absent in double or triple bonds (due to the changed bond angles; see Chapter 7), so the pi-bond energies of N, O, and F are not anomalously weak.

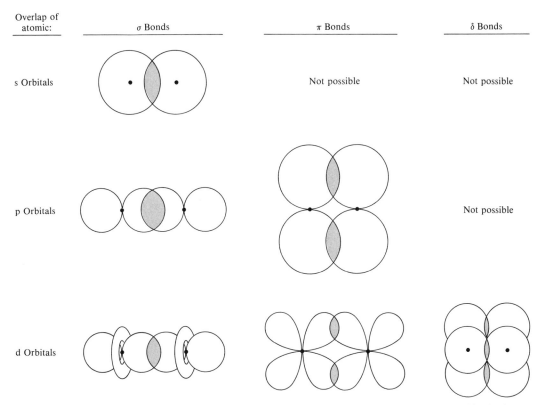

| Overlap of atomic: | σ Bonds | π Bonds | δ Bonds |
|---|---|---|---|
| s Orbitals | | Not possible | Not possible |
| p Orbitals | | | Not possible |
| d Orbitals | | | |

**Figure 6.7**  Overlap of atomic orbitals of two atoms (the nuclei of which are represented by dots) to give covalent bonds. Regions of atomic orbital overlap are indicated by shading. At left, two *s*, *p*, or *d* atomic orbitals overlap in one shaded region to give sigma bonds. At center, two *p* or *d* orbitals overlap in two regions to give pi bonds. At right, two *d* orbitals overlap in four regions (one is hidden from view) to give a delta bond.

Although firm data are much harder to come by in the *d*-block of elements, it is generally apparent that pi bonds *increase* in stability down a group of *d*-block metals, in accord with the increased availability of 4*d* and 5*d* orbitals for bonding. In addition, *d* orbitals present the possibility of a new type of bonding not found in the *p*-block: **delta bonding** (Figure 6.7). Delta bonds seem to be weaker than pi bonds, but their strength also increases down a group of *d*-block metals. Delta bonds, along with two pi bonds and a sigma bond, are found in a famous series of acetates and other derivatives of quadruply bonded diatomic ions of early *d*-block metals: $Cr_2^{4+}$, $Mo_2^{4+}$, $W_2^{4+}$, $Re_2^{6+}$, and so forth.

Some allotropes of nonmetallic elements differ from each other in the number of sigma and pi bonds formed by the atoms. For example, the atoms of Group 15/VA have three valence electrons available to participate in three covalent bonds. These three bonds may be three sigma bonds to three different atoms (i.e., three single bonds), or they may be two sigma bonds to two atoms plus a pi bond to one of them (a single and a double bond). Or they may be one sigma and two pi bonds to one other atom (a triple bond).

Since each atom contributes half of each of these bonds, we can compute that each Group 15/VA atom releases $\frac{3}{2}$ times the sigma-bond energy upon forming an allotrope involving only sigma bonds. The same atom releases $\frac{1}{2}$ the sigma-bond energy plus the pi-bond energy upon forming a diatomic allotrope having one triple bond. For nitrogen, the formation of three single bonds releases $\frac{3}{2} \times 167$ kJ $=$ 250 kJ per mole of nitrogen atoms, while the formation of one triple bond releases $\frac{1}{2} \times 167 + 387 = 470$ kJ per mole of nitrogen atoms. The latter is clearly the preferable alternative for nitrogen. But for phosphorus the single-bonding alternative releases more energy (301 kJ per mole of atoms) than does the triple-bonding alternative (240 kJ per mole of atoms); hence a single-bonded allotrope is preferred.

This example can be generalized: Multiply bonded allotropes (graphite, $N_2$, $O_2$, $O_3$) are favored for $p$-block elements in the second period. Singly bonded allotropes are favored for $p$-block elements in the third period and below. Indeed, it is only in recent years that any stable compounds involving multiple bonds between like $p$-block atoms of the third or later periods have been synthesized [5].

We previously encountered multiple bonds of several of the elements to oxygen in the oxides and oxo anions of the elements (Chapters 2 and 4). We noted that these were most predominant for elements in the second period (e.g., in $CO_3{}^{2-}$, $NO_3{}^-$, $NO_2{}^-$, $CO_2$, CO, and the oxides of nitrogen), for which they are more stable than the sigma-bonded alternatives. In oxygen compounds, double bonds are not entirely absent in the third period, however. Any Lewis structure of monomeric $SO_2$, for example, requires the presence of a double bond. (Many other oxides and oxo anions may well involve some resonance structures that include double bonds.) But as we go down the periodic table to $SeO_2$ and $TeO_2$, we find that the most stable form is not the monomeric gaseous molecule, but single-bonded polymeric forms. Thus not only the effect of increasing size of the central atom but of increasing preference of that atom for sigma over pi bonding must be involved.

Finally, it is interesting to note that the early $d$-block elements that are capable of forming double, triple, or quadruple bonds to each other are also capable of forming them to $p$-block elements. A number of complex organometallic compounds are known that contain $Mo{=}O$, $W{=}N{-}H$, $W{\equiv}N$, $W{\equiv}C{-}H$, and the like, units. This may or may not be related to the fact that these $d$-block elements are rather readily oxidized by the usually inert gas $N_2$ and that some of them are involved in the biochemical or artificial *fixation* of inert $N_2$ (conversion to biochemically or industrially useful forms such as $NH_3$).

## 6.6

## The Metals; Conductors and Semiconductors

Our description of the covalent bond in the last section is not adequate for describing the bonding in elemental boron and in the metals. Boron has only three valence electrons and, even in the prepared-for-bonding electron configuration of $2s^1 2p^2$, would appear to be capable of forming only three covalent bonds—not enough to fulfull its potential maximum coordination number and thus achieve a maximum bonding energy. Indeed, as shown for only 1 of the 12 equivalent B atoms in Figure 6.3a, each boron atom has 5 nearest neighbors within the icosahedron and (typically) one nearest neighbor outside the icosahedron, for a coordination number of 6. To form this many covalent bonds as we normally picture them, boron would need 6 valence electrons, which it does not have.

We can satisfactorily treat the bonding in boron compounds with the **molecular orbital theory**, which does not make our common assumption that a bonding pair of electrons is shared by only *two* atoms. The molecular orbital theory (which is basically beyond the scope of this course) allows a bonding pair of electrons to be shared by all the atoms in the whole boron macromolecule. One pair of electrons by itself would not give much bonding, but the use of $\frac{3}{2}$ pairs of electrons per boron atom in a large molecule to form many of these bonding **molecular orbitals** can and does give rise to quite high bonding energies.

The metallic elements to the left of or below boron face the same problem as boron: They have insufficient valence electrons to form ordinary (**two-centered, two-electron**) chemical bonds to all their neighbor atoms. Furthermore, they have higher coordination numbers than does boron. Most metals have one of three types of metallic lattice. Two of these are close-packed lattices (Section 3.9), the cubic close-packed and the hexagonal close-packed lattices, in which each metal atom has a coordination number of 12. (Unlike the close-packed anions discussed in Section 3.9, there are no counterions present in metallic lattices.) The third common form of metallic lattice is the body-centered cubic lattice, in which each metal atom sits at the center of a cube of nearest-neighbor metal atoms, thus having a coordination number of 8.

Using classical two-center, two-electron covalent bonds, each metal atom would thus need 8 or 12 valence electrons to bond to its neighbors, but the metal atoms have a maximum of 6 usable valence electrons (as indicated by the maximum atomization energies at W and Mo). These valence electrons are thus, in general, put into up to $\frac{6}{2}$ times $N$ (Avogadro's number) of bonding molecular orbitals in a 1-mol crystal of a metal. The total number of bonding molecular orbitals *available* for electrons in such a crystal will in general be one-half per each valence orbital. In addition to the $ns$ valence orbitals available in the $s$-block and the $ns$ and $(n-1)d$ orbitals available in the $d$-block, the $np$ orbitals generally contribute to the collection of bonding molecular orbitals, giving a total of $\frac{4}{2}$ times $N$ bonding valence orbitals available in the $s$- and $p$-blocks, and $\frac{9}{2}$ times $N$ bonding molecular orbitals available in the $d$-block.

Characteristically, then, metal atoms have too few valence electrons to fill all of their numerous bonding molecular orbitals. These trillions of molecular orbitals differ in energy only by infinitesimal gaps and thus give rise to a **band** of molecular orbitals. Thus for any electron in an occupied molecular orbital there is available (at only a slightly higher energy) an alternative, nonoccupied molecular orbital. Any frequency of light will suffice to promote an electron from some occupied to some unoccupied orbital; hence metals have a characteristic metallic luster resulting from

199

the interaction with light of all frequencies. Some of these accessible orbitals will bring that electron closer to a positive electrode at one edge of a crystal; hence metals are **conductors** of electricity, even in the solid state when the metal atoms themselves cannot move. The electrical conductivity of metals characteristically decreases at higher temperatures, however, as thermal vibrations of the metal atoms gives rise to increasing resistance. The mobile conduction electrons also make metals good conductors of heat.

Some properties of metals are quite variable, however, and depend markedly on the number of valence electrons available to the metal. Among these are their heats of atomization and related physical properties such as melting and boiling points. Thus although most metals are high-melting solids, there is one metal that is a liquid at room temperature (mercury), and there are some that melt just above room temperature (cesium, gallium). Except for gallium, these have few valence electrons and hence low bonding energies.

The strengths of metals are also related to the number of valence electrons available for metallic bonding. Group 1 metals, Group 11 metals (copper, silver, and gold), and some of the subsequent *base metals* such as thallium and lead have few electrons prepared for bonding. These are soft and easily cut and are also quite malleable (capable of being hammered into thin sheets) and ductile (capable of being drawn into thin wires). On the other hand, metals with many available valence electrons (from the center of the *d*-block) have much greater strength and are more useful for structural purposes, although they often tend to break upon hammering (they may be brittle). Of course one may modify the strength and related properties of either group of metals by preparing alloys, since incorporating other metals alters the number of valence electrons available for bonding.

The density of metals is a function both of the number of bonding electrons (stronger bonding results in shorter metallic bond distances, hence greater density) and of the atomic weight of the metal. Because of the former factor, it is difficult to get a strong structural metal that is also very light. Metals of atomic weight less than that of iron should generally be lighter, but light metals such as magnesium and aluminum lack many bonding electrons and are rather weak. A very good metal for such purposes (e.g., aircraft manufacture) is titanium, which is very abundant, is lighter than steel, and yet is strong. But its reactivity even with atmospheric nitrogen at high temperatures makes it very difficult to produce and to weld (argon atmospheres are required). There is little demand for dense metals per se, but it is interesting to note that osmium is the densest of known metals ($22.48$ g/cm$^3$).

One other variable property of metals is magnetism. As we mentioned previously, overlap of $3d$ orbitals with each other is relatively poor, as is overlap of $4f$ orbitals with each other. Consequently electrons in these orbitals are not as effective in forming molecular orbitals as are electrons in $4d$, $5d$, $5f$, or $ns$ or $np$ valence orbitals; instead many or most of these electrons may remain unpaired in $3d$ or $4f$ orbitals of individual atoms. This accounts for the relatively easy atomization of elements using these valence orbitals (Table 6.1). Furthermore, these unpaired electrons tend to align with magnetic fields, giving rise to unusual magnetic properties for some of these metals. If the electrons in $3d$ or $4f$ orbitals of different metal atoms align with each other, their combined interaction with a magnetic field is strong and is known as **ferromagnetism**. Ferromagnetism is found in the $3d$ metals iron, cobalt, nickel, and the $4f$ metals europium through thulium. In contrast, most of the electrons in $4d$ and $5d$ and (at least early) $5f$ metals pair up in molecular orbitals (a few remain unpaired), so that only a much weaker interaction with applied magnetic fields occurs. (Magnetic properties of $d$-block metals and their compounds will be covered in more detail in a later course in inorganic chemistry.)

In the $p$-block of elements generally only the $ns$ and $np$ orbitals are available for forming molecular orbitals, giving rise to $\frac{4}{2}$ times $N$ bonding molecular orbitals. When we reach Group 14/IVA, we have 4 times $N$ valence electrons, enough to fill these orbitals; hence diamond, for example, is a **nonconductor** of electricity. In the molecular orbital theory, however, there are also **antibonding** molecular orbitals that are not normally filled with electrons; these are usually much higher in energy (580 kJ/mol in diamond) than the bonding molecular orbitals. As we go down a group of the $p$-block, however, bonding becomes weaker (Table 6.5), and the **energy gap** between the band of bonding molecular orbitals and the band of antibonding molecular orbitals (and perhaps bonding molecular orbitals based on $nd$ atomic orbitals) becomes smaller. At the bottom of many of the $p$-block groups this gap becomes zero, so that metallic conduction becomes possible for lead, for example.

In between the nonconductor graphite and the conductor lead we have an intermediate gap between the two bands of molecular orbitals. (The band gap between the two sets of molecular orbitals is 105 kJ/mol for silicon and 58 kJ/mol for germanium.) This gap is large enough that (at low temperatures) few electrons can get into the band of unoccupied molecular orbitals (the *conduction band*) and migrate to an electrode, but at higher temperatures more electrons acquire enough energy to be able to do so; hence these elements are known as **semiconductors**. Their conductivity is limited at low temperature but *increases* with increasing temperature. Their conductivity also increases greatly upon irradiation with light (which promotes electrons to the conduction band). These elements (B, Si, Ge, As, Sb, Se, and Te) are commonly called the **metalloids**, as their electrical (and chemical) properties lie between those of the metals and the nonmetals.

Some of these elements are extremely important for their semiconducting properties in modern solid-state electronic devices (transistors, etc.). Very high purity is mandatory, since introduction of even small levels of impurities has great effects on semiconduction. For example, a small level of As in a Ge semiconductor introduces one extra electron per As atom. Since the occupied molecular orbitals of Ge are filled, these electrons must go up into the unoccupied molecular orbitals, which are consequently no longer unoccupied, and thus can give rise to more ready conduction. (Such a semiconductor is known as an **n-type** (negative) semiconductor.) Similarly, impurities of Ga in a Ge semiconductor lead to a vacancy or **hole** in the band of occupied molecular orbitals, into which, upon the input of a relatively small amount of energy, another electron can jump, also increasing the electrical conductivity and producing a **p-type** (positive) semiconductor. Consequently Si and Ge must be prepared in an extraordinary state of purity (at least 99.9999999% pure) for the electronics industry.

For many solid-state electronic devices, impurities such as Ga or As are deliberately *doped* at controlled levels into the Si or Ge to produce *n*- and *p*-type semiconductors. When two such semiconductors are put in contact, for example, the resulting *diode* will conduct electricity better in one direction than in the other, hence rectifying an alternating current.

Finally, we note that any substance can exist in the metallic state if its band of antibonding orbitals can be brought down to overlap its band of filled bonding molecular orbitals. In theory, this can be achieved by the application of very high pressure to the substance, which causes increased overlap of high-energy orbitals such as antibonding molecular orbitals and atomic orbitals that are higher in energy than the valence orbitals (for example, the $3d$ orbitals of Si). Such broadening of the band of high-energy orbitals eventually causes it to overlap the band of filled bonding orbitals; hence the energy gap becomes zero, and the substance becomes metallic. This phenomenon has been observed in practice with xenon at very high pressures, such as are normally found only in the centers of explosions. However, very high gravitational fields also result in high pressures, and data from satellites flying by the planets Jupiter and Saturn suggest that the cores of these very heavy planets may contain metallic hydrogen.

## 6.7

## Catenation of the Elements

Covalent bonds between like atoms are not confined to the elements themselves, but also occur in many compounds of the elements. The formation of bonds between like atoms is referred to as the **catenation** of elements. When catenation occurs for a given element, it greatly expands the number of possible derivatives of a given class that exist for that element. The element showing the greatest tendency to catenation is carbon. Were it otherwise, carbon would form only one hydrogen compound, $CH_4$. But given that carbon can form any number of C—C bonds, there is an infinite number of possible C—H compounds $C_nH_{2n+2}$, plus other C—H compounds containing double and triple bonds, catenated rings of carbon compounds, and so forth. So many compounds result that a thorough study of the chemistry of just the element carbon (the field of organic chemistry) takes about as long as the study of the chemistry of all 107 other elements (the field of inorganic chemistry).

The C—C single bond is the second-strongest single bond (Table 6.5). Thus one might think that this is the reason that other elements show a lesser tendency to catenation than does carbon. (Although the H—H bond is stronger, hydrogen does not have any additional valence electrons with which to form longer chains.) Although very low sigma-bond energies are barriers to catenation, probably a greater problem is the fact that bond energies of most atoms to oxygen are substantially greater than bond energies of those atoms to other like atoms (this will be discussed further in Chapter 8); hence catenated compounds of the elements tend to be thermodynamically unstable to oxidation and may have to be prepared in

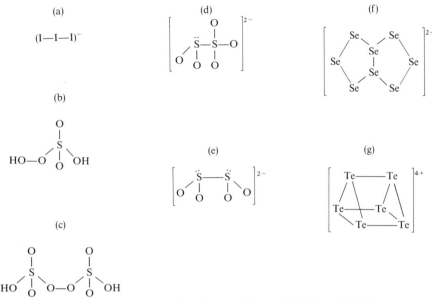

**Figure 6.8**   Some catenated compounds of the elements of Groups 17/VIIA and 16/VIA: (a) triiodide ion; (b) peroxomonosulfuric acid; (c) peroxodisulfuric acid; (d) disulfite ion; (e) dithionite ion; (f) $Se_4^{2+}$ cation; (g) $Te_6^{4+}$ cation.

inert atmospheres with rigorous exclusion of air. But if the central atom of the element is small and/or the outer atoms of the compound are large, there may be sufficient steric hindrance to the approach of oxygen or other reactive molecules to allow decent stability for certain catenated compounds. (There may also be other reasons why a particular catenated compound is slow to react.)

GROUP 17/VIIA   Halogen atoms, like hydrogen, generally cannot form more than one bond to other atoms of the same kind, and so do not catenate. However, there are some polyhalide ions in which a halogen atom uses more than an octet of valence electrons; these occur especially for iodine but not at all for fluorine. The best known of these is the *triiodide* ion, $I_3^-$ (Figure 6.8); other polyiodide ions such as $I_5^-$ and $I_8^{2-}$ are also known as salts of nonacidic cations.

GROUP 16/VIA   A number of catenated oxygen compounds exist, but due to the relative weakness of the O—O single bond these are generally strong oxidizing agents and are often explosives (Section 5.6). Two catenated anions with odd numbers of valence electrons exist, the ozonide ion $O_3^-$ and the superoxide ion $O_2^-$. (These large, relatively nonbasic anions are most easily obtained as $Cs^+$ salts.) Most catenated oxygen compounds are derivatives of hydrogen peroxide, $H_2O_2$, or of its dianion, the peroxide ion, $O_2^{2-}$. There are a number of oxo anions in which one or more oxo groups are replaced by peroxo groups. The casually written formulas and names of these are confusing because they look like impossible oxo anions. For example, $H_2SO_5$ and $H_2S_2O_8$ (Figure 6.8), commonly called *Caro's acid* and *persulfuric acid* respectively, really contain sulfur in its common oxidation number of $+6$. These are better named peroxomonosulfuric and peroxodisulfuric acid, respectively; they are very strong oxidizing agents.

Sulfur shows a very great tendency to catenate in all sorts of compounds, consistent with the large number of allotropes that the element itself exhibits. There are polysulfide ions, including $S_2^{2-}$, $S_4^{2-}$, and $S_6^{2-}$, which have open-chain structures; the first of these is perhaps most familiar in the form of the mineral iron pyrite or *fool's gold*, $FeS_2$. Mixtures of such ions are formed by dissolving sulfur in a sulfide-ion solution. The corresponding hydrogen compounds, $H_2S_x$, and halides, $S_xCl_2$, etc., will be discussed along with the simple hydrides and halides of the elements later in the book.

A number of oxo anions of sulfur are catenated; the nomenclature of these leaves a great deal to be desired. We have already mentioned the familiar thiosulfate ion, $S_2O_3^{2-}$ (Figure 5.8). Somewhat related are the series of polythionate ions $S_nO_6^{2-}$ ($n = 2$ to 6), which actually have the structure $^-O_3S—S_{n-2}—SO_3^-$, incorporating a catenated sulfur chain between two terminal $SO_3^-$ groups. Two other catenated sulfur oxo anions have more reduced terminal groups: the disulfite ion $S_2O_5^{2-}$ (Figure 6.8), obtained by loss of water from solutions of hydrogen sulfite ion ($HSO_3^-$), and the dithionite ion (formerly called hydrosulfite), $S_2O_4^{2-}$ (Figure 6.8). The latter is a useful powerful reducing agent.

Catenated compounds of selenium and tellurium are less abundant than those of sulfur. One class that has attracted attention is a group of catenated cations (also found for S), obtained by dissolving elemental S, Se, or Te in disulfuric acid containing excess $SO_3$. These include bright yellow $S_4^{2+}$ and bright red $Se_4^{2+}$ and $Te_4^{2+}$, which have a square arrangement of atoms. Deep blue $S_8^{2+}$ and green $Se_8^{2+}$ have a cross-linked ring structure (Figure 6.8); brown $Te_6^{4+}$ has a trigonal-prismatic structure (Figure 6.8).

GROUP 15/VA  There are several catenated compounds of nitrogen, but due to the low N—N single-bond energy they are often explosive or otherwise unstable, since formation of $N_2$ is so thermodynamically advantageous. Some of the oxides mentioned in Chapter 4 technically are catenated; the catenated hydrogen compound *hydrazine*, $N_2H_4$, is an important reducing agent and rocket fuel. We have already mentioned the very powerfully reducing azide ion, $N_3^-$ (Figure 5.8). Higher degrees of catenation are usually achieved in organic derivatives, for example of tetrazene, $H_2N—N{=}N—NH_2$. Catenated hydrides and halides of phosphorous are also known (but containing only single bonds); there are also somewhat more stable organometallic derivatives. Numerous catenated polyphosphorus oxo anions exist, but these are of little importance. Even less catenation is observed for arsenic, antimony, and bismuth, although Bi forms interesting polyhedral **cluster compound** cations such as $Bi_5^{3+}$ and $Bi_9^{5+}$ (Figure 6.9).

The anions formed by these elements are usually catenated. The heavier Group 15/VA and other electronegative nonmetal atoms are not electronegative enough to form simple monoatomic anions (such as $X^{3-}$) very readily and instead

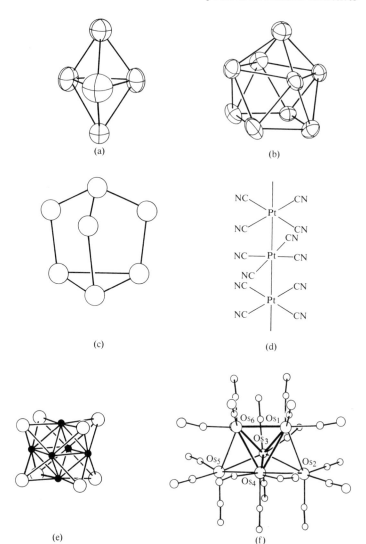

**Figure 6.9** Geometries of some cluster compounds of Groups 15/VA and 14/IVA: (a) the isoelectronic ions $Bi_5^{3+}$ and $Pb_5^{2-}$; (b) $Sn_9^{4-}$ — the isoelectronic ion $Bi_9^{5+}$ is similar in structure; (c) $As_7^{3-}$; (d) the polymeric $[Pt(CN)_4]_n^{-1.7n}$ ion; (e) the $Mo_6X_8^{4+}$ cation (dark circles = Mo); (f) $Os_6(CO)_{18}$ (large circles = Os).

retain some of the catenation found in the elemental form in their very complex, less reduced anions. For example, $(As^-)_n$ in LiAs adopts a spiral chain structure like that taken by elemental Se and Te, with which it is formally isoelectronic. Many of these compounds are also semiconductors. There are also nonpolymeric cluster anions $P_7^{3-}$, $As_7^{3-}$, and $Sb_7^{3-}$ (Figure 6.9). All these polyatomic anions avoid building up as great a density of negative charge on a weakly electronegative nonmetallic atom as we find in $X^{3-}$.

GROUP 14/IVA    You will see enough examples of catenation of carbon when you take organic chemistry; we will say no more about this topic here. But chemists, using periodic reasoning, have long tried to duplicate the catenating ability of

205

carbon in the element just below it, silicon. Early attempts focused on the hydrogen compounds of silicon. A number of catenated silicon hydrides, $Si_nH_{2n+2}$, can be made, but these are spontaneously flammable in air. They could not be studied adequately until Alfred Stock, in his studies, invented the vacuum-line techniques for handling volatile materials in the complete absence of air. Greater stability is found in the catenated halides of silicon, which are known up to $Si_{16}F_{34}$, $Si_6Cl_{14}$, and $Si_4Br_{10}$. The greatest stability of all, however, is found in the organometallic polysilanes (such as $Si_n(CH_3)_{2n+2}$), which are as stable as the hydrocarbons. In this case it appears that the larger bulk of a methyl group is needed to cover the larger silicon atom as effectively as the small hydrogen atom can shield the small carbon atom from attack by oxygen and other reactive species.

As one might expect from the increasing size of the Group 14/IVA elements and the decreasing M—M bond energies, catenation falls off slowly on going farther down the group; it is most extensively observed, again, in the organometallic compounds. As in Group 15/VA, catenation is usually observed in the "anions" of the Group 14/IVA elements. Thus in calcium silicide, $CaSi_2$, the $Si^-$ structural units are isoelectronic with P or As atoms and form infinite puckered layers as in elemental As or in black phosphorus. But other nonpolymeric anions have recently been discovered that have finite three-dimensional structures. These metal cluster anions include such species as $Ge_9^{4-}$ and $Sn_9^{4-}$, and $Sn_5^{2-}$ and $Pb_5^{2-}$ (Figure 6.9), which are isoelectronic with the $Bi_9^{5+}$ and $Bi_5^{3+}$ cations. Predicting the formulas and geometries of metal clusters such as these provides an interesting challenge for the molecular-orbital theory.

GROUP 13/IIIA    Although a few compounds containing single Ga—Ga or In—In bonds are known, catenation of Group 13/IIIA atoms is by far the most characteristic for boron. A large body of chemistry dealing with the hydrides of boron will be discussed in Chapter 10 and in more detail in further courses in inorganic chemistry. We point out here that most of these compounds involve finite three-dimensional clusters of boron atoms. Many of the structures involve fragments of the icosahedral structural unit found in elemental boron or involve other polyhedra of boron atoms. Characteristically, too few valence electrons are available to allow the construction of reasonable Lewis dot structures (representing two-center, two-electron bonds) for these compounds.

SUMMARY FOR NONMETALS    Catenation is not a strong feature of the halogens, which normally lack the ability to share more than one electron, or of nitrogen and oxygen, which form abnormally weak sigma bonds. Instead it is most frequently found among the lightest of the remaining nonmetal atoms (B, C and Si, P, S), which have the strongest sigma bonds and which, being small, are most easily shielded from oxidizing agents by outer atoms.

d-BLOCK METALS    The most familiar case of catenation involving a metal involves the Hg—Hg bond in the mercury(I) cation, $Hg_2^{2+}$, which occurs commonly in aqueous solution, although it is susceptible to fairly easy oxidation. More recently, longer chains have been prepared: $Hg_3^{2+}$, $Hg_4^{2+}$, and an infinite-chain $Hg_{2.86x}^{x+}$, for which a large $-1$ anion occurs in the lattice after each 2.86 mercury atoms of the chain.

Among the more typical $d$-block elements, many catenated compounds and ions have been prepared, although these are characteristically quite sensitive to oxidation. We have already mentioned the quadruply bonded dimeric metal ions such as $Cr_2^{4+}$ and $Mo_2^{4+}$. These are, of course, catenated; similar ions exist that are triply or doubly bonded. Metal-metal bonds are sometimes created by partial oxidation of compounds of the $Pt^{2+}$, $Pd^{2+}$, and $Ni^{2+}$ ions in which the metal ion has a coordination number of 4, but in which the nearest-neighbor atoms are arranged in a square around the metal ion. These ions have unshared pairs of electrons extending above and below the metal; upon partial removal of these electrons by oxidation these orbitals can overlap with their counterparts on nearby metal ions, giving rise to a one-dimensional polymer (Figure 6.9). With nonstoichiometric oxidation (as in $K_2Pt(CN)_4Br_{0.3}$) a conduction band is created that is partially filled; such compounds are conductors of electricity in one dimension only. Two-dimensional infinite catenation of metal atoms also occurs in some low-oxidation-number halides, etc., such as ScCl; Cl—Sc—Sc—Cl units are linked in sheets via metallic bonding of the Sc atoms in neighboring units.

Recently there has been a great deal of interest in the creation of finite three-dimensional cluster compounds involving catenation of $d$-block elements. An older example of such a cluster compound is the cation $[Mo_6X_8]^{4+}$ (Figure 6.9; $X =$ a halogen), which involves an octahedral cluster of molybdenum atoms. More recent research has focused on organometallic clusters such as $Os_6(CO)_{18}$ (Figure 6.9). $d$-block metals have long been important **heterogeneous catalysts**: They catalyze numerous important organic reactions and are, of course, insoluble in the solutions involved. This property allows them to be recovered easily by filtration, but it is hard to know just how they carry out the catalysis. $d$-block organometallic compounds have recently become very important as **homogeneous catalysts** of other organic reactions. Since these compounds are soluble, one can study the influence of their concentration on reaction kinetics and thus determine the mechanism of catalysis. But these compounds do not duplicate all the catalytic abilities of bulk metals. Consequently, inorganic chemists find it interesting (from both theoretical and practical points of view) to see what minimum number of metal-metal bonds are needed in order to develop the characteristics of metallic bonding, especially as it manifests itself in catalysis.

## 6.8

## Thermochemical Analysis of Redox Reactions: The Born-Haber Cycle

Now that we have some understanding of the bonding in the elements themselves, we will attempt to analyze in detail some of the redox reactions of the elements that we discussed in Chapter 5. Even such a simple redox reaction as the burning of sodium in chlorine (equation (6.1)) is complex enough that it is best broken down into a sum of simpler reactions that can easily be analyzed thermodynamically. Then, by Hess's law, the enthalpy change for reaction (6.1) ($\Delta H_f^\circ$, the enthalpy of formation of NaCl), will be equal to the sum of the enthalpy changes for the different simpler reactions that add up to reaction (6.1).

$$Na(s) + \tfrac{1}{2}Cl_2(g) \rightarrow NaCl(s) \qquad \Delta H_f^\circ = -411 \text{ kJ/mol} \qquad (6.1)$$

The thermochemical cycle of simpler reactions we construct should relate this complex redox reaction to simpler gas-phase reactions; hence we use the following general strategy, involving three steps:

1. Take all reactants to the gaseous state; identify and assign all energy changes involved.
2. Let the reactants react in the gaseous state to give gaseous products; identify and assign all energy changes involved.
3. Take the products to their final state; identify and assign all energy changes involved.

Then, by Hess's law, the energy change for the overall reaction is equal to the sum of the energy changes for each of the steps.

Analysis of the reactions by which ionic compounds are formed from their elements in the standard state was pioneered by Born and Haber [6], so such an analysis is now known as a **Born-Haber cycle**. The classical Born-Haber cycle for the formation of NaCl from Na metal and $Cl_2$ gas is shown in Figure 6.10. (Recall that we have already carried out one thermochemical cycle, for dissolution of ionic solids, in Section 3.5).

In step 1 of this thermochemical cycle we convert sodium metal (a solid) and chlorine gas (which consists of diatomic molecules) to gaseous atoms. The energy required to produce 1 mol of gaseous atoms is, of course, the heat [7] of atomization

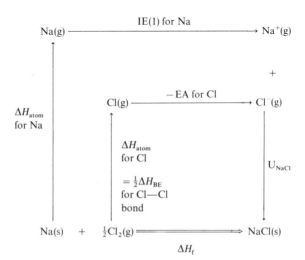

**Figure 6.10** The Born-Haber cycle for the formation of NaCl from the elements in their standard states. For interpretation of the symbols see the text.

of that element (Table 6.1). For Na this is $+107$ kJ/mol; for Cl this is $+122$ kJ/mol [8].

In step 2 we allow the gaseous atoms to react to give the products in gaseous form. In this case we remove an electron from Na to form the gaseous $Na^+$ ion, and we give the electron to the Cl atom to form the gaseous $Cl^-$ ion. The energy required to remove the outermost (most loosely bound) electron from a gaseous atom to produce a gaseous cation of $+1$ charge is known as its **first ionization energy** (often called ionization potential, and abbreviated IE or IP).

$$Na(g) \rightarrow Na^+(g) + e^- \qquad \Delta H = IE(1) = +495.8 \text{ kJ/mol} \qquad (6.2)$$

Such a process is inherently endothermic, as separating positive and negative charges (cations and electrons) in the gas phase requires an input of energy. The first ionization energies of the elements are tabulated in Table 6.6.

We may then remove a second electron to produce a gaseous cation of $+2$ charge. This requires a **second ionization energy** (IE(2), Table 6.7), which is substantially larger than the first ionization energy, since an electron must be pulled away from an already positively charged ion. Likewise, third ionization energies (Table 6.8) are larger yet. These energy terms would need to be included in the Born-Haber cycle for the formation of a metal di- or trihalide.

The energy change involved in *adding* an electron to an atom to form an anion is known as the **electron affinity** of that element (Table 6.9). Due to an unfortunate tradition regarding the signs of electron affinities, they are better regarded as the energies required to *remove* the electron of a gaseous anion of $-1$ charge to produce a gaseous atom of that element:

$$Cl^-(g) \rightarrow Cl(g) + e^- \qquad \Delta H = EA = +348.8 \text{ kJ/mol} \qquad (6.3)$$

As we see from Table 6.9, not all gaseous anions hold on to their electrons well; many have zero or negative electron affinities. No gaseous anion holds onto a second extra electron well: The second electron affinity even of oxygen (as $O^{2-}$ ion) is $-780$ kJ/mol.

Thus in step 2, we must add the first ionization energy of sodium, $+496$ kJ/mol, but *subtract* the electron affinity of chlorine, since the chloride ion is not being stripped of its electron but is instead being formed. Note that adding all the energy terms listed for the first two steps gives $+376$ kJ/mol; the reaction of two of the most active elements would be hopelessly endothermic were it not for the final step.

Finally, in step 3, we allow the gaseous ions to come together to form the final solid product. As we recall from Chapter 3, a very large amount of energy, the *lattice energy*, $U$, is released at this step. Adding all these energy terms should then give us the observed enthalpy of formation of a metal halide such as NaCl:

$$\Delta H_f(MX_n) = \Delta H_{atom}(M) + n\,\Delta H_{atom}(X) + \sum IE(n) \text{ for M} \qquad (6.4)$$
$$- n\,EA(X) + U$$

In fact, one of the practical uses of the Born-Haber cycle is to obtain experimental values of lattice energies, to which calculated values such as the $-751$ kJ/mol for NaCl can be compared. Subtracting the energy terms obtained for steps 1 and

**Table 6.6**  First Ionization Energies of the Elements (kJ/mol)

| | $s^1$ | $s^2$ | $s^2d^1$ | $s^2d^2$ | $s^2d^3$ | $s^2d^4$ | $s^2d^5$ | $s^2d^6$ | $s^2d^7$ | $s^2d^8$ | $s^2d^9$ | $s^2d^{10}$ | $s^2p^1$ | $s^2p^2$ | $s^2p^3$ | $s^2p^4$ | $s^2p^5$ | $s^2p^6$ |
|---|---|---|---|---|---|---|---|---|---|---|---|---|---|---|---|---|---|---|
| 1 | H 1312 | | | | | | | | | | | | | | | | | He 2372 |
| 2 | Li 520 | Be 899 | | | | | | | | | | | B 801 | C 1086 | N 1402 | O 1314 | F 1681 | Ne 2081 |
| 3 | Na 496 | Mg 738 | | | | | | | | | | | Al 578 | Si 786 | P 1012 | S 1000 | Cl 1251 | Ar 1520 |
| 4 | K 419 | Ca 590 | Sc 631 | Ti 658 | V 650 | Cr 653 | Mn 717 | Fe 759 | Co 758 | Ni 737 | Cu 746 | Zn 906 | Ga 579 | Ge 762 | As 944 | Se 941 | Br 1140 | Kr 1351 |
| 5 | Rb 403 | Sr 550 | Y 616 | Zr 660 | Nb 664 | Mo 685 | Tc 702 | Ru 711 | Rh 720 | Pd 805 | Ag 731 | Cd 868 | In 558 | Sn 709 | Sb 832 | Te 869 | I 1008 | Xe 1170 |
| 6 | Cs 376 | Ba 503 | Lu 524 | Hf 654 | Ta 761 | W 770 | Re 760 | Os 840 | Ir 880 | Pt 870 | Au 890 | Hg 1007 | Tl 589 | Pb 716 | Bi 703 | Po 812 | At | Rn 1037 |
| 7 | Fr | Ra 509 | Lr 490 | Rf | Ha | | | | | | | | | | | | | |

| | $s^2f^1$ | $s^2f^2$ | $s^2f^3$ | $s^2f^4$ | $s^2f^5$ | $s^2f^6$ | $s^2f^7$ | $s^2f^8$ | $s^2f^9$ | $s^2f^{10}$ | $s^2f^{11}$ | $s^2f^{12}$ | $s^2f^{13}$ | $s^2f^{14}$ |
|---|---|---|---|---|---|---|---|---|---|---|---|---|---|---|
| 6 | La 538 | Ce 528 | Pr 523 | Nd 530 | Pm 536 | Sm 543 | Eu 547 | Gd 592 | Tb 564 | Dy 572 | Ho 581 | Er 589 | Tm 597 | Yb 603 |
| 7 | Ac 490 | Th 590 | Pa 570 | U 590 | Np 600 | Pu 585 | Am 578 | Cm 581 | Bk 601 | Cf 608 | Es 619 | Fm 627 | Md 635 | No 642 |

SOURCE: First ionization energies, IE(1), of the elements listed are taken from James E. Huheey, *Inorganic Chemistry: Principles of Structure and Reactivity*, 3d ed., Harper and Row, New York, 1983, pp. 42–44.
NOTE: Group headings indicate the characteristic valence-electron configurations of the atoms being ionized.

**Table 6.7** Second Ionization Energies of (+1 Ions of) the Elements (kJ/mol)

| Period | $s^1$ | | | | | | | | | | | $s^2$ | $s^2p^1$ | $s^2p^2$ | $s^2p^3$ | $s^2p^4$ | $s^2p^5$ | $s^2p^6$ |
|---|---|---|---|---|---|---|---|---|---|---|---|---|---|---|---|---|---|---|
| 1 | | | | | | | | | | | | | | | | | He 5250 | Li 7298 |
| 2 | Be 1757 | | | | | | | | | | | B 2427 | C 2353 | N 2856 | O 3388 | F 3374 | Ne 3952 | Na 4562 |
| 3 | Mg 1451 | | | | | | | | | | | Al 1817 | Si 1577 | P 1903 | S 2251 | Cl 2297 | Ar 2666 | K 3051 |
| 4 | Ca 1145 | Sc 1235 | Ti 1310 | V 1414 | Cr 1496 | Mn 1509 | Fe 1561 | Co 1646 | Ni 1753 | Cu 1958 | Zn 1733 | Ga 1979 | Ge 1537 | As 1798 | Se 2045 | Br 2100 | Kr 2350 | Rb 2633 |
| 5 | Sr 1064 | Y 1181 | Zr 1267 | Nb 1382 | Mo 1558 | Tc 1472 | Ru 1617 | Rh 1744 | Pd 1875 | Ag 2074 | Cd 1631 | In 1821 | Sn 1412 | Sb 1595 | Te 1790 | I 1846 | Xe 2046 | Cs 2230 |
| 6 | Ba 965 | Lu 1340 | Hf 1440 | Ta | W | Re | Os | Ir | Pt 1791 | Au 1980 | Hg 1810 | Tl 1971 | Pb 1450 | Bi 1610 | Po | At | Rn | Fr |
| 7 | Ra 979 | Lr | Rf | Ha | 106 | 107 | | 109 | | | | | | | | | | |

| 6 | La 1067 | Ce 1047 | Pr 1018 | Nd 1034 | Pm 1052 | Sm 1068 | Eu 1085 | Gd 1170 | Tb 1112 | Dy 1126 | Ho 1139 | Er 1151 | Tm 1163 | Yb 1175 |
|---|---|---|---|---|---|---|---|---|---|---|---|---|---|---|
| 7 | Ac 1170 | Th 1110 | Pa | U | Np | Pu | Am | Cm | Bk | Cf | Es | Fm | Md | No |

SOURCE: Second ionization energies, IE(2), of the +1 ions of the elements listed are taken from James E. Huheey, *Inorganic Chemistry: Principles of Structure and Reactivity*, 3d ed., Harper and Row, New York, 1983, pp. 42–44.
NOTE: Group headings indicate the valence electron configurations of the ions being ionized a second time.

**Table 6.8**  Third Ionization Energies of (+2 Ions of) the Elements (kJ/mol)

| | $d^1$ | $d^2$ | $d^3$ | $d^4$ | $d^5$ | $d^6$ | $d^7$ | $d^8$ | $d^9$ | $d^{10}$ | $s^1$ | $s^2$ | $s^2p^1$ | $s^2p^2$ | $s^2p^3$ | $s^2p^4$ | $s^2p^5$ | $s^2p^6$ |
|---|---|---|---|---|---|---|---|---|---|---|---|---|---|---|---|---|---|---|
| 1 | | | | | | | | | | | | | | | | | Li 11815 | Be 14849 |
| 2 | | | | | | | | | | | B 3660 | C 4621 | N 4578 | O 5300 | F 6050 | Ne 6122 | Na 6912 | Mg 7733 |
| 3 | | | | | | | | | | | Al 2745 | Si 3232 | P 2912 | S 3361 | Cl 3822 | Ar 3931 | K 4411 | Ca 4912 |
| 4 | Sc 2389 | Ti 2652 | V 2828 | Cr 2987 | Mn 3248 | Fe 2957 | Co 3232 | Ni 3393 | Cu 3554 | Zn 3833 | Ga 2963 | Ge 3302 | As 2736 | Se 2973 | Br 3500 | Kr 3565 | Rb 3900 | Sr 4210 |
| 5 | Y 1980 | Zr 2218 | Nb 2416 | Mo 2621 | Tc 2850 | Ru 2747 | Rh 2997 | Pd 3177 | Ag 3361 | Cd 3616 | In 2705 | Sn 2943 | Sb 2440 | Te 2698 | I 3200 | Xe 3100 | Cs | Ba |
| 6 | Lu 2022 | Hf 2250 | Ta | W | Re | Os | Ir | Pt | Au | Hg 3300 | Tl 2878 | Pb 2082 | Bi 2466 | Po | At | Rn | | |

| | $f^1$ | $f^2$ | $f^3$ | $f^4$ | $f^5$ | $f^6$ | $f^7$ | $f^8$ | $f^9$ | $f^{10}$ | $f^{11}$ | $f^{12}$ | $f^{13}$ | $f^{14}$ |
|---|---|---|---|---|---|---|---|---|---|---|---|---|---|---|
| 6 | La 1850 | Ce 1949 | Pr 2086 | Nd 2130 | Pm 2150 | Sm 2260 | Eu 2400 | Gd 1990 | Tb 2110 | Dy 2200 | Ho 2200 | Er 2190 | Tm 2284 | Yb 2415 |
| 7 | Ac | Th 1930 | Pa | U | Np | Pu | Am | Cm | Bk | Cf | Es | Fm | Md | No |

SOURCE: Third ionization energies, IE(3), of the +1 ions of the elements listed are taken from James E. Huheey, *Inorganic Chemistry: Principles of Structure and Reactivity*, 3d ed., Harper and Row, New York, 1983, pp. 42–44.
NOTE: Group headings indicate the valence electron configurations of the ions being ionized a third time.

**Table 6.9**  Electron Affinities of the Elements (kJ/mol) (Zeroth Ionization Energies of the −1 Ions of the Elements)

| | $s^1$ | $s^2$ | $s^2p^1$ | | | | | | | | | | | $s^2p^2$ | $s^2p^3$ | $s^2p^4$ | $s^2p^5$ | $s^2p^6$ |
|---|---|---|---|---|---|---|---|---|---|---|---|---|---|---|---|---|---|---|
| 1 | | H 73 | | | | | | | | | | | | | | | | |
| 2 | He 0 | Li 60 | Be 0 | | | | | | | | | | | B 27 | C 122 | N −7 | O 141 | F 328 |
| 3 | Ne 0 | Na 53 | Mg 0 | | | | | | | | | | | Al 44 | Si 134 | P 72 | S 200 | Cl 349 |
| 4 | Ar 0 | K 48 | Ca 0 | Sc 0 | Ti 20 | V 50 | Cr 64 | Mn 0 | Fe 24 | Co 70 | Ni 111 | Cu 118 | Zn 0 | Ga 29 | Ge 120 | As 77 | Se 195 | Br 325 |
| 5 | Kr 0 | Rb 47 | Sr 0 | Y 0 | Zr 50 | Nb 100 | Mo 100 | Tc 70 | Ru 110 | Rh 120 | Pd 60 | Ag 126 | Cd 0 | In 29 | Sn 121 | Sb 101 | Te 190 | I 295 |
| 6 | Xe 0 | Cs 46 | Ba 0 | Lu 50 | Hf 0 | Ta 60 | W 60 | Re 15 | Os 110 | Ir 160 | Pt 205 | Au 223 | Hg 0 | Tl 30 | Pb 110 | Bi 110 | Po 180 | At 270 |

SOURCE: Electron affinities (EA) of the elements listed are taken from James E. Huheey, *Inorganic Chemistry: Principles of Structure and Reactivity*, 3d ed., Harper and Row, New York, 1983, pp. 42–44.

NOTE: Group headings indicate the characteristic valence electron configurations of the anions being ionized.

2 (+376 kJ/mol) from the enthalpy of formation of NaCl (−411 kJ/mol) gives us an experimental lattice energy of −787 kJ/mol. The calculated value of the lattice energy for NaCl (−751 kJ/mol, Section 3.6) is close enough to this to give us confidence in the essential correctness of our model for the Coulombic attractive forces in ionic compounds (Chapter 3).

Not all such calculations agree so well with experimental lattice energies, however. For example, the enthalpy of formation of silver bromide is −96 kJ/mol [9]. Subtracting from this the other component energy terms analogous to those in Figure 6.10, we obtain an experimental lattice energy of −899 kJ/mol.

AgBr crystallizes in the NaCl lattice; substituting the appropriate ionic radii, Madelung constant, Born exponents, and charge into equation (3.11) gives us a calculated lattice energy of −702 kJ/mol. This is seriously short of the experimental value. In fact, had we taken this calculated value as valid and added it to the other terms in Figure 6.10, we would have predicted an enthalpy of formation for AgBr of +101 kJ/mol. This value suggests that AgBr is thermodynamically unstable, and that if it exists it could be explosive! Since cameras using photographic processes based on AgBr seldom if ever explode, we may conclude that the Coulombic model for the attraction of $Ag^+$ and $Br^-$ fails to account completely for the lattice energy of this compound. If we analyzed a number of other such compounds, we would find that this result (a lattice energy more exothermic than calculated) was characteristic when the *electronegativities* of the cationic and anionic elements were reasonably close. Thus equation (3.11) is deficient insofar as it does not take electronegativity into account. When the two ions are of comparable electronegativity their interaction will no longer be purely Coulombic but will also involve covalent bonding.

Another use of Born-Haber calculations is in predicting whether unknown compounds will be thermodynamically stable or not. When Bartlett prepared the surprising ionic compound $O_2^+PtF_6^-$, he noticed that the first ionization energy of the "noble gas" xenon was lower than the first ionization energy of the $O_2$ molecule. He made a rough Born-Haber calculation to predict that there might be a stable compound $Xe^+PtF_6^-$. So (in a vacuum system) he mixed xenon gas and $PtF_6$ gas and obtained a solid ionic product [10] that created quite a sensation in inorganic chemistry, since at the time it was thought that noble gases could form no compounds.

EXAMPLE Extrapolate the appropriate properties of the element below mercury, "eka-mercury," and use these to predict whether this element will form a fluoride analogous to $HgF_2$.

Solution ■ Extrapolation of numerical values down a short and irregular group such as 12 is of course a risky business, but let us try anyway just for fun. We extrapolate the following values for eka-Hg: $\Delta H_{atom} = +20$ kJ/mol; IE(1) = +1130 kJ/mol; IE(2) = +2000 kJ/mol. We also need the known terms for fluorine: $2(\Delta H_{atom}) = +158$ kJ/mol; −2(EA) = −656 kJ/mol. Thus the enthalpy of formation of eka-$HgF_2$ should be roughly +2652 kJ/mol plus the lattice energy of this compound.

■ Since the electronegativity of F is much higher than that of Hg, we will assume that equation (3.11) will also give us a reasonably accurate lattice energy for eka-$HgF_2$. We extrapolate an ionic radius of 123 pm for eka-$Hg^{2+}$, which is virtually identical with that of $F^-$ (119 pm), so the fluorite lattice type should be adopted (Madelung constant 2.51939). We will take the Born exponent of eka-$Hg^{2+}$ to be 14; averaging this with the Born exponent for $F^-$, 7, gives us 10.5. Substituting these numbers into equation (3.11), we predict a lattice energy for eka-$HgF_2$ of about −2617 kJ/mol. □

If we believe these numbers (and in view of their uncertainties we should *not*), $\Delta H_f$ for the fluoride of eka-Hg will be roughly +35 kJ/mol. Unless covalent bonding stabilizes this compound, it will be thermodynamically unstable. Since fluorine is the strongest of oxidizing agents and gives some of the greatest lattice energies, this suggests the possibility that the element below mercury may not form *any* compounds at all—i.e., that it may be a noble gas or liquid (the small heat of atomization may not suffice to keep it a solid). As we will see in the last chapter, there are sounder reasons than the calculations we have just carried out for believing that this might turn out to be the case.

## 6.9

## Thermochemical Analysis of the Activity Series of the Elements

In Experiment 4 we examined the activity of the metals as reducing agents (for water or hydrogen ion), and found a better correlation of their activity with their Pauling electronegativities than with their (first) ionization energies. This may well have surprised you, since the reaction of the metals with water or hydrogen ion produces metal ions and obviously does involve ionization of the metals.

One problem in correlating activity and first ionization energies is that most of the metals form, not +1, but +2 or +3 (or even +4) ions. Hence the second and third (and perhaps fourth) ionization energies should have been involved in the correlation. But there is still another problem: Ionization energies apply to *gaseous* metal *atoms* giving *gaseous* metal ions, whereas the reactions of metals with water or hydrogen ion involve *solid* (bulk) metals giving *hydrated* metal ions. To find the relationship of activity to ionization energies, we need to construct a thermochemical cycle for the relatively complicated process of dissolving a metal in standard (1 M) acid:

$$M(s) + z\,H^+(aq) \rightarrow M^{+z}(aq) + \frac{z}{2}H_2(g) \tag{6.5}$$

We can generalize the thermochemical cycle for the reaction of any metal with 1 M hydrogen ion; the generalized cycle is shown schematically in Figure 6.11.

Let us apply this process to the (dangerously exothermic) reaction of sodium metal with 1 M aqueous hydrogen ion. In the first step we take the two reactants to the gaseous state. For the hydrogen ion we must remove its water of hydration;

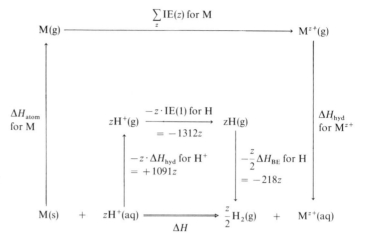

**Figure 6.11** Thermochemical cycle for the reaction of a metal in its standard state with 1 $M$ aqueous hydrogen ion to give an aqueous metal ion of charge $+z$ and gaseous hydrogen. For interpretation of the symbols for the different enthalpy changes see the text.

to do this we must *supply* the energy of hydration of $H^+$, $-\Delta H_{hyd} = +1091$ kJ/mol (Table 2.1). The sodium metal must be atomized; its heat of atomization $\Delta H_{atom}$ is $+107$ kJ/mol.

In the second step we allow the gaseous sodium atom and hydrogen ion to react—an electron is ionized from sodium (giving gaseous $Na^+$) and is acquired by the hydrogen ion (giving a gaseous H atom). To do this we must supply the first ionization energy of Na, $\Delta H_{IE(1)} = +496$ kJ/mol from Table 6.6, but we allow the *reverse* of the first ionization of hydrogen to occur, $-\Delta H_{IE(1)} = -1312$ kJ/mol.

In the third step we allow the gaseous sodium ion to hydrate itself, $\Delta H_{hyd} = -405$ kJ/mol. We allow the individual hydrogen atoms to come to their final state, as gaseous diatomic molecules. Energy is released when $\frac{1}{2}$ mol of covalent H—H bonds is formed per mole of H atoms. This energy is the negative of the heat of atomization of H (or one-half the reverse of the conventional H—H bond dissociation energy, $-\frac{1}{2}\Delta H_{BE}$), $-218$ kJ/mol.

Finally, we apply Hess's law and add the energy changes in all these steps. The terms involving the hydrogen add up to $-439$ kJ/mol, while those involving sodium give $+198$ kJ/mol; overall the theoretically calculated enthalpy change for the reaction of sodium metal with 1 $M$ acid is $-241$ kJ/mol. Experimentally, we know that the standard reduction potential of sodium ion is $-2.71$ V (Table 5.4). This value can be converted to thermodynamic units using the relationship

$$\Delta G^\circ = -nFE^\circ \qquad (6.6)$$

where $n$ is the number of electrons involved in the redox reaction (one in this case) and $F$, the faraday, is 96.5 kJ per electron volt. (In the thermochemical cycle of Figure 6.11, the metal ions are not being reduced but are being formed by oxidation; hence we also need to reverse the sign of the standard reduction potential.) If we neglect entropy changes (i.e., formation of gaseous $H_2$, hydration and dehydration of hydrated ions), we can equate $\Delta G$ and $\Delta H$ and obtain $\Delta G$ for the oxidation of sodium, $-262$ kJ/mol. In view of our neglect of entropy changes, the agreement with the calculated enthalpy change, $-241$ kJ/mol, is good.

The calculations for the activity of sodium can be applied to the activity of any metal by substituting the appropriate atomization and ionization energies. We may note that the overall theoretical enthalpy changes involving hydrogen are constant at $-439$ kJ per mole of $H^+$ involved. Thus the generalized $\Delta H$ for the reaction is

$$\Delta H = \Delta H_{atom} + \sum IE(z) + \Delta H_{hyd} - 439z \qquad (6.7)$$

where $z$ is the number of electrons and hydrogen ions involved and is the charge on the metal cation formed. Finally, assuming the equality of $\Delta H$ and $\Delta G$, then replacing $\Delta G$ by the negative of the standard reduction potential, we obtain in general:

$$96.5\ E^\circ = \frac{\Delta H_{atom}}{z} + \frac{\sum IE(z)}{z} + \frac{\Delta H_{hyd}}{z} - 439 \text{ kJ/mol} \qquad (6.8)$$

## 6.10
## Periodicity and Significance of Ionization Energies and Electron Affinities

SUCCESSIVE IONIZATION ENERGIES  As we see by paging through the tables of electron affinities (*zeroth ionization energies*), followed by the first, second, and third ionization energies, for a given element successive ionizations generally become more difficult in a relatively regular pattern. But if the $n$th ionization is one that begins removing core electrons from an atom, that ionization will naturally be much more difficult than would otherwise have been the case.

HORIZONTAL PERIODIC TRENDS  We generally find that ionization energies and electron affinities increase across a period. Among atoms using $2s$ and $2p$ valence electrons, for example, ionization is easiest for the $2s^1$ electron configuration (IE(1) of Li, IE(2) of Be$^+$, IE(3) of B$^{2+}$, EA of He$^-$; Figure 6.12). Ionization becomes more difficult farther to the right for atoms or ions of the same charge. The highest ionization energy is required for atoms or ions of the $2s^2 2p^6$ electron configuration (IE(1) of Ne, IE(2) of Na$^+$, IE(3) of Mg$^{2+}$, EA of F$^-$). When the electron being lost changes from being an $s$ valence electron to a $p$ valence electron, however, ionization becomes easier (e.g., compare the first IE's of B and Be). When the electron being removed first comes from a more than half-filled subshell, ionization is also some-what easier, since an electron in a filled orbital is repelled by the other electron occupying the same orbital. Thus the first IE of O ($2s^2 2p^4$, with one paired-up $p$ electron) is less than that of N ($2s^2 2p^3$, with each $p$ electron in its own orbital).

VERTICAL PERIODIC TRENDS  Generally speaking, we find that the $n$th ionization energy or electron affinity *decreases* down a group: The electron being ionized is in a larger orbital of higher principal quantum number and hence is not attracted as

| Valence electron configuration: | $s^1$ | $s^2$ | $s^2p^1$ | $s^2p^2$ | $s^2p^3$ | $s^2p^4$ | $s^2p^5$ | $s^2p^6$ |
|---|---|---|---|---|---|---|---|---|
| Examples: | $He^-$ | $Li^-$ | $Be^-$ | $B^-$ | $C^-$ | $N^-$ | $O^-$ | $F^-$ |
| | Li | Be | B | C | N | O | F | Ne |
| | $Be^+$ | $B^+$ | $C^+$ | $N^+$ | $O^+$ | $F^+$ | $Ne^+$ | $Na^+$ |
| | $B^{2+}$ | $C^{2+}$ | $N^{2+}$ | $O^{2+}$ | $F^{2+}$ | $Ne^{2+}$ | $Na^{2+}$ | $Mg^{2+}$ |

**Figure 6.12**  Horizontal periodic trends in ionization energies and electron affinities.

strongly to the nucleus. A couple of irregularities are noteworthy, however: (1) The electron affinities of the *p*-block elements of the second period are unexpectedly less than those of the third (or later) periods because of the great electron-electron repulsions found in such small anions as $F^-$ and $O^-$; (2) the ionization energies of some of the later *d*-block elements of the sixth period and early *f*-block elements of the seventh period are higher than expected. (Gold has the highest electron affinity of any element other than a halogen.) The reason for this is not immediately apparent but will be explored in Chapter 11.

As an illustration of the significance of ionization energies, let us consider the question of the relative stabilities of the $+2$ and $+3$ ions of the *d*-block metals of the fourth period. As we see from Table B, there is a rather irregular alternation in stability of these two oxidation states across the fourth period. In the production of a $+3$ ion by dissolving a *d*-block metal in acid, we may consider the final stage to be the following:

$$M^{2+} + H^+ \rightarrow M^{3+} + \tfrac{1}{2}H_2 \qquad (6.9)$$

A thermochemical cycle may be set up for this reaction for which the energy terms add up as follows:

$$96.5E^\circ = IE(3) + \Delta H_{hyd}(+3 \text{ ion}) - \Delta H_{hyd}(+2 \text{ ion}) - 439 \qquad (6.10)$$

Using the Latimer equation (2.1), we can estimate the hydration energies of the $+3$ and $+2$ ions for which radii are given in Table C. Taking the difference between these hydration energies gives us a contribution ranging from $-2560$ kJ/mol for Ti to $-2693$ kJ/mol for Cr. (The hydration energies are always more favorable for the $+3$ ions, of course.) This variation is, however, much smaller than the variation in IE(3), which ranges from $+2389$ kJ/mol for Sc to $+3833$ kJ/mol for Zn.

Thus the variation in the third ionization energies of the $d$-block elements of the fourth period is largely responsible for the varying stability of the $+3$ ions as compared with the $+2$ ions. In general, IE(3) increases across the fourth period in the $d$-block; hence $+3$ ions generally are energetically unfavorable after Cr. However, the $+3$ ion of Fe is more easily formed, since it involves the relatively easy ionization of the sixth $d$ electron of $Fe^{2+}$, which is one electron above a half-filled subshell.

Similarly, the third ionization energy increases across the sixth period in the $f$-block, resulting in reduced instability of the $+2$ oxidation state—until $Gd^{2+}$ is reached. $Gd^{2+}$ has one electron above a half-filled subshell and is therefore very readily oxidized to $Gd^{3+}$. But the ease of the third ionization then decreases again to the end of the $f$-block; hence although the $+3$ oxidation state is the most stable for all the $f$-block elements in the sixth period, the $+2$ oxidation state is least unfavorable for the element before Gd ($Eu^{2+}$, with a half-filled subshell) and at the end of the $f$-block ($Yb^{2+}$, with a filled subshell).

Finally, it is important to remember the distinction between the terms electronegativity and electron affinity. *Electronegativity* refers to the attraction of an atom *in a compound* for its *shared electron pairs. Ionization energy* and *electron affinity* refer to attractions of *isolated gaseous* atoms or ions for their (unshared) electrons. The distinction is not just semantic: If you refer to the anomalies in the periodic trends in Pauling electronegativities (discussed in Chapter 1), you will see that they are often dissimilar to the anomalies discussed above.

It has been suggested that there should be some relationship between the two properties, however. There is a scale of **Mulliken** electronegativities that are obtained by scaling the *average of the electron affinity and first ionization energy* of each element. But the electron affinities and ionization energies used are not those from Tables 6.6 to 6.9 but are for electrons coming from orbitals that are averages of the valence orbitals used by that atom (*hybrid orbitals*) in its compounds. For the $p$-block elements, in which $s$ and $p$ valence orbitals are used, the Mulliken electronegativities do indeed agree satisfactorily with the Pauling electronegativities that we have been using. Unfortunately Mulliken electronegativities are not available for the $d$-block elements, since it is very difficult to determine to what extent each of the three types of available orbitals ($s$, $p$, and $d$) are being used in the compounds of $d$-block elements.

## 6.11

## Significance of Atomization Energies in Redox Chemistry

To see the importance of the atomization energies of the metals, let us calculate standard reduction potentials from equation (6.8) with the $\Delta H_{atom}$ term deleted— i.e., for the reaction of *gaseous* metal atoms with acids. This calculation amounts to subtracting a factor of $\Delta H_{atom}/96.5z$ from the recorded standard reduction potentials (Table 5.4); in the case of sodium this changes its standard reduction potential from $-2.71$ V to $-3.82$ V. More interestingly, it changes the reduction potential of the least active metal, gold, from $+1.68$ V to $-2.13$ V—i.e., makes it almost as active as magnesium actually is [11].

In fact, it *is* now possible to start reactions with gaseous metal atoms, by providing the atomization energy separately with a high-temperature resistance heater in a high-vacuum apparatus known as a *metal-atom reactor*. The metal atoms produced by this means are in fact much more reactive than bulk metals as we know them. They must be handled in high vacuum at very low temperatures to prevent them from reacting with each other, with other gaseous substances, or with the walls of the reaction vessel. With the enhanced activity possessed by gaseous metal atoms, it is possible to carry out many reactions that are completely impossible with solid or liquid metals [12].

EXAMPLE

Compare the activity of the metals Hg, Pt, Pd, and Ba (all of which form $+2$ ions) in the solid state; as gaseous atoms.

Solution

■ We may use the standard reduction potentials of these metals as indicators of their solid-state activities. Referring to Table 5.4, we can set up an activity series: Activity of Ba $(-2.91$ V$) \gg$ Pb $(-0.13$ V$) \gg$ Hg $(+0.85$ V$) >$ Pt $(+1.2$ V$)$. This is the order we would have expected based on the electronegativities of the elements.

■ To calculate the gas-phase activities either we can use equation (6.8), omitting the atomization term, or we can subtract the atomization energy (in volts) from the standard reduction potentials. Either way of calculating may introduce some errors resulting from entropy changes, but we will disregard these. If we take the latter approach, we find that we must subtract the following amounts from the standard reduction potentials: Ba, 0.94 V; Pb, 1.01 V; Hg, 0.33 V; Pt, 2.92 V. This causes platinum to jump up dramatically in the activity series: Ba $(-3.85$ V$) \gg$ Pt $(-1.72$ V$) >$ Pb $(-1.14$ V$) \gg$ Hg $(+0.52$ V$)$. □

There is one problem in trying to calculate the gas-phase activities of these metals from equation (6.8) without the atomization term: The hydration energies of $Hg^{2+}$ and $Pt^{2+}$ are not known. On the other hand, we can use equation (6.8) in its complete form to calculate those hydration energies, although the resulting values necessarily assume a zero entropy change in the activity reaction, which is not likely the case. Doing so anyway gives estimated hydration energies of $-1839$ kJ/mol for $Hg^{2+}$ and $-2115$ kJ/mol for $Pt^{2+}$, which are qualitatively in line with those given in Table 2.1. Alternatively, the hydration energies can be estimated using the Latimer equation (2.1). But the Latimer equation does not include a term for the effects of the high electronegativities of Hg and Pt, and it cannot correct for the fact that these metal ions (especially $Pt^{2+}$) have anomalously small coordination numbers.

As another illustration of the significance of atomization energies, let us consider further some of the redox chemistry of the elements with especially high atomization energies (e.g., carbon and the metals in the vicinity of tungsten). Certainly we can understand how the high atomization energies of these elements reduce their activities. But although these elements are relatively reluctant to assume positive oxidation states, when they do so they prefer to assume *very high* positive oxidation states. Thus although Mo and W are much less active metals than Cr, they show a strong preference for the $+6$ oxidation state when they are oxidized, while Cr prefers $+3$ (Table B).

To analyze this situation, let us suppose that we have 2 mol of $Mo^{3+}$ or $W^{3+}$ ions. Let us consider the possibility of a *disproportionation* reaction:

$$2\,M^{3+}(aq) \rightleftharpoons M^{6+}(aq) + M(s) \tag{6.11}$$

This disproportionation reaction will be more favorable for Mo and W than for Cr, since producing the solid metal will recover the heat of atomization, which is more significant for Mo and W than for Cr. Looking at this from another angle, we note that the $3d$ orbitals of Cr are not so readily available for bonding; in the $Cr^{3+}(aq)$ ion these hold three electrons that are not used in bonding to the waters of hydration. In the products $M^{6+}(aq)$ (or, more realistically, $MO_3(s)$), bonding to oxygen does occur; in M(s) bonding of metal atoms to each other does occur. The $4d$ orbitals of Mo and the $5d$ orbitals of W are more suitable for such extensive bonding; it is energetically unfavorable to let these orbitals be "wasted" as in $Cr^{3+}(aq)$.

## 6.12

## Study Objectives

1. Compare (qualitatively) the relative heats of atomization and the boiling points of two elements, using periodic trends. Explain these values in terms of the valence electron configurations of the two elements.
2. Describe the structures and physical properties (allotropes, physical state, color, ease of boiling, conductivity) of any nonmetallic elements.
3. Distinguish the following types of atomic radii and compare them in magnitude for a given atom: covalent radius, metallic radius, cationic radius, anionic radius, van der Waals radius.
4. Draw and distinguish sigma, pi, and delta bonds. Know and explain the periodic trends in abilities of atoms to form these kinds of bonds and in the strengths of these bonds when formed.
5. Use sigma- and pi-bond energies to calculate the relative stabilities of allotropes of elements that differ in their number of sigma and pi bonds.
6. Give examples of elements that will be most useful (a) as oxidizing agents, (b) as inert gases, (c) as semiconductors, (d) as structural metals, and (e) for applications requiring malleability or ductility.
7. Know periodic trends in the ability of nonmetals to form catenated compounds; give examples for nonmetals of catenated compounds or ions.

8. Given thermodynamic and electrochemical data, set up a Born-Haber cycle and an equation for the oxidation of a metal, and calculate an energy term from it.

9. Set up thermochemical cycles for other redox reactions (such as that of the activity series of elements) and calculate energy terms or explain periodic trends in activity using them.

10. Distinguish clearly among ionization energy, electron affinity, and electronegativity, and know the main periodic trends in each.

## 6.13

## Exercises

1. Without referring to Table 6.1, choose the element from each set that should have the highest atomization energy and the element that should be easiest to atomize, and explain why you made the choice you did:
   **a.** C, Si, Ge, Sn, Pb; **b.** V, Nb, Ta;
   **c.** Li, Be, B, C, N, O, F, Ne;
   **d.** Sr, Zr, Mo, Ru, Pd, Cd, Sn, Te, Xe.

*2. Which element in each set shows the highest boiling point?
   **a.** C, Si, Ge, Sn, Pb;
   **b.** Sr, Zr, Mo, Ru, Pd, Cd, Sn, Te, Xe.

3. Draw the structure of each of the following elements, and characterize each as either a *gas, liquid, easily vaporized solid*, or *hard-to-vaporize solid*:
   **a.** boron, **b.** white phosphorus, **c.** black phosphorus,
   **d.** molybdenum.

4. Give the symbols of two elements to which each of the following apply: **a.** occur in the diamond structure; **b.** occur in polymeric helical chains; **c.** occur in a black, sheetlike allotrope; **d.** occur as diatomic molecules; **e.** occur as tetrahedral $X_4$ molecules; **f.** occur as $X_8$ molecules; **g.** occur at room temperature in allotropes with multiple bonds; **h.** occur as monoatomic gases.

5. What element occurs as: **a.** linked icosahedra of atoms; **b.** in both a diamond-type lattice and a metallic lattice?

*6. Many allotropes of the nonmetals do not melt and/or boil but convert to other allotropes on heating (or cooling). Pretending for the moment that this does not happen, classify each allotrope in Figure 6.1 as one of the following: **a.** likely to have very high melting and boiling points; **b.** likely to be a gas or a liquid at room temperature; or **c.** likely to be a solid with a relatively low melting point. Explain, in general, why classes **a**, **b**, and **c** contain the allotropes they do.

7. **a.** Draw the structures of the three major allotropes of phosphorus. **b.** Which of the three forms should have the highest vapor pressure at room temperature? **c.** Which form is chemically the most reactive? **d.** Which has the least stable sigma bonds, and why? **e.** Which has no counterpart in arsenic or antimony chemistry?

*8. Choose the largest and the smallest of each set: **a.** Li, C, F, Ne; **b.** Be, Ca, Ba, Ra; **c.** cationic radius of Na, van der Waals radius of Na, covalent radius of Na.

9. The single-bonded metallic radius of Na is 157 pm. Give a reasonable estimate of the following radii, assuming that the increment between radii of different magnitudes is 60 pm: **a.** the covalent radius of Na; **b.** the cationic radius of Na; **c.** the anionic radius of Na; **d.** the van der Waals radius of Na.

10. Which element shows the greatest tendency to form double or triple bonds? C, Si, Ge, Sn, Pb.

*11. Give the name (sigma, etc.) of each of the kinds of bonds drawn below, and arrange them in order of *increasing* bond dissociation energy.

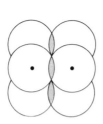 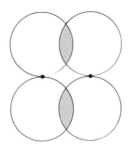 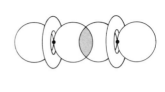

*12. From the sigma and pi O—O bond energies, calculate $\Delta H$ for the conversion of 3 mol $O_2$ to 2 mol $O_3$. Which is more stable?

13. For each of the Group 15/VA and Group 16/VIA elements except Bi and Po, write balanced chemical equations illustrating the process

    multiply bonded allotrope $\rightleftharpoons$ singly bonded allotrope

    (One of these allotropes may have to be an imaginary analogue of a known allotrope of another element in the group.) Using the average sigma- and pi-bond energies from Table 6.5, calculate $\Delta H$ for this process in each case. What is the general trend in the stability of multiply bonded allotropes? Can you explain this in terms of overlap of orbitals and bond lengths?

14. Calculate $\Delta H$ for the conversion of 6 mol of sulfur *atoms* from the allotropic form $S_2$ to the allotropic form $S_3$ (with a structure like that of ozone). Is this reaction favored?

15. What conditions of temperature and pressure should favor the formation of multiply bonded allotropes for Group 15/VA and 16/VIA and why?

16. Which is a better conclusion? Pi-bonded allotropes are found for the second-period elements because: **a.** pi bonds are especially stable among the second-period elements, or **b.** sigma bonds between certain second-period elements are especially unstable. Explain your conclusion, using data from Table 6.5.

*17. Which element shows the greatest tendency to show the properties of a semiconductor? B, Al, Ga, In, Tl.

18. Consider the following elements: Cs, Ti, Co, Ir, As, Bi. **a.** Which one of the metals in this list would be softest? **b.** Which of the metals would be most useful for structural purposes, such as building airplanes? **c.** Which of the elements would be a semiconductor? **d.** Which of the d-block metals in the list would give off the most heat on being alloyed with the metal Zr? **e.** Which of the metals shows ferromagnetism?

19. Briefly explain why, among the three compounds $(C_6H_5)_6C_2$, $(C_6H_5)_6Si_2$, and $(C_6H_5)_6Ge_2$, you may expect the silicon compound to be the most stable.

20. Set up and evaluate a thermochemical cycle to evaluate the enthalpy change for the reaction $2\,Na(s) \rightarrow Na_2(g)$.

*21. Set up a Born-Haber cycle for each of the following reactions, and calculate the experimental lattice energies of the products:
   **21.1** $Ca(s) + F_2(g) \rightarrow CaF_2(s)$;  $\Delta H_f = -1215$ kJ/mol
   **21.2** $Ca(s) + \frac{1}{2}O_2(g) \rightarrow CaO(s)$;  $\Delta H_f = -636$ kJ/mol
   **21.3** Calculate the theoretical lattice energies for these two compounds and compare them with your experimental values.

22. **a.** Set up and evaluate a thermochemical cycle to evaluate the enthalpy change for the reaction $2\,Cr^{3+}(aq) + H_2 \rightarrow 2\,Cr^{2+}(aq) + 2\,H^+(aq)$. **b.** Could $H_2$ be used as a reducing agent to prepare $Cr^{2+}$? **c.** Ignoring entropy effects, calculate $E°$ for this reaction.

23. The first and second ionization energies of Mg and Pb do not differ greatly, but their activities (in reacting with acid to form hydrated +2 ions) do. **a.** Calculate $\Delta H$ for the reaction of Mg metal with $1\,M\,H^+$ ion to produce hydrated $Mg^{2+}$ ions and $H_2$. **b.** Make the same energy calculation for Pb metal. **c.** Which metal is more active? Calculate the approximate standard reduction potential $E°$ for each metal from your $\Delta H$ value. **d.** Fundamentally, what is the cause of the difference in activity between Mg and Pb?

*24. Consider the possibility of a reaction between tin metal and aqueous strontium ion:

$$Sr^{2+}(aq) + Sn(s) \rightleftharpoons Sn^{2+}(aq) + Sr(s)$$

**a.** Draw a diagram of the thermochemical cycle for this reaction, and (using appropriate data) calculate $\Delta H$ for this reaction. **b.** Does this reaction go as written to products? Neglecting entropy effects, calculate the cell potential $E°$ for the reaction as written.

25. (Thought question) Suggest a better way to attempt to correlate activities of metals (Experiment 4) with ionization energy than simply correlating with the first ionization energies of metals. (*Hint*: Look at equation (6.8).) Try this correlation (on graph paper) with the $E°$'s from Table 5.4. Do you get a more

satisfactory correlation with activity using your ionization-energy function than the correlation of activity with electronegativity (Figure 5.6)? If not, why not?

## Notes

[1]  See, for example, N. N. Greenwood and A. Earnshaw, *Chemistry of the Elements*, Pergamon Press, Oxford, 1984.

[2]  Brewer, L., in *Electronic Structure and Alloy Chemistry of the Transition Metals*, P. A. Beck, ed., Interscience, New York, 1963, as cited by W. L. Jolly, *Modern Inorganic Chemistry*, McGraw-Hill, New York, 1984, p. 293.

[3]  Phillips, C. S. G., and R. J. P. Williams, *Inorganic Chemistry*, Oxford University Press, New York, 1966, vol. II, p. 5.

[4]  Pauling, L., *The Nature of the Chemical Bond*, 3d ed., Cornell University Press, Ithaca, N.Y., 1960, Ch. 11.

[5]  See, for example, A. H. Cowley, *Acc. Chem. Res.*, 17, 386 (1984).

[6]  Born, M., *Verhandl. Deut. Physik. Ges.*, 21, 13 (1919); F. Haber, *Verhandl. Deut. Physik. Ges.*, 21, 750 (1919).

[7]  If we do this under constant pressure, this heat is actually an enthalpy change.

[8]  The atomization of a nonmetal may be regarded as the sum of two other steps: the *vaporization* of the nonmetallic molecule $X_n$, requiring its *heat of vaporization*, and the breaking of the bonds in the gaseous $X_n$ molecule, requiring the input of the *bond dissociation energy*. Thus to obtain 1 mol of Cl atoms, we would not need a heat of vaporization, since $Cl_2$ is already gaseous, but we would need to invest one-half the Cl—Cl bond dissociation energy in order to break $\frac{1}{2}$ mol of Cl—Cl bonds.

[9]  Useful compilations of enthalpies of formation and of lattice energies are in M. C. Ball and A. H. Norbury, *Physical Data for Inorganic Chemists*, Longman, London, 1974, pp. 59–96.

[10]  N. Bartlett, *Proc. Chem. Soc. (London)*, 1962, 218; N. Bartlett, *Endeavor*, 88, 3 (1963).

[11]  This neglects the fact that converting a hot gaseous metal to cold liquid-phase hydrated ions will have a new unfavorable entropy term, but this is presumably a relatively minor correction.

[12]  Klabunde, K. J., *Chemistry of Free Atoms and Particles*, Academic Press, New York, 1980.

# Coordination Compounds and the Lewis Acid-Base Concept

Most metal ions are present in the body in very low concentrations, yet they are often extremely important in the function of the body's catalysts—the enzymes. Similarly, many modern processes of industrial organic chemistry have incorporated small amounts of certain metal compounds as powerful catalysts. In this chapter we will see that many of these catalytic functions of metals can be related to the acid-base chemistry of cations and anions that we examined in earlier chapters. To do this we must first broaden our concept of acids and bases. The broadened concept is more abstract than our familiar notion of acids as substances that taste sour and turn litmus paper pink. In order to give you more concrete experience with the nature of Lewis acid-base reactions, your instructor may now assign Experiment 5 from Chapter 13.

## 7.1

## Coordination Compounds and the Lewis Acid-Base Concept: Terms Used

In Chapter 2 we observed that metal cations can distort the electron distribution in water to the point that hydrogen (hydronium) ions are released, so we classified metal cations as acids. Similarly, we classified oxo (and other) anions as bases due to their ability to accept protons from the solvent, water, thus releasing hydroxide ions. But in a sense the presence of water is not essential to the acidic and basic functions of these ions. Indeed, too much water actually *diminishes* the acidity of acidic substances such as oxo acids and metal ions. In very concentrated solutions, such species (as well as their basic counterparts) behave as if they are even more concentrated than they actually are. A characteristic of such concentrated solutions is that there are not enough water molecules present to leave many water molecules free to act as solvent or, often, to complete the secondary (or even the primary) hydration spheres. For example, a concentrated (19 $M$) solution of KOH behaves in many respects in a manner that we would expect from a 10,000 $M$ solution of hydrated hydroxide ions!

Analytical and physical chemists are often concerned with this phenomenon. Concentrated solutions of ions do not show ideal colligative properties (freezing-point depression, etc.) or ideal compliance to the **law of chemical equilibrium** unless account is taken of the enhanced activity of such ions. This activity is often measured in terms of the **activity coefficient** of a dissolved ion, which is the ratio of its apparent or effective concentration to its actual concentration. In sufficiently dilute solutions, in which nearly all the ions are completely hydrated, the activity coefficients of ions approach 1.00. In solutions of moderate concentrations, activity coefficients are less than 1.00 due to the greater tendency for pairing of cations and anions to occur. But in highly concentrated solutions activity coefficients become much greater than 1.00.

This suggests that the ultimate source of acidity and basicity lies not in the water but in the metal cation or the anion; the water actually serves to dampen out the activity of the metal ion or the anion. Inorganic chemists nowadays do many of their reactions in the absence of water. As one example, the reactions of acidic and basic oxides with each other to give salts (Chapter 4) can occur with water absent. If you did Experiment 5, you observed additional examples of reactions that seem very much like acid-base reactions, but in which water does not participate. A scheme for classifying and predicting nonaqueous reaction patterns along the lines that we are familiar with in aqueous solutions thus would be very helpful.

Among his many contributions to chemistry [1], G. N. Lewis, one of the greatest of American chemists, noted many reactions and properties that resembled acid-base reactions and properties, although they occurred in compounds in the complete absence of water and even in compounds that did not contain the elements hydrogen and oxygen. Organic amines such as pyridine (Figure 7.1) act like bases in that they react with (neutralize) acids, yielding solid salts such as pyridinium chloride, $[C_5H_5NH]^+Cl^-$. But Lewis thought it significant that bases such as pyridine also react with numerous metal ions and metal compounds to give solids and solid salts such as $[Cd(C_5H_5N)_2Cl_2]$ and $[Ni(C_5H_5N)_4]^{2+}(ClO_4^-)_2$. In these reactions, the metal ions and salts seem to be reacting very much like familiar, proton-containing acids.

*Acid* and *base* are collective terms defined by chemists in such a way as to include all compounds having similar chemical properties. Thus a statement that "acids do thus and so" economically describes hundreds or thousands of reactions of the compounds included in the definition. Several times in the history of chemistry, chemists have noted additional groups of compounds that show many properties or reactions analogous to those of recognized acids and bases; hence several new, broader definitions of acids and bases have appeared.

G. N. Lewis, who was also involved in work on the octet rule and in pioneering the drawing of dot structures, applied the drawing of these structures to the reactions he felt should be considered as acid-base reactions. In such reactions the product was found to possess a new covalent bond (two shared electrons) not present in either reactant. The two electrons in this bond both came from the *same* reactant, which was in all cases the species that had been accepted as the *base* in the reaction (Figure 7.1).

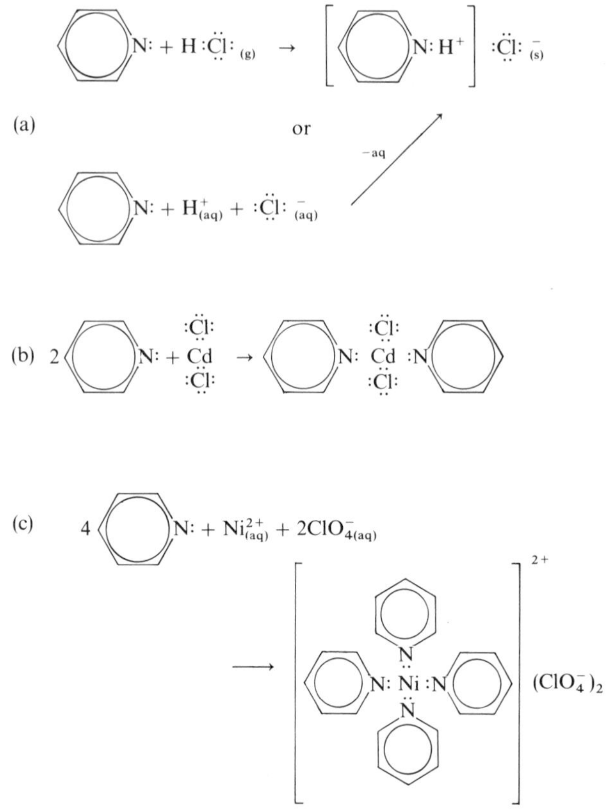

**Figure 7.1** Representation of different Lewis acid-base neutralization reactions, all involving the same Lewis base, pyridine. (a) Reaction with gaseous or aqueous HCl; (b) reaction with $CdCl_2$ to give a neutral, nonionic coordination compound; (c) reaction with $Ni(ClO_4)_2$ to give a product containing a complex ion and unattached perchlorate ions.

Thus Lewis proposed broadening the chemists' definition of bases to include *any species that can donate a pair of electrons.* Such species are now commonly called **Lewis bases, donors,** or **ligands**; organic chemists commonly call them **nucleophiles.** Correspondingly, Lewis proposed broadening the definition of acids to include not only hydrogen-ion-producing materials, but such species as metal ions, since they are also *species that can accept a pair of electrons.* These species are now often called **Lewis acids** or **acceptors**; organic chemists commonly call these **electrophiles** ("electron lovers"). In the product (usually a solid) the two are held together by the sharing of the pair of electrons. The resulting bond is called a **coordinate covalent bond** and is distinguished from an "ordinary" covalent bond by the fact that both electrons in the bond came from the same atom in the Lewis base, the **donor atom** of the base. The species in which the Lewis acid and base are joined together in chemical matrimony is known by various names: It may be called an **acid-base adduct,** a **coordination complex** or **coordination compound,** or (if it is charged) a **complex ion.**

Any atom in the Lewis base that possesses unshared electron pairs capable of being donated is a potential donor atom. For example, in all oxo anions the oxygen atoms are potential donor atoms. The central atoms usually are not, unless they are in their group oxidation number minus two (or lower) and thus have an unshared pair of electrons. Except in the hydride ion ($H^-$), the hydrogen atom does not have an unshared pair of electrons in its compounds and thus does not seem capable of functioning as a donor atom. There are compounds and ions in which any nonmetal atom other than hydrogen (or the noble gases [2]) may have unshared pairs of electrons and thus be a potential donor atom; even metal atoms occasionally can do so. When carbon atoms (such as the C atom in the methide ion, $:CH_3^-$) act as donor atoms, metal-carbon bonds are formed; such compounds are known as **organometallic compounds** [3].

Not only inorganic oxo and other anions, and many inorganic molecules, but also all classes of organic compounds containing oxygen, nitrogen, sulfur, or halogen atoms with unshared electron pairs can function as ligands. Thus we may see that the number of potential Lewis acid-base reactions in the world, or even in our bodies, is enormous. (From an inorganic chemist's point of view, cell fluids are "soups full of ligands" waiting to attach to metal-ion Lewis acids.) If we can develop the acid-base concepts we already have to apply to this greatly expanded list of acids and bases, our power to predict chemical reactions and properties will certainly be enhanced.

A complex ion, when isolated from solution, must be accompanied by ion(s) of opposite charge (**counterions**). To show that there is no sharing of electrons between these ions and the Lewis acid or base in the complex ion, the complex ion itself is traditionally enclosed in brackets. For example, we earlier showed formulas and structures (Figure 7.1) of two coordination complexes, $[Cd(C_5H_5N)_2Cl_2]$ and $[Ni(C_5H_5N)_4]^{2+}(ClO_4^-)_2$. The cadmium compound is one uncharged molecule, enclosed by the brackets, with no charge. But the nickel compound contains a complex ion enclosed by brackets, the $+2$ charge of which is balanced by two perchlorate ions not joined to the nickel ion by any coordinate covalent bond. This distinction is real: Only the latter compound, when dissolved in an appropriate solvent, will conduct an electric current.

The distinction did not come easily to chemists, however: Alfred Werner, a Swiss chemist, won the first Nobel prize awarded for work in inorganic chemistry (1913) for his study of the chemistry of coordination complexes. He prepared numerous such complexes, among them a series of complexes of the $Co^{3+}$ ion with the ligand ammonia, which differed in composition, color, and reactivity. He found that the yellow compound $CoCl_3 \cdot 6NH_3$ conducted electricity as well in solution as, say, $LaCl_3$, and reacted with 3 mol $AgNO_3$ rapidly to precipitate 3 mol $AgCl$. The purple compound $CoCl_3 \cdot 5NH_3$ had a lesser conductivity, comparable to that of $BaCl_2$, and yielded only 2 mol $AgCl$ immediately on treatment with excess $AgNO_3$. He also found two isomers (green and violet, respectively) of $CoCl_3 \cdot 4NH_3$, each of which had a conductivity comparable to that of $NaCl$, and each of which released 1 mol $AgCl$. Later he prepared the compound $CoCl_3 \cdot 3NH_3$, which did not conduct electricity and did not react immediately with $AgNO_3$. Werner saw that these data could be explained by assuming that there were two kinds of chloride in these compounds. The chloride not present as free chloride ions could not conduct an electric current or react with $Ag^+$ because it was attached to the

$Co^{3+}$—i.e., was a ligand. He rewrote the above formulas to make this distinction: $[Co(NH_3)_6]Cl_3$, $[Co(NH_3)_5Cl]Cl_2$, $[Co(NH_3)_4Cl_2]Cl$, and $[Co(NH_3)_3Cl_3]$, respectively.

The term *coordination number*, which we have been using all along to refer to the number of nearest neighbor atoms or ions to a given atom or ion, is especially useful here. The coordination number of the Lewis acid or acceptor atom or ion in coordination compounds is an important property and is in most circumstances equal to the number of donor atoms attached to the acceptor atom. Provided that there is no multiple bonding to the acceptor atom, it is also equal to the number of "ordinary" and coordinate covalent bonds to the acceptor atom.

Chemists of Werner's time thought only in terms of the number of ordinary covalent bonds characteristically formed by an element—its **valence**. Thus they drew some strange-looking (to modern eyes) structures to account for the complexes of $CoCl_3$ with up to six ammonia molecules, since they assumed that, with a valence of 3, cobalt could form only three bonds. With our modern appreciation of the relative sizes of cations, anions, and donor atoms, we can see that cobalt would be expected to have a coordination number greater than 3. Werner noticed that in each of the cobalt complexes mentioned above and in many others he studied, the experimental data could best be explained by assuming a constant coordination number of 6 for the $Co^{3+}$ ion.

Many or most metal ions can be quite variable in their coordination numbers in coordination complexes. Werner was fortunate that certain *d*-block metal ions such as $Co^{3+}$ with certain valence electron configurations have strong preferences for certain coordination numbers and certain angles between the coordinate covalent bonds. These preferences cannot be explained just by considerations of ionic radii and maximum coordination numbers, but are explained by theories such as the crystal field theory, which will be studied later in inorganic chemistry.

The definition of a coordinate covalent bond is somewhat artificial. Once a bond is assembled, there is no certain way to tell from where the electrons came, since all electrons are indistinguishable. But we can mentally disassemble a bond in two ways: (1) by returning the two electrons, one to each reactant, or (2) by returning both to the likely base and none to the likely acid. If the reactants seem more familiar and plausible to us with one electron returned to each, we assume that this is how the bond was formed and consider the bond to be an ordinary covalent bond, formed (probably) by an oxidation-reduction reaction. If the reactants seem more familiar and plausible when both electrons go to the donor atom of a base, we call the bond a coordinate covalent bond and the reaction a Lewis acid-base reaction. Sometimes either way seems about equally likely, and we then have our choice of classifications.

In this chapter and the next, however, we will classify all bonds to a central metal atom as coordinate covalent bonds so that we can take advantage of certain very useful principles. It will thus be quite important to develop the ability to treat a given compound as the product of a Lewis acid-base reaction, and to be able to dissect its structure to identify the Lewis acids and bases from which it might have been assembled.

EXAMPLE    Draw structures of the complex ions or coordination compounds in the following, and identify, for each one, the ligands, the coordinate covalent bonds, the Lewis bases, the Lewis acid, the donor atoms, and the coordination number of the Lewis acid: (a) $[Al(H_2O)_6]^{3+}$; (b) $[Co(NH_3)_5Cl]SO_4$; (c) $K_3[FeF_4(OH)_2]$.

Solution    ■ Our eyes should first be drawn to the brackets in these equations, since these separate the coordination compound or complex ion from any counterions. Thus in (b) we can disregard the sulfate ion, $SO_4{}^{2-}$, and in (c) we can omit the three potassium ions, $K^+$. When these are omitted, we can see the true charges on the complex ions: $+2$ in (b) to counterbalance the sulfate ion, and $-3$ in (c) to counterbalance the three potassium ions.

■ Within the brackets, we look for the Lewis acid, which will normally be a metal atom or ion. However, nonmetal ions can fill this function too; more often than not the Lewis acid is written just inside the left bracket. The other species will be ligands or Lewis bases, attached to the Lewis acids. It is customary to enclose each polyatomic ligand inside a set of parentheses; monoatomic ligands such as $Cl^-$ are not enclosed. With this information, we can draw structures (Figure 7.2) in which each ligand is connected to the metal ion through a likely donor atom (a nonmetal, nonhydrogen atom that might have an unshared pair of electrons). In Figure 7.2a each line represents a coordinate covalent bond. The donor atoms in each case are the nonmetal atoms at the end of the coordinate covalent bond: O in the aluminum complex ion, N and Cl in the cobalt complex, and O and F in the iron complex. The coordination number of the metal ion (acceptor) equals the total number of donor atoms or the total number of covalent bonds: six in each case.

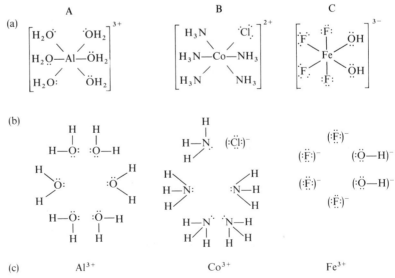

**Figure 7.2** (a) Structures of the coordination complexes or complex ions in the example; (b) the ligands removed from each complex; (c) the Lewis acids that remain after removal of the ligands.

■ Next we dissect each complex ion into its component ligands and the metal-ion acceptor, as shown in Figure 7.2b; the two electrons in each coordinate covalent bond are put back onto the donor atoms as now-unshared pairs of electrons. Then we must complete the Lewis structure of each ligand according to the rules of Chapter 1 in order to determine whether the ligand is charged or not, and if so, what its negative charge is. (Positive ions are almost never ligands.) Thus we have drawn octet-rule-obeying dot structures for $H_2O$, $NH_3$, Cl, F, and OH. But if these neutral species are the true ligands, then the total number of valence electrons in each one (eight, by chance, in each case) must equal the sums of the group numbers of the elements. This is true for $H_2O$ and $NH_3$; these are the true, neutral ligands. But for Cl, F, and OH, we have drawn one more electron in each case than the sum of the group numbers. This one extra electron gives the species a negative charge: The true ligands (Lewis bases) in each case are $Cl^-$, $F^-$, and $OH^-$.

■ Finally, we identify the Lewis acids: In each case this is a metal ion, but we must know the charge on the metal ion. This is simply equal to the charge on the whole complex ion minus any charges on anionic ligands that we have removed. Thus for $[Al(H_2O)_6]^{3+}$, removal of six neutral water molecules still leaves a $+3$ charge; the Lewis acid is $Al^{3+}$. Removal of some neutral ammonia molecules and a $Cl^-$ from $[Co(NH_3)_5Cl]^{2+}$ leaves a $Co^{3+}$ ion as the Lewis acid. Removal of six $-1$-charged ligands from the $-3$-charged iron complex ion leaves a $Fe^{3+}$ ion as the Lewis acid.

□

Note that this last procedure is quite similar to that given in Section 1.7.2 for deriving oxidation numbers from Lewis structures. Thus the $+3$ charges assigned to Al, Co, and Fe are also their assigned oxidation numbers in these complex ions. In the rare case in which the donor atom is *less* electronegative than the central metal atom, however, the two procedures are *not* equivalent. In such situations the assignment of oxidation numbers is a bit ambiguous. The usual procedure seems to be to follow the rules of Section 1.7.2 if the central atom is from the *p*-block, but to follow the rules of this section if it is from the *d*- or *f*-blocks.

## 7.2

## Coordination Compounds in Biochemistry and in Industrial Catalysis

In the last three decades the study of coordination compounds has grown tremendously in scope. Chemists began to realize that coordination compounds play very important roles in catalyzing many biochemical processes. Furthermore, coordination and organometallic compounds can be used to catalyze industrial organic chemical reactions to produce higher yields of products contaminated with fewer other products that have to be removed and discarded at great expense and hazard to the environment. Since many students using this text will not have had an organic or biochemistry course, we will not go into details on the structures and reactions of these biochemical and industrial catalysts. But we will illustrate some of the species involved and see how the concept of the metal ion as a Lewis acid is involved.

Many very important biochemicals are coordination compounds involving one or more metal ions coordinated to an organic group or groups of great size and complexity. In many cases the metal ions are so outnumbered by the atoms of C, H, N, O, S, and P that their presence was not even realized until modern methods of analytical chemistry came along that were capable of detecting very low concentrations of metals. Two very important and famous compounds—namely, chlorophyll and hemoglobin—have relatively small organic portions, so their metal content has long been known (Figure 7.3). Chlorophyll is the key to photosynthesis

**Figure 7.3** Structures of some biologically significant coordination compounds: (a) chlorophyll, containing a $Mg^{2+}$ ion coordinated to four nitrogen donor atoms of an organic (porphyrin) ligand; (b) heme, containing a $Fe^{2+}$ ion coordinated to four nitrogen donor atoms of a porphyrin ligand and a nitrogen donor atom from a protein; (c) methylcobalamin, one derivative of vitamin $B_{12}$; (d) methylcobaloxime, a synthetic mimic of methylcobalamin. Structures reproduced from F. A. Cotton and G. Wilkinson, *Basic Inorganic Chemistry*. Copyright © 1976 by John Wiley & Sons, Inc. Reprinted by permission.

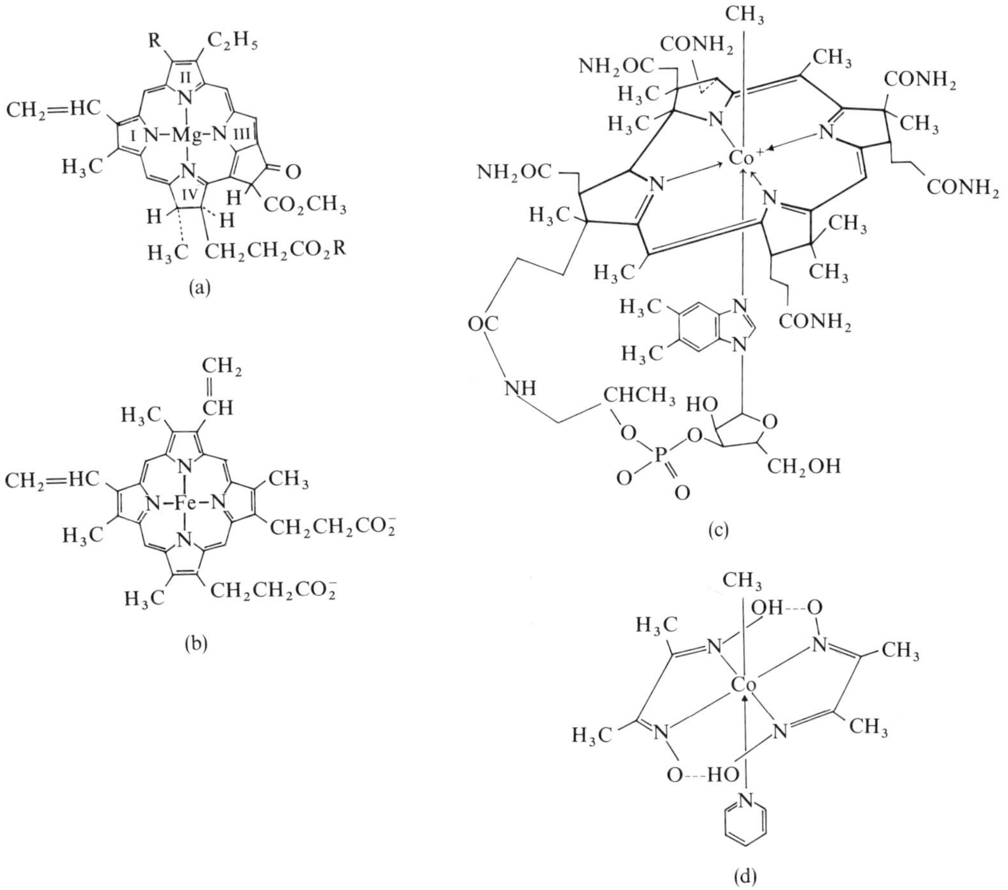

in plants. The heme unit in hemoglobin transports the oxygen molecule in the body: The $O_2$ molecule acts as a ligand and is coordinated to the iron atom of heme during its transport. One fairly complex vitamin, vitamin $B_{12}$ or cobalamin (Figure 7.3c), incorporates a cobalt atom that, in the various forms of this vitamin, is identifiable as a $Co^{3+}$, a $Co^{2+}$, or a $Co^+$ ion. The $Co^{3+}$ ion functions by coordinating ligands with carbon donor atoms (such as $CH_3^-$), which it then transfers to other compounds. (This is perhaps the only naturally occurring organometallic compound.) Interestingly, inorganic chemists have found that many of the functions of vitamin $B_{12}$ can be mimicked using the cobalt complex of a much simpler organic ligand, shown in Figure 7.3d [4]. The search for simpler coordination complexes that mimic the behavior of complicated biochemical complexes is one of the activities of the field of research called **bioinorganic chemistry**. With simpler ligands present, there is more hope of being able to sort out or even calculate the reasons for the functions of the catalyst.

**Enzymes** are nature's catalysts for speeding up chemical reactions in the body to rates as much as 100,000 times as fast as they would occur without catalysis. Many enzymes contain from one to several metal ions firmly incorporated into the protein structure of the enzyme; these are known as **metalloenzymes**. Other enzymes, although not incorporating the metals irreversibly into their structures, do require the reversible coordination of metal ions in order to become active; these are called **metal-activated enzymes**. Most of the metal ions of the third and fourth periods of the periodic table function in one or more metalloenzymes or in activating numerous enzymes. These metals (Na, Mg, K, Ca, Cr through Zn, Mo, Sn, and V for sea squirts) are essential for animal life. An example of one of the many such enzymes is carboxypeptidase A, shown in Figure 7.4, which helps cleave protein molecules by hydrolyzing the amide group linking the carboxyl-terminal amino acid to the polypeptide or protein. In the center of the 307 amino acids linked together to form this enzyme is one $Zn^{2+}$ ion, which is intimately involved in the mechanism of action of this enzyme [5].

In recent years the single most active area of research in inorganic chemistry has dealt with the use of $d$-block organometallic complex compounds to mimic the efficiency and selectivity of enzymes for the assembly of organic compounds not in the body, but also in the laboratory or (especially) in industrial plants. For illustrative purposes we mention a few of these reactions and the organometallic or complex compounds used to catalyze them:

■ The polymerization of alkenes to give polymers such as polyethylene, which uses the Ziegler-Natta catalyst, $TiCl_4$ and $Al(C_2H_5)_3$.

■ The production of acetic acid ($CH_3COOH$) from methanol ($CH_3OH$) and carbon monoxide, using $[Rh(CO)_2I_2]^-$.

■ Hydroformylation of (addition of $H_2$ and CO to) alkenes $RCH=CH_2$ to give aldehydes $RCH_2-CH_2-CH=O$, using $[RhH(CO)\ (triphenylphosphine)_3]$.

■ Alkene metathesis, which converts 2 mol $RCH=CH_2$ to 1 mol $CH_2=CH_2$ and 1 mol $RCH=CHR$, using $W(CO)_6$.

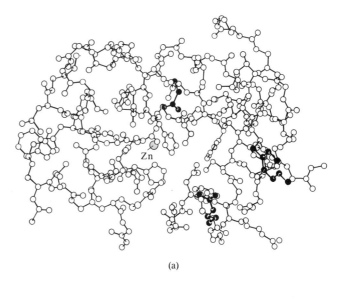

(a)

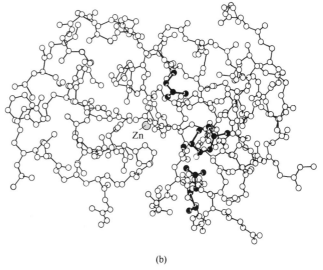

(b)

**Figure 7.4** About one-fourth of the structure of the zinc-containing metalloenzyme carboxypeptidase A (a) in its "resting" state, and (b) after incorporating glycyl-*L*-tyrosine. Reproduced with permission from W. N. Lipscomb, *Chem. Soc. Rev.*, 1, 319 (1972).

All these reactions are thought to involve the reacting organic molecules functioning as organometallic ligands using carbon donor atoms [6].

Although the detailed chemistry of either metalloenzymes or organometallic catalysts is beyond the scope of this book, we may note the conclusions of workers in these fields regarding the overall functions of metal ions in biochemical and other catalytic systems [5]:

■ $K^+$, $Na^+$, $Ca^{2+}$, and $Mg^{2+}$ trigger and control certain biochemical mechanisms; the passage of $Na^+$ ions across nerve cell walls constitutes an electrical current involved in nerve impulse transmission.

235

■ *d*-Block metals with several oxidation states are particularly useful in catalyzing biological oxidation-reduction reactions: In biochemistry, copper, iron, cobalt, and molybdenum are frequently involved.

Many other functions of metal ions in organometallic and biochemical catalysis clearly involve their function as Lewis acids:

■ If two molecules that are to react with each other can each act as a ligand and be coordinated to the same metal ion, the statistical odds of their finding each other and colliding may be greatly enhanced. Often the geometries of the ligands and of the coordination complex may be such that the reactive atoms of the two molecules may be brought close together. This is called *the template effect*.

■ Metal ions can change the whole conformation of an enzyme or biomolecule containing negatively charged functional groups that are close together (such as triphosphate ion groups in DNA and RNA). In the absence of $+2$-charged metal ions, DNA and RNA double helices tend to unwind due to the repulsion of the triphosphate groups. The mutual attraction of the negatively charged triphosphate groups to (especially) the weakly acidic $Mg^{2+}$ ion helps retain the double helix.

■ Just as the acidity of a metal ion polarizes the water molecule in the complex ion $[M(H_2O)_x]^{y+}$ and causes it to release hydrogen ions, an acidic metal ion will enhance the acidity of other coordinated molecules. Then, for example, bonds other than O—H may also be attacked by water (hydrolyzed), producing products more readily than would otherwise be the case.

■ More controversially, it has been suggested [7] that the geometric shape and the bond angles about the metal ion in the enzyme may be distorted from the ideal in a direction that makes it look more like the transition state in the reaction of the organic molecules. If this is the case, then the activation energy of the reaction will be lowered and the reaction will be speeded up. This is known as the *entatic-state hypothesis*.

Since metal ions can perform more of the above functions than hydrogen ions, metal complexes (particularly of the *d*-block metals) are now replacing the older hydrogen-ion acid catalysts, such as sulfuric acid, in many industrial processes. Because the geometry or shape of ligands within complexes and of complex compounds as a whole plays an important part in the function of Lewis-acid catalysts, we next turn to the prediction of the shapes of chemical species, followed by the classification of ligands (in part) on the basis of their shape.

## 7.3

# The Shapes of *p*-Block Molecules, Ions, and Coordination Compounds (VSEPR)

The geometric shapes of molecules, ions, and coordination compounds of the *p*-block elements, and the approximate angles between their covalent bonds, may be predicted remarkably well using a simple scheme known as the **valence shell electron-pair repulsion (VSEPR)** model. This model assumes that the shapes of molecules and ions are determined primarily by the electrostatic repulsions of the negatively charged covalent-bond electron pairs and unshared pairs of valence *p* electrons about the central atom. The geometry and bond angles adopted are those that allow these pairs of electrons to move as far apart as possible.

We begin the determination of molecular geometry with a correct Lewis dot structure of the species. If it is at all large there will be more than one central atom, and each central atom may have its own geometry and bond angles that must be determined separately. For each central atom in this structure we determine the *total coordination number* [8], which is the sum of the number of other atoms attached to the central atom in question and the number of unshared valence *p*-electron pairs about it. Whether the atoms are attached by single, double, or triple bonds does not matter in this count, since all the electrons in double or triple bonds must occupy the same general region of space—between the two atoms in the double or triple bond. The only kind of unshared valence-electron pairs we will consider are *p*-electron pairs. The results with *d*-electron pairs often must be predicted using the crystal field theory, which will be covered in a later course in inorganic chemistry.

### 7.3.1  Compounds with No Unshared Electron Pairs on the Central Atom

We will first take the simpler case, in which there are no unshared valence *p*-electron pairs on the central atoms. (For these compounds the *total* coordination number of each central atom equals its coordination number.) The least repulsive arrangements of electron pairs for total coordination numbers between 2 and 9 are shown in Figure 7.5.

A total coordination number of 2 about a given atom (as in $HgF_2$ or $O{=}C{=}O$) means that there are two independent pairs of electrons repelling each other. Repulsion is minimized with the two electron pairs at a bond angle of 180°, giving a *linear* geometry about that central atom. The total coordination number of 3 gives rise to an equilateral triangular arrangement of bonds, with bond angles of 120° between electron pairs.

For a total coordination number of 4 or higher, three-dimensional structures give less repulsion than structures that confine all the atoms and bond electrons to a single plane. Thus a square array of four electron pairs about a central atom separates the pairs by 90°, but a tetrahedral array gives a greater separation of 109.5°. Chemists had become so used to drawing molecules on planar sheets of paper and blackboards that this concept was surprisingly hard to accept at first.

For a total coordination number of 5, it is impossible for all the electron pairs to be equivalently situated about the central atom, even if all the outer atoms are identical. Instead there is a three-layered *trigonal bipyramidal* arrangement of outer atoms or unshared electron pairs into two different types: Those above and below the central atom are said to be in **axial** positions (designated by the subscript "ax" in Figure 7.5). Those in the same plane as the central atom are said to be in **equatorial**

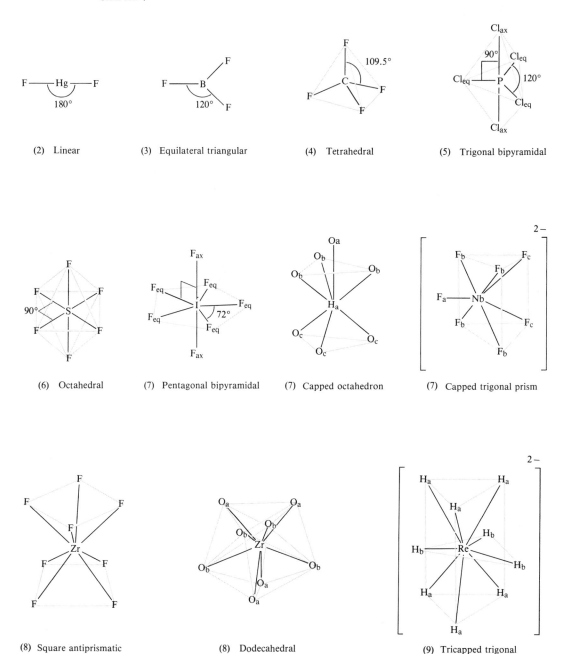

**Figure 7.5** Idealized arrangements of electron pairs that minimize repulsions among two to nine electron pairs (the total coordination number, shown in parentheses) about a central atom. The examples shown are complete molecules or complex ions except for (7) the capped octahedron, part of the structure of $[Ho(H_2O)(C_{15}H_{11}O_2)_3]$; (8) the square antiprism, part of the complex ion $[Zr_6F_{31}]^{7-}$; (8) the dodecahedron, part of the complex $[Zr(C_5H_7O_2)_4]$.

positions (designated by "eq"). The angles between equatorial electron pairs are 120°; the angle between any of these pairs and any axial electron pair is 90°. (Note that one axial pair and the three equatorial pairs together form a pyramid with a triangular (trigonal) base; the name of the geometry arises from the presence of two such pyramids sharing a base.)

Although it may not be apparent from the figure (it is a good idea to examine as many of these arrangements as possible using molecular models), the preferred arrangement of six electron pairs, the *octahedron*, is highly symmetrical: No two pairs are any more axial than any other two pairs. All bond angles are 90°.

For a total coordination number of 7 (as well as for most numbers over 9), there is more than one geometry of nearly equal efficiency in separating electron pairs. No clear prediction among these geometries (pentagonal bipyramidal, capped octahedral, and capped trigonal prismatic) is possible. Except in the case of the pentagonal bipyramid, the bond angles are variable. A total coordination number of 8 gives rise to two arrangements of minimal repulsion, the square antiprism and the dodecahedron. For total coordination number 9, a unique geometry, the tricapped trigonal prism, is predicted to minimize the electron-pair repulsions.

Although total coordination numbers up to 14 are known for some very large central atoms with tiny outer atoms, real examples are not common and predictions of the favored geometry are not very clear.

## 7.3.2    The Case of Unshared *p*-Electron Pairs on the Central Atom

Until recently our methods of determining the geometry of molecules and ions (such as X-ray crystallography) could not locate unshared electron pairs; this is still usually the case. Hence we describe what we can observe, the geometry of the *central* and *outer* or *donor* atoms only, which results in some new geometries (Figure 7.6).

Although the bond angles are approximately the same for a given total co-ordination number whether or not any of the central-atom electron pairs are unshared, VSEPR does predict slight distortions of the ideal bond angles of Figure 7.5. It is assumed that an outer atom "pulls out" an electron pair and makes it require less angular space about a central atom than does an unshared pair of electrons (Figure 7.7) [9]. We expect this assumption to have two kinds of consequences: (1) The bond angles will tend to distort to give more room to unshared electron pairs; (2) in geometries with positions that are not equivalent (such as the trigonal bipyramidal) the unshared electron pairs will occupy the more spacious positions.

Since unshared electron pairs are expected to take more space around the central atoms, the angles we can measure, those between bonds to different outer atoms, should generally be less than shown in Figure 7.5. For example, the ideal tetrahedral angle of 109.5° in $CF_4$ is expected to be reduced in other molecules with a total coordination number of 4 but which include unshared electron pairs. Bond-angle data given in Table 7.1 confirm this prediction. Similar trends are found among the angles between the axial and equatorial bond pairs in molecules of total coordination number 6 with one unshared electron pair. However, the angle

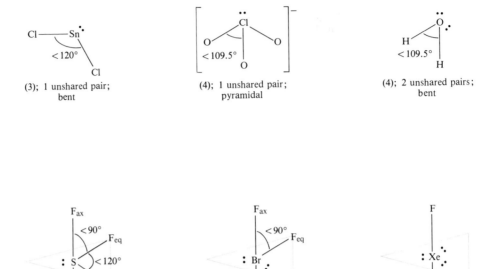

(3); 1 unshared pair;
bent

(4); 1 unshared pair;
pyramidal

(4); 2 unshared pairs;
bent

(5); 1 unshared pair;
see-saw

(5); 2 unshared pairs;
T-shaped

(5); 3 unshared pairs;
linear

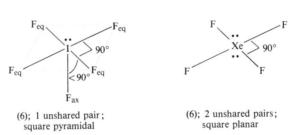

(6); 1 unshared pair;
square pyramidal

(6); 2 unshared pairs;
square planar

**Figure 7.6** Geometries described by the central and outer atoms in molecules and ions also having unshared $p$-electron pairs about the central atom. Total coordination numbers of central atoms are shown in parentheses.

between different equatorial pairs remains 90°, since the space-demanding axial unshared pair pushes equatorial bond pairs toward the axial bond pair, but not toward each other.

In the trigonal bipyramid, the second consequence comes into play. Electron pairs in the axial positions have three near neighbors—the three equatorial pairs,

**Figure 7.7** A schematic illustration of two assumptions of VSEPR. The first and second drawings illustrate an unshared electron pair occupying more space (taking a greater *cone angle*) about a central atom than an electron pair shared with an outer atom. The second and third drawings illustrate a bond pair to an atom of low electronegativity occupying more space about the central atom than a bond pair to a highly electronegative atom.

**Table 7.1** Bond-Angle Distortions Due to Unshared Electron Pairs in Molecules with a Total Coordination Number of 4

| Molecule | Angle | Molecule | Angle | Molecule | Angle | Molecule | Angle |
|---|---|---|---|---|---|---|---|
| $NH_3$ | 107.3 | | | $NF_3$ | 102.1 | $OH_2$ | 104.5 |
| $PH_3$ | 93.3 | $PCl_3$ | 100.3 | $PF_3$ | 97.8 | $SH_2$ | 92.2 |
| $AsH_3$ | 91.8 | $AsCl_3$ | 98.7 | $AsF_3$ | 96.2 | $SeH_2$ | 91 |
| $SbH_3$ | 91.3 | $SbCl_3$ | 99.5 | $SbF_3$ | 88 | $TeH_2$ | 89.5 |

Source: Data from F. A. Cotton and G. Wilkinson, *Advanced Inorganic Chemistry: A Comprehensive Treatise*, 4th ed., Wiley-Interscience, New York, 1980.
Note: Bond angles in degrees.

only 90° away—so they instead adopt the equatorial position in order to have only two near neighbors (the two axial pairs at 90°). (The two equatorial pairs at 120° are far enough away not to matter so much.) This results in the unusual geometries seen in Figure 7.6, such as the "seesaw" or "teeter-totter" geometry that is reminiscent of a child's balancing toy. Additionally, the remaining bond angles are reduced from 90° or 120° due to the spatial demands of the unshared electron pairs, except in the case of three equatorial pairs, the "pushes" of which cancel each other out.

With a total coordination number of 6 and two lone pairs, the second principle leads us to expect the two lone pairs to go to opposite sides of the octahedron (as far from each other as possible), giving rise to the square planar geometry. Again bond angles will not distort, as the pushes cancel each other out.

A most interesting case is that of total coordination number 7 with one unshared electron pair. Since we cannot predict the geometry even in the absence of unshared electron pairs, it is not easy to do so if an unpaired electron pair is present. Various distorted geometries often result, but often a completely unexpected result turns up: an undistorted octahedron, in which the unpaired electron pair occupies no angular space at all. To explain this geometry, it is necessary to use more sophisticated theories than VSEPR. We have paid no attention to the types of valence orbitals in which the bond or unshared electrons are located. You will study bonding theories that do take this into account in an organic or a later inorganic chemistry course. For example, one of these theories, the valence bond theory, assumes that the orbitals used by the bond and unshared pairs are not pure *s, p,* or *d* orbitals but are *hybrids* or mixtures of these orbital types. The undistorted octahedral geometry can then be rationalized by assuming that, for some reason, in this case the unshared electron pair remains in a pure valence *s* orbital, which is spherical and takes no angular space.

One final consequence of our assumptions illustrated in Figure 7.7 is that a bond pair to a more electronegative outer atom should occupy less space than a bond pair to a less electronegative outer atom. At least one example of this theory generally works out: In trigonal bipyramidal molecules of the *p*-block elements having more than one kind of outer atom, it is found that the more electronegative substituent preferentially occupies axial positions. Thus in the mixed halide $PF_2Cl_3$ the fluorine atoms are axial and the less electronegative chlorine atoms are equatorial.

EXAMPLE  Predict the geometric shapes and the bond angles of the following species: the pseudohalide ligands pictured in Figure 5.9a–f; (g) $NO_2^-$; (h) $ClO_2^-$; (i) $CO_3^{2-}$; (j) $SO_3^{2-}$; (k) $PCl_2F_3$; (l) $HN_3$.

Solutions  ■ (1) Draw a correct Lewis structure of the species. (2) Determine the total co-ordination number about each central atom. (3) Count the number of unshared *p*-electron pairs on each central atom; locate these in the most uncrowded positions. (4) Describe the resulting geometry (Figure 7.5 or 7.6) and identify the undistorted bond angles (Figure 7.5). (5) Predict any bond angle distortions due to repulsions from unshared electron pairs. (6) If two different types of outer atoms are involved in a trigonal bipyramidal structure, locate the most electronegative atoms in axial positions if possible.

■ Figure 5.9 shows the Lewis structures for the pseudohalide and phenyl anions. (a) The·phenyl group has six central atoms. Each carbon in turn can be considered separately as having a total coordination number of 3, consisting (for five of the six carbon atoms) of two other carbon atoms and the hydrogen atom as outer atoms; hence each carbon will have an equilateral triangular geometry about it except the unique carbon, which will have an apparent bent geometry. The bond angles should all be approximately 120°; there might be a slight distortion to a smaller angle at the unique carbon atom.

■ (b) The central nitrogen atom of the azide ion has a total coordination number of 2, since double bonds count the same as single. Hence the azide ion is linear, with a 180° bond angle. (c) The two atoms of the cyanide ion by definition are linear. (d, e) As in (b), the total coordination numbers of the central (C) atoms of the cyanate and thiocyanate ions are 2, so these are also linear. (f) The central S atom of the thiocyanate ion has a total coordination number of 4, so this ion is tetrahedral, although some deviation from 109.5° bond angles might be expected (larger around S, smaller between oxygens).

■ (g–j) You may want to refer back to Chapters 1 and 2 to obtain correct Lewis structures of the oxo anions; you are unlikely to get a correct geometry from an incorrect Lewis dot structure. You should obtain total coordination numbers and numbers of unshared electron pairs for these oxo anions as follows: for $NO_2^-$, 3 and 1, respectively; 3 and 2 for $ClO_2^-$; 3 and 0 for $CO_3^{2-}$; 4 and 1 for $SO_3^{2-}$. From these, the following geometries and bond angles follow: for $NO_2^-$, bent and $<120°$; for $ClO_2^-$, bent and $<109.5°$; for $CO_3^{2-}$, equilateral triangular and 120°; for $SO_3^{2-}$, pyramidal and $<109.5°$.

■ (k) $PCl_2F_3$ has a total coordination number of 5 and a trigonal bipyramidal structure. The more electronegative fluorines will prefer the axial sites, but as there are three fluorines and only two axial sites, the third fluorine will have to go equatorial.

■ (l) The acid corresponding to the linear azide ion (hydrazoic acid) will *not* be linear. The $H^+$ must attach to one of the two outer nitrogen atoms, since only these have unshared electron pairs, but then that nitrogen atom becomes a new central atom with its own geometry. It has a total coordination number of 3 with one unshared pair; hence the hydrazoic acid molecule will be bent at that nitrogen atom, with a bond angle $<120°$ (found 110.5°). □

---

## 7.4

## Classification of Ligands

Since so many organic and inorganic molecules have so many unshared pairs of electrons in so many geometric arrangements, it is quite useful to develop some general classes of ligands. There are several ways of doing this; the first we will examine is structural, based on the number of unshared electron pairs geometrically available for donation to a Lewis acid or to different Lewis acids.

The simplest Lewis bases have only one donor atom and only one unshared pair of electrons; clearly such ligands can form only one bond to one metal ion or other Lewis acid. Such ligands are called **monodentate** ligands.

The halide and other monoatomic anions have more than one unshared electron pair located on only one donor atom. Sometimes two or three pairs may be donated to a metal ion, forming a double or triple bond. But since pi bonds are weaker than sigma bonds, the second (and third) pair(s) of electrons are more commonly donated to a second (and a third) metal ion. A ligand that donates more than one electron pair to more than one metal ion or other Lewis acid is known as a **bridging ligand**.

Many important ligands have more than one donor atom. Such ligands are capable of donating electron pairs to different atoms and thus are *potential* bridging ligands. However, in many complexes this potential is unfulfilled, and only one donor atom is used, or only one metal ion is coordinated. In such cases we do not call the ligand bridging (although it is useful to observe the possibility of forming a bridge in a subsequent reaction).

Due to their linear geometries, the pseudohalide ions are good examples of potentially bridging ligands. Nonetheless they often use only one donor atom. Most of the pseudohalide ions have atoms of different elements that can serve as donor atoms. (For example, when it is not bridging, the thiocyanate ion can donate the unshared electron pair from either its sulfur atom or its nitrogen atom.) Such ligands are called **ambidentate**.

Many ligands with more than one donor atom have geometries such that they can form more than one coordinate covalent bond to the same metal ion. Such ligands are known as potential **chelating ligands**. Such ligands, of course, must be nonlinear, and they must form bonds with reasonable angles (i.e., somewhere near 90° or 109.5°) at the metal atom. This normally requires that the different donor atoms be far enough apart in the ligand to form *five- or six-membered rings*, counting the metal ion. Thus there should be two or three atoms intervening between the two donor atoms (Figure 7.8). Note that neither the pseudohalide ions nor the simple oxo anions meet these requirements: They normally act either as monodentate or as bridging ligands.

Chelating ligands are classified according to the number of donor atoms they possess that are suitably positioned to be simultaneously donated to the same metal ion. (Often they have additional donor atoms that are "out of position" and that could bridge to another metal ion, but usually don't.) Ligands with two suitably positioned donor atoms (such as ethylenediamine, $NH_2CH_2CH_2NH_2$) are called

**Figure 7.8** Geometric requirements for chelation by a ligand. (a, b) Bond angles at the metal are unfavorable if chelation is attempted with $O_2^{2-}$ or $CO_3^{2-}$, giving three- or four-membered rings. (c, d) Bond angles are favorable upon chelation with ethylenediamine or the acetylacetonate anion, giving five- and six-membered rings.

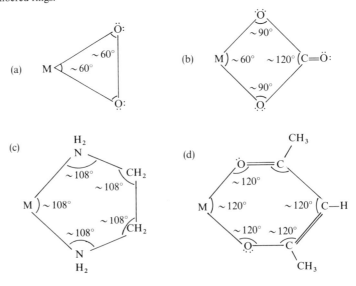

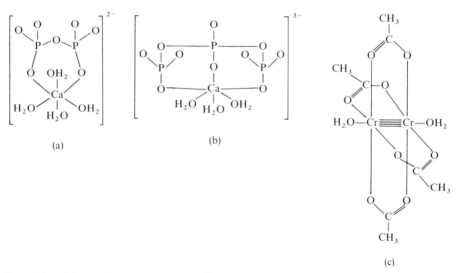

**Figure 7.9** (a) A complex formed between $Ca^{2+}$ and a pyrophosphate ion ($P_2O_7^{4-}$); (b) a complex formed between $Ca^{2+}$ and a tripolyphosphate ion ($P_3O_{10}^{5-}$); (c) the complex formed between the catenated, quadruply bonded $Cr_2^{4+}$ ion and acetate ion ($CH_3CO_2^-$) and water.

(potential) **bidentate chelating ligands**; those with three (such as diethylenetriamine, $NH_2CH_2CH_2NHCH_2CH_2NH_2$) are called **tridentate chelating ligands**, and so on. In general, such ligands are called **polydentate ligands**. As we will see in the next section, ligands that *can* form chelate rings with metal ions usually *do* so.

The linear polynuclear oxo anions (such as pyrophosphate and tripolyphosphate) can act as chelating ligands. Their function in detergents is to form complexes with the metal ions ($Ca^{2+}$, etc.) in tap water (Figure 7.9) so that the metal ions do not form insoluble salts with the detergent. Note that although the polynuclear oxo anions have very large numbers of potential donor atoms, their geometry restricts the number that can be used to donate electron pairs to a given metal ion. Thus, in Figure 7.9, pyrophosphate acts as a bidentate chelating ligand, and tripolyphosphate acts as a tridentate chelating ligand.

Although oxo anions do not form five- or six-membered rings with single metal ions, they can do so with dimeric catenated metal ions such as the (singly bonded) $Cu_2^{2+}$ ion and the (quadruply bonded) $Mo_2^{4+}$ and $Cr_2^{4+}$ ions (Figure 7.9c; Chapter 6). Although the ligands in such complexes are, in a sense, both bridging and chelating, they are usually termed bridging ligands.

A special class of chelating ligand includes those that are large ring compounds even without a metal atom present and that can position several donor atoms inside their ring to donate to a metal ion. Such ligands are called **macrocyclic** ligands. Chlorophyll, heme, and vitamin $B_{12}$ all contain tetradentate macrocyclic ligands (Figure 7.3).

There are certain compounds (mostly organic) containing double or triple bonds and hence containing *pi* bonds that can actually donate the electrons in a pi

245

bond to a metal ion, even though these are already shared. The existence of such **pi-donor** ligands requires an extension of our definitions: Not only unshared but also loosely shared electrons can be donated. In such cases it is hard to define the number of donor atoms and hence the coordination number of the metal and the geometry of the complex. Since such compounds have carbon donor atoms, they are classified as organometallic compounds; they will get only brief mention in this text.

Although not often listed as a separate classification, it is sometimes important to distinguish between *charged* complexes (i.e., complex *ions*) and *uncharged* complexes. This classification is important because it affects the solubility of complexes. Charged complex ions and their counterions will attract hydration spheres; since they are quite large, complex ions will be nonacidic cations or nonbasic anions. Often they may be *hydrophobic structure makers* (Section 3.4). (There will be exceptions, since the ligand may contain extra functional groups that are acidic or basic, and the metal ion will enhance the acidity of these.)

Generally speaking, then, complex *cations* are expected to form soluble salts with basic anions and are best isolated from solution using nonbasic anions. Perchlorates of these complex cations crystallize nicely, but it must be remembered that the ligands in the cations are often organic or otherwise easily oxidized, so the resulting crystals are dangerously explosive! Similar dangers attach to the other affordable nonbasic oxo anions, so the nonbasic anion of choice is often a *complex anion*. **Halo anions** (such as $TiCl_6^{2-}$, $BiCl_4^-$, $SnCl_6^{2-}$, or $SbCl_5^{2-}$, which you may have prepared in the last part of Experiment 5) are often used for this purpose. The chloro, bromo, and iodo anions are still often subject to hydrolysis in excess water but can be used for this purpose in acidic solutions containing excess halide ion. The fluoro anions are especially useful, since all fluoro anions are nonbasic, and many of them resist hydrolysis. The ions $BF_4^-$ and $PF_6^-$ resemble perchlorate in size and nonbasicity, but these ions do not involve unusually high oxidation states and hence are nonoxidizing. Similarly, $SiF_6^{2-}$ somewhat resembles $SO_4^{2-}$ in forming precipitates with feebly acidic cations.

In contrast, *uncharged complexes* do not have a charge to attract waters of hydration, so these are generally *insoluble in water*, unless the ligand has groups such as —OH or —$NH_2$ that can hydrogen-bond to water. If the ligands are organic and lack hydrogen-bonding groups, the complex is hydrophobic: It will not only be insoluble in water, but will also be soluble in nonpolar solvents such as hydrocarbons, fats, oils, and lipids. This property is made use of very often in analytical chemistry. A metal ion present in very low concentration in a natural water is complexed with an organic ligand to give an uncharged complex. This complex is then extracted into a much smaller volume of a nonpolar solvent and is thus separated from many impurities and concentrated at the same time.

Complexes of this sort, which naturally include many organometallic compounds, tend to accumulate in fatty tissue and lipids in the body and are not easily excreted in the watery urine; hence toxic effects of metals in these complexes persist longer. When an animal dies and is eaten by another animal higher up the food chain, the uncharged complex is passed on to it. Since such animals eat many times their own weight of the lower-chain animal, the concentration of the uncharged metal complex builds up; this process is known as **bioamplification** of the toxin.

## 7.5

## The Chelate and Macrocyclic Effects

Clearly the number of possible complex ions and compounds in a natural system (such as a polluted body of water, or the living cell) is very large indeed. It is often important to know the specific complex in which each metal ion is present in such systems. Determining this experimentally requires some very difficult analytical separations and detections. Predicting this quantitatively requires the knowledge of numerous equilibrium constants and the use of much computer time. Fortunately there are two very important qualitative generalizations that will enable us to predict some of the most likely complexes in a complicated mixture of metal ions and ligands. The first generalization is that *chelate ligands form more stable complexes than analogous monodentate ligands; macrocyclic ligands form more stable complexes than chelate ligands.* (The second generalization is so useful that the entire next chapter is devoted to it.)

To see the **chelate effect**, we need to set up a competition (for a metal ion) between 1 mol of an *n*-dentate chelating ligand and *n* mol of a very similar monodentate ligand. For example, if we mix 1 mol $Ni^{2+}$ ion with 3 mol ethylenediamine and 6 mol ammonia, the equilibrium will favor the formation of the complex with ethylenediamine:

$$[Ni(NH_3)_6]^{3+} + 3\,NH_2CH_2CH_2NH_2 \rightleftharpoons \tag{7.1}$$
$$[Ni(NH_2CH_2CH_2NH_2)_3]^{3+} + 6\,NH_3$$

The free energy change driving this reaction to the right is $-67$ kJ/mol, which breaks down into a very small enthalpy change of $-13$ kJ/mol and a larger entropy change term ($-T\Delta S = -54$ kJ/mol). The entropy effect is responsible for the shift of this and similar equilibria in favor of the complex with the chelated ligand attached to the metal ion. The enthalpy change is small, since each complex ion in equation (7.1) involves six very similar Ni—N coordinate covalent bonds. The reason for the favorable entropy change is more complex and involves several factors [10], the largest of which arises from the fact that there are 4 mol of solute particles on the left side of equation (7.1) and 7 mol on the right; the increasing degrees of freedom (disorder) among the products favor complex formation by the chelate ligand.

Similarly, a competition between a noncyclic chelating ligand and a macrocyclic (chelating) ligand having the same number and type of donor atoms will generally lead to complex formation predominantly by the macrocyclic ligand. This is known as the **macrocyclic effect**.

$$[K(CH_3O(CH_2CH_2O)_5CH_3)]^+ + cyclo\text{-}(CH_2CH_2O)_6 \rightleftharpoons \tag{7.2}$$
$$[K(CH_2CH_2O)_6]^+ + CH_3O(CH_2CH_2O)_5CH_3 \qquad K_{eq} = 10^4$$

Even though the same number of moles and reactants are involved in this reaction, the entropy term also drives this type of reaction to the right. This is attributed to the fact that a long noncyclic ligand such as $CH_3O(CH_2CH_2O)_5CH_3$ is more flexible than the corresponding macrocyclic ligand $(CH_2CH_2O)_6$ and can adopt many more conformations than the macrocycle when it is *not* coordinated.

Thus for a given number and type of donor atom, the macrocyclic ligand will generally form the most stable complex. This is certainly reflected in the biochemically significant complexes shown in Figure 7.3, all of which involve macrocyclic ligands. One important feature of the chemistry of macrocyclic ligands, however, is that the size of the "cavity" in the center of all the donor atoms is limited by van der Waals repulsions of the donor atoms with each other. Often the size is limited by the rigidity of the ligand itself; hence such ligands are often very selective for a particular size of metal ion. Among Group 1 metal cations, for example, the ligand $(CH_2CH_2O)_6$ selectively coordinates to $K^+$ in preference to both larger and smaller cations from the group.

## 7.6

## Nomenclature of Coordination and Organometallic Compounds

Given that several complex organic ligands may be present in a coordination compound and that geometric isomers may be present, the name of a coordination compound can be quite formidable, sometimes taking more than a line of type. We will cover the naming of only some of the simpler coordination and organometallic compounds, omitting complex organic ligands and consideration of geometric and stereoisomers, etc.

NAMING OF LIGANDS

**1.** Neutral ligands take the names they normally use as neutral molecules. There are four specific exceptions: $H_2O$ as a ligand is known as *aqua* (formerly *aquo*), $NH_3$ is named *ammine*, CO is named *carbonyl*, and NO is named *nitrosyl*.

**2.** The names of anionic ligands have *-o* in place of the final *-e*. (Thus *carbonate* as a ligand is known as *carbonato*.) Halide ions and some pseudohalide or similar anions ending in *-ide* drop the whole *-ide* ending: Thus $F^-$ is *fluoro*, $Cl^-$ is *chloro*, $Br^-$ is *bromo*, $I^-$ is *iodo*, $O^{2-}$ is *oxo*, $S^{2-}$ is *thio* (an exception), $OH^-$ is *hydroxo*, $CN^-$ is *cyano*, and $O_2^{2-}$ is *peroxo*. (But $N_3^-$ is *azido* and $H^-$ is usually *hydrido* to avoid confusion with other usages of the terms azo and hydro.) The *-o* ending is *not* used, however, with carbon anions, which keep their *-yl* names: Thus $CH_3^-$ as a ligand is just named *methyl*, and $C_6H_5^-$ is still named *phenyl*.

NAMING THE CENTRAL (METAL) ATOM:

**3.** In a complex cation or a neutral coordination compound, the name of the central (metal) atom is used unchanged. If the ion is a complex *anion*, the name of the central atom is converted to its Latin form (if there is one), and the suffix -*ate* is added. This is analogous to the naming of oxo anions; thus an iron-containing complex anion is named -*ferrate*.

**4.** If the central (metal) atom is capable of having more than one oxidation number, the oxidation number of the central atom is shown last, in Roman numerals enclosed in parentheses (the *Stock convention*). (There is an alternative convention, the *Ewens-Bassett convention*, by which the net charge on the *whole* complex ion, in Arabic numerals, is enclosed in the parentheses instead.) Thus, $FeO_4{}^{2-}$ can be called *ferrate(VI)* (Stock convention) or *ferrate (2−)* (Ewens-Bassett). It should be noted that in coordination and especially in *d*-block organometallic chemistry, metal-atom oxidation numbers can be unusual—i.e., they may not appear in Tables B or C, and can, in some cases, be zero or even negative [11].

**5.** An alternative practice is commonplace with *p-block organometallic compounds*: If *neutral*, the central atom takes the name of its *hydride*; if *cationic*, the -*ane* or -*ine* ending of the hydride becomes -*onium*. (If anionic, rule 3 still holds.) Thus the arsenic in $R_3As$ is named *arsine*, while in $R_4As^+$ it is named *arsonium*, and in $R_6As^-$ it is named *arsenate (1 −)* (by the Ewens-Bassett convention).

ORDER OF NAMING

**6.** The *cation* of a complex compound is named first, then the *anion*—as with ordinary salts, and regardless of which one is a complex ion, or whether both are.

**7.** Within the name of the coordination complex, the *ligands* are named first, then the *metal* or *central* atom or ion.

**8.** The ligands within a given complex ion are *listed in alphabetical order*. (Before 1971, anionic ligands were listed before neutral ligands, but now their listing is integrated.)

USE OF PARENTHESES AND INDICATION OF NUMBERS OF LIGANDS PRESENT

**9.** The names of *neutral ligands* are enclosed in parentheses, except for the four ligands $H_2O$, $NH_3$, CO, and NO, which have short common names. The official rules on whether to use parentheses around the names of anionic ligands seem often not to be followed. The usual practice in U.S. literature seems to be to enclose long anionic names (say, four syllables or more) in parentheses. Shorter names are not enclosed in parentheses unless confusion could result by the running together of a ligand's name and that of the previous one. Thus if the complex contains a chloro group, followed alphabetically by a methyl group, it is better to use each of these names with parentheses, since "chloromethyl" also refers to the $ClCH_2{}^-$ ligand. Similarly, the ligand pyridine is usually enclosed in parentheses to avoid potential confusion of two of these ligands with one bipyridyl ligand, another organic compound.

**10.** The *number of each type of ligand* present is indicated by prefixes. (a) If the ligand name is short and hence not enclosed in parentheses, *di-*, *tri-*, *tetra-*, *penta-*, and *hexa-* are added as prefixes to the name of the ligand to indicate the presence of 2, 3, 4, 5, and 6 ligands. (b) If the ligand name is long and hence enclosed in parentheses, the prefixes *bis-*, *tris-*, *tetrakis-*, *pentakis-*, *hexakis-*, etc., are used before the parentheses enclosing the name of the ligand. If no prefix is used, the presence of one ligand is implied; but if it is necessary to emphasize this, *mono-* may be used.

**EXAMPLE**     Name the following compounds: (a) $B(CH_3)_3$; (b) $K[B(C_6H_5)_4]$; (c) $(CH_3)_3P$; (d) $[(CH_3)_4P]Cl$.

**Solution**     ■ The compounds listed all have carbon donor atoms in their methyl and phenyl ligands and are thus *p*-block organometallic compounds. Hence the special provisions of rules 2 and 5 come into play.

■ (a, c) The ligand $CH_3{}^-$ is named *methyl* (without parentheses) by rule 2. By the common practice of rule 5, the boron is named *borane* and the phosphorus *phosphine* (without oxidation numbers). Rules 7 and 10 tell us to assemble these names as follows: (a) trimethylborane; (c) trimethylphosphine.

■ (b) By rule 2 the ligand is *phenyl*; by rules 5, 3, and 4 the central atom is named *borate(III)* (Stock convention) or *borate* $(1-)$ (Ewens-Bassett convention). Rules 6, 7, and 10 lead to the final name: potassium tetraphenylborate(III) (or $1-$). (The oxidation state could be omitted since boron has no other common oxidation number.)

■ (d) This is named similarly except that rule 5 tells us to name the phosphorus as *phosphonium*. The compound thus is named tetramethylphosphonium chloride.     □

As we will see in more detail in Chapter 10, many organometallic compounds can themselves act as Lewis acids and bases. The Lewis dot structure of trimethyl-borane shows only six valence electrons around the central boron atom; consequently this (and other Group 13/IIIA methyls, phenyls, etc.) are Lewis acids. The Group 14/IVA methyl and phenyl compounds have a complete octet of electrons and do not act either as Lewis acids or Lewis bases. The neutral methyl and phenyl compounds of Groups 15/VA, 16/VIA, and 17/VIIA show unshared electron pairs on the Group 15/VA, 16/VIA, and 17/VIIA atoms; these compounds act as Lewis bases or ligands in forming additional complex compounds. (The tendency to act as a Lewis base is almost but not quite absent from the Group 17/VIIA methyls and phenyls [12].) The boron and phosphorus atoms in the tetraphenylborate and tetraphenylphosphonium ions have octets of electrons, however, and show little or no Lewis acid or Lewis base tendencies.

**EXAMPLE**     Name the following coordination complexes: (a) $K_2[SiF_6]$; (b) $[Co(NH_3)_4Cl_2]Cl$; (c) $[Pt(S_2O_3)(SCN)_2(H_2O)]$; (d) $[ReH_3(P(C_6H_5)_3)_5]$.

**Solution**     ■ (a) By rule 2 the ligand of this complex anion is *fluoro*; by rules 3 and 4 the central atom is *silicate(IV)* (oxidation number IV is so common for Si that this may be omitted) or *silicate*(2−). Rule 6 tells us to name the cation, $K^+$, first: potassium hexafluorosilicate (with nothing or IV or 2− in parentheses).

■ (b) Rule 1 tells us to name $NH_3$ *ammine*; by rule 2 $Cl^-$ is *chloro*. By rules 3 and 4 the Co is *cobalt(III)* or *cobalt(1+)*. By rules 6, 7 and 8 we name the components in this order: ammine, then chloro, then the cobalt, then the counteranion, chloride. Adding the numbering of the ligands as in rule 10, we finally obtain tetramminedichlorocobalt(III) chloride (using the Stock convention).

■ (c) The anionic ligands $S_2O_3^{2-}$ and $SCN^-$ are named (*thiosulfato*) and (*thiocyanato*) by rule 2, with the parentheses courtesy of rule 9. $H_2O$ is *aqua* by rule 1. The Pt is *platinum(IV)* in the Stock convention (rule 4). Applying rules 7 and 8, we name the ligands first, in alphabetical order, and indicate the presence of two $SCN^-$ ligands with *bis*. The compound is named aquabis(thiocyanato)(thiosulfato)platinum(IV).

■ (d) This compound contains a common organophosphorus compound, triphenylphosphine, as a ligand. Once that is named, the complex is named trihydridopentakis(triphenylphosphine)rhenium(III). ☐

Additional rules are also used to designate bridging ligands, to tell which donor atom is used by an ambidentate ligand, etc. For these rules refer to an advanced text or to the latest rules for inorganic nomenclature of the International Union of Pure and Applied Chemistry (IUPAC) [13].

## 7.7

# Thermodynamics of the Lewis Acid-Base Interaction: Drago's $E$ and $C$ Parameters

We saw in Chapter 3 that the very large energies of attraction of positive and negative ions—their lattice energies—can be measured indirectly via thermochemical cycles. In Chapter 6 we saw that the large, pure covalent bond-dissociation energies can be measured directly in the gas phase. Naturally, the energies of coordinate covalent bond formation are also of interest. These enthalpy changes can be measured directly while Lewis acids and bases react with each other. The reactions should be carried out *in the gas phase*, since nearly all solvents are themselves Lewis acids or bases. Since many Lewis acids and bases are nonvolatile, it is often necessary to carry out these reactions instead in very nonpolar (nonacidic and nonbasic) solvents and assume that none of the measured enthalpy changes are affected by solute-solvent interactions.

A few hundred such thermodynamic measurements have been made. Drago and co-workers [13] have found that these measurements can be quite adequately predicted using the *Drago-Wayland equation*:

$$\Delta H \text{ (in kJ/mol)} = -4.184(C_A C_B + E_A E_B) \tag{7.3}$$

In this equation there are two parameters for each acid, $C_A$ and $E_A$, and two parameters for each base, which must be found by fitting experimental enthalpy-change data for reactions involving that acid or that base. (The parameter 4.184 converts the energy units from the original kilocalories per mole to kilojoules per mole.) Once these parameters have been obtained, however, they can be used to predict the enthalpy changes for thousands of acid-base combinations that have not been tested. If such predictions are tested later, the predictions are usually found to be quite accurate unless that particular acid-base combination involves steric hindrance (spatial interference of groups attached to the donor atom with other groups attached to the acceptor atom). Table 7.2 contains Drago-Wayland $E$ and $C$ parameters for a number of Lewis acids and bases.

EXAMPLE   Calculate the enthalpy changes of the following Lewis acid-base reactions: (a) $I_2$ + ethyl acetate; (b) $SbCl_5$ + tetrahydrofuran.

Solution   Taking the appropriate $E$ and $C$ parameters from Table 7.2, we find that
(a) $\Delta H = -4.184(1.000 \times 1.74 + 1.000 \times 0.975) = -11.4 \text{ kJ/mol}$;
(b) $\Delta H = -4.184(5.13 \times 4.27 + 7.38 \times 0.978) = -122 \text{ kJ/mol}$.   □

These two calculations alone show that there can be quite a substantial variation in the exothermicity of Lewis acid-base reactions, from the very weak to the quite exothermic. We would be tempted to conclude that antimony pentachloride is a stronger Lewis acid than iodine, and that tetrahydrofuran is a stronger Lewis base than ethyl acetate.

Before we make any such classifications, we must notice that equation (7.3) contains *two* parameters for each acid and two for each base. Neither parameter can be clearly identified with strength in quite the way that we identified $Z^2/r$ (or $Z^2/r + 0.096[\chi_P - 1.50]$) with acid strength in Chapter 2. *Both* are strength parameters, but they measure strength of different types: the $E$ parameters are interpreted as measures of strength in *electrostatic bonding* and the $C$ parameters as measures of strength in *covalent bonding*. We can legitimately compare the $C_A$ parameter of $I_2$ to the $C_A$ parameter of $SbCl_5$ and say that $SbCl_5$ is *better at covalent bonding* than $I_2$. Likewise, we can compare their $E_A$ values to say that $SbCl_5$ is *better at ionic bonding* than $I_2$. Although in this particular case it is reasonable to conclude that $SbCl_5$ is a stronger Lewis acid overall than $I_2$, there are many pairs of Lewis acids in Table 7.2 for which one is better at ionic and the other better at covalent bonding.

In the derivation of the Drago-Wayland parameters, there are always two more parameters than there are measurable enthalpy changes, so that two of these parameters must be arbitrarily set. These are the $E_A$ and the $C_A$ terms for $I_2$, both of which were arbitrarily set at 1.000. However, we know that the nonpolar, uncharged molecule $I_2$ really cannot be as good at electrostatic bonding as it is at covalent bonding. Thus clearly the equality of $E_A$ and $C_A$ for iodine is meaningless. This meaninglessness carries through the calculations for other acids and bases; we cannot meaningfully compare the $E$ and $C$ values to each other—we can only compare $E_A$ values to other $E_A$ values, and so on.

**Table 7.2** Drago-Wayland $E$ and $C$ Parameters for Uncharged Lewis Acids and Bases

| Acid | Acceptor atom | $E_A$ | $C_A$ | $C/E$ |
|---|---|---|---|---|
| Iodine | I | 1.00 | 1.00 | 1.00 |
| Iodine monobromide | I | 2.41 | 1.56 | 0.65 |
| Iodine monochloride | I | 5.10 | 0.830 | 0.16 |
| Sulfur dioxide | S | 0.92 | 0.808 | 0.88 |
| Antimony pentachloride | Sb | 7.38 | 5.13 | 0.70 |
| Chlorotrimethylstannane | Sn | 5.76 | 0.03 | 0.01 |
| Boron trifluoride | B | 9.88 | 1.62 | 0.16 |
| Trimethylborane | B | 6.14 | 1.70 | 0.28 |
| Trimethylalane | Al | 16.9 | 1.43 | 0.08 |
| Trimethylgallium | Ga | 13.3 | 0.881 | 0.07 |
| Trimethylindium | In | 15.3 | 0.654 | 0.04 |
| Water | H | 1.64 | 0.571 | 0.35 |
| Phenol | H | 4.33 | 0.422 | 0.10 |
| Thiophenol | H | 0.99 | 0.198 | 0.20 |
| Ethanol | H | 3.88 | 0.451 | 0.12 |
| (Tetraphenylporphyrinato)zinc(II) | Zn | 5.15 | 0.620 | 0.12 |
| (Tetraphenylporphyrinato)cobalt(II) | Co | 4.44 | 0.58 | 0.13 |

| Base | Donor atom | $E_B$ | $C_B$ | $C/E$ |
|---|---|---|---|---|
| Dimethyl selenide | Se | 0.217 | 8.33 | 38.4 |
| Dimethyl sulfide | S | 0.343 | 7.46 | 21.8 |
| Diethyl sulfide | S | 0.339 | 7.40 | 21.8 |
| Diethyl ether | O | 0.936 | 3.25 | 3.5 |
| Acetone | O | 0.937 | 2.33 | 2.5 |
| Ethyl acetate | O | 0.975 | 1.74 | 1.8 |
| Tetrahydrofuran | O | 0.978 | 4.27 | 4.4 |
| Trimethylphosphine | P | 0.838 | 6.55 | 7.8 |
| Trimethylamine | N | 0.808 | 11.54 | 14.2 |
| Pyridine | N | 1.17 | 6.40 | 5.5 |
| Dimethylamine | N | 1.09 | 8.73 | 8.0 |
| Methylamine | N | 1.30 | 5.88 | 4.5 |
| Ammonia | N | 1.15 | 4.75 | 4.1 |
| Acetonitrile | N | 0.886 | 1.34 | 1.5 |
| 1-Azabicyclo[2.2.1]octane | N | 0.700 | 13.2 | 18.9 |
| Benzene | C | 0.280 | 0.590 | 2.1 |

SOURCES: Data from R. S. Drago, *Coord. Chem. Rev.* 33, 251 (1980); R. S. Drago, *Struct. Bonding* 15, 73 (1973); and J. E. Huheey, *Inorganic Chemistry: Principles of Structure and Reactivity*, 2d ed., Harper and Row, New York, 1979, pp. 275–276.

*In general, it is impossible to characterize the Lewis acid-base interaction by a single parameter* such as strength. Our characterizations of relative acidities of metal ions in Chapter 2 worked relatively well because we always had the same Lewis base present, $H_2O$. (The bonding of water is mainly electrostatic to cations of metals of low electronegativity.) But, in general, the Lewis acid-base interaction involves varying mixes of electrostatic/ionic and covalent bonding and requires at least two parameters to characterize for each species. In fact, Kroeger and Drago [14] found that the two-parameter equation (7.3) works only for *uncharged* Lewis acids and

**Table 7.3** Bond Lengths, M—X, in Acceptor Molecules and in Their Donor-Acceptor Complexes

| Acceptor | M—X (pm) | Complex ion | M—X (pm) |
|---|---|---|---|
| $CdCl_2$ | 223.5 | $CdCl_6^{4-}$ | 253 |
| $SiF_4$ | 154 | $SiF_6^{2-}$ | 171 |
| $TiCl_4$ | 218–221 | $TiCl_6^{2-}$ | 235 |
| $ZrCl_4$ | 233 | $ZrCl_6^{2-}$ | 245 |
| $GeCl_4$ | 208–210 | $GeCl_6^{2-}$ | 235 |
| $GeF_4$ | 167 | $GeF_6^{2-}$ | 177 |
| $SnBr_4$ | 244 | $SnBr_6^{2-}$ | 259–264 |
| $SnCl_4$ | 230–233 | $SnCl_6^{2-}$ | 241–245 |
| $SnI_4$ | 264 | $SnI_6^{2-}$ | 285 |
| $PbCl_4$ | 243 | $PbCl_6^{2-}$ | 248–250 |
| $PF_5$ | 154–157 | $PF_6^{-}$ | 173 |
| $SbCl_5$ | 231 | $SbCl_6^{-}$ | 247 |
| $SO_2$ | 143 | $SO_3^{2-}$ | 150 |
| $SeO_2$ | 161 | $SeO_3^{2-}$ | 174 |
| ICl | 230 | $ICl_2^{-}$ | 236 |
| $I_2$ | 266 | $I_3^{-}$ | 283 |

SOURCE: From V. Gutmann, *The Donor-Acceptor Approach to Molecular Interactions*, Plenum, New York, 1978. Used with permission.

bases. To account for the generally more exothermic interactions of gaseous *ionic* Lewis acids and bases, we need yet a third parameter for each acid and base. (This parameter relates the degree to which the negative charge of the anion is transferred to the positive cation in the adduct and is related to the electron affinities and ionization energies of the species involved.) In the next chapter we will continue our study of Lewis acid-base interactions, but we will return to aqueous solution, in which we can still perform reasonably well, at least in a qualitative sense, with two parameters.

We briefly mention one other consequence of Lewis acid-base interactions: Bond lengths of adjacent bonds are altered as the coordination number of the donor and acceptor atoms increase [15]. We saw this happen in purely ionic compounds as coordination numbers increased (Table 3.5); the lengthenings are perhaps more dramatic in coordination complexes (Table 7.3). It is generally true that the stronger the Lewis acid-base interaction, the more the adjacent bonds are lengthened.

## 7.8
## Nonaqueous Solvents

Many ionic compounds are insoluble in water, and many of those that are soluble undergo hydrolysis in water. Therefore it is often very useful to find **nonaqueous solvents** for ionic (as well as polar and nonpolar covalent) compounds. The Lewis acid-base properties of many liquid substances allow them to be used as solvents for selected ionic and other compounds. There is a wide variety of such solvents; here we can only indicate categories and give a few examples.

Nonaqueous solvents are usually divided into two broad categories: **protic solvents**, which have ionizable hydrogen atoms, and **aprotic solvents**, which have either no hydrogen atoms or only very nonacidic hydrogen atoms.

## 7.8.1    Aprotic Solvents

The aprotic solvent category is usually subdivided into three groups. The first of these is the group of *nonpolar, weakly solvating* solvents. These include the saturated hydrocarbons such as pentane, hexane, heptane, "petroleum ether" (a mixture of saturated hydrocarbons), and cyclohexane. Such solvents cannot interact effectively with either the cation or the anion of an ionic salt, and thus there is no energy of solvation to offset the lattice energy; hence ionic salts are insoluble in such solvents. But these can be good solvents for nonpolar *hydrophobic* solutes such as neutral organometallic compounds and neutral complexes of some carbon- and hydrogen-rich alkoxides ($RO^-$) and amides ($R_2N^-$), etc. Their advantage in such cases is that they are unlikely to interfere in reactions of the solute with Lewis acids or bases.

Also included in this group are aromatic hydrocarbons such as benzene and toluene, which are usually weakly solvating, but which can act as pi-donor ligands for some metal ions. Such solvents are generally very poor for ionic compounds of most metals but can dissolve certain salts of selected metal ions (e.g., $Ag^+$). The very slightly polar and weakly Lewis-basic chlorocarbons such as $CH_2Cl_2$ and $CHCl_3$ are similarly useful.

The second group is that of *polar, strongly solvating but nonionized solvents*. These solvents may dissolve a number of ionic solutes by virtue of their properties either as Lewis acids or Lewis bases. It is more important in overcoming the lattice energy of an ionic compound to solvate the small cation rather than the large anion, since the cation will generally provide the larger solvation energy; hence most of these solvents are good *Lewis bases*. Numerous organic compounds fit into this group: for example, the ethers ($R_2O$), the ketones ($R_2C{=}O$) such as acetone, the esters ($RCO_2R'$), various sorts of organic nitrogen compounds (amines, nitriles, etc.), amides ($RCONR'_2$), and organic nitro compounds ($RNO_2$). Some of these can be made chelating and hence even more effective solvents (e.g., the polyethers, $RO(CH_2CH_2O)_nR$).

The choice among these solvents may be based partly on the appropriateness of the melting and boiling point of the solvent (some of these have quite high boiling points and may be difficult to remove in order to recover solutes). The Lewis-base strength of these solvents is also a consideration. Unfortunately, for reasons indicated in the last section, a single number characterizing the Lewis-base strength of a solvent, applicable to all solutes, is not possible.

A few useful solvents are predominantly Lewis acids. Sulfur dioxide can be liquefied by mild cooling and is often used as a solvent. Some nonmetal halides such as $SbF_5$ and $SbCl_5$ are very strong Lewis acids and can be used to promote ionization of very weak Lewis bases:

$$(C_6H_5)_3CCl + SbCl_5 \rightarrow (C_6H_5)_3C^+ + SbCl_6^- \tag{7.4}$$

The third group of aprotic solvents have both Lewis-acid and Lewis-base properties (i.e., are amphoteric). These solvents, such as bromine trifluoride, can dissolve both types of solutes:

$$KBr + BrF_3 \rightarrow K^+ + BrF_4^- \tag{7.5}$$

$$SbF_5 + BrF_3 \rightarrow BrF_2^+ + SbF_6^- \tag{7.6}$$

In a manner analogous to the autoionization of the amphoteric solvent water (into $H^+$ and $OH^-$), these are also *autoionizing solvents*:

$$2 BrF_3 \rightleftharpoons BrF_2^+ + BrF_4^- \tag{7.7}$$

In this solvent the reverse of equation (7.7) can be called a neutralization reaction (analogous to the reaction of $H^+$ and $OH^-$); titrations can be carried out using solutions of $SbF_5$ as the Lewis acid and KBr as the Lewis base. Due to their combined Lewis acid-base properties, the autoionizing solvents are quite reactive, for example, with the water of the air, and are rather difficult to handle without contamination.

## 7.8.2    Protic Solvents

Protic solvents are fundamentally like the aprotic solvents, but because of the presence of acidic hydrogen atoms they possess two additional features that we may utilize. Since some form of hydrogen ion is released, it is possible to use the concept of pH. Also, reactions analogous to hydrolysis can occur: Metal ions may enhance the acidity of such solutions to the point that protons are lost, and nonmetal anions may enhance the basicity to the point that protonation of the anion may occur. Such reactions, known as **solvolyses**, often give rise to new and useful chemical products; these will be discussed in Chapter 9.

The protic equilibria may be illustrated with a very important *basic protic solvent*, liquid ammonia (b.p. $-33\,°C$). Like water, these solvents are amphoteric and undergo autoionization:

$$2 NH_3 \rightleftharpoons NH_4^+ + NH_2^- \tag{7.8}$$

In such a solution, the hydrogen ion is attached (via a coordinate covalent bond) to an ammonia molecule to give the ammonium ion, which is the analogue (in ammonia) to the hydronium ion $H_3O^+$ (in water). The equilibrium constant for this autoionization is about $10^{-27}$, so it would be possible to define a pH scale for liquid ammonia in which $pH = -\log[NH_4^+]$. With this definition a neutral solution (one in which the concentrations of $NH_4^+$ and $NH_2^-$ were equal) would have a pH of 13.5 rather than 7 as in water. It would be difficult, however, to meaningfully compare a pH measured in liquid ammonia with one measured in water.

The concept of "effective pH" has been devised to render such comparisons

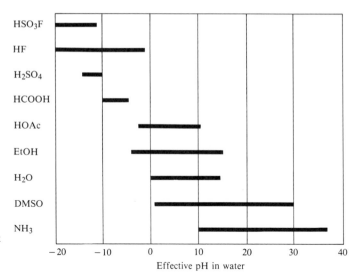

**Figure 7.10** Effective pH ranges of various protic solvents. Reproduced with permission from W. L. Jolly, *Modern Inorganic Chemistry*. Copyright 1984 by McGraw-Hill Book Company. Reprinted by permission.

Effective pH in water

possible. The **effective pH** of a nonaqueous solution is the pH of a hypothetical aqueous solution in which a dissolved test acid would have the same equilibrium cation-anion ratio as found in the nonaqueous solution [16]. An equilibrium mixture of $NH_4^+$ and $NH_2^-$ could not be achieved in water due to the complete reaction (leveling) of $NH_2^-$ with water to give $OH^-$ and $NH_3$, but we can compute that this would happen if a pH of 23.5 could be achieved in aqueous solution; hence the "neutral" solution in liquid ammonia has an effective pH of 23.5. Since by aqueous standards this is very basic, ammonia is categorized as a basic solvent.

Figure 7.10 shows the effective pH ranges of various protic solvents. In each case the neutral effective pH of that solvent is in the middle of the range shown; its acidic pH's are to the left, and its basic pH's are to the right. If the middle of the effective pH range of a solvent falls to the right of the middle of the pH range of water (pH = 7), we classify the solvent as a basic solvent (e.g., $NH_3$ and DMSO, $(CH_3)_2SO$). Basic solvents such as $NH_3$ are useful for investigating the properties of very strongly basic anions, since they may not be very strongly basic relative to the solvent and will undergo less solvolysis than they would in water.

The solvents above ethanol (EtOH) in Figure 7.10 are among those in the group of *acidic protic solvents*; these are useful for studying the properties of very acidic cations. The most acidic solvent in Figure 7.10, fluorosulfonic acid ($HSO_3F$), may be further enhanced in acidity by the addition of the aprotic Lewis acid $SbF_5$. This acid combines with the fluorosulfonate anion (a *very* weak Lewis base indeed), shifting the ionization equilibrium farther to the right:

$$HSO_3F + SbF_5 \rightleftharpoons FSO_3SbF_5^- + H^+ \qquad \text{(solvated to give } H_2SO_3F^+\text{)} \quad (7.9)$$

Such a solution is called a *superacid*, since it is capable of protonating any known Lewis base, no matter how feeble (e.g., Xe is converted, in part, to $XeH^+$).

It is even capable of protonating some species that lack either unshared electron pairs (classical Lewis bases) or shared pi-bond electrons (nonclassical ligands or Lewis bases). For example, $(CH_3)_4C$ lacks either, but is protonated to form an intermediate cation, $(CH_3)_4CH^+$, in which a *shared sigma-bond pair* is protonated. Such an electron pair is then shared among *three* atoms, in a situation reminiscent of the bonding in metals or elemental boron (Chapter 6). This bonding is best treated using the molecular orbital theory that you will learn later.

Nonaqueous solvents are not only useful in studying the chemistry of species that are too acidic or too basic to study in water, but they can also be used to study species that are so strongly oxidizing that they would oxidize water, or so strongly reducing that they would reduce water.

For example, liquid ammonia is more difficult to reduce (to $H_2$ and the $NH_2^-$ ion) than water (to $H_2$ and the $OH^-$ ion). Thus the very electropositive metals (such as the Group 1 metals and Ca, Sr, Ba, Eu, and Yb) react violently with water (Experiment 4), releasing $H_2$. But unless a catalyst is present, these metals do not react with liquid ammonia, but instead dissolve to give intensely blue solutions of remarkably low density. These solutions contain the ammonia equivalents of hydrated cations, $M(NH_3)_n^{2+ \text{ or } 1+}$, and solvated *electrons*, $e^-(NH_3)_m$. The deep blue color results from excitement of the electron by light; this has led to the facetious proposal that electrons are blue! The low density results from the fact that the ammonia molecules solvating the electron are remarkably far apart. The solvated electron acts as if it has a surprisingly large "ionic radius" of about 150 to 170 pm.

If additional Group 1 metal is added to one of the above blue solutions, a second, even less dense liquid of bronze color forms, which is insoluble in the first and floats on it. These properties, which have been known since 1864, have fascinated chemists ever since, but the nature of the bronze solution proved difficult to determine, in part because evaporating either the blue or bronze solution to dryness generally gave back the metal. But this problem was overcome by the application of the principles of Section 3.8: The large anion in the blue solution (the "electride ion") and the unknown (probably large) anion in the bronze solution should best be stabilized in the solid state by using larger cations than simple Group 1 cations. Relatively stable crystals were produced [17] by adding macrocyclic ligands (the *crown* or *crypt* polyethers), which form more stable and larger complex cations with the Group 1 metal than does ammonia. By this method a cesium salt of the electride ion was isolated.

Similar treatment of the bronze solutions gave golden crystals of a material containing one molecule of crown or crypt polyether and *two* atoms of Group 1 metal ion. X-ray crystallography of the product with Na showed that one Na atom sits at the center of the polyether as a typical complex cation (Figure 7.11). The other sodium atom sits outside by itself, and hence must be present as the $Na^-$ ion. The combination of a sodium cation and a sodide anion is normally less stable than metallic sodium, but evidently either the extra energy of coordinate covalent bond formation with the polyether or its physical barrier to electron transfer (or both) gives marginal stability to these remarkable compounds.

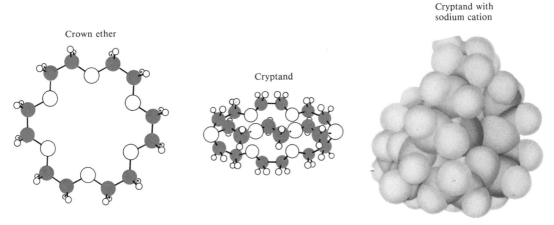

Cryptand with
sodium cation

Crown ether

Cryptand

**Figure 7.11** Molecular models of a crown ether, a *cryptand* or crypt polyether (which also contains two nitrogen atoms), and a Group 1 cation (in black) encapsulated by a crypt polyether ligand. From "Anions of the Alkali Metals," by James L. Dye. In *Scientific American*, (July 1977). Copyright © 1977 by Scientific American, Inc. All rights reserved.

## 7.9

## Study Objectives

1. Know the use of the common terms for Lewis acid-base (coordination) chemistry; identify examples for these terms in the formulas or structures of given coordination compounds.
2. Know the functions of metal ions in catalysis and in enzymes.
3. Draw Lewis dot structures of and predict the geometry of molecules and ions, including bond angles and distortions.
4. Classify ligands.
5. Predict the solubility of a given coordination compound or complex ion in (a) water; (b) nonpolar solvents.
6. Predict the positions of equilibria involving monodentate, chelating, and macrocyclic ligands; give possible explanations of the chelate and macrocyclic effects.
7. Name coordination compounds.
8. Calculate the enthalpy change of formation of a coordinate covalent bond from $E$ and $C$ parameters, or vice versa.
9. Interpret the $E$ and $C$·parameters of different Lewis acids or bases to compare their strengths at covalent bonding or at electrostatic bonding.
10. Classify different solvents as protic or aprotic, then also within the subgroups of these classifications; choose an appropriate solvent to study the properties of a strongly acidic cation or a strongly basic anion.

**11.** Predict possible products of Lewis acid-base reactions like those of Experiment 5.

## 7.10

**Exercises**

†**1.** Draw the Lewis dot structures of the following molecules or ions, and identify the potential donor atoms in each:
**a.** $NH_3$; **b.** $NH_2$—$CH_2$—COOH (an amino acid); **c.** $SO_4^{2-}$;
**d.** $SO_3^{2-}$.

**2. a.** Draw Lewis dot structures of the following complex ions:
$[Au(CH_3)_2(NH_3)_2]^+$; $[Na(CH_3$—O—$CH_2$—$CH_2$—O—$CH_3)_3]^+$;
$[SnI_6]^{2-}$; $[HgCl_2]$; $[Hg(CH_3)_2]$. Treat all the bonds to the metal as co-ordinate covalent bonds. **b.** Write the formulas (including charges, if any) of the ligands found in each complex ion. **c.** Underline the donor atoms in those ligands. **d.** Write the formulas of the Lewis acids (including charges, if any) found in these complex ions. **e.** Give the coordination number of the metal atoms in each of these complex ions.

*\***3.** Below are drawn Lewis structures of two coordination compounds. Treat all bonds to the central metal atom in each as coordinate covalent bonds. For each complex: **a.** Write the formulas (including charges) of each ligand. **b.** Tell which ligand (if any) is a chelating ligand. **c.** Write the formula (including charge) of the central metal ion. **d.** Write the formula of the counterion for the complex ion (if any).

**4.** For the complexes $(NH_4)_2[Pd(SCN)_4]$ and $[Co(NH_3)_5(SeO_4)]ClO_4$: **a.** Write the electron-dot structure for each kind of ligand, showing unshared electron pairs and underlining its donor atom(s). **b.** Give the charge on each free ligand. **c.** Give the coordination number of the central (metal) atom. **d.** Give the oxidation number of the central (metal) atom.

**5.** Pretend you are Alfred Werner. You have, by painstaking work, isolated the following set of compounds:

| | |
|---|---|
| $PdCl_2 : 4(CH_3)_3P$ | (contains 3 ions; $2\,Cl^-$ ions pptd. by $Ag^+$) |
| $PdCl_2 : 3(CH_3)_3P$ | (contains 2 ions; $1\,Cl^-$ ion pptd. by $Ag^+$) |
| $PdCl_2 : 2(CH_3)_3P$ | (contains 0 ions; $0\,Cl^-$ ion pptd. by $Ag^+$) |

$$PdCl_2 : (CH_3)_3 P : KCl \quad \text{(contains 2 ions; 0 Cl}^- \text{ ion pptd. by Ag}^+)$$
$$PdCl_2 : 2 KCl \quad \text{(contains 3 ions; 0 Cl}^- \text{ ion pptd. by Ag}^+)$$

    **a.** Write the formulas of each of these compounds in the modern form, with complex ions enclosed in brackets. **b.** What is the coordination number of palladium in this series of compounds? (It is constant.)

**6.** List the functions of metal ions in enzymes.

**7.** (Thought question) Complexes containing coordinated $NH_3$ and $H_2O$ are common, as are complexes containing coordinated $R_3N$, $R_2O$, $R_3P$, $R_2S$, etc. But complexes containing coordinated $PH_3$ and $H_2S$ (or $RPH_2$ and $RSH$) are very seldom seen. Suggest a possible reason why, based on the concepts given in Sections 5.5 and 7.2.

**†8.** Draw the Lewis dot structures of each of the following molecules or ions. Predict the bond angles (including distortions) about the central atom in each species. Give the name of the resulting geometry (atoms only, not unshared electron pairs): **a.** $NO_2^+$;    **b.** $XeF_4$;    **c.** $(CH_3)_2TeCl_2$;    **d.** $SbCl_5$;   **e.** $IF_5$;    **f.** $ClNO$;    **g.** $XeF_3^+$;    **h.** $SbCl_5^{2-}$;    **i.** $XeO_4$;   **j.** $XeF_5^+$;    **k.** $(CH_3)_2ICl$;    **l.** $NCN^{2-}$.

**9.** Draw the structure of the following, showing their geometric arrangement of groups (cis or trans, axial or equatorial, etc.): **a.** $(CH_3)_2TeCl_2$;  **b.** $SbBr_2F_2^-$;    **c.** $(CH_3)_2SbCl_3$.

**\*10.** Consider the following four forms of mercury: $Hg(l)$, $Hg(CH_3)_2(l)$, $HgCl_2(aq)$, $HgS(s, insoluble)$. Which form or forms will: **a.** pass unchanged through the digestive system (digestion requires solubility in lipids or water)? **b.** be most easily eliminated through the kidneys? **c.** most readily cross the (nonpolar) blood-brain barrier? **d.** most readily undergo bioamplification; **e.** be the greatest air-pollution hazard(s)?

**11.** Tell whether the complexes in questions 3, 4, and 5 are likely to be soluble in water; in nonpolar solvents.

**12.** Which of the following are *chelating* ligands? **a.** $PO_4^{3-}$;    **b.** $Si_2O_7^{6-}$;  **c.** $CH_3$—$O$—$CH_2$—$O$—$CH_3$;    **d.** $(CH_3)_3N$;    **e.** $N(CH_2$—$CO_2^-)_3$;  **f.** $[NH_3CH_2CH_2NH_3]^{2+}$.

**13.** **a.** Which of the ligands drawn below are chelating ligands? **b.** The $Pd^{2+}$ ion normally has a coordination number of 4. Predict the formula (including overall charge) of the complex ion formed by the $Pd^{2+}$ ion with each ligand.

A          B         C         D

**\*14.** Below is drawn the structure of a (hypothetical) truly complex complex ion.

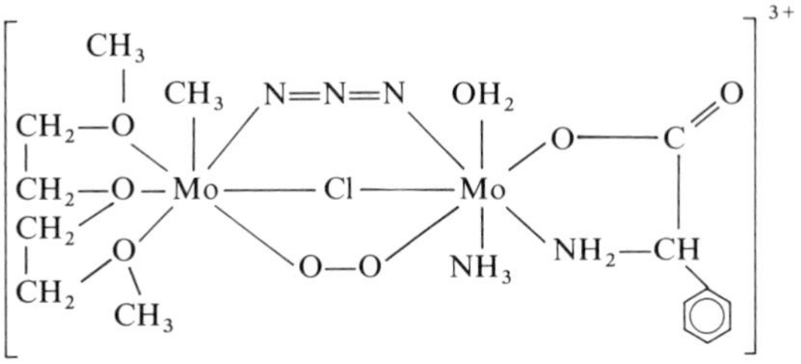

**a.** Write the formulas, including charges if any, of the ligands found in this complex. Underline the donor atoms on each ligand. **b.** List all chelating ligands in the above complex, if any. Identify each as bidentate, etc. **c.** List all bridging ligands in the above complex, if any. **d.** Give the charge of each Mo ion in the above complex, assuming that they either are the same or differ by only one unit of charge. **e.** Give the coordination number of each Mo ion in the above complex. (They may or may not be the same.)

**\*15.** Some ligands are drawn below. Which of these is/are: **a.** terdentate ligands; **b.** macrocyclic ligands; **c.** ligands that can chelate; **d.** ligands that can bridge metal ions but not chelate them; **e.** the ligand that will form the *least* stable complex with a metal ion?

$$NH_2-CH_2-CH_2-NH_2$$

A

$$H_2N-NH_2$$

B

$$(NH_2CH_2CH_2)_2NH$$

C

$$(NH_2CH_2CH_2)_3N$$

D

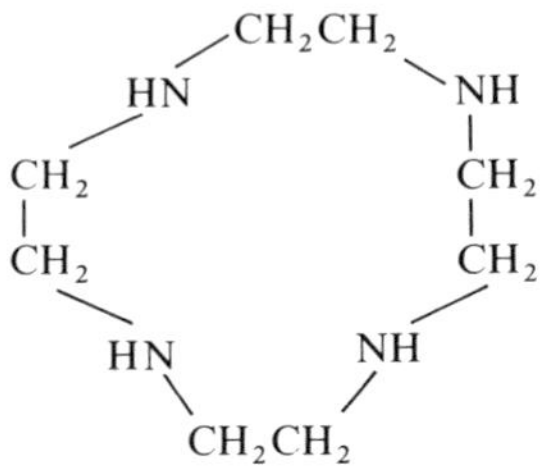

E

**16.** Shown below are some molecules and ions that *may* be capable of acting as ligands. **a.** Pick out all that are capable of acting as chelating ligands. **b.** Classify each potentially chelating ligand as (potentially) *bidentate, tridentate*, etc. (Choose the highest possible classification for each.)

A. $\overset{\oplus}{H_3}N-CH_2-CH_2-\overset{\oplus}{N}H_3$

C. $(^{\ominus}O_2C-CH_2)_2NCH_2CH_2N(CH_2CO_2^{\ominus})_2$

B.

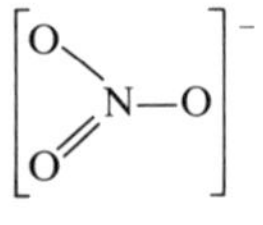

D.

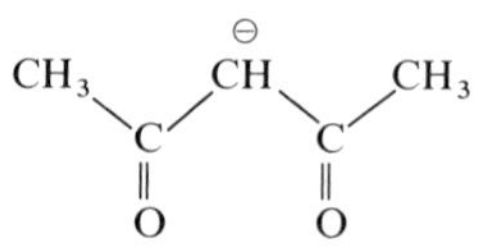

E. $:N\equiv C-\overset{..}{\underset{..}{S}}-\overset{..}{\underset{..}{S}}-C\equiv N:$

**17.** Below is drawn a neutral coordination compound of the metal niobium, Nb. Assume each bond to Nb to be a coordinate covalent bond. **a.** What is the coordination number of Nb in this complex? **b.** What is the likely geometric shape about the Nb atom? **c.** What are the likely approximate bond angles about the Nb atom? **d.** What *charge* is present on each of the ligands before they coordinate to the Nb? **e.** What charge is present on the Nb atom before it coordinates all these ligands? **f.** List all chelating ligands present in this complex. **g.** List all ambidentate ligands present in this complex.

18. Give the name of each ligand present in the Nb complex of the previous question *as it will be named in the nomenclature for coordination compounds.*

19. Give the *name* of each compound given in questions 3, 4, and 5.

20. Name the complexes given as examples of those used as industrial catalysts (Section 7.2).

†21. Name each of the following complex compounds: **a.** $Ca_3[VCl_6]_2$; **b.** $[Co(H_2O)_4(OH)_2][AlCl_4]$; **c.** $Cs[V(CO)_3(P(C_6H_5)_3)_3]$; **d.** $K_2[ReH_9]$; **e.** $[Co(CO)_5][BF_4]$; **f.** $K[Te(CH_3)_2Cl_3]$; **g.** $[(CH_3)_4P][BrF_4]$; **h.** $[Ag(C_6H_6)_2][Nb(CO)_6]$.

22. Look up one of the following complex compounds in the latest Cumulated Index of *Chemical Abstracts*. (The booklet *How to Search Printed CA* will be helpful; note that *Chemical Abstracts* uses Ewens-Bassett nomenclature.) **a.** Make a photocopy of the one or two pages of the index that contain all the references to the complex ion and *all* its salts (or the neutral complex). Mark where these citations begin and where they end. **b.** Circle the part of these pages that contains the citations of the *specific* complexes listed below (or their hydrates). **c.** Look up the same complex ion in *Inorganic Syntheses*, in the indexes at the ends of volumes 10, 20, and the latest issue. Is a synthesis of your complex ion present? Complex compounds: (1) $[Cr(CH_3COCHCOCH_3)_3]$; (2) $K_3[Cr(SCN)_6]$; (3) $K_3[Cr(C_2O_4)_3]$; (4) $[Cr(NH_2CH_2CO_2)_3]$; (5) $[Cr(NH_2CH_2CH_2NH_2)_3]Cl_3$; (6) $Na_3[CrCl_6]$.

23. Consider the two complexes (A) dicyanobis(1,2-diaminoethane)platinum(IV) perchlorate and (B) tetramminedicyanoplatinum(IV) sulfate. **a.** Write the *formulas* of these two complexes. **b.** Consider also complex (C), which is drawn below. List these three complexes in order of *increasing* stability. **c.** Which complex (A, B, C, or none) contains a chelating ligand? Which complex contains a macrocyclic ligand? Which complex contains a ligand that is bridging?

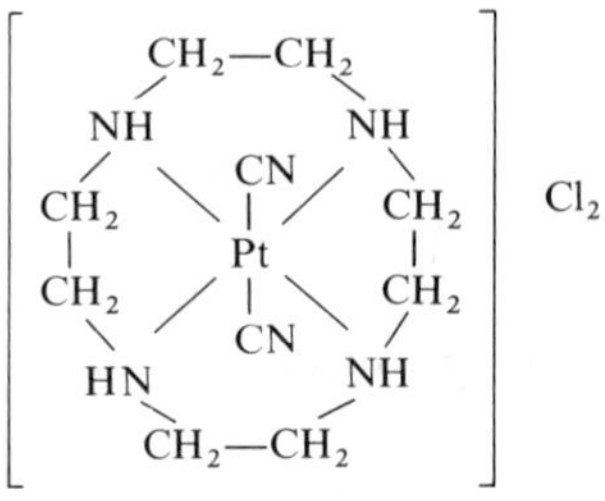

24. Calculate the enthalpy changes for the following reactions:
    **24.1** Diethyl ether + trimethylaluminum →
    diethyl ether : trimethylaluminum
    **24.2** dimethyl selenide + trimethylaluminum →
    dimethyl selenide : trimethylaluminum
    Which Lewis base is more effective at covalent bonding? (Use the $E$ and $C$ parameters to show this.)

*25. Consider the following reaction between different Lewis acid-base adducts:

    1-azabicyclo[2.2.1]octane : $SbCl_5$ + acetone : $Al(CH_3)_3$ →

    1-azabicyclo[2.2.1]octane : $Al(CH_3)_3$ + acetone : $SbCl_5$

    **a.** Calculate the enthalpy of formation of each of the adducts from its constituent Lewis acid and base. **b.** Calculate the overall $\Delta H$ for the above reaction. Does the reaction go to the left or the right?

26. Using Figure 7.11, select a suitable nonaqueous solvent in which to do an acid-base titration of: **a.** the nonmetal anions listed in Table 5.8; **b.** the +4 ions in Table 2.2; **c.** the first ionizable proton of telluric acid; **d.** the first ionizable proton of selenic acid; **e.** the third ionizable proton of telluric acid.

27. What solvents would you choose to enhance: **a.** the acidic properties of acetone, $(CH_3)_2C{=}O$; **b.** the basic properties of $HNO_3$; **c.** the basic properties of $SiH_4$; **d.** the acidic properties of $SiH_4$? Write an equation showing a reasonable reaction of each solute with its solvent.

28. Give plausible products (or just one product) for each of the following reactions, or tell if no reaction is expected. **a.** $CsCl + AlCl_3 →$ ?

    **b.** $SiCl_4$ +  → ?     (A conductor of electricity in solution)

    **c.** $PbCl_4$ +  → ?     (A nonconductor of electricity in solution)

†29. (Question for thought and review.) Iron forms a number of well-known complexes with the $CN^-$ ion. The complexes $K_4[Fe(CN)_6]$, $K_3[Fe(CN)_6]$, $Fe[Fe(CN)_6]$, and $Fe_2[Fe(CN)_6]$ are not as intensely colored as $KFe[Fe(CN)_6]$, commonly known as Prussian blue.
    **29.1** Insofar as possible, name each of these five complexes.
    **29.2** In each complex in which an iron atom is written outside of the brackets,

it is also acting as a Lewis acid. What is the Lewis base, and what donor atom(s) is it using? What is the classification of this kind of ligand?

**29.3** Explain why Prussian blue is more intensely colored than the other complex compounds.

**Notes**

[1]   See the series of articles in *J. Chem. Educ.*, 61, nos. 1–3, Jan.–Mar. 1984.

[2]   The xenon atom has recently been reported as a ligand in the short-lived coordination complex $[Cr(CO)_5Xe]$: M. B. Simpson, M. Poliakoff, J. J. Turner, W. B. Maier III, and J. G. McLaughlin, *J. Chem. Soc., Chem. Commun.*, 1983, 1355.

[3]   Traditionally, coordination complexes of the cyanide ion, :CN⁻, and often of carbon monoxide, :CO, are not classified as organometallic compounds.

[4]   Schrauzer, G. N., *Acc. Chem. Res.*, 1, 97 (1968).

[5]   A number of books cover the role of metals in biochemistry, e.g., M. N. Hughes, *The Inorganic Chemistry of Biological Processes*, 2d ed., John Wiley, Chichester, 1981.

[6]   Several books cover the chemistry of organometallic catalysts, e.g., C. Lukehart, *Fundamental Transition Metal Organometallic Chemistry*, Brooks/Cole, Monterey, Calif., 1985.

[7]   Vallee, B. L., and R. J. P. Williams, *Proc. Natl. Acad. Sci. U.S.A.*, 59, 498 (1968).

[8]   Jolly, W. L., *Modern Inorganic Chemistry*, McGraw-Hill, New York, 1984, Ch. 3.

[9]   Although recent theoretical calculations indicate that this assumption is incorrect (M. B. Hall, *J. Amer. Chem. Soc.*, 100, 6333 (1978); *Inorg. Chem.*, 17, 2261 (1978)), it is so much easier to use than more "correct" models is predicting observed results that chemists continue to use it.

[10]   Chung, C.-S., *J. Chem. Educ.*, 61, 1062 (1984); *Inorg. Chem.*, 18, 1321 (1979); Myers, R. T., *Inorg. Chem.*, 17, 952 (1978).

[11]   If, as discussed in Section 7.1, a ligand is less electronegative than the metal to which it is attached, there can be an ambiguity in the oxidation number of the metal. This ambiguity is avoided by the Ewens-Bassett convention.

[12]   Wulfsberg, G., J. Yanisch, R. Meyer, J. Bowers, and M. Essig, *Inorg. Chem.*, 23, 715 (1984) and references therein.

[13]   International Union of Pure and Applied Chemistry, *How to Name an Inorganic Substance*, Pergamon Press, Oxford, 1977.

[14]   Drago, R. S., and B. B. Wayland, *J. Amer. Chem. Soc.*, 87, 3571 (1965); R. S. Drago, *Struct. Bond.*, 15, 73 (1973); R. S. Drago et al., *J. Amer. Chem. Soc.*, 93, 6014 (1971); ibid., 99, 3203 (1977).

[15]   Kroeger, M. K., and R. S. Drago, *J. Amer. Chem. Soc.*, 103, 3250 (1981).

[16]   Gutmann, V., *The Donor-Acceptor Approach to Molecular Interactions*, Plenum, New York, 1978.

[17]   See, for example, W. L. Jolly, *Modern Inorganic Chemistry*, McGraw-Hill, New York, 1984, p. 202.

[18]   Issa, D., and J. L. Dye, *J. Amer. Chem. Soc.*, 104, 3781 (1982), and other papers by Dye.

# The Hard and Soft Acid-Base (HSAB) Principle and Its Applications

The large majority of organic compounds and inorganic anions possess donor atoms and can act as ligands. Since there are so many different metal ions, there are a limitless number of coordination compounds that may be formed in natural waters, in biological fluids, or in industrial process waters or wastewaters. Important properties of metal ions such as their solubilities in fat or water and their toxicities will depend on which of the countless number of possible ligands ends up attached to the metal ion.

From the mathematical point of view, the question, Which metal ion will combine with which ligand? [1] is rather formidable. In principle, we must measure the equilibrium constants relating to the stepwise or overall formation of each possible combination of metal ion and ligand. We must know the initial concentration of each metal ion and each ligand in a given natural water, biological fluid, or analytical sample, and the pH and the reducing potential of the sample. All this information, and a large computer, are needed to calculate the equilibrium concentration of each possible complex. Simplified calculations of the *speciation* of metals have actually been done, involving, for example [2], 12 metals (Fe, Mn, Cu, Ba, Cd, Zn, Ni, Hg, Pb, Co, Ag, and Al) and 8 ligands ($CO_3^{2-}$, $S^{2-}$, $Cl^-$, $F^-$, $NH_3$, $PO_4^{3-}$, "$SiO_3^{2-}$", and $OH^-$).

Fortunately we can, to a limited extent, predict the results of such computations without any calculations whatsoever. Your instructor may assign Experiment 6 at this point; if you did the qualitative analysis scheme in your general chemistry course, he or she may assign only part 7 of this experiment. In this experiment you can discover the principle that enables us to make qualitative predictions of the predominant products that result when Lewis acids and Lewis bases compete with each other to form precipitates or complex ions. After we develop this principle, we will use it to organize the chemistry of such diverse phenomena as (1) the qualitative analysis scheme that you may have done in general chemistry, (2) the activity series experiment that you also may have done, (3) additional solubility rules for inorganic compounds, (4) the geochemistry of the elements and why they occur in the ores in which they are found, (5) the nutritional and toxic effects of many metals and nonmetals in the body (e.g., micronutrients and toxic effects of heavy metals), and (6) the ways in which medicinal chemists devise drugs to counteract the effects of heavy metal poisoning.

## 8.1

## The Hard and Soft Acid-Base (HSAB) Principle

In Experiment 6 you may have made the observation that in mixtures of several Lewis acids and bases, a certain set of metal ions tends to combine with the iodide ion in preference to the fluoride ion. This same set of metal ions prefers the sulfide ion to the hydroxide ion, thiourea to urea, and the sulfide ion to the silicate ion. This last experiment is, of course, related to the **differentiation** of the elements in nature. Berzelius first noted in 1796 that certain metal ions tend to occur in nature as sulfides, while others tend to occur as oxides, carbonates, sulfates, or silicates. Only about 20 years ago was it realized just how broadly these observations could be generalized and how many thousands of chemical reactions and phenomena could be (reasonably well) predicted using the generalized observation. The general answer to the question, Which metal ion will tend preferentially to form a complex ion with which ligand? was best summarized several years ago by Pearson [3]: *Hard (Lewis) acids tend to combine with hard (Lewis) bases; soft acids prefer soft bases.*

This statement is now known as the **hard and soft acid-base**, or **HSAB**, **Principle**. It is simply a summary of the observed results of thousands of chemical reactions in inorganic and organic chemistry, biochemistry, aquatic chemistry, medicinal chemistry, geochemistry, and so on. Put into equation form, for the following chemical reaction among complex ions, the equilibrium will tend to favor the products on the right side of equation (8.1):

$$[\text{hard acid(:soft base)}_n] + [\text{soft acid(:hard base)}_n] \rightleftharpoons \quad (8.1)$$
$$[\text{hard acid(:hard base)}_n] + [\text{soft acid(:soft base)}_n]$$

To be able to use the HSAB Principle, we must find a way to divide metal ions into the two classes *hard* and *soft* (other authors use the terms *class a* and *class b* instead). And we must divide all ligands into the two classes called hard and soft (or class a and class b, respectively). We can indeed do this reasonably well, although some acids and bases will not fall clearly into either class; we will call them *borderline*. The classification scheme that is usually devised is shown in Table 8.1, superimposed upon a table of Pauling electronegativities.

SOFT ACIDS OR CLASS B METAL IONS    Despite their presence in different groups of the periodic table, these metal ions, popularly known as the *heavy metal* ions, have a number of chemical similarities. They have insoluble chlorides, so they tend to occur in Group I of the qualitative analysis scheme. They are chemically rather inert and thus are low on the activity series of metals. They occur in nature as the free elements or in sulfide and related minerals. In terms of fundamental atomic properties (as you may have noted in the lab experiment), these are the *electro-*

**Table 8.1**  Hard and Soft Acids and Bases

Most often seen as Lewis bases

Hard bases · Borderline bases · softest · soft bases · bases

| H 2.2 | | | | | | | | | | | | | | | | | He |
|---|---|---|---|---|---|---|---|---|---|---|---|---|---|---|---|---|---|
| Li 0.98 | Be 1.57 | | | | | | | | | | | | | | | | Ne |
| Na 0.93 | Mg 1.31 | | | | | | | | | | | | | | | | Ar |
| K 0.82 | Ca 1.00 | Sc 1.36 | Ti 1.54 | V 1.63 | Cr 1.66 | Mn 1.55 | Fe(+3)1.83(+2) | Co(+3)1.88(+2) | Ni 1.91 | Cu(+1)2.0 (+2) | Zn 1.65 | | | | | | Kr 3.0 |
| Rb 0.82 | Sr 0.95 | Y 1.22 | Zr 1.33 | Nb 1.6 | Mo 2.16? | Tc 1.9? | Ru 2.2 | Rh (+3)2.28(+1) | Pd 2.20 | Ag 1.93 | Cd 1.69 | | | | | | Xe 2.6 |
| Cs 0.79 | Ba 0.89 | Lu 1.27 | Hf 1.3 | Ta 1.5 | W 2.36? | Re 1.9? | Os 2.2 | Ir (+3)2.2(+1) | Pt 2.28 | Au 2.54 | Hg 2.0 | | | | | | |
| Fr 0.7 | Ra 0.9 | | | | | | | | | | | | | | | | |

Right-hand p-block (groups 13–17):

| B 2.04 | C 2.55 | N 3.04 | O 3.44 | F 3.98 |
|---|---|---|---|---|
| Al 1.61 | Si 1.90 | P 2.19 | S 2.58 | Cl 3.16 |
| Ga 1.81 | Ge 2.01 | As 2.18 | Se 2.55 | Br 2.96 |
| In(+3)1.78(+1) | Sn(+4)1.96(+2) | Sb 2.05 | Te 2.1 | I 2.66 |
| Tl (+1)1.60 (+3)2.04 | Pb (+2)1.87 (+4)2.33 | Bi 2.02 | | |

Borderline acids · (+2) · Hard acids · Soft acids · Borderline acids · Hard acids

| La 1.10 | Ce 1.12 | Pr 1.13 | Nd 1.14 | Pm | Sm 1.17 | Eu | Gd 1.20 | Tb | Dy 1.22 | Ho 1.25 | Er 1.24 | Tm 1.25 | Yb |
|---|---|---|---|---|---|---|---|---|---|---|---|---|---|
| Ac 1.1 | Th 1.3 | Pa 1.5 | U 1.38 | Np 1.36 | Pu 1.28 | Am 1.3 | Cm 1.3 | Bk 1.3 | Cf 1.3 | Es 1.3 | Fm 1.3 | Md 1.3 | No 1.3 |

NOTES: Numbers in parentheses are oxidation numbers. The number below each atomic symbol is the Pauling electronegativity of that element.

*negative metals*. They are characterized by quite *high* Pauling electronegativities for metals, generally in the range of 1.9 to 2.54. Other characteristics that they share to a lesser extent are large size (ionic radii in excess of 90 pm) and low charge (usually $+1$ or $+2$). Simply put, they form a triangle of metals around the element gold, which is the most electronegative of all metals.

SOFT BASES OR CLASS B LIGANDS    According to the HSAB Principle, the group of ligands with which these metal ions tend to associate will be called *soft bases* or *class b* ligands. The soft-acid metals are found in nature either in the form of the native (free) elements or as chlorides, bromides, iodides, sulfides, selenides, tellurides, arsenides, etc. They are not found as oxides or fluorides or as the salts of various oxo anions such as sulfate, silicate, carbonate, etc. If we have the misfortune to swallow the ions of the soft-acid metals and live through the experience, we will find the ions bonded to certain donor atoms in the proteins and enzymes of our body: to sulfur and selenium but *not* to oxygen, fluorine, or (probably) nitrogen. Table 8.1 indicates that the following donor atoms are characteristic of soft bases: C, P, As, S, Se, Te, Br, and I. (We may also add H when it occurs as a donor atom in the hydride ion, $H^-$.) Fundamentally, these are the *electronegative nonmetals*, with Pauling electronegativities of 2.1 to 2.96. These are also the largest nonmetal atoms, with anionic radii in excess of 170 pm.

HARD BASES OR CLASS A LIGANDS    Contrasting most strongly with the soft non-metal donor atoms are *oxygen* and *fluorine*, which are seldom found associated in nature with the soft acids. These donor atoms are of very high electronegativity (3.44 and 3.98) and are the smallest of the nonmetal atoms (anionic radii of about 120 pm). Although this seems like a small group of atoms, they (oxygen in particular) are the donor atoms in countless ligands—the oxo anions such as sulfate, carbonate, silicate; the anions of organic acids such as acetate; and several classes of organic molecules such as alcohols and ketones.

HARD ACIDS OR TYPE A METAL IONS    Although there are only two unambiguously hard donor atoms, in nature a large number of metal ions occur associated with ligands containing these two. These *hard acids* occupy a good share of the periodic table (Table 8.1); they have in common *low* electronegativities of (usually) 0.7 to 1.6 and are thus mostly the *very electropositive* metals. They frequently (but not invariably) have relatively small cationic radii (less than perhaps 90 pm) and may often have high charges ($+3$ or higher) [4].

As you may have noticed in the lab experiment, the most important atomic variable in classifying hard and soft acids and bases is the Pauling electronegativity of the donor or acceptor atom. Sometimes the other variables (charge and size) override considerations of electronegativity. Thus $H^+$ is classified as a hard acid despite its high electronegativity on the basis of its extremely small size (0 pm). $B^{3+}$ and $C^{4+}$ are considered hard acids despite their electronegativities both because of their very small size and their high charges.

Once we have classified the two Lewis acids and the two Lewis bases, we apply the HSAB Principle to determine whether the equilibrium in question favors products or reactants. Note that if there are less than two Lewis acids or less than two Lewis bases, the principle does not apply.

EXAMPLE   Predict whether reactants or products are favored in each of the following equilibria:
(a) $Nb_2S_5 + 5\,HgO \rightleftharpoons Nb_2O_5 + 5\,HgS$;
(b) $La_2(CO_3)_3 + Tl_2S_3 \rightleftharpoons La_2S_3 + Tl_2(CO_3)_3$;
(c) $2\,CH_3MgF + HgF_2 \rightleftharpoons (CH_3)_2Hg + 2\,MgF_2$.

Solution    ■ (a) The two cations are $Nb^{5+}$ and $Hg^{2+}$. On the basis of its electronegativity and charge, the former is clearly a hard acid, whereas the latter is a soft acid. The two anions are $O^{2-}$ and $S^{2-}$, which are hard and soft bases, respectively. It is convenient to write "HA," "HB," "SA," and "SB" under the species as they occur in the equation:

$$Nb_2S_5 + 5\,HgO \rightleftharpoons Nb_2O_5 + 5\,HgS$$

HA SB     SA HB     HA HB     SA SB

The side of the equation under which the combinations HAHB and SASB occur will be favored.

■ (b) Based on electronegativity considerations, we classify $La^{3+}$ as a hard acid and $Tl^{3+}$ as a soft acid. We classify $S^{2-}$ as a soft base. In $CO_3{}^{2-}$ it is important to identify the donor atom: It is oxygen, hence this is a hard base. The combinations HAHB and SASB occur under the reactants in this case.

■ (c) $Mg^{2+}$ is clearly a hard acid and $Hg^{2+}$ is a soft acid; $F^-$ is a hard base and $CH_3{}^-$, with a carbon donor atom, is a soft base. Confusingly, the species $CH_3MgF$ will have written under it SB-HA-HB. But you will notice that on the right side, the $F^-$ is still coordinated to the $Mg^{2+}$. It has not participated in the reaction, so the relevant exchange is of the soft base $CH_3{}^-$ away from the hard acid $Mg^{2+}$ to the soft acid $Hg^{2+}$. This equilibrium favors products.     □

## 8.2

## Relative and Borderline Hardness and Softness

Suppose that we are considering a reaction in which both Lewis acids are soft and both Lewis bases are soft, e.g,

$$CdSe + HgS \rightleftharpoons CdS + HgSe \tag{8.2}$$

The HSAB Principle can even be used to predict the result here (the reaction goes to the right), because it is possible to assign *relative* softnesses to different soft acids and bases. Among soft bases the following orders of softness have been observed:

Group 17/VIIA: $I > Br > Cl > F$

Group 16/VIA: $Te = Se > S \gg O$

Group 15/VA: $Sb < As = P > N$

As we see in Table 8.1, the softest donor atoms are those of the lowest electronegativity, and they form a "ridge" along the metal-nonmetal boundary. Thus we find, in a competitive precipitation experiment involving all the halide ions, that AgI will precipitate preferentially (is least soluble). AgBr is not quite so insoluble as AgI; AgCl is not quite so insoluble as AgBr; AgF, which involves a hard base, is much more soluble than any of the others.

Evaluating the relative softness of the metal ions has proved somewhat harder. There are some minor discrepancies among different competitive equilibria involving different soft metal ions with different soft bases. For our purposes we can say that the farther the metal is from *gold* (which all seem to agree is the softest acid), the harder it is. (Gold is, of course, the metal with the *highest* Pauling electronegativity.)

For use with competitive equilibria in which softness is present in more than one Lewis acid and more than one Lewis base, we may restate the HSAB Principle as follows: **Less soft acids tend to combine with less soft bases; softer acids prefer softer bases.** Thus in reaction (8.2), mercury, being closer to gold in the periodic table, should be softer, and should prefer the softer base, the selenide ion.

So far it does not seem to be profitable to attempt to assign relative hardnesses to different hard acids and bases, although Pearson is currently working on this problem [5]. If a competitive equilibrium involves only one species with any degree of softness, it is best to use, not the HSAB Principle, but the concepts in the first few chapters of this book. For example, the aqueous equilibrium

$$Hg(ClO_4)_2 + 2\,KOH \rightleftharpoons Hg(OH)_2 + 2\,KClO_4 \tag{8.3}$$

involves only one soft acid and no soft bases. We can predict that products will be favored, and $Hg(OH)_2$ and $KClO_4$ will precipitate, by evaluating the *strengths* of the Lewis acids and bases. $K^+$ is a nonacidic cation and should give a precipitate with the nonbasic $ClO_4^-$ anion; $Hg^{2+}$ is a weakly to moderately acidic cation and should give a precipitate with the strongly basic hydroxide anion.

BORDERLINE ACIDS AND BASES   Given that there are degrees of softness, it is not surprising that there are some metal and nonmetal atoms that show such a small degree of softness that softness will not consistently govern the results of their competitive equilibria. As Table 8.1 shows, there are several such metal ions, which are labeled borderline acids. (Many of these have insoluble sulfides but soluble halides, for example.) Likewise two of the nonmetal atoms, chlorine and nitrogen, are classified as borderline bases. In these cases the electronegativities of the atoms or their sizes may fall between the ranges that are typical for soft and for hard species. Although the chemistry of borderline acids and bases is a little harder to predict, we can still use the HSAB Principle: A borderline base (such as Cl) will be *softer* than F but *not so soft* as I.

MODIFYING THE SOFTNESS OF AN ATOM   Since high oxidation numbers are one characteristic of a hard acid, and low oxidation numbers are a characteristic of a soft acid, it is possible to alter the softness of a metal ion by changing its oxidation number. This is a particularly important characteristic of the later $d$-block metals of the fourth period, which are fundamentally borderline acids and thus particularly susceptible to a change in categories. Among the metals Fe to Zn the $+1$ oxidation state (which occurs only for Cu) is definitely soft, whereas the $+2$ oxidation state is borderline, and the $+3$ ion (such as for Fe) is characteristically hard. Some other examples of this effect can also be seen in Table 8.1. We also note that, in terms of their catalytic effects, metallic forms of elements (oxidation number $= 0$) are generally soft acids.

Because of the increased softness associated with low oxidation numbers, we generally find that low oxidation states of metals are stabilized by the presence of soft-base ligands. High oxidation states of metals are usually stabilized by the presence of hard-base ligands. Thus the maximum oxidation states of metals are usually found in their fluorides, oxides, fluoro anions, or oxo anions. (Of course $F^-$ and $O^{2-}$ are the two anions that would be the most difficult for the metal in its high oxidation state to oxidize.) Oxidation states for metals of zero or below are mainly found in their organometallic compounds, in which the soft donor atom carbon is attached to the metal atom.

The softness of a donor or acceptor atom can also be altered by changing the substituents on the atom. Attaching soft bases to a Lewis acid generally softens it as a Lewis acid in any further reactions with additional Lewis bases; the converse holds true for the attachment of hard bases. For example, in the form of hydrated ions (water is a hard base), all the Group 13/IIIA $+3$ metal ions except $Tl^{3+}$ are hard acids; in the form of their methyl compounds $M(CH_3)_3$, all are soft acids except $B(CH_3)_3$ and $Al(CH_3)_3$.

EXAMPLE    Predict whether reactants of products are favored in the following equilibria:
(a) $ZnI_2 + HgCl_2 \rightleftharpoons ZnCl_2 + HgI_2$;
(b) $(CH_3)_2O : BF_3 + (CH_3)_2S : BH_3 \rightleftharpoons (CH_3)_2S : BF_3 + (CH_3)_2O : BH_3$;
(c) $CuI_2 + Cu_2O \rightleftharpoons 2\,CuI + CuO$.

Solution    ■ (a) $I^-$ is a soft base and $Hg^{2+}$ is a soft acid. Although $Zn^{2+}$ is a borderline acid (Table 8.1) and $Cl^-$ is a borderline base, relative to $Hg^{2+}$ and $I^-$, they are each the harder species. Thus the SA–SB combination is $HgI_2$, and the harder acid–harder base combination is $ZnCl_2$; products are favored.

■ (b) In the four coordination compounds chosen, the change that occurs is a switch of B—O and B—S coordinate covalent bonds. Breaking apart the compounds at these bonds, we identify the hard base, $(CH_3)_2O$, and the soft base, $(CH_3)_2S$. The two Lewis acids $BH_3$ and $BF_3$ differ in their substituents; $BF_3$ should be the harder. Hence the reactants should be preferred.

■ (c) The bases are easily classified; $I^-$ is soft and $O^{2-}$ is hard. The copper occurs in two oxidation states: $Cu^+$ in $Cu_2O$ and $CuI$, and $Cu^{2+}$ in $CuO$ and $CuI_2$. Since $Cu^+$ is softer, the SA–SB combination is $CuI$, and products are favored.    □

. Second, the strongest bonds in the gas state are for the halides
far left of the periodic table. These gaseous halides, of course,
, triplets, etc. Thus it seems that the bond energies of gaseous
irectly related to their degree of polarity: Ion pairs show higher
an polar covalent bonds, which show higher bond energies than
ent bonds.

hus suggested that bond energies can be analyzed as a sum of contri-
covalent bonding and ionic bonding. Pauling chose to take the
ntribution to the bond energy as the average of the halogen-halogen
nt-element covalent bond energies (Table 6.5). We may suppose that the
ntribution should depend on the magnitude of the partial positive charge
element times the magnitude of the partial negative charge of the halide
d by the bond distance, i.e., $Z^2/r$. If we neglect the variability of the bond
nce, an equation for the element-halogen bond energy, $\Delta H(E—X)$, begins to
erge:

$$\Delta H(E—X) = \tfrac{1}{2}[\Delta H(X—X) + \Delta H(E—E)] \tag{8.6}$$
$$+ k(\text{partial pos. charge})(\text{partial neg. charge})$$

The magnitudes of the partial positive and negative charges in polar covalent
bonds are even now difficult to determine unambiguously. But logically each should
depend on the *difference* between the relative electron-attracting abilities of the two
elements: The more unequal this ability, the more charge will build up at each end
of the bond. This electron-attracting ability Pauling called the **electronegativity** of
that element; he defined it as the attraction of an atom in a molecule (or polyatomic
ion) for the electrons in its covalent bonds. Thus each of the partial charges would
be expected to build up in proportion to the difference between the electronegativities
of the two atoms, until the charge separation became as complete as it could be in
the ion pairs:

$$\text{Partial pos. charge} = -\text{partial neg. charge} = k'[\chi_P(X) - \chi_P(E)] \tag{8.7}$$

Substituting this expression into equation (8.6), collecting the constants $-kk'^2$
together, and setting them equal to 96.5 kJ/eV gives us the relationship between
bond energies and differences between Pauling electronegativities $\chi_P$:

$$\Delta H(E—X) = \tfrac{1}{2}[\Delta H(X—X) + \Delta H(E—E)] + 96.5[\chi_P(X) - \chi_P(E)]^2 \tag{8.8}$$

Provided that we arbitrarily set one electronegativity value (Pauling chose 4
for F; we now use 3.98), equation (8.8) can be used, along with modern experimental
gaseous bond dissociation energies, to determine the values of the Pauling electronegativities given in Table A [8]. Values have been obtained separately for each
kind of gaseous halide and for other single-bonded compounds of the elements,
such as hydrides. The values in Table A represent mean values for these different
compounds, and have standard deviations of $\pm 0.05$. For the $d$- and $f$-block
metals the element-element single-bond energies are generally unknown, so the
electronegativity difference $\chi_P(E) - \chi_P(X)$ is obtained approximately as the heat
of formation of the halide $EX_n$ divided by $n$. (Faulty thermodynamic data will, of

**Table 8.2** Characteristic Properties of Hard and Soft Acids and Bases

| Property | Hard Acids | Hard Bases | Soft Acids | Soft Bases |
|---|---|---|---|---|
| Electronegativity | 0.7–1.6 | 3.4–4.0 | 1.9–2.5 | 2.1–3.0 |
| Ionic radius | Small | Small | Large | Large |
| Ionic charge | High | — | Low | — |

## 8.3

## The HSAB Principle, Polar Covalent Bonding, and Pauling Electronegativities

The HSAB Principle is what we call *empirical*: It is based on observation. Your
conclusions upon doing the lab experiments have also been empirical. Although
empirical relationships are very valuable, science also seeks to find theoretical
reasons for the relationships. (We have generally done this in the chapters following
the lab experiments.) For the HSAB Principle this has been more difficult than usual,
in part because it is a qualitative relationship: We do not have good numerical
values for softness of species (but see [5]). Sophisticated methods and arguments
(such as perturbation molecular-orbital theory [6]) have been applied to the
question of why the principle works. But for now let us analyze the principle at a
very simple level, looking at the fundamental properties of hard and soft acids and
bases (Table 8.2).

Of these properties the most important in determining hardness or softness is
the Pauling electronegativity [3a]. We note that hard acids and hard bases have
Pauling electronegativities that differ by a large amount (at least 1.8 units). We recall
from general chemistry that a large electronegativity difference between bonded
atoms favors a highly ionic (electrostatic) character to the bond. Hard acids and
hard bases also tend to be small, and hard acids to be highly charged. All these
factors contribute to high lattice energies and hence stability for ionic hard acid–
hard base reactants or products.

On the other hand, soft acids and soft bases have very similar Pauling electronegativities; the small difference between them favors the formation of nonpolar
covalent bonds. The large size and low charges of the atoms, although not favorable
for covalent bonding, are even less favorable for ionic bonding.

On examining the combination of a hard base and a soft acid (or a hard acid
and a soft base), we find that the relative electronegativities and sizes are not optimal
for either ionic or covalent bonding. The electronegativity differences of about one
are characteristic of **polar covalent bonds**. So the HSAB Principle, in a sense, is
restating the observation (common in organic chemistry, for example) that polar
covalent compounds tend to be rather reactive if they can react to give an ionic
product and a nonpolar covalent product. Many of the more reactive species in
organic chemistry (such as Grignard reagents and alkyl halides) have polar covalent
bonds (C—Mg and C—halogen, respectively).

We also observe a thermodynamic difference between the hard acid–hard base and the soft acid–soft base interactions. We noted in Chapter 3 that the most common hard acid–hard base precipitation reaction, that of an acidic cation and a basic anion, is driven by its *entropy change*, as many solvent water molecules are released. For example, the typical hard acid–hard base reaction (8.4) involves a small, unfavorable $\Delta H$ of $+51$ kJ/mol and is driven by its $-T\Delta S$ of $-121$ kJ/mol.

$$H^+(aq) + F^-(aq) \rightleftharpoons HF(aq) \tag{8.4}$$

Reactions of the large, low-charged soft acids and bases generally do not involve much of an entropy change but are associated with a favorable *enthalpy change*. Reaction (8.5), a typical soft acid–soft base reaction, involves a negligible $-T\Delta S$ of $+10$ kJ/mol and is driven by the magnitude of its $\Delta H$, $-315$ kJ/mol.

$$Hg^{2+}(aq) + I^-(aq) \rightleftharpoons HgI^+(aq) \tag{8.5}$$

This enthalpy change is the result of the formation of a *covalent bond*. Thus the bond energies of polar and nonpolar covalent compounds are related to the driving force behind the HSAB Principle.

Bond dissociation energies for gaseous elements were considered in Chapter 6. These can also be evaluated for gaseous single-bonded polar covalent compounds provided that there is only one type of bond present. (If more than one type of bond is present, all of them may dissociate; there is then no unambiguous way to divide the enthalpy change among the different bond types.) The bond energies of the gaseous halides of many of the elements are tabulated in Tables 8.3 to 8.6.

The chemist Linus Pauling noted some interesting periodic trends in these data [7]. First, the polar bonds of elements to halogens are almost always stronger than the nonpolar bonds of the elements to themselves or the bonds of the halogens to

**Table 8.3**  Element–Fluorine Bond Dissociation Energies (kJ/mol)

| HF | 565 | | | | | | | | | | | | | | |
|---|---|---|---|---|---|---|---|---|---|---|---|---|---|---|---|
| LiF | 573 | BeF$_2$ | 632 | BF$_3$ | 613 | CF$_4$ | 485 | NF$_3$ | 283 | OF$_2$ | 189 | F$_2$ | 155 | | |
| NaF | 477 | MgF$_2$ | 513 | AlF$_3$ | 583 | SiF$_4$ | 565 | PF$_3$ | 490 | SF$_6$ | 284 | ClF$_5$ | 142 | | |
| | | | | | | | | | | | | ClF$_3$ | 172 | | |
| | | | | | | | | | | | | ClF | 249 | | |
| KF | 490 | CaF$_2$ | 550 | GaF$_3$ | 469 | GeF$_4$ | 452 | AsF$_5$ | 406 | SeF$_6$ | 285 | BrF$_5$ | 187 | KrF$_2$ | 50 |
| | | | | | | GeF$_2$ | 481 | AsF$_3$ | 484 | SeF$_4$ | 310 | BrF$_3$ | 201 | | |
| | | | | | | | | | | SeF$_2$ | 351 | BrF | 249 | | |
| RbF | 490 | SrF$_2$ | 553 | InF$_3$ | 444 | SnF$_4$ | 414 | SbF$_5$ | 402 | TeF$_6$ | 330 | IF$_7$ | 231 | XeF$_6$ | 126 |
| | | | | InF | 523 | SnF$_2$ | 481 | SbF$_3$ | 440 | TeF$_4$ | 335 | IF$_5$ | 268 | XeF$_4$ | 130 |
| | | | | | | | | | | TeF$_2$ | 393 | IF$_3$ | 272 | XeF$_2$ | 131 |
| CsF | 502 | BaF$_2$ | 578 | TlF | 439 | PbF$_4$ | 331 | BiF$_5$ | 297 | | | | | | |
| | | | | | | PbF$_2$ | 394 | BiF$_3$ | 393 | | | | | | |

SOURCE: Data are from J. E. Huheey, *Inorganic Chemistry: Principles of Structure and Reactivity*, 3d ed., Harper and Row, New York, 1983, pp. A-28 to A-40.
NOTE: Compounds are shown for which each bond dissociation energy applies.

**Table 8.4**  Element–Chlorine Bond Dissoci...

| HCl | 428 | | | | |
|---|---|---|---|---|---|
| LiCl | 464 | BeCl$_2$ | 461 | BCl$_3$ | 4. |
| NaCl | 408 | MgCl$_2$ | 406 | AlCl$_3$ | 421 |
| KCl | 423 | CaCl$_2$ | 429 | GaCl$_3$ | 354 |
| | | | | G. | |
| RbCl | 444 | SrCl$_2$ | 469 | InCl$_3$ | 328 | SnC |
| | | | | InCl | 435 | SnCl$_2$ |
| CsCl | 435 | BaCl$_2$ | 475 | TlCl | 364 | PbCl$_4$ |
| | | | | | | PbCl$_2$ | 304 |

SOURCE: Data are from J. E. Huheey, *Inorganic Chemistry: Principles of Structure a...* pp. A-28 to A-40.
NOTE: Compounds are shown for which each bond dissociation energy applies.

**Table 8.5**  Element–Bromine Bond Dissociation Energies (kJ/mol)

| HBr | 362 | | | | | | | | | | |
|---|---|---|---|---|---|---|---|---|---|---|---|
| LiBr | 418 | BeBr$_2$ | 372 | BBr$_3$ | 377 | CBr$_4$ | 285 | | | H. | |
| NaBr | 363 | MgBr$_2$ | 339 | | | SiBr$_4$ | 310 | PBr$_3$ | 264 | S$_2$Br |
| KBr | 379 | CaBr$_2$ | 402 | GaBr$_3$ | 302 | GeBr$_4$ | 276 | AsBr$_3$ | 258 | SeBr |
| | | | | | | GeBr$_2$ | 325 | | | SeBr$_2$ |
| RbBr | 385 | SrBr$_2$ | 405 | InBr$_3$ | 279 | SnBr$_4$ | 273 | SbBr$_5$ | 184 | TeBr$_4$ | 176 |
| | | | | InBr | 406 | SnBr$_2$ | 329 | SbBr$_3$ | 260 | TeBr$_2$ | 243 |
| CsBr | 416 | BaBr$_2$ | 427 | TlBr | 326 | PbBr$_4$ | 201 | BiBr$_3$ | 232 | |
| | | | | | | PbBr$_2$ | 260 | | | |

SOURCE: Data are from J. E. Huheey, *Inorganic Chemistry: Principles of Structure and Reactivity*, 3d ed., Harper and Row, New York, pp. A-28 to A-40.
NOTE: Compounds are shown for which each bond dissociation energy applies.

**Table 8.6**  Element–Iodine Bond Dissociation Energies (kJ/mol)

| HI | 295 | | | | | | | | | | | | |
|---|---|---|---|---|---|---|---|---|---|---|---|---|---|
| LiI | 347 | BeI$_2$ | 289 | | | CI$_4$ | 213 | | | HOI | 201 | FI | 278 |
| NaI | 304 | MgI$_2$ | 264 | | | SiI$_4$ | 234 | PI$_3$ | 184 | | | ClI | 208 |
| KI | 326 | CaI$_2$ | 326 | GaI$_3$ | 237 | GeI$_4$ | 212 | AsI$_3$ | 200 | SeI$_2$ | 151 | BrI | 175 |
| | | | | | | GeI$_2$ | 264 | | | | | | |
| RbI | 331 | SrI$_2$ | 335 | InI$_3$ | 225 | SnI$_4$ | 205 | SbI$_3$ | 195 | TeI$_4$ | 121 | I$_2$ | 149 |
| | | | | | | SnI$_2$ | 261 | | | TeI$_2$ | 192 | | |
| CsI | 335 | BaI$_2$ | 360 | TlI | 280 | PbI$_2$ | 205 | BiI$_3$ | 168 | | | | |

SOURCE: Data are from J. E. Huheey, *Inorganic Chemistry: Principles of Structure and Reactivity*, 3d ed., Harper and Row, New York, 1983, pp. A-28 to A-40.
NOTE: Compounds are shown for which each bond dissociation energy applies.

course, lead to faulty electronegativities; this is thought to be the case, for example, for lead(IV).)

Equation (8.8) is unreliable for any electronegativity differences greater than about 1.8. As we note from Tables 8.3 to 8.6, at the left of the periodic table the bond energies are approximately constant despite changes in Pauling electronegativities. This observation is one source of the suggestion that the bonding in a halide or other compound becomes as ionic as it can get with electronegativity differences much over 1.8.

We can now apply the Pauling electronegativity equation (8.8) to the analysis of a simple HSAB equilibrium occurring in the gas phase:

$$HA:SB(g) + SA:HB(g) \rightleftharpoons HA:HB(g) + SA:SB(g) \qquad (8.9)$$

We evaluate the bond energies of each of the four reactants and products using equation (8.8). We then take the sum of the bond dissociation energies of the *reactants* minus the sum of the bond dissociation energies of the *products* (in which bonds are formed, not dissociated). When we do this, all the terms involving element-element bond energies ($\Delta H(HA{-}HA)$, etc.) drop out, leaving four terms:

$$\Delta H = 96.5\{[\chi_P(HA) - \chi_P(SB)]^2 + [\chi_P(SA) - \chi_P(HB)]^2 \qquad (8.10)$$
$$- [\chi_P(HA) - \chi_P(HB)]^2 - [\chi_P(SA) - \chi_P(SB)]^2\}$$

Let us now take some typical electronegativity values for the donor and acceptor atoms of hard and soft acids and bases: $\chi_P(HA) = 1.6$; $\chi_P(HB) = 3.4$; $\chi_P(SA) = \chi_P(SB) = 2.5$. Substituting these values in the previous equation gives us

$$\Delta H = 96.5[0.81 + 0.81 - 3.24 - 0] = -156 \text{ kJ/mol} \qquad (8.11)$$

This calculation indicates, as expected, that the products HA—HB and SA—SB are favored, but it also indicates why: *The driving force of the HSAB reaction lies in the formation of the very stable ionic hard acid–hard base product.*

---

## 8.4

## Effects of Changing Substituents and Oxidation Numbers on Electronegativities

Just as we have seen that changing substituents and oxidation numbers can change the softness of the atom, we also find that changing these factors can change the Pauling electronegativity of an atom.

EXAMPLE    The I—F bond energy differs in the different iodine fluorides; it is 231 kJ/mol in $IF_7$, 268 kJ/mol in $IF_5$, 272 kJ/mol in $IF_3$, and 278 kJ/mol in IF. Compute the Pauling electronegativity of iodine in each of these compounds.

Solution    We assume the electronegativity of fluorine to be 3.98 and take the I—I and F—F bond energies from Table 6.5; these are 149 and 155 kJ/mol, respectively, for an average of 152 kJ/mol. Referring to the Pauling electronegativity equation, if we deduct this average from the bond energies, then divide by 96.5, we get

$[\chi_P(\text{F}) - \chi_P(\text{I})]^2$. For I(VII) in IF$_7$, we obtain 0.82; for I(V) in IF$_5$ we obtain 1.20; for I(III) we obtain 1.24; for I(I) we obtain 1.30. Taking square roots and deducting the square root from the electronegativity of F, 3.98, we obtain the following Pauling electronegativities: 3.07 for I(VII), 2.88 for I(V), 2.87 for I(III), and 2.84 for I(I).

$\square$

The example shows that the Pauling electronegativity of an element increases with increasing oxidation number, which makes sense, since a more electron-poor atom should have a greater attraction for shared electrons in its covalent bonds. The example also shows that the change in electronegativity with changing oxidation state is usually not great enough to justify tabulating separate values, especially since the values have inherent uncertainties of $\pm 0.05$. In the few cases where such data are available, the variation is not much greater than 0.20, even over a range of several oxidation numbers. There are two exceptions, however; the electronegativities of thallium and lead are very sensitive to their oxidation number (changes of over 0.40 for a change of only 2 in their oxidation states). Hence Table A contains separate values for the two oxidation states indicated for these two elements. (The oxidation numbers implied for the other elements in Table A are $\pm$ the Roman numerals indicated at the head of each group—but these values are normally used for any oxidation state.)

The effects of changing substituents on Pauling electronegativities cannot be investigated by looking at bond dissociation energies; the enthalpy change for the dissociation of a compound containing two different substituents cannot cleanly be divided into enthalpy changes for the two different kinds of bonds; however, other methods can be used to estimate Pauling electronegativities on such compounds. Results obtained by one such method on different compounds of C, Si, Ge, B, and N are shown in Table 8.7. They indicate that the electronegativity of an atom can vary by up to 0.4 Pauling units depending on its substituents; even greater variability is suggested by other methods [9].

Although such differences are perhaps worth tabulating, the size of the table that would be needed to include all such *group electronegativities* is impractical. Instead, both inorganic and organic chemists, working from the "base" Pauling electronegativities in Table A, then rationalize variations in terms of the **inductive effects** of the substituents. From the examples in Table 8.7, we see that *electron-donating substituents* (those of low electronegativities themselves) *lower the electronegativity of an atom; electron-withdrawing substituents on an atom increase its electronegativity.* This is eminently logical: An atom losing bond electrons to some of its substituents may be expected to try to make up its losses by withdrawing bond electrons from other substituents.

There is a general agreement between Pauling's and Pearson's (HSAB) approaches to the results of competitive Lewis acid-base equilibria. But it has been noted [10] that, under certain circumstances, an apparent paradox results (the *Pearson-Pauling paradox*).

Let the acceptor atom be the cation of an electronegative metal. Let us now either increase the oxidation number of the cation or put more electron-withdrawing ligands on it. Either step ought to increase the electronegativity of the cation.

**Table 8.7**  Effects of Substituents on the Pauling Electronegativity of an Atom

| Compound | Electronegativity of Element | Tabulated $\chi_P$ |
|---|---|---|
| **A. Carbon Compounds** | | 2.55 |
| $(CH_3)_3C—Cl$ | 2.29 | |
| $(CH_3)_2CH—Cl$ | 2.33 | |
| $CH_3CH_2—Cl$ | 2.36 | |
| $CH_3—Cl$ | 2.40 | |
| $ClCH_2—Cl$ | 2.47 | |
| $Cl_2CH—Cl$ | 2.56 | |
| $Cl_3C—Cl$ | 2.64 | |
| **B. Extreme Values for Atoms Other Than Carbon** | | |
| $(CH_3)_3Si—Cl$ | 1.76 ⎫ | 1.90 |
| $Cl_3Si—Cl$ | 1.92 ⎭ | |
| $(CH_3)_3Ge—Cl$ | 1.82 ⎫ | 2.01 |
| $Cl_3Ge—Cl$ | 2.11 ⎭ | |
| $(HNBCl)_3$ | 1.90 ⎫ | 2.04 |
| $HCB_{10}Cl_{10}CH$ | 2.13 ⎭ | |
| $\overline{CH_2CH_2N}—Cl$ | 2.88 ⎫ | 3.04 |
| $(CH_3)_2C(CO)_2(NCl)_2$ | 3.28 ⎭ | |

SOURCES: The source of details on these, and of the NQR frequencies themselves, is G. K. Semin, T. A. Babushkina, and G. G. Yakobson, *Nuclear Quadrupole Resonance in Chemistry*, Israel Program for Scientific Translations (Jerusalem) and John Wiley (New York), 1975. For a review of methods of determining group electronegativities, see P. R. Wells, *Prog. Phys. Org. Chem.*, 6, 111 (1968).

NOTE: Electronegativities were computed by the author from the $^{35}Cl$ nuclear quadrupole resonance (NQR) frequencies of the chlorines in the compounds indicated, using the relationship found among simple singly bonded chlorides of the elements, NQR frequency = 26.0 $(\chi_P - 1.13)$ (author's unpublished correlation), which may also be derived from the Townes-Dailey theory for NQR frequencies using the Gordy relationship of ionic character to electronegativity differences.

According to the Pauling electronegativity equation (8.8), this increase will diminish the cation's relative affinity for very electronegative donor atoms. Thus we may say that this metal atom has become a softer acid. However, according to Section 8.2, increasing oxidation numbers and attaching very electronegative (hard) substituents make an acceptor atom a *harder* acid. Pearson [10] notes that when these contradictory predictions arise, the HSAB Principle almost invariably gives the correct prediction. This may be true partly because the influence of oxidation state on electronegativity is really quite small. In the two cases in which it is *not* small, Tl and Pb, it is generally felt that the unusual trend prevails: Tl(III) seems to be softer than Tl(I), and Pb(IV) seems softer than Pb(II).

## 8.5

## The Need for Two Parameters: Softness and Strength

Despite its many successful predictions (many of which will be illustrated in the remainder of the chapter), we are by no means ready to replace all the principles from earlier in the book with the HSAB Principle. We noted in the last chapter when dealing with Drago's $E$ and $C$ parameters that at least two parameters are necessary to characterize each acid and two are needed for each base. Softness is just one parameter, which must be considered in addition to another parameter or concept, the *strength* of the acid or base. When we evaluate the $Z^2/r$ ratio for an acid or base, or count its oxo groups, etc., we are evaluating its strength. We may, if we wish, set up a counterpart statement dealing with strength: *Stronger acids tend to react with stronger bases; weaker acids prefer weaker bases.* (This, for example, summarizes our solubility rules in a nutshell.)

Since a competitive Lewis acid-base reaction is governed by at least two independent principles, we may expect to find frequent cases in which the two principles come into conflict and in which the HSAB Principle by itself fails to give a correct prediction. Reaction (8.12) is an example of such a case:

$$CH_3HgSO_3^- \quad + \quad OH_2:OH^- \quad \rightleftharpoons \quad CH_3HgOH \quad + \quad OH_2:SO_3^{2-}$$

$$\text{SASB} \qquad\qquad \text{HAHB} \qquad\qquad \text{SAHB} \qquad\qquad \text{HASB}$$

Strong A–weak B    weak A–strong B    strong A–strong B    weak A–weak B

$$(8.12)$$

(Note that we have included the solvent, water, as a hard acid—it has a hard acceptor atom, H, engaged in hydrogen bonding to the anion.) The analyses beneath the equation show that the HSAB Principle favors reactants, while considerations of acid and base strengths favor products. The observed equilibrium constant of only 10 indicates that neither is strongly preferred over the other.

Fortunately, in everyday experience there are relatively few cases in which the principle of strength overrides or even counterbalances the HSAB Principle. Probably this is due to the fact that the typical enthalpy changes of covalent-bond formation are much larger than the typical entropy changes driving aqueous electrostatic reactions of strong acids and strong bases.

As our discussion of reaction (8.12) indicates, it is often important to include the solvent explicitly in the reaction. Most solvents are Lewis acids and bases, and can be classified as hard, borderline, or soft; their contribution to the direction of the reaction may be crucial. For the reaction

$$KCl + AgNO_3 \rightleftharpoons AgCl(s) + KNO_3 \qquad\qquad (8.13)$$

we are, in aqueous solution, actually dealing with hydrated ions. Water is a hard Lewis base (oxygen donor atom) and a hard Lewis acid (hydrogen acceptor atom). In net ionic form in aqueous solution the above reaction may be represented as

$$Cl^-:H_2O + Ag^+:OH_2 \rightleftharpoons AgCl(s) + H_2O:H_2O \qquad\qquad (8.14)$$

Clearly the HSAB Principle favors products. But if the solvent is liquid ammonia, which has a borderline nitrogen donor atom, the prediction is not so clear, since the borderline base $NH_3$ is competing with the borderline base $Cl^-$. In point of fact reaction (8.13) goes in the opposite direction in liquid ammonia, and KCl precipitates. It should be noted that the HSAB Principle is most commonly used for competitive equilibria in aqueous solutions.

Since Drago's $E$ and $C$ parameters (discussed in Chapter 7) are also two-parameter equations covering Lewis acid-base reactions, it is of interest to see how they relate to the HSAB concept. It would appear that the Drago $E$ and $C$ parameters deal with the two concepts of strength and softness in a different (but equally valid) way: Comparing the $E$ parameters of two Lewis acids will tell us which one has greater *strength as a hard acid*; comparing the $C$ parameters of two Lewis bases will tell us which has greater *strength as a soft base*. Taking some ratio of these parameters (e.g., $C/E$ in Table 7.2) may partially factor out the strength factor. Lewis bases with high $C/E$ ratios may be expected to be softer than those with low $C/E$ ratios; thus we find the $C/E$ ratios for ligands with Se donor atoms to be higher than those with S atoms, which are higher than those with O donor atoms.

But this does not always work: The $C/E$ ratios for N donors are sometimes higher than those for the P donor ligand trimethylphosphine. Drago's $E$ and $C$ parameters are developed from measurements in the gas phase or in nonpolar solvents, while the HSAB conclusions commonly come from observations in aqueous or other protic solvents. Again we see that solvent effects are involved in HSAB; hence it may be prudent to use each set of parameters ($E$ and $C$ or HSAB) in the medium in which it was developed (gaseous or nonpolar versus polar or protic).

As we also indicated in the last chapter, two parameters do not always suffice to characterize a Lewis acid-base reaction completely; the same is true of the concepts of softness and strength. Thus although soft bases generally prefer to combine with soft acids in competitive acid-base equilibria, there are some soft bases that distinctly prefer *d-block* soft acids and will not combine with *p-block* soft acids. These bases, such as carbon monoxide, contain empty pi-bonding (antibonding molecular) orbitals that can form a pi bond only with filled pi-bonding $d$ orbitals. Such a specific interaction is, of course, not included in the general concept either of softness or of strength of CO as a base.

We hope that the nit-picking in the past few sections will have convinced you that the HSAB Principle is not infallible and does not invariably give correct predictions. But it would be just as bad to leave you with the impression that the HSAB Principle is just another abstract theory, too far removed from reality to give any useful predictions; hence the last half of this chapter is devoted to the numerous and varied phenomena in real life that can be successfully predicted and understood using the HSAB Principle.

## 8.6

## Applications of HSAB: Solubility of Halides and Chalcogenides

In Chapter 3 we developed solubility rules, but these were only for oxo anions—hard bases. Using the HSAB Principle, we can now supplement these rules with solubility rules for salts of soft bases.

At first glance it may seem that a precipitation equilibrium reaction such as $Ag^+ + Cl^- \rightleftharpoons AgCl(s)$ cannot be treated by the HSAB Principle, since the equation appears to contain only one acid and one base. But when we write a metal ion such as $Ag^+$ and are speaking of an aqueous solution, we really mean a *hydrated* ion of the type $[Ag(H_2O)_n]^+$, in which the soft silver ion is coordinated by the hard oxygen donor atoms of several water molecules. Likewise, by $Cl^-$ we really mean $[Cl(H_2O)_m]^-$, in which the borderline base, the chloride ion, is hydrogen-bonded to the hard-acid hydrogen atoms of several water molecules. If we rewrite this precipitation reaction in the form of equation (8.14), we have an equation that is indeed of the form of equation (8.1). Thus we can predict that *the chlorides, bromides, and iodides of the soft acids are insoluble.* Indeed, this turns out generally to be true for the $+1$ and $+2$ ions of the soft-acid metals: The insoluble chlorides are $CuCl$ (but not $CuCl_2$), $AgCl$, $AuCl$ (but not $AuCl_3$), $TlCl$, $Hg_2Cl_2$ (but not $HgCl_2$), $OsCl_2$, $IrCl_2$, $PdCl_2$, $PtCl_2$ (but not $PtCl_4$), and $PbCl_2$. The lists of insoluble bromides and insoluble iodides are similar but slightly larger.

We may anticipate a similar solubility rule for sulfides, selenides, and tellurides. As it turns out, the sulfides, selenides, and tellurides of the soft *and borderline* acids are insoluble. This we may rationalize by noting that the $S^{2-}$, $Se^{2-}$, and $Te^{2-}$ ions are *stronger* bases than the $Cl^-$, $Br^-$, and $I^-$ ions. This combination of softness and some strength enables them to combine with borderline acids, which although not so soft are generally smaller and/or more highly charged than the purely soft acids, and thus are *stronger* acids.

We can also make use of relative softness here: We find that for a given soft acid, the iodide is less soluble than the bromide, which is less soluble than the chloride; similarly, the telluride and selenide are less soluble than the sulfide. In fact, a relationship has been noted [11] between the solubility products of sulfides, selenides, and tellurides of 1 : 1 stoichiometry (i.e., of MS, MSe, and MTe) and the electronegativity difference between the cation and the anion (Figure 8.1).

We may make similar predictions of the solubilities of *pseudohalide* salts (Section 5.5). The cyanide ion (using its carbon donor atom) and the thiocyanate ion (using its sulfur donor atom) qualify as soft bases, while the azide ion can be identified as a borderline base. We may predict, and generally do find, that the cyanides, thiocyanates, and azides of the soft acids are insoluble.

In contrast, the HSAB Principle is not relevant to the prediction of the water solubilities of the salts of the hard bases $F^-$, $OH^-$, and $O^{2-}$, since there is no soft base involved in their precipitation equilibria. Instead we must note that the fluoride ion is a weak base (Table 5.6), while the hydroxide and oxide ions are very strong bases. Consequently *the fluorides, oxides, and hydroxides of acidic cations are insoluble.* Since there is a substantial difference in strength between the fluoride and the other two anions, it is not too surprising that there are some differences in detail between fluorides and oxides or hydroxides. Thus feebly acidic cations have soluble oxides and hydroxides but insoluble fluorides, while some of the soft-acid cations ($Ag^+$, $Hg^{2+}$) have soluble fluorides but insoluble oxides.

Technically, the HSAB Principle cannot be used to predict the solubility of the sulfides, selenides, and tellurides of the hard acids, since no soft acid is involved in their solubility equilibria. The concept of strength can be used, however. These anions are basic anions, and thus may be expected to give insoluble salts with hard

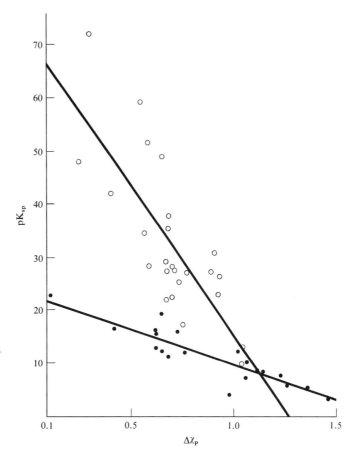

**Figure 8.1** Negative logarithms of solubility products ($pK_{sp}$) of MS, MSe, and MTe (open circles), and of MCl, MBr, MI, $MN_3$, MCN, and MSCN (closed circles), versus differences in Pauling electronegativities of the Lewis acids and bases. $pK_{sp}$ data are from A. F. Clifford, *Inorganic Chemistry of Qualitative Analysis*, Prentice-Hall, Englewood Cliffs, N.J., 1961, Appendix 7.

acidic cations, such as $La^{3+}$, $Ti^{4+}$, or $Al^{3+}$. To some extent this is the case. But as we may expect intuitively from HSAB, the oxides of such metal ions are much less soluble than the sulfides, selenides, and tellurides. Consequently the sulfides, selenides, and tellurides of the hard acids undergo *hydrolysis* to give the oxides or hydroxides of the cation, which may precipitate, while the basic anions hydrolyze to give $HS^-$ or $H_2S$ (or the selenium or tellurium analogues). Hydrolyses of salts of these types will be discussed further in Chapter 9.

## 8.7

## Applications of HSAB: The Qualitative Analysis Scheme for Metal Ions

In your general chemistry course you may have spent many weeks in the laboratory learning to separate and identify the metal ions present in a solution, via the **qualitative analysis scheme** ("qual scheme") for the cations [12]. You may recall that all the metal ions are first separated into five groups:

- Group I, precipitated with 0.3 $M$ HCl

- Group II, precipitated with $H_2S$ from acidic solution

- Group III, precipitated with $(NH_4)_2S$ from basic solution

- Group IV, precipitated with $(NH_4)_2CO_3$

- Group V, which remains in solution throughout

Perhaps at the time you wondered how the principle of periodicity applied to the results of this series of experiments. (Obviously the members of qual scheme Group I ($Ag^+$, $Hg_2^{2+}$, $Pb^{2+}$) do not belong to Group I of the periodic table or to any other group; neither are they in the same period.) It helps, in trying to see the periodic pattern, to know the pattern that results when we do a separation and analysis of *all* the metal ions. The results of this herculean task have been described [13] and are summarized in Table 8.8.

As you may have foreseen, the metals of qual scheme Group I are the soft acids. These metal ions combine strongly with the borderline base, the chloride ion, to give insoluble chlorides. But in the qual scheme a large excess of chloride ion is present, and many of these ions combine with the excess chloride present to give complex ions, $[MCl_n]^{x-}$, which remain in solution. The $Ag^+$, $Hg_2^{2+}$, $Tl^+$, and $Pb^{2+}$ ions precipitate, while the ions of the metals Cu, Pd, Os, Ir, Pt, and Au form soluble complex ions.

The group precipitating reagent for qual scheme Group II, the sulfide ion, is very soft and is a stronger base than the chloride ion, so it combines with additional metal ions to give more insoluble sulfides. Had we neglected to put the chloride ion in, the sulfide ion would now precipitate all the soft acids of qual scheme Group I, plus the largest number of the borderline acids. If we do the experiment properly and first delete the metals of qual scheme Group I, we are left with the metals not quite so near to gold—approximately the borderline acids, minus those of the first row of the transition metals.

The group precipitation of qual scheme Group III involves a basic solution of the sulfide ion. The sulfide ion is now present in much higher concentration, but it is also competing with another strong base, the much harder hydroxide ion. In solution we now have only hard-acid metal ions and a small group of borderline acids from the +2 ions of the first row of the $d$-block metals, $Mn^{2+}$, $Fe^{2+}$, $Co^{2+}$, $Ni^{2+}$, and $Zn^{2+}$. These precipitate as *sulfides*, while the numerous *strong* hard acids present (Figure 8.1, including also all the $f$-block elements) combine with the *strong* hard base, $OH^-$, to precipitate the hydroxides or oxides of these metals.

At this point only the *nonacidic* and *feebly acidic* hard acids of the first two groups of the periodic table are present. In qual scheme Group IV the moderately basic carbonate ion is added, since it precipitates the feebly acidic cations only. (Of course, had we not carried out the earlier separations, it would precipitate almost all the preceding metals too.) Remaining in solution to the end are the nonacidic cations of qual scheme Group V, which prefer association with the (nonbasic) hard base water, in hydrated-ion form. (Some of the confirmatory tests involve adding nonbasic anions to get precipitates of these metals.)

**Table 8.8** The Complete Qualitative Analysis Scheme of the Cations

Nonmetals present as *anions* (not included in scheme)

GROUP III (as hydroxides except * as sulfides)

| 1 | 2 | 3 | 4 | 5 | 6 | 7 | 8 | 9 | 10 | 11 | 12 | 13 | 14 | 15 | 16 | 17 | 18 |
|---|---|---|---|---|---|---|---|---|---|---|---|---|---|---|---|---|---|
| H 2.2 | | | | | | | | | | | | | | | | | He |
| Li 0.98 | Be 1.57 | | | | | | | | | | | B 2.04 | C 2.55 | N 3.04 | O 3.44 | F 3.98 | Ne |
| Na 0.93 | Mg 1.31 | | | | | | | | | | | Al 1.61 | Si 1.90 | P 2.19 | S 2.58 | Cl 3.16 | Ar |
| K 0.82 | Ca 1.00 | Sc 1.36 | Ti 1.54 | V 1.63 | Cr 1.66 | Mn* 1.55 | Fe* 1.83 | Co* 1.88 | Ni* 1.91 | Cu 2.0 | Zn* 1.65 | Ga* 1.81 | Ge 2.01 | As 2.18 | Se 2.55 | Br 2.96 | Kr 3.0 |
| Rb 0.82 | Sr 0.95 | Y 1.22 | Zr 1.33 | Nb 1.6 | Mo 2.16 | Tc 1.9 | Ru 2.2 | Rh 2.28 | Pd 2.20 | Ag 1.93 | Cd 1.69 | In* 1.78 | Sn 1.80 | Sb 2.05 | Te 2.1 | I 2.66 | Xe 2.6 |
| Cs 0.79 | Ba 0.89 | Lu 1.27 | Hf 1.3 | Ta 1.5 | W 2.36 | Re 1.9 | Os 2.2 | Ir 2.2 | Pt 2.28 | Au 2.54 | Hg 2.0 | Tl 1.60 | Pb 1.87 | Bi 2.02 | Po 2.0 | At 2.2 | Rn |
| Fr 0.7 | Ra 0.9 | | | | | | | | | | | | | | | | |

W 2.36 — insoluble oxide

Lanthanides and Actinides:

| | | | | | | | | | | | | | |
|---|---|---|---|---|---|---|---|---|---|---|---|---|---|
| La 1.10 | Ce 1.12 | Pr 1.13 | Nd 1.14 | Pm | Sm 1.17 | Eu | Gd 1.20 | Tb | Dy 1.22 | Ho 1.25 | Er 1.24 | Tm 1.25 | Yb |
| Ac 1.1 | Th 1.3 | Pa 1.5 | U 1.38 | Np 1.36 | Pu 1.28 | Am 1.3 | Cm 1.3 | Bk 1.3 | Cf 1.3 | Es 1.3 | Fm 1.3 | Md 1.3 | No 1.3 |

Group labels: GROUP I, GROUP II, GROUP III, GROUP IV, GROUP V

## Applications of HSAB: Redox Chemistry

As we have seen, the versatile Lewis acid-base concept encompasses much of chemistry: It includes not only the earlier, more restricted types of acid-base reactions but also precipitation reactions and what we usually call coordination chemistry. Only oxidation-reduction (redox) chemistry is excluded. A reducing agent does "donate" (or transfer) an electron to an electron acceptor (oxidizing agent), but of course this donation or transfer does not fit the Lewis definition because an electron *pair* is not involved.

As we suggested at the beginning of Chapter 7, acids and bases are what we define them to be. Soviet chemist M. Usanovich has proposed (in effect) removing the word *pair* from the Lewis definition, thus including redox chemistry within the scope of acid-base chemistry [14]. This definition has not caught on, but this is not due to any intrinsic flaw in such a definition. Within such a definition the electron itself could be thought of as a (Usanovich) base. Would such an extension be worthwhile—for example, could we thus extend the HSAB Principle to include redox reactions?

There is at least one set of experimental results that follows a pattern like that of the HSAB Principle: the listing of the *activity series of metals* (Experiment 4) according to their decreasing redox reactivity with water or hydrochloric acid. In Chapter 5 we divided the metals into three groups according to their Pauling electronegativities. As hardness and softness mainly depend on Pauling electronegativities, it is not surprising that we can also organize the activity series of metals according to the hardness or softness of the metal ions produced in the half-reaction

$$M(s) + H_2O : H_2O \rightleftharpoons M^{n+}(aq) + e^-(aq) \tag{8.15}$$

The most active metals, the *very electropositive metals*, are largely the metals that give *hard-acid* metal ions in solution. The metals of intermediate activity, the *electropositive metals*, are, in general, the borderline acids, while the inactive *electronegative metals* correspond well to the *soft acids*.

In order to fit reaction (8.15) to the HSAB pattern, we must identify the *electron* as a *soft base*. Then we can think of the solid metal as a combination of the (hard- or soft-acid) metal ion with the soft base, the electron. This combination will be preferred if the metal ion is a soft acid.

Can we justify the classification of the electron as a soft base? (Indeed, it has been suggested as the "ultimate soft base.") With a great deal of dubious rationalization we can. Thus soft bases have donor atoms of low electronegativity. A free electron is on no atom. No atom has no attraction for bond electrons, hence must have a very low electronegativity. Soft bases are relatively large. As we noted in Section 7.8, solvated electrons in liquid ammonia seem to have rather large "anionic radii" of 150 to 170 pm. The logic of this rationalization would no doubt make Aristotle turn over in his grave, but we *do* note empirically a relationship between the HSAB scheme and the activity series of metals.

The formation of complex ions is often used to increase the activity of a metal in redox reactions. We noted in Chapter 5 that nitric acid is a good oxidizing agent but that many electropositive metals are passivated by the formation of a tough insoluble film of oxide on their surfaces upon treatment with $HNO_3$. If the metal is a hard acid, this problem may often be overcome by the use of a mixture of nitric and hydrofluoric acids; the surface oxide then dissolves to give a soluble fluoro complex:

$$Nb + 5\,HNO_3 + 6\,HF \rightarrow H^+ + NbF_6^- + 5\,NO_2 + 5\,H_2O \qquad (8.16)$$

On the other hand, nitric acid is not strong enough to oxidize the softest-acid metal, gold. However, a mixture of hydrochloric and nitric acids, known as *aqua regia*, is able to dissolve gold, since the gold ion is complexed by the chloride ion:

$$2\,Au + 11\,HCl + 3\,HNO_3 \rightarrow 2\,H^+ + 2\,AuCl_4^- + 3\,NOCl + 6\,H_2O \qquad (8.17)$$

The even softer base, the cyanide ion, forms such stable complexes with the soft acids that the parent metals may be oxidized with atmospheric oxygen. This reaction is exploited in the extraction of these precious metals from deposits having very low percentages of metal:

$$4\,Au + 8\,CN^- + O_2 + 2\,H_2O \rightarrow 4\,[Au(CN)_2]^- + 4\,OH^- \qquad (8.18)$$

Related to this reaction is the well-known observation that although silver is quite unreactive to the strong oxidizing agent in the air, $O_2$, it is readily tarnished by sulfur or by $H_2S$ and air to give black silver sulfide.

One other general relationship of the HSAB Principle to redox chemistry may be noted: The cations of the (inactive) electronegative metals, which are soft acids, are also strong oxidizing agents. Similarly, we may note that the anions of the (inactive) electronegative nonmetals, which are soft bases, are also strong reducing agents. (These characteristics are sometimes added to tables such as Table 8.2 as general characteristics of soft acids and bases.) Thus it may seem surprising that soft-acid metal ions do not generally oxidize soft-base nonmetal anions. The soft acid–soft base combination is usually stabilized sufficiently, however, by the formation of a covalent bond (as in reaction (8.5)). Nonetheless, the heats of formation of simple soft acid–soft base salts are usually rather low, and some of them (such as the azides previously noted) are explosives.

## 8.9

# Applications of HSAB: The Geochemical Classification and Differentiation of the Elements

In studying the types of minerals in which different elements are found [15], Berzelius long ago noted that certain metals tend to occur as sulfides and others as carbonates or oxides. Nowadays geochemists classify the elements into four classes, according to their predominant geological pattern of behavior (Table 8.9):

**Table 8.9**  Geochemical Classification of the Elements

Lithophiles

| Li | Be |    |    |    |    |    |    |    |    |    |    | B  | C  | N  |    |    | He |
|----|----|----|----|----|----|----|----|----|----|----|----|----|----|----|----|----|----|
| Na | Mg |    |    |    |    |    |    |    |    |    |    | Al | Si | P  | O  | F  | Ne |
| K  | Ca | Sc | Ti | V  | Cr | Mn | Fe | Co | Ni | Cu | Zn | Ga | Ge | As | S  | Cl | Ar |
| Rb | Sr | Y  | Zr | Nb | Mo |    | Ru | Rh | Pd | Ag | Cd | In | Sn | Sb | Se | Br | Kr |
| Cs | Ba | Lu | Hf | Ta | W  | Re | Os | Ir | Pt | Au | Hg | Tl | Pb | Bi | Te | I  | Xe |

Atmophiles

Siderophiles

Chalcophiles

NOTE: This classification emphasizes behavior under conditions at the surface of the earth; many elements can display other classes of behavior as well.

**1.** The *atmophiles*, the chemically unreactive nonmetals that occur in the atmosphere in elemental form ($N_2$, the noble gases).

**2.** The *lithophiles*, the metals and nonmetals that tend to occur in oxides, silicates, sulfates, or carbonates. Since these anions all possess oxygen as a donor atom, the metal ions involved are the hard acids. The nonmetals that are classified as lithophiles either are inherently hard bases or have been oxidized by atmospheric oxygen to oxo anions, which are hard bases.

**3.** The *siderophiles*, the metals that tend to occur native (in the elemental form). Since the oxidizing agent $O_2$ is present in the environment of the crust of the earth but does not affect the core of the earth, the listing of siderophiles differs for these two parts of the planet. In the crust of the earth the siderophiles include the most noble (least active) metals (Table 8.9), those which are the combination of the soft-acid metal ions with the softest base, the electron. In the core of the earth the siderophiles include any metal less active than iron (the most abundant metal in the core), so such metals as Fe, Co, Ni, Mo, W, and Re are added to this classification.

**4.** The *chalcophiles*, which occur in nature as sulfides (less commonly with other soft bases, since sulfide is much the most common soft base). The chalcophiles are mostly the borderline acids, although soft acids occur not only in native form, but also as tellurides, arsenides, etc. Logically, many of the soft bases are also listed as chalcophiles, since they are found in such minerals.

The process by which the elements separated into these classes is known as the *primary differentiation of the elements*. It is believed that early in the history of the earth it was hot, so that many materials were molten. Overall, metals are in excess over nonmetals, so only the more active metals were present as salts (silicon is assumed also to have been present as silica and silicates). The abundant, only moderately active metal iron is presumed to have been present in molten form, dissolving nickel and the less active siderophile metals as it sank (due to its high density) to form the *core* of the earth. Hard-acid metal ions tended to associate with the silicate ions and formed a type of slag that became the *mantle* of the earth. (The whole process is thought to have been much like that occurring in a blast furnace, Chapter 5.) The *crust* of the earth contains other silicates (see Chapter 4) and borderline- and soft-acid metal ion sulfides.

These original mineral deposits have not remained unchanged, however. A variety of chemical processes (weathering, Chapter 4; metamorphism, hydrothermal synthesis, and sedimentation) have continued to alter them, giving rise to new minerals. Sometimes particular minerals were concentrated in deposits that are rich enough to mine; such minerals are known as *ores*. Table 8.10 lists the commercially most important ores of some of the elements that are utilized in the greatest quantities. The hard-acid metals occur only as hydrated ions or as salts of hard bases, while the soft acids occur very predominantly either as salts of soft bases or as the elements themselves. The sulfides of the borderline acids, not being as insoluble as the sulfides of the soft acids, are more susceptible to weathering—such as oxidation by $O_2$—to give soluble sulfates or other secondary minerals that

**Table 8.10**  Major Mineral Sources of Selected Elements

| Element | Mineral Names, Formulas | Tonnes Used/Year | Uses of Elements |
|---|---|---|---|
| **A. Hard-Acid Metals** | | | |
| Sodium | Halite, NaCl | 90,700,000 | To make NaOH |
| | Natron, $Na_2CO_3 \cdot 10H_2O$ | | |
| Calcium | Limestone, $CaCO_3$ | 54,400,000 | To make CaO, cement |
| | Gypsum, $CaSO_4 \cdot 2H_2O$ | | |
| Potassium | Sylvite, KCl | 9,060,000 | Fertilizer |
| Magnesium | Seawater, $Mg^{2+}$(aq) | 8,160,000 | Light metal |
| | Magnesite, $MgCO_3$ | | |
| Aluminum | Bauxite, $Al_2O_3$ | 5,500,000 | Light metal |
| Chromium | Chromite, $FeCr_2O_4$ | 1,270,000 | Alloys with Fe |
| Titanium | Ilmenite, $FeTiO_3$ | 907,000 | Light metal |
| | Rutile, $TiO_2$ | | |
| **B. Borderline-Acid Metals** | | | |
| Iron | Hematite, $Fe_2O_3$ | 281,000,000 | Structural metal |
| | Magnetite, $Fe_3O_4$ | | |
| | Siderite, $FeCO_3$ | | |
| Manganese | Pyrolusite, $MnO_2$ | 5,440,000 | Alloys with Fe |
| | Psilomelane, $BaMn_5O_{11}$ | | |
| Zinc | Sphalerite, ZnS | 3,400,000 | Coating metals |
| Nickel | Pentlandite, $(Ni, Fe)_9S_8$ | 362,000 | Alloys with Fe |
| | Garnierite, $Ni_3Si_2O_5(OH)_4$ | | |
| Tin | Cassiterite, $SnO_2$ | 172,000 | Coating metals |
| Antimony | Stibnite, $Sb_2S_3$ | 54,500 | Alloys with Pb |
| Molybdenum | Molybdenite, $MoS_2$ | 40,800 | Alloys with Fe |
| **C. Soft-Acid Metals** | | | |
| Copper | Chalcopyrite, $CuFeS_2$ | 4,800,000 | Electrical |
| | Chalcocite, $Cu_2S$ | | |
| Lead | Galena, PbS | 2,500,000 | Metal in batteries |
| Mercury | Cinnabar, HgS | 8,160 | Electrical, to make NaOH |
| Silver | Argentite, $Ag_2S$ | 7,250 | Photography |
| Gold | Gold, Au | 1,450 | Coinage, jewelry |
| | Calaverite, $AuTe_2$ | | |
| Platinum | Platinum, Pt | 27 | Catalysts |
| | Sperrylite, $PtAs_2$ | | |
| **D. Soft-Base Nonmetals** | | | |
| Carbon | Coal | 1,600,000,000 | Reducing agent, fuel |
| Sulfur | Sulfur, S | 18,100,000 | To make $H_2SO_4$ |
| | Pyrite, $FeS_2$ | | |
| Phosphorus | Apatite, $Ca_5(PO_4)_3OH$ | 6,300,000 | Fertilizer |
| Bromine | Seawater, brines, $Br^-$(aq) | 99,700 | Organobromine compounds |
| Arsenic | Arsenopyrite, FeAsS | 36,300 | Alloys with Pb |
| | Enargite, $Cu_3AsS_4$ | | |
| Iodine | Brines, $I^-$(aq) | 3,630 | Photography, medicine |
| | Caliche, $NaIO_3$ | | |

| Element | Mineral Names, Formulas | Tonnes Used/Year | Uses of Elements |
|---------|------------------------|------------------|------------------|
| **E. Borderline-Base Nonmetals** | | | |
| Nitrogen | Air, as $N_2$ | 15,000,000 | Fertilizer |
| | Soda niter, $NaNO_3$ | | |
| Chlorine | Halite, NaCl | 4,500,000 | Organochlorine compounds, HCl |
| **F. Hard-Base Nonmetals** | | | |
| Oxygen | Air, as $O_2$ | 18,100,000 | Steel manufacture |
| Fluorine | Fluorite, $CaF_2$ | 997,000 | To make fluorides |

SOURCES: Data from J. W. Moore and E. A. Moore, *Environmental Chemistry*, Academic Press, New York, 1976, and F. H. Day, *The Chemical Elements in Nature*, Reinhold, New York, 1964.

involve hard bases as anions; hence the minerals bearing these elements are more diverse than those of the hard and soft acids.

Recent oceanographic research has revealed that these secondary processes that generate minerals are still going on beneath the ocean, at the midoceanic ridges [16]. Here magma from the mantle at a temperature of 1200 °C continuously rises and is cooled to form the oceanic crust. This rock cracks as it moves off the midoceanic ridges, allowing seawater to seep in close to the hot magma. In this superheated water (350 °C but still liquid, due to the very high pressures at the bottom of the ocean) chemical processes are speeded up. In particular, $Mg^{2+}$ and $SO_4^{2-}$ are removed from the ocean water. The $Mg^{2+}$ reacts with the basaltic rock (which acts as a source of $SiO_2$) and water to give an insoluble magnesium hydroxy-silicate and hydrogen ion. The sulfate is reduced by iron(II) silicate found in the mantle rock to give sulfide (as pyrite):

$$2\,SO_4^{2-} + 4\,H^+ + 11\,Fe_2SiO_4 \rightarrow FeS_2 + 7\,Fe_3O_4 + 11\,SiO_2 + 2\,H_2O \quad (8.19)$$

The superheated water finally emerges from vents at the top of the midoceanic ridge at 350°C, containing no $Mg^{2+}$ or $SO_4^{2-}$ but containing a great deal of dissolved $H_2S$ (from the hydrogen and sulfide ions produced) and dissolved *d*-block metal ions (released from the basalt by the action of the hydrogen ions). As this superheated solution spews into the cold ocean water, the qualitative analysis scheme of the anions is replayed: The softer *d*-block metal ions ($Fe^{2+}, Ni^{2+}, Cu^{2+}, Zn^{2+}$) give a precipitate of their sulfides as a black *smoker* that gives rise to a black *chimney*. The sulfide of the harder $Mn^{2+}$ ion, however, tends to be carried away from the smoker, where it is oxidized by dissolved $O_2$ to give $MnO_2$. Possibly this process is connected with the manganese-rich nodules known to be present on the ocean floor.

As we see from Table 8.10, the soft-acid metals and many of the borderline-acid metals are generally used in much lower quantities than are the hard-acid metals because their ores are much less abundant, both in the universe and in the crust of the earth. They are, however, extremely useful, and consequently, in general, command very high prices. The known worldwide reserves of some of them are nearly exhausted. Their scarce minerals are found in only a few regions of the world, which makes their availability particularly dependent on political factors [17]. For example, cobalt is needed to make jet engines; it comes principally from Zaire. But when a civil war hit Zaire, the price and availability of cobalt changed dramatically. Such situations have focused a lot of attention on the resources (such as manganese nodules) that lie at the bottom of the sea and have made quite important the question of ownership of such resources.

Since the soft-acid metals are siderophiles, they are believed to be perhaps a thousand times more abundant in the earth's core (dissolved in iron) than in its crust. They are also more abundant in meteorites, which have not gone through a molten phase that would allow these elements to concentrate in an iron core. Recently it was discovered that sediments deposited around the earth about 65 million years ago are a thousand times enriched in (at least) two of these elements, osmium and iridium, as compared with normal sediments. This enrichment has been taken as evidence that an extraterrestrial body (perhaps an asteroid) may have collided with the earth at that time, creating a shower of dust over the earth that was (incidentally) iridium-rich and (more important) may have shut out a good deal of sunlight for many years. This was the period in which the dinosaurs and all other large animals of the time became extinct.

## 8.10

## Applications of HSAB to Coordination Chemistry: Toxicology of the Elements and Medicinal Chemistry

### 8.10.1 Elements That Are Essential and/or Toxic

The HSAB Principle is also very useful in organizing the chemistry of the metal ions in ecology, biochemistry, and medicine [18]. Many of the metal ions are essential for life; these are listed, along with a summary of their functions, in Table 8.11. The donor atoms of biochemical ligands to which these metal ions prefer to bind are also indicated. These follow the general pattern we would expect from the HSAB Principle. Other metal and nonmetal ions are toxic components of natural waters and wastewaters; they are listed in Table 8.12 along with their sources and effects. We note immediately that soft acids are seldom essential for life and are often toxic. The HSAB Principle also tells us the kind of biochemical sites at which toxic metal ions are likely to bind.

HARD ACIDS As we see from Table 8.11, several hard acids are essential for life. $Na^+$, $K^+$, $Mg^{2+}$, and $Ca^{2+}$ often function in the form of their hydrated ions, which the body allows to pass selectively across various barriers; this passage of charged ions constitutes an electric current, which can transmit a nerve impulse or trigger some response. $Mg^{2+}$ and $Ca^{2+}$ also serve structural purposes: the latter in the form of calcium phosphate or carbonate as bones, teeth, shells, etc; the former by attracting the negatively charged triphosphate groups of DNA together in the

**Table 8.11**  Functions and Preferred Ligand Binding Groups for Essential Metal Ions

| Metal ion | Function | Ligand groups, with donor atoms in parentheses |
|---|---|---|
| **A. Hard Acids** | | |
| $Na^+$ | As charged ion | Hydrated ions (O) |
| $K^+$ | As charged ion | Singly charged oxygen donor atoms or neutral oxygen ligands (O) |
| $Mg^{2+}$ | As charged ion, structural | Carboxylate (O), phosphate (O), nitrogen donors (N) |
| $Ca^{2+}$ | As charged ion, structural | Like $Mg^{2+}$ but less affinity for nitrogen donors, phosphate, and other multidentate anions |
| $Fe^{3+}$ | Redox reactions | Carboxylate (O), tyrosine (O), $—NH_2$ (N), porphyrin ("hard" N) |
| $Co^{3+}$ | Redox reactions | Similar to $Fe^{3+}$ |
| $V^{n+}$ | Essential to sea squirts | |
| $Cr^{3+}$ | Glucose tolerance factor | |
| $Ni^{2+}$ | Stabilizes coiled ribosomes | |
| $Al^{3+}$ | May activate two enzymes | |
| **B. Borderline Acids** | | |
| $Mn^{2+}$ | Lewis acid | Similar to $Mg^{2+}$ |
| $Fe^{2+}$ | Redox reactions | $—SH$ (S), $—NH_2$ (N) > carboxylates (O) |
| $Zn^{2+}$ | Lewis acid | Imidazole (N), cysteine (S) |
| $Cu^{2+}$ | Redox reactions | Amines (N) $\gg$ carboxylates (O) |
| $Mo^{2+}$ | Redox reactions | $—SH$ (S) |
| **C. Soft Acids** | | |
| $Cu^+$ | Redox reactions | Cysteine (S) |

Sources: Data from M. N. Hughes, *The Inorganic Chemistry of Biological Processes*, 2d ed., John Wiley, Chichester, 1981; J. Huheey, *Inorganic Chemistry: Principles of Structure and Reactivity*, 3d ed., Harper and Row, Cambridge, 1983.

same region of space, thus helping preserve the three-dimensional arrangement (conformation) of the DNA helix.

Most hard-acid metal ions are small and/or highly charged and thus are unavailable in natural waters (Chapter 2), since they precipitate as hydroxides; hence most hard-acid metal ions are neither essential nor toxic. Beryllium is one exception—it is very dangerously poisonous. Apparently it substitutes for $Mg^{2+}$ in certain enzymes in the body, but being much more acidic, it alters the functions of those enzymes so that they can no longer function. Barium is also toxic if soluble, but in the form of insoluble $BaSO_4$ it is safely given as "barium milkshakes" to make the stomach opaque to X-rays. Some of the $f$-block elements (such as plutonium) of the seventh period are exceedingly toxic due to their radioactivity. Plutonium tends to concentrate in the bones where it emits dangerous alpha-particles, which irradiate the bone marrow and cause leukemia.

**Table 8.12**  Occurrence and Significance of Trace Elements in Natural Waters and Wastewaters

| Element | Sources | Effects and significance | USPHS Limit (mg/L)[a] | Occurrence: % of samples, high, mean ($\mu$g/L)[b] |
|---------|---------|--------------------------|-----------------------|--------------------------------------------------|
| **A. Hard Acids** | | | | |
| Beryllium | Coal, nuclear power and space industries | Acute and chronic toxicity, possibly carcinogenic | Not given | Not given |
| Boron | Coal, detergent formulations, industrial wastes | Toxic to some plants | 1.0 | 98% (above 1 $\mu$g/L), 5000, 101 |
| **B. Borderline Acids** | | | | |
| Chromium | Metal plating, cooling-tower water additive (chromate), normally found as Cr(VI) in polluted water | Essential trace element (glucose tolerance factor), possibly carcinogenic as Cr(VI) | 0.05 | 24.5 %, 112, 9.7 |
| Manganese | Mining, industrial waste, acid mine drainage, microbial action on manganese minerals at low pE | Relatively nontoxic to animals, toxic to plants at higher levels, stains materials (bathroom fixtures and clothing) | 0.05 | 51.4% (above 0.3 $\mu$g/L), 3230, 58 |
| Iron | Corroded metal, industrial wastes, acid mine drainage, low pE water in contact with iron minerals | Essential nutrient (component of hemoglobin), not very toxic, damages materials (bathroom fixtures and clothing) | 0.05 | 75.6%, 4600, 52 |
| Copper | Metal plating, industrial and domestic wastes, mining, mineral leaching | Essential trace element, not very toxic to animals, toxic to plants and algae at moderate levels | 1.0 | 74.4%, 280, 15 |
| Molybdenum | Industrial waste, natural sources | Possibly toxic to animals, essential for plants | Not given | 32.7 (above 2 $\mu$g/L), 5400, 120 |
| Zinc | Industrial waste, metal plating, plumbing | Essential element in many metalloenzymes, aids wound healing, toxic to plants at higher levels, major component of sewage sludge limiting land disposal of sludge | 5.0 | 76.5% (above 2 $\mu$g/L), 1180, 64 |
| **C. Soft Acids and the Softer Borderline Acids ("Heavy Metals")** | | | | |
| Cadmium | Industrial discharge, mining waste, metal plating, water pipes | Replaces zinc biochemically, causes high blood pressure, kidney damage, destruction of testicular tissue and red blood cells, toxic to aquatic biota | 0.01 | 2.5%, not given, 9.5 |
| Lead | Industry, mining, plumbing, coal, gasoline | Toxic (anemia, kidney disease, nervous system), wildlife destruction | 0.05 | 19.3% (above 2 $\mu$g/L), 140, 23 |

**Table 8.12** (continued)

| Element | Sources | Effects and significance | USPHS Limit (mg/L)[a] | Occurrence: % of samples, high, mean ($\mu$g/L)[b] |
|---|---|---|---|---|
| **C. Soft Acids and the Softer Borderline Acids ("Heavy Metals")** (continued) | | | | |
| Mercury | Industrial waste, mining, pesticides, coal | Acute and chronic toxicity | Not given | Not given |
| Silver | Natural geological sources, mining, electroplating, film-processing wastes, disinfection of water | Causes blue-gray discoloration of skin, mucous membranes, eyes | 0.05 | 6.6% (above 0.1 $\mu$g/L), 38, 2.6 |
| **D. Soft Bases** | | | | |
| Iodine (iodide) | Industrial waste, natural brines, seawater intrusion | Prevents goiter | Not given | Rare in fresh water |
| Arsenic | Mining byproduct, pesticides, chemical waste | Toxic, possibly carcinogenic | 0.05 | 5.5% (above 5 $\mu$g/L) 336, 64 |
| Selenium | Natural geological sources, sulfur, coal | Essential at low levels, toxic at higher levels, causes "alkali disease" and "blind staggers" in cattle, possibly carcinogenic | 0.01 | Not given |
| **E. Hard Base** | | | | |
| Fluorine (fluoride ion) | Natural geological sources, industrial waste, water additive | Prevents tooth decay at about 1 mg/L, causes mottled teeth and bone damage at around 5 mg/L in water | 0.8–1.7 depending on temperature | Not given |

SOURCE: Adapted From S. E. Manahan, *Environmental Chemistry*, 3d ed., p. 147. Copyright © 1979 by Willard Grant Press, Boston. Reprinted by permission of Brooks/Cole Publishing, Monterey, California.

NOTES: [a] *Public Health Service Drinking Water Standards*, U.S. Public Health Service, 1962. [b] John F. Kopp and Robert C. Kroner, *Trace Metals in Waters of the United States*, United States Environmental Protection Agency, 1969. The first figure is the percentage of samples showing the element; the second is the highest *value* found; the third is the mean value in positive samples.

BORDERLINE ACIDS  Many of the lighter borderline-acid metal ions, although needed only in tiny quantities, have critical biochemical functions. Their appreciable Lewis acidity enables them to bind certain ligands (e.g., $O_2$) or to enhance the acidity and reactivity of those ligands as a step in synthesizing or metabolizing them. Many of these metal ions also have two or more accessible oxidation states differing by only one electron, which allows them (in appropriate enzymes) to catalyze important redox reactions. In excessive quantities these metal ions can be toxic.

SOFT ACIDS  In order to function properly, enzymes must maintain their correct three-dimensional conformational structure, which is held in position by comparatively weak bonds: hydrogen bonds and the sulfur-sulfur single bonds found in the component amino acid cystine. The latter bond, which is formed by oxidation of the S—H bonds of two cysteine amino acid components, is easily ruptured by the action of soft-acid metal ions (giving S—metal bonds instead). (Compounds such as cystine that contain the —SH functional group are often called *mercaptans*. This term is short for "mercury capturer.") Critical enzymes, being very efficient at catalyzing biochemical reactions, are often present in quite low concentrations and are quite susceptible to deactivation by soft-metal ions; hence all soft-metal ions are toxic when present in the body at concentrations of tens of milligrams per kilogram of body weight (tens of parts per million, ppm); none is essential (other than $Cu^+$ while present during redox reactions of $Cu^{2+}$).

Perhaps the best known of these is mercury. The toxic effects of this element, as well as many others, depends on the chemical form in which it is found. As metallic mercury it is insoluble and passes through the digestive system unchanged. (It was once used as a laxative!) However, it is volatile, and the vapors are exceedingly toxic. (In general, the lungs lack the kinds of defense mechanisms against toxins that the digestive system possesses.) Spills of metallic mercury need to be cleaned up carefully, since mercury spilled in a laboratory can poison the scientists working there over many years [19].

Soluble inorganic mercury salts (such as $HgCl_2$) are absorbed by the gastro-intestinal tract and are toxic. The expression "mad as a hatter" refers to the long-term effects of inorganic mercury ingestion by persons who engaged in the old practice of using $HgCl_2$ in treating beaver pelts to make hats. If a single dose of a soluble inorganic mercury salt is ingested and does not cause death, the body can excrete the mercury over a period of months, since it is water soluble. In contrast, very insoluble mercury salts such as $Hg_2Cl_2$ (*calomel*) and $HgS$ are relatively nontoxic to begin with, since they are nonvolatile and insoluble and cannot be absorbed by the body.

Of particular danger are the derivatives of the *methylmercury* cation, $CH_3Hg^+$, which is perhaps the softest of all cations. Its derivatives, such as $CH_3HgCl$ and $(CH_3)_2Hg$, are both volatile and absorbable through the digestive system. They are much more soluble in fats and lipids than in water, however, so they cannot readily be excreted. Due to this lipid solubility, these compounds can penetrate the blood-brain and placental barriers, causing brain damage and birth defects. Cases of methylmercury poisoning have occurred in areas where fish were eaten that came from waters into which only inorganic mercury was being dumped. It seems that certain microorganisms are able to methylate inorganic mercury via the natural methylating agent vitamin $B_{12}$ (Figure 7.3).

*Lead* is also a notorious soft-acid poison, occurring in the modern environment as a result of the former practice of using a lead hydroxy-carbonate as a white paint pigment and more recently from the use of tetraethyllead, $(CH_3CH_2)_4Pb$, as an antiknock agent in gasoline. As expected, lead reacts with enzyme —SH groups; its most serious effects arise from the consequent deactivation of two enzymes needed for the biosynthesis of heme and from adverse effects upon the function of the brain. Biochemically, *cadmium* behaves as a soft acid and is also quite toxic. In excess it causes itai-itai disease, in which the bones become brittle and break easily and painfully. Its levels also correlate with increased levels in the population of diseases related to high blood pressure. In the body, cadmium tends to concentrate in the kidneys, where it is bound to sulfur donor atoms in the protein metallothionein. *Thallium* is used principally as a rat poison, which speaks for its toxicity.

*Gold* and *platinum* are much too valuable to allow them to escape into waste-waters, but they are toxic. Both have found medicinal uses. Gold is used in antiarthritic drugs such as chloro(triethylphosphine)gold(I), $(C_2H_5)_3P:AuCl$, or Myocrisin, the gold(I) thiomalate $[Au—S—CH(COO^-Na^+)(CH_2COO^-Na^+)]_n$. Although quite effective, there are often toxic side effects due to the long-term accumulation of gold in the body. Cisplatin, an important anticancer drug, is the complex $[PtCl_2(NH_3)_2]$. Somewhat surprisingly from the HSAB point of view, this binds to the nitrogen donor atoms of DNA. As we may expect, toxic side effects must be guarded against.

SOFT BASES    Many of the soft-base nonmetals are beneficial in small doses and may even be essential. *Sulfur* and *phosphorus* are, of course, major components of amino acids and nucleic acids, etc. *Chlorine* is essential as $Cl^-$; *iodine* is essential in the form of the hormone thyroxin. Of particular interest are the very soft *arsenic* and *selenium*. Both are essential, but only the functions of selenium are known. It is a component of glutathione peroxidase, and as a result of its reducing properties it protects against free radicals, which may be implicated in aging and causing cancer. Since it is softer than sulfur, it attracts soft metal ions preferentially and thus protects against their toxic effects (such as carcinogenesis). Thus despite the fact that tuna fish have been found with elevated mercury levels, no one has been found to suffer mercury poisoning from high consumption of tuna. Tuna is also high in selenium, which may tie up the mercury in the form of very stable complexes.

On the other hand, both of these elements are *very* toxic. Arsenic poisoning, of course, has been well known for millennia. Selenium occurs in nature at high levels in certain soils in New Zealand and the western United States, where it is concentrated in certain plants such as locoweed. Cattle eating this weed develop a disease known as the *blind staggers* due to the selenium. Using HSAB, one may speculate that accumulation of Se much in excess of that needed to tie up soft-acid metals may result in binding to borderline-acid metals that are essential in enzyme functions. Whether or not this happens, it is not the contradiction that it appears to be to say that selenium (or other substances) can be both *essential* and (in slightly higher doses) also *highly toxic*. With medication, it is certainly true that more is not necessarily better.

Other soft bases are also well-known toxics: carbon monoxide, $:CO$, and the cyanide ion, $:CN^-$. With both, the soft carbon donor atom binds strongly to borderline-acid metal ions such as $Fe^{2+}$ in heme, preventing the proper functioning of the enzyme. Thus the hard base $O_2$ is unable to compete equitably with $:CO$ for the iron ion in hemoglobin. Many nonmetal hydrogen or methyl compounds (such as phosphine, arsine, hydrogen sulfide, hydrogen selenide, and their methylated forms) have soft donor atoms and are very poisonous.

### 8.10.2 Removal of Unwanted Metal Ions: Chelation Therapy

Given the potency and variety of metal ions in biochemical systems, it can easily be appreciated that the removal of excess or unwanted metal ions selectively from the body (or even from analytical samples or from wastewaters) is no simple task. Any number of ligands can be imagined that may form a complex with a given metal ion, but a number of additional constraints are usually imposed. Concentrations of the offending metal ion and the reagent to be used may both need to be low, which does not favor complex formation unless a very stable complex results. At the same time the ligand should not form too stable complexes with other metal ions, since these may need to be undisturbed. In analytical chemistry, it may be desirable for the complex to be soluble in nonpolar solvents so that it can be extracted and concentrated. On the other hand, this is generally undesirable in medicinal chemistry, since the complex then cannot be excreted. A charged (or perhaps a hydrogen-bonding) complex is desired, which is more likely to be water-soluble. The complex itself, of course, must not be too toxic.

In designing reagents for these tasks, the medicinal, analytical, or environmental chemist has two principles to guide him or her: the HSAB Principle and the fact that chelating ligands form more stable complexes than do monodentate ones (Chapter 7). Figure 8.2a shows some drugs used to remove excess or toxic metal ions from the body.

The chelate principle is best illustrated by the drug EDTA (ethylenediaminetetraacetic acid), which when deprotonated at the four —COOH groups is a hexadentate ligand and is a very powerful chelating ligand indeed (it is used extensively in analytical chemistry to titrate metal ions). The problem lies in its nonselectivity—it prefers highly charged metal ions but will form very stable complexes with almost any metal ion except for a $+1$-charged one. It is the treatment of choice for plutonium poisoning, being one of the few ligands capable of dislodging the $Pu^{4+}$ ion from its precipitated form in the bones, but excesses will then remove the calcium from the bone. This problem is overcome by administering EDTA with calcium ions already chelated in it. EDTA is also used as a food preservative, since it chelates metal ion impurities in foods that would otherwise catalyze air oxidation (spoiling) of the food.

In other ligands that are less overwhelmingly chelating, greater selectivity is possible using the HSAB Principle. For example, the drug British Anti-Lewisite (BAL), when deprotonated at the —SH groups, is chelating with soft-base sulfur donor atoms, and thus selectively chelates soft acids in the body. Originally developed during World War I as an antidote for the war gas Lewisite, $ClCH{=}CH_2AsCl_2$, it is now used in treating poisoning by other soft acids such as mercury and thallium.

As we previously mentioned, some borderline acids, though essential, are toxic at higher levels. Wilson's disease, for example, is a metabolic disorder that results in the inability to excrete excess $Cu^{2+}$ ion, which accumulates to toxic levels. It is treated with the ligand penicillamine, containing a soft and a borderline donor atom in a chelate ring. In human history, iron has often been a deficiency problem (anemia), since $Fe^{2+}$ is normally unstable to oxidation and $Fe^{3+}$ is too acidic to be

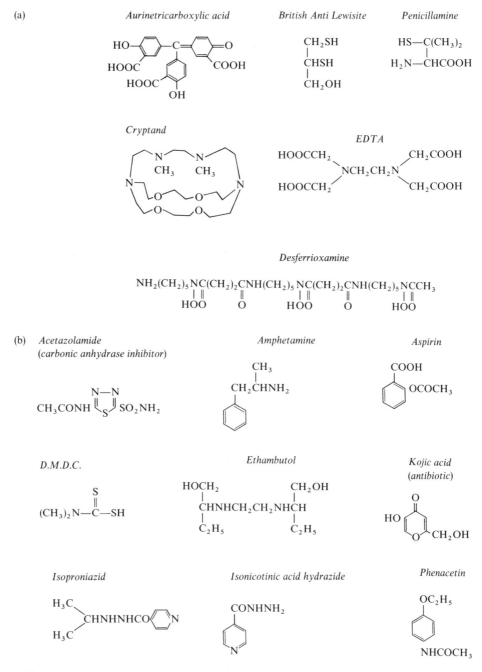

**Figure 8.2** (a) Some chelating ligands used as drugs to remove metal ions from the body; (b) other pharmaceuticals, the actions of which *may* involve metal ions. From M. N. Hughes, *The Inorganic Chemistry of Biological Processes*, 2d ed. Copyright © 1981 by John Wiley and Sons Ltd., Chichester, England. Reprinted by permission.

*Nialamide*

$$\text{CH}_2\text{NHCOCH}_2\text{CH}_2\text{NHNHCO}$$

*disulfiram*

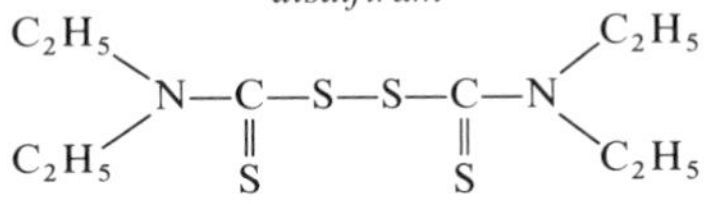

$$\text{C}_2\text{H}_5 \quad \text{C}_2\text{H}_5$$
$$\text{N—C—S—S—C—N}$$
$$\text{C}_2\text{H}_5 \quad \| \quad \| \quad \text{C}_2\text{H}_5$$
$$\text{S} \quad \text{S}$$

*Thiacetazone*

$$\text{S}$$
$$\|$$
$$\text{CH}=\text{NNHCNH}_2$$

$$\text{NHCOCH}_3$$

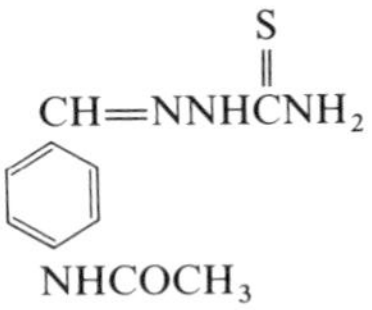

**Figure 8.2** (continued)

available at the pH of the intestine, where absorption must take place; hence in vitamins, simple iron salts are not used. Instead, salts with chelating organic anions are commonly included. But the body has evolved *no* mechanism (other than bleeding) for excreting excess iron, since historically this situation hardly ever arose. But iron poisoning does now occur and is treated with the hard-base chelating ligand desferrioxamine. Another hard-base chelating ligand, aurinetricarboxylic acid, is used in analytical chemistry as a reagent for aluminum; medicinally it is used to treat poisoning by the $Be^{2+}$ ion (similar size and acidity).

The macrocyclic cryptand ligand shown in Figures 8.2 and 7.11 forms more stable complexes than chelating ligands but also selects on the basis of the restricted size of the cavity formed by its donor atoms. It selects $Cd^{2+}$ over $Ca^{2+}$ and $Zn^{2+}$ by a factor of a million and thus can be used for the selective removal of cadmium.

A number of drugs used to treat various conditions not caused by metal poisoning per se are nonetheless good candidates to act as ligands, and their mechanisms of action in the body may involve metal ions (for example, aspirin action may involve copper). Some of these are shown in Figure 8.2b.

The use of chelate ligands with hard or soft donor atoms to concentrate and to detect specific metal ions is a vast topic that you will study extensively in your course on quantitative analysis, so we will not attempt to summarize this area here except to point out that the same principles are involved.

## 8.11

## Study Objectives

1. List a group of Lewis acids or of Lewis bases in order of increasing softness.
2. Use HSAB to predict whether a given reaction will go to the left or to the right.
3. Use HSAB to predict whether a given halide, sulfide, selenide, or telluride will be soluble or insoluble in water.
4. Use HSAB to predict in which group of the qualitative analysis scheme a given metal ion will fall.

5. Use HSAB to classify an element as an atmophile, a lithophile, or a chalcophile or siderophile. Identify likely mineral sources of given elements and forms in which they would occur in natural waters.

6. Use HSAB to predict the binding sites of different metal ions to biological molecules and to select likely ligands for medicinal use. Identify and discuss the toxicities of elements using the HSAB Principle.

7. Tell how the hardness or softness of a donor or acceptor atom is modified (a) by changing its oxidation number, or (b) by changing the hardness or softness of other groups attached to it.

8. Know that the HSAB Principle does not work well if there is a great difference in strength in the acids or bases involved.

9. Discuss explanations (in terms of thermodynamics and bonding types) of why the HSAB Principle works.

10. Calculate bond energies from Pauling electronegativities, or vice versa.

11. Know how the Pauling electronegativity of an element is affected by changing substituents on it or by changing its oxidation number.

## 8.12

### Exercises

1. In each of the following complexes, classify both the metal ions and each ligand as *hard*, *soft*, or *borderline*: **a.** $(NH_4)_2[Pd(-SCN)_4]$; **b.** $[Co(NH_3)_4(SeO_4)]ClO_4$.

*2. Draw the Lewis dot structures of the following molecules or ions, identify the potential donor atoms in each, and classify the donor atoms as hard, soft, or borderline bases: **a.** $NH_3$; **b.** $NH_2-CH_2-COOH$ (an amino acid); **c.** $SO_4^{2-}$; **d.** $SO_3^{2-}$.

3. Classify each Lewis acid and each Lewis base in the following complexes as hard, soft, or borderline. **a.** $[Na(CH_3-O-CH_2-CH_2-O-CH_3)_3]^+$ **b.** $[SnI_6]^{2-}$ **c.** $[HgCl_2]$ **d.** $[Hg(CH_3)_2]$

†4. Rewrite each of these lists of Lewis acids or bases so that each is in increasing order of softness: **a.** $Cu^+$, $Au^+$, $Ag^+$, $K^+$; **b.** $Br^-$, $I^-$, $F^-$, $Cl^-$; **c.** $Mo^{2+}$, $Mo^{6+}$, $Mo^{4+}$; **d.** $BF_3$, $B(OCH_3)_3$, $B(CH_3)_3$; **e.** $FHg^+$, $CH_3Hg^+$, $Fe^{2+}$, $Fe^{3+}$; **f.** $(CH_3)_2S$, $(CH_3)_2Se$, $(CH_3)_2O$.

5. Most oxo anions are hard bases, but some are ambidentate and may also act as soft bases. List some of these.

†6. Predict whether reactants (left side) or products (right side) will be favored in each of the following equilibria:
6.1 $As_2S_5 + 5HgO \rightleftharpoons As_2O_5 + 5HgS$
6.2 $ZnI_2 + HgCl_2 \rightleftharpoons ZnCl_2 + HgI_2$
6.3 $La_2(CO_3)_3 + Bi_2S_3 \rightleftharpoons La_2S_3 + Bi_2(CO_3)_3$
6.4 $2CH_3MgF + HgF_2 \rightleftharpoons (CH_3)_2Hg + 2MgF_2$
6.5 $AgF + LiI \rightleftharpoons AgI + LiF$
6.6 $[Cu(thiourea)_4]^+ + [Cu(urea)_4]^{2+} \rightleftharpoons [Cu(thiourea)_4]^{2+} + [Cu(urea)_4]^+$

**6.7** $PbSe + HgS \rightleftharpoons HgSe + PbS$

**6.8** $3\,FeO + Fe_2S_3 \rightleftharpoons Fe_2O_3 + 3\,FeS$

**6.9** $CdSO_4 + CaS \rightleftharpoons CdS + CaSO_4$

**6.10** $Zn(SCH_3)_2 + Hg(SeCH_3)_2 \rightleftharpoons Hg(SCH_3)_2 + Zn(SeCH_3)_2$

**6.11** $2\,Fe(SCN)_3 + 3\,Fe(OCN)_2 \rightleftharpoons 2\,Fe(OCN)_3 + 3\,Fe(SCN)_2$

**6.12** $CaS + ZnSeO_4 \rightleftharpoons ZnS + CaSeO_4$

7. Give plausible products (or just one product) for each of the following reactions, or tell if no reaction is expected.

   **7.1** $CdSO_4 + MgS \rightarrow$

   **7.2** $(CH_3)_2Hg + CaF_2 \rightarrow$

8. Predict any products that will precipitate in each of the following mixtures in water:

   **8.1** $HgCl_2 + KI + KF$

   **8.2** $PrCl_3 + Na_2S$

   **8.3** $TlNO_3 + KI + KF$

9. What do relative electronegativities suggest about the nature of the bond between a hard acid and a hard base? a soft acid and a soft base?

10. Are the ionic radii and charges of hard acids and bases also favorable to the kind of bonding you cited in the previous question? Explain.

*11. Hydrogen and boron do not have electronegativities characteristic of hard acids, yet the $H^+$ and $B^{3+}$ ions are classified as hard acids. Which of their properties other than electronegativity might justify such a classification?

12. What other characteristic of acids and bases—besides their hardness and softness—determines whether a reaction will go to the left or to the right?

*13. Circle all *insoluble* salts: CdTe; AgI; AgF; KI; EuSe; $TiO_2$; $TiTe_2$; $PtAs_2$.

14. Give the symbols for two metal ions that would fall into *each* of the groups of the qualitative analysis scheme.

15. Into which group of the qualitative analysis scheme will each of the following ions fall? **a.** $Tl^+$; **b.** $Cu^+$; **c.** $Rb^+$; **d.** $Cr^{3+}$; **e.** $Sn^{2+}$.

†16. Predict in which group of the qualitative analysis scheme each of the following metal ions would occur. If the metal ion gives a precipitate, write the formula of the precipitate: **a.** $Sn^{2+}$; **b.** $Pr^{3+}$; **c.** $Cu^+$; **d.** $Sr^{2+}$; **e.** $Sb^{3+}$; **f.** $Eu^{3+}$; **g.** $Au^+$; **h.** $Ra^{2+}$; **i.** $Rb^+$.

†17. Answer the following questions about the following metal ions: **a.** $Zr^{4+}$; **b.** $Ag^+$; **c.** $Sb^{3+}$.

   **17.1** Is its bromide soluble or insoluble?

   **17.2** Is its selenide soluble or insoluble?

   **17.3** Is its oxide soluble or insoluble?

   **17.4** In which group of the qualitative analysis scheme does this metal ion occur?

**17.5** Is this metal a lithophile?

**17.6** Which is the more likely mineral source: a silicate, a sulfide, or seawater?

†**18.** Classify each element as (*A*) a *lithophile* or (*B*) either a *chalcophile* or a *siderophile*. Then choose its most likely mineral source.

**18.1** La; source: $LaPO_4$, LaAs, $LaI_3$, or $LaCl_3$.

**18.2** Pt; source: $PtAs_2$, $PtN_2$, $PtSiO_4$, or $PtF_2$.

**18.3** Zr; source: $ZrSiO_4$, $ZrPbS_4$, or $ZrCl_4$.

**18.4** Sb; source: $Sb_2(SiO_3)_3$, $Sb_2S_3$, $SbF_3$, or $SbCl_3$.

**18.5** Te; source: $TeF_4$, $Na_2Te$, PbTe, or $TeI_4$.

**18.6** Be; source: $Be^{2+}(aq)$, $Be_3Al_2Si_6O_{18}$, or $BeSeS_2$.

**18.7** F; source: $F^-(aq)$, $CaF_2$, or $HgF_2$.

**18.8** Co; source: $CoCO_3$, $CoAs_2$, or $Co^{2+}(aq)$.

**18.9** Th; source: $ThS_2$, $ThO_2$, or $Th^{4+}(aq)$.

**19.** Which are more generally toxic, soft-acid metal ions or hard-acid metal ions? To which type of biochemical ligand would soft-acid metal ions bind most strongly? **a.** phosphate groups; **b.** porphyrin groups (nitrogen donor); **c.** cysteine groups (sulfur donor).

*****20.** Which of the following medicinal chemicals would be most effective in combating poisoning by a soft-acid metal ion:

$$
\begin{array}{ccccc}
CH_3 & CH_3 & CH_2{-}OH & CH_2{-}OH & CH_2{-}SH \\
| & | & | & | & | \\
CH{-}SH & CH{-}OH & CH{-}OH & CH{-}OH & CH{-}SH \\
| & | & | & | & | \\
CH_3 & CH_3 & CH_3 & CH_2{-}SH & CH_2{-}OH
\end{array}
$$

**21.** Consider the following six ligands. **a.** List them in order of increasing softness (note that some will be equally soft). **b.** Which one of these might make the best medicine to combat poisoning by the $Pt^{2+}$ ion? What principles did you use to choose this ligand?

**22.** Briefly explain why the element selenium is necessary for life and protects against mercury poisoning but is also highly poisonous itself.

**23.** Which of the following manners of disrupting protein structure would you expect to be most uniquely associated with the heavy-metal ions? **a.** Hydrolysis of the CO—NH (peptide) linkages; **b.** disruption of hydrogen bonding between different CO—NH groups; **c.** disruption of S—S bond formation between different cysteine amino acid S—H groups; **d.** disruption of the electrostatic attraction of oppositely charged amino acid side chains.

24. Compare the compounds $(CH_3)_3SnCl$ and $F_3SnCl$: **a.** Which should show the highest electronegativity for tin? **b.** Which should show the strongest Sn—Cl bond?

*25. The electronegativity of carbon in $CCl_4$ is estimated as 2.64; the electronegativity of carbon in $(CH_3)_3CCl$ is estimated as 2.29. Compute the expected C—Cl bond energy in each compound.

26. You are given the following bond energies: Ge—Ge = 188 kJ/mol; Cl—Cl = 240 kJ/mol; Ge—Cl (in $GeCl_2$) = 385 kJ/mol. Calculate the electronegativity of Ge in $GeCl_2$. Does your answer differ from the Pauling electronegativity of Ge in $GeCl_4$ (2.01) in the expected manner? Explain.

27. For which elements (other than iodine, used in an example) are there appropriate data in Tables 6.5 and 8.3 to 8.6 for the calculation of the effects of changing oxidation numbers on the Pauling electronegativities of elements? Carry out such calculations for two of those elements.

28. Some new elements have just been discovered in the author's laboratory. The following atomic parameters have been obtained:

| Element: | Sheltonium | Bennerine | Coatsium | Kamelogen |
|---|---|---|---|---|
| Electronegativity: | 1.06 | 3.82 | 1.97 | 2.34 |
| Ionic radius (pm): | 107 | 123 | 105 | 189 |
| Element-element single-bond energy (kJ/mol): | 78 | 166 | 88 | 183 |

**28.1** Will products or reactants be favored in the equilibrium Sheltonium Benneride + Coatsium Kamelide $\rightleftharpoons$ Sheltonium Kamelide + Coatsium Benneride?

**28.2** Calculate the covalent bond energies expected for the Coatsium-Kamelogen bond; for the Sheltonium-Bennerine bond.

Notes

[1] Williams, R. J. P., *The Metals of Life*, Van Nostrand Reinhold, New York, 1971, Ch. 4.

[2] Morel, F., R. E. McDuff, and J. J. Morgan, in *Trace Metals and Metal-Organic Interactions in Natural Waters*, ed. Philip C. Singer, Ann Arbor Science, Ann Arbor, Mich., 1973, p. 157.

[3] Pearson, R. G., *Hard and Soft Acids and Bases*, Dowden, Hutchinson & Ross, Stroudsburg, Penn., 1973; (a) op. cit., p. 6.

[4]  Note that many of the nonmetals may function as "cations" or Lewis-acidic sites in some compounds and as donor atoms in others. For example, arsenic as $As^{3+}$ is a hard acid, but as a donor ($As^{3-}$, etc.) it is a soft base.

[5]  Parr, R. G., and R. G. Pearson, *J. Amer. Chem. Soc.*, 105, 7512 (1983).

[6]  Klopman, G., *J. Amer. Chem. Soc.*, 90, 223 (1968).

[7]  Pauling, L., *The Nature of the Chemical Bond*, 3d ed., Cornell University Press, Ithaca, N.Y., 1960, Ch. 3.

[8]  Allred, A. A., *J. Inorg. Nucl. Chem.*, 17, 215 (1961).

[9]  Wells, P. R., *Prog. Phys. Org. Chem.*, 6, 111 (1968).

[10]  Pearson, R. G., *J. Chem. Soc., Chem. Commun.*, 65 (1968).

[11]  Clifford, A. F., *et al., J. Amer. Chem. Soc.*, 79, 5404 (1957).

[12]  Clifford, A. F., *Inorganic Chemistry of Qualitative Analysis*, Prentice-Hall, Englewood Cliffs, N.J., 1961.

[13]  Phillips, C. S. G., and R. J. P. Williams, *Inorganic Chemistry*, Oxford University Press, New York, 1965, vol. 2, Ch. 34.

[14]  Usanovich, M., *Zhur. Obschei. Khim.*, 9, 182 (1939).

[15]  See, for example, F. H. Day, *The Chemical Elements in Nature*, Reinhold, New York, 1964.

[16]  Edmond, J. M., K. L. Von Damm, R. E. McDuff, and C. I. Measures, *Nature*, 297, 187 (1982); J. M. Edmond and K. Von Damm, *Scientific American*, 248 (4), 78 (April 1983); P. A. Rona, *Scientific American*, 254 (1), 84 (January 1986).

[17]  Lepkowski, W., *Chem. & Eng. News*, June 4, 1979, p. 14.

[18]  See, for instance, M. N. Hughes, *The Inorganic Chemistry of Biological Processes*, 2d ed., John Wiley, Chichester, 1981.

[19]  Although it seems logical from the point of view of the HSAB Principle, the traditional practice of sprinkling powdered sulfur over spilled mercury is not a good clean-up procedure since the reaction of elemental mercury and sulfur is *very* slow.

# The Halides, Nitrides, and Sulfides of the Elements

At this point in the course we have covered the four basic types of inorganic reactions: acid-base, precipitation, oxidation-reduction, and complexation (Lewis acid-base); hence Chapters 9 and 10 introduce no new principles. The descriptive chemistry of the remaining classes of compounds is rich and varied; it cannot be completely predicted using the principles we have developed. But these principles should be helpful in *organizing* and *understanding* the physical properties and chemical reactions of these important classes of compounds.

In this chapter we will first examine the *halides* of the elements, which are widely used in synthesizing other inorganic, organometallic, and organic compounds, including polymers. In the latter half of this chapter we will focus on the products obtained from the halides principally by (Lewis) acid-base reactions: the *sulfides*, *nitrides*, and *esters* or *alkoxides* of the elements. In the next chapter we will look at other classes of compounds that, although often made from the halides, have more pronounced chemistries as reducing agents: the *hydrides* and simple *organometallic compounds* of the elements.

## 9.1

### Existence or Nonexistence of Given Halides

The halides of the elements often are not confined to those corresponding to the common oxidation numbers. For example, the known sulfur fluorides include the two we may expect, $SF_6$ and $SF_4$. But sulfur also forms fluorides in which sulfur has oxidation numbers of $+5$ and $+1$, $S_2F_{10}$ and two isomers of $S_2F_2$. Although the reason for the odd oxidation numbers in these compounds is that they are catenated (contain S—S bonds), the unexpected oxidation number $+2$ also occurs in $SCl_2$.

Not all halides of an element in a given oxidation state exist. For example, there is no sulfur hexachloride, hexabromide, or hexaiodide. Although $SCl_4$ exists, $SBr_4$ does not. Indeed, there is no stable iodide of sulfur at all.

One reason for this is that elements such as sulfur in high oxidation states have oxidizing properties, while the halide ions have quite varying reducing properties $(F^- < Cl^- < Br^- < I^-)$. Although sulfur(VI) is not capable of oxidizing the very stable fluoride ion, it does oxidize the iodide ion, as in attempts to prepare hydrogen iodide from sodium iodide and concentrated sulfuric acid:

$$2\,NaI + 3\,H_2SO_4 \rightarrow I_2 + SO_2 + 2\,H_2O + 2\,NaHSO_4 \qquad (9.1)$$

The corresponding experiment with the less easily oxidized bromide ion (as NaBr) gives some HBr and some $Br_2$; NaCl gives HCl exclusively.

$$NaCl + H_2SO_4 \rightarrow HCl + NaHSO_4 \qquad (9.2)$$

Familiar examples of redox reactions that occur during attempts to prepare the halides themselves include the nonsyntheses of $FeI_3$ and $CuI_2$:

$$2\,Fe^{3+} + 6\,I^- \rightarrow 2\,FeI_2 + I_2 \qquad (9.3)$$

$$2\,Cu^{2+} + 4\,I^- \rightarrow 2\,CuI + I_2 \qquad (9.4)$$

The corresponding bromides, chlorides, and fluorides of iron(III) and copper(II) can be prepared without difficulty. Similar redox reactions occur with other reducing ions, such as sulfide or some of the pseudohalide ions.

Note, however, that such redox reactions generally do *not* occur if the oxidizing cation is a soft acid, even if its reduction potential is similar to those of $Fe^{3+}$ or $Cu^{2+}$. In these combinations (e.g., AgI) the enthalpy of formation of the product and its lattice energy are substantially more favorable than we would anticipate. The stabilizing energy in such cases is connected with the formation of a good covalent bond. But the hard cation $Fe^{3+}$, the borderline cation $Cu^{2+}$, and the small (hence evidently not soft) imaginary $S^{n+}$ cations evidently do not form good enough covalent bonds to iodine to give stable products.

Since fluorine and oxygen are the most active nonmetals, the fluoride and the oxide (or hydroxide) ions are the most resistant to oxidation. These ions are hard bases; hence we concluded in Section 8.2 that hard bases tend to stabilize elements in their highest oxidation states (in which they are harder acids). Since fluorine is the most electronegative element and the strongest oxidizing agent, we may well expect the elements generally to show their highest oxidation state in fluorides (or in fluoro anions) above any other kind of derivative, even oxides or oxo anions. Often this is the case, as for example in $PtF_6$, the only compound of platinum(VI), but as we can see from Table 9.1, frequently the oxide or oxo anion exhibits the highest oxidation number.

To explain this, let us consider the problem of forming chlorine(VII), for example, as a fluoride or as an oxide. The oxide $Cl_2O_7$ (Figure 4.2) requires that chlorine have the reasonable coordination number of 4, but the corresponding fluoride $ClF_7$ would have a coordination number of 7, which exceeds the maximum possible for a third-period element. But in the presence of a Lewis acid the cation $ClF_6^+$ can be formed, and the larger iodine atom does form a heptafluoride.

## 9.2

## Color of Halides: Charge-Transfer Bands and Internal Redox Reactions

Many of the halides of the elements are quite colorful compounds, with very intense colors, but this property is by no means characteristic of all halides. If you per-

**Table 9.1** Highest Fluorides or Fluoro Anions of the Elements (or Oxides if These Have Higher Oxidation Numbers)

| 1 | 2 | 3 | 4 | 5 | 6 | 7 | 8 | 9 | 10 | 11 | 12 | 13 | 14 | 15 | 16 | 17 | 18 |
|---|---|---|---|---|---|---|---|---|---|---|---|---|---|---|---|---|---|
| HF | | | | | | | | | | | | | | | | | |
| LiF | $BeF_2$ | | | | | | | | | | | $BF_3$ | $CF_4$ | $NF_4^+$ | $OF_2$ | $F_2$ | |
| NaF | $MgF_2$ | | | | | | | | | | | $AlF_3$ | $SiF_4$ | $PF_5$ | $SF_6$ | $Cl_2O_7$ | |
| KF | $CaF_2$ | $ScF_3$ | $TiF_4$ | $VF_5$ | $CrF_6$ | $Mn_2O_7$ | $FeO_4^{2-}$ | $CoO_4^{3-}$ | $NiF_6^{2-}$ | $CuF_6^{2-}$ | $ZnF_2$ | $GaF_3$ | $GeF_4$ | $AsF_5$ | $SeF_6$ | $BrO_4^-$ | $KrF_2$ |
| RbF | $SrF_2$ | $YF_3$ | $ZrF_4$ | $NbF_5$ | $MoF_6$ | $Tc_2O_7$ | $RuO_4$ | $RhF_6$ | $PdF_4$ | $AgF_4^-$ | $CdF_2$ | $InF_3$ | $SnF_4$ | $SbF_5$ | $TeF_6$ | $IF_7$ | $XeO_4$ |
| CsF | $BaF_2$ | $LuF_3$ | $HfF_4$ | $TaF_5$ | $WF_6$ | $ReF_7$ | $OsO_4$ | $IrF_6$ | $PtF_6$ | $AuF_5$ | $HgF_2$ | $TlF_3$ | $PbF_4$ | $BiF_5$ | $PoF_6^{2-}$ | $AtO_3^-$ | $RnF_2$ |
| FrF | $RaF_2$ | | | | | | | | | | | | | | | | |

Lanthanides and Actinides:

| | | | | | | | | | | | | | |
|---|---|---|---|---|---|---|---|---|---|---|---|---|---|
| $LaF_3$ | $CeF_4$ | $PrF_4$ | $NdF_7^{3-}$ | $PmF_3$ | $SmF_3$ | $EuF_3$ | $GdF_3$ | $TbF_4$ | $DyF_7^{3-}$ | $HoF_3$ | $ErF_3$ | $TmF_3$ | $YbF_3$ |
| $AcF_3$ | $ThF_4$ | $PaF_5$ | $UF_6$ | $NpO_6^{5-}$ | $PuO_6^{5-}$ | $AmO_2^{2+}$ | $CmF_4$ | $BkF_4$ | $CfF_4$ | $EsF_3$ | $FmF_3$ | $MdF_3$ | $NoF_3$ |

formed Experiment 6, you might well have noticed that soft acid–soft base adducts (complexes or salts) often have vivid colors. Table 9.2 summarizes the colors (or lack of color, if none is indicated) for a selection of $s$- and $p$-block halides.

Two characteristics of colored halides are apparent. (1) Quite a number of iodides of the elements are intensely colored; few fluorides of the elements are intensely colored. (If fluorides contain $d$- or $f$-block metals, they contain colors due to these cations, although such colors are relatively pale.) (2) The halides of the elements at the left of the $s$- and $p$-blocks, in low oxidation states, are colorless; those at the right, occurring just before the halides become nonexistent, are colored. The cations forming the colored halides are generally better oxidizing agents than the cations forming the colorless halides. These characteristics hint that this intense color is related to the redox properties of the Lewis acid and base forming the halide or other adduct.

The combination of a cation that is a good oxidizing agent with an anion that is a good reducing agent often results in a redox reaction—the expected halide is nonexistent. But if the oxidizing power of the cation and the reducing power of the anion are just insufficient to produce a redox reaction, the halide can exist (although it will have a low heat of formation), and it will likely be intensely colored. Such compounds are often marginally stable; this characteristic is the source of a commonplace observation that intensely colored compounds are sometimes quite reactive (in a redox sense) and should be treated with respect.

The color results from what is known as an electronic **charge-transfer absorption**. Although the redox reaction of the cation and anion is nonspontaneous, light energy is capable of promoting the electron out of the valence orbital of the anion and into the valence orbital of the cation. Normally the electron subsequently returns to the original atom, but sometimes the compound is permanently decomposed, so we say it is *photosensitive*. Silver bromide is an example of a photosensitive halide; its decomposition after exposure to light is the basis of the process of photography.

A particular frequency of light is absorbed during the charge-transfer process; the remaining light that passes through or is reflected no longer contains all frequencies of visible light and hence is colored. Because it can be moving in the same direction as the electron being transferred, the light wave has an unusually high probability of being absorbed in the charge-transfer process; hence the color depletion of the reflected light is strong, and we perceive intense color. In contrast, the process of light absorption in $d$- and $f$-block metal ions (Section 2.5) does not involve motion of the electron—it stays on the same atom in the same type of orbital—and a given light wave is much less likely to be absorbed; hence the colors of these metal ions are relatively faint.

Charge-transfer absorptions are not confined to halides, of course: recall the sulfides of the qualitative analysis scheme. They can occur even with poorly reducing Lewis bases if the Lewis acid is powerfully oxidizing (but not oxidizing enough to make the compound nonexistent). Examples from earlier in the book include some of the powerfully oxidizing oxo anions: $MnO_4^-$ is purple; $CrO_4^{2-}$ is yellow; $FeO_4^{2-}$ is red-violet. These intense colors contrast not only with the usual colorless or white

**Table 9.2**  Colors of the Highest Halides of Some Elements

### A. Fluorides

| | | | | | | | | | |
|---|---|---|---|---|---|---|---|---|---|
| LiF | BeF$_2$ | | | BF$_3$ | CF$_4$ | NF$_3$ | (OF$_2$) | (F$_2$) yellow | |
| NaF | MgF$_2$ | | | AlF$_3$ | SiF$_4$ | PF$_5$ | SF$_6$ | ClF$_5$ | |
| KF | CaF$_2$ | ScF$_3$ | ZnF$_2$ | GaF$_3$ | GeF$_4$ | AsF$_5$ | SeF$_6$ | BrF$_5$ | (KrF$_2$) |
| RbF | SrF$_2$ | YF$_3$ | CdF$_2$ | InF$_3$ | SnF$_4$ | SbF$_5$ | TeF$_6$ | IF$_7$ | XeF$_6$ |
| CsF | BaF$_2$ | LuF$_3$ | HgF$_2$ | TlF$_3$ | PbF$_4$ | BiF$_5$ | | | |

### B. Chlorides

| | | | | | | | | | |
|---|---|---|---|---|---|---|---|---|---|
| LiCl | BeCl$_2$ | | | BCl$_3$ | CCl$_4$ | NCl$_3$ yellow | | | |
| NaCl | MgCl$_2$ | | | AlCl$_3$ | SiCl$_4$ | PCl$_5$ | SCl$_4$ yellow | (Cl$_2$) yellow-green | |
| KCl | CaCl$_2$ | ScCl$_3$ | ZnCl$_2$ | GaCl$_3$ | GeCl$_4$ | AsCl$_5$ yellow | SeCl$_4$ yellow | (BrCl) red | |
| RbCl | SrCl$_2$ | YCl$_3$ | CdCl$_2$ | InCl$_3$ | SnCl$_4$ | SbCl$_5$ | TeCl$_4$ yellow | (ICl$_3$) orange | (XeCl$_2$) ? |
| CsCl | BaCl$_2$ | LuCl$_3$ | HgCl$_2$ | TlCl$_3$ | PbCl$_4$ yellow | BiCl$_3$ | | | |

### C. Bromides

| | | | | | | | | | |
|---|---|---|---|---|---|---|---|---|---|
| LiBr | BeBr$_2$ | | | BBr$_3$ | CBr$_4$ | NBr$_3$ ? | | | |
| NaBr | MgBr$_2$ | | | AlBr$_3$ | SiBr$_4$ | PBr$_5$ yellow | S$_2$Br$_2$ red | | |
| KBr | CaBr$_2$ | ScBr$_3$ | ZnBr$_2$ | GaBr$_3$ | GeBr$_4$ | AsBr$_3$ yellow | SeBr$_4$ orange | (Br$_2$) red | |
| RbBr | SrBr$_2$ | YBr$_3$ | CdBr$_2$ | InBr$_3$ yellow | SnBr$_4$ | SbBr$_3$ | TeBr$_4$ orange | (IBr) red | |
| CsBr | BaBr$_2$ | LuBr$_3$ | HgBr$_2$ | TlBr$_3$ yellow | PbBr$_2$ | BiBr$_3$ yellow | | | |

### D. Iodides

| | | | | | | | | | |
|---|---|---|---|---|---|---|---|---|---|
| LiI | BeI$_2$ | | | BI$_3$ | CI$_4$ red | | | | |
| NaI | MgI$_2$ | | | AlI$_3$ | SiI$_4$ | PI$_5$ black | | | |
| KI | CaI$_2$ | ScI$_3$ | ZnI$_2$ | GaI$_3$ yellow | GeI$_4$ orange | AsI$_3$ red | | | |
| RbI | SrI$_2$ | YI$_3$ | CdI$_2$ | InI$_3$ yellow, red | SnI$_4$ yellow | SbI$_3$ yellow, red | TeI$_4$ gray | (I$_2$) purple | |
| CsI | BaI$_2$ | LuI$_3$ | HgI$_2$ yellow, red | TlI yellow | PbI$_2$ yellow | BiI$_3$ red | | | |

NOTE: If no color is listed, the compound is colorless or white. Compounds to the left of the bar are in the group oxidation state; those to the right are in the (group number minus 2) oxidation state; those in parentheses are in still lower oxidation states.

oxo anion, but also with the pale colors of the nonoxidizing $Mn^{2+}$, $Cr^{3+}$, and $Fe^{2+}$ ions.

## Methods of Synthesis of Halides

Although many of the more important halides of the elements can be purchased commercially, many of them are too reactive to be available, and some of those that are available may have decomposed or become contaminated by the time the purchaser receives them; hence chemists often have to synthesize halides. A variety of methods are available [1], depending on the properties of the desired halide. Generally the simpler ones are satisfactory only for the less reactive halides.

**1.** Perhaps the simplest method of synthesis is *precipitation* of an insoluble halide by reacting a soluble halide with a soluble salt of the desired cation. This is generally a satisfactory synthesis for the chlorides, bromides, and iodides of the soft acids (Section 8.6). The fluoride ion is a hard weak base that gives fairly insoluble fluoride salts with most hard or borderline feebly to strongly acidic metal ions. Many of these, however, are hydrated. Anhydrous fluorides do precipitate with many of the hard feebly acidic cations such as $Ca^{2+}$.

**2.** Soluble halides of the nonacidic or feebly acidic cations can be obtained by acid-base neutralization reactions of the hydrohalic acid with the hydroxide, oxide, or carbonate of the cation involved, followed by evaporation of the resulting solution. The chlorides, bromides, and iodides of the nonacidic cations usually crystallize out in anhydrous form (since both cation and anion are large). In the other cases hydrated salts usually crystallize out, but the water can normally be driven off by applying heat to the crystals in a vacuum.

**3.** With weakly or moderately acidic cations, hydrated chlorides, bromides, and iodides can be obtained by evaporation of the solution resulting from an acid-base reaction of the hydrohalic acid with an oxide, hydroxide, or carbonate of the cation. But attempts to drive off the water of hydration by heating lead to hydrolysis (Chapter 2):

$$NiCO_3(s) + 2\,HBr(aq) + heat \rightarrow [Ni(H_2O)_6]Br_2(s) + CO_2(g) \qquad (9.5)$$

$$[Ni(H_2O)_6]Br_2(s) + heat \rightleftharpoons Ni(OH)Br(s) + HBr(g) + 5\,H_2O(g) \qquad (9.6)$$

In some cases the anhydrous halide can be obtained if the dehydration is carried out in a stream of the anhydrous hydrogen halide, which tends to reverse the above equilibrium.

**4.** Alternatively, after a neutralization reaction the water may be removed from many of the hydrated salts of weakly or moderately acidic cations by chemical reaction with the chemical dehydrating agents 2,2-dimethoxypropane (equation 9.7), triethyl orthoformate (equation 9.8), or thionyl chloride (equation 9.9). The first two of these are used to produce nonaqueous solutions of metal halides that contain coordinated methanol or ethanol. (Since these are weaker ligands than water, their presence may be unobjectionable.)

$$[Co(H_2O)_6]Cl_2 + 6(CH_3)_2C(OCH_3)_2 \rightarrow \qquad (9.7)$$
$$[Co(CH_3OH)_6]Cl_2 + 6CH_3OH + 6(CH_3)_2C{=}O$$

$$[Co(H_2O)_6]Br_2 + 6HC(OC_2H_5)_3 \rightarrow \qquad (9.8)$$
$$[Co(C_2H_5OH)_6]Br_2 + 6C_2H_5OH + 6HCOOC_2H_5$$

If a completely solvent-free and solid halide is needed, thionyl chloride is useful, since it produces gaseous nonbasic byproducts and is itself volatile:

$$[Cr(H_2O)_6]Cl_3(s) + 6SOCl_2(l) \rightarrow CrCl_3(s) + 6SO_2(g) + 12HCl(g) \qquad (9.9)$$

**5.** With more acidic cations than the preceding ones, it is necessary to completely avoid the presence of water. (This is sometimes a good idea even with weakly and moderately acidic cations.) Two general strategies are then available, one based on exchange of halide groups for other groups (halogenation) and the other based on redox reactions (some syntheses involve both). Exchange reactions include exchange of one halogen for another:

$$FeCl_3 + excess\ BBr_3 \rightleftharpoons FeBr_3 + BCl_3 \qquad (9.10)$$

There are also exchange reactions of halide for oxide:

$$2MO + SF_4 \rightarrow 2MF_2 + SO_2(g) \qquad (9.11)$$

$$Cr_2O_3 + 3CCl_4 \rightarrow 2CrCl_3 + 3COCl_2 \qquad (600\,°C) \qquad (9.12)$$

**6.** Redox reactions can, in principle, be used for most halides, but may be too vigorous for the more active metals. Fluorine is such a powerful oxidizing agent that its oxidations are difficult to control except with the most resistant elements:

$$Xe + 3F_2 \rightarrow XeF_6 \qquad (400\,°C) \qquad (9.13)$$

Often the fluoride of an element in a high oxidation state (such as $CoF_3$ or $XeF_2$ or simply HF) may be used in place of elemental fluorine. This method is particularly convenient for making the lower halide of an element from a more common higher halide:

$$14TaI_5 + 16Ta \rightarrow 5Ta_6I_{14} \qquad (9.14)$$

Oxidation with the halogen itself is more convenient with the heavier halogens, but the reaction is still often quite vigorous:

$$2Sb + 3Cl_2 \rightarrow 2SbCl_3 \qquad (9.15)$$

If a metal with a high heat of atomization (e.g., W) is being oxidized with a halogen, it is often necessary to use temperatures of several hundred degrees Celsius. This reaction can be carried out in a *tube furnace*: Vapors of the halogen are passed over the hot metal; the halide formed vaporizes at the high temperature and passes out of the furnace to condense as crystals:

$$2\,Cr(s) + 3\,I_2(g) \rightarrow 2\,CrI_3(g) \qquad (450\,°C) \tag{9.16}$$

**7.** Finally, there are reactions that combine reduction with halogen exchange at high temperatures. These often involve carbon as a reducing agent to remove the oxygen. Such reactions are economically favorable:

$$Ta_2O_5 + 5\,C + 5\,Br_2 \rightarrow 5\,CO + 2\,TaBr_5 \qquad (460\,°C) \tag{9.17}$$

EXAMPLE   Choose reasonable syntheses of the following halides: $SF_6$, LiF, $CoF_2$, and $PbI_2$.

Solution   ■ The later methods given in the above list tend to be more difficult than the earlier, so if the softness or the low acidity of the cation allows the use of an earlier method, it should be chosen. $Li^+$ is a hard, feebly acidic cation that gives an insoluble anhydrous fluoride, so route 1 is the simplest. $Pb^{2+}$ is a soft acid, so its iodide should be (and is) insoluble; route 1 can also be used. (Recrystallization of the yellow $PbI_2$ from large volumes of hot water gives crystals that sparkle like mosaic gold.)

■ $Co^{2+}$ is a weakly acidic cation that would be expected to (and does) give an insoluble hydrated fluoride. Dehydration of the fluoride by heat would be expected to cause some hydrolysis. Direct fluorination of cobalt would likely lead to a higher oxidation state, so an exchange reaction might be preferable. $CoF_2$ is made by heating $CoCl_2$ in HF.

■ $SF_6$ is the fluoride of a very acidic "cation" in a very high oxidation state, so direct oxidation with a powerful oxidizing agent is likely to be needed. This gas is prepared by burning sulfur in fluorine. □

For experimental details in preparing a given halide (especially if one of the later, more complex methods is to be used), you should consult the original literature or a collected volume [2].

9.4

# Physical States of Halides and Their Ionic, Covalent, or Polymeric Covalent Structures

The physical properties of the halides (melting and boiling points; whether each is a solid, liquid, or gas at room temperature) are governed by the principles discussed in Section 4.2 and applied there to the oxides. Table 9.3, for example, shows the melting points of the highest fluoride of each element. At the far left of the table we have metallic fluorides with the high melting points characteristic of substances in ionic lattices in which both metal ion and fluoride ion have coordination numbers higher than 2. At the far right of the table we have monomeric covalent nonmetal fluorides with melting points generally well below room temperature—most of these are gases at room temperature. Similarly, the high-valent fluorides in the middle of the d-block are gases or at least are quite volatile.

**Table 9.3  Melting Points of the Highest Fluoride of Each Element**

| 1 | 2 | 3 | 4 | 5 | 6 | 7 | 8 | 9 | 10 | 11 | 12 | 13 | 14 | 15 | 16 | 17 | 18 |
|---|---|---|---|---|---|---|---|---|---|---|---|---|---|---|---|---|---|
| HF −83 | | | | | | | | | | | | | | | | | |
| LiF 848 | BeF$_2$ 535 | | | | | | | | | | | BF$_3$ −127 | CF$_4$ −184 | NF$_3$ −206 | OF$_2$ −223 | F$_2$ −219 | |
| NaF 1012 | MgF$_2$ 1263 | | | | | | | | | | | AlF$_3$ (1272) | SiF$_4$ −86 | PF$_5$ −75 | SF$_6$ −64 | ClF$_5$ −113 | |
| KF 857 | CaF$_2$ 1423 | ScF$_3$ 1227 | TiF$_4$ (283) | VF$_5$ 110 | CrF$_4$ 277 | MnF$_4$ dec. | FeF$_3$ 1102 | CoF$_3$ | NiF$_3$ dec. | CuF$_2$ 755 | ZnF$_2$ 927 | GaF$_3$ (800) | GeF$_4$ −37 | AsF$_5$ −63 | SeF$_6$ −40 | BrF$_5$ −61 | KrF$_2$ dec. |
| RbF 775 | SrF$_2$ 1400 | YF$_3$ 1152 | ZrF$_4$ (908) | NbF$_5$ 78 | MoF$_6$ 17 | TcF$_6$ 33 | RuF$_6$ 54 | RhF$_6$ | PdF$_4$ | AgF$_2$ 690 | CdF$_2$ 1100 | InF$_3$ 1150 | SnF$_4$ 705 | SbF$_5$ 7 | TeF$_6$ −35 | IF$_7$ 5 | XeF$_6$ 50 |
| CsF 682 | BaF$_2$ 1290 | LuF$_3$ 1182 | HfF$_4$ | TaF$_5$ 97 | WF$_6$ 2 | ReF$_7$ | OsF$_6$ 32 | IrF$_6$ 44 | PtF$_6$ 57 | AuF$_3$ 727 | HgF$_2$ 645 | TlF$_3$ 550 | PbF$_4$ 500 | BiF$_5$ dec. | | | |
| | | AcF$_3$ | ThF$_4$ 900 | PaF$_5$ (500) | UF$_6$ 64 | NpF$_6$ 53 | PuF$_6$ 50 | | | | | | | | | | |

Sources: Melting points in degrees Celsius taken from the *Handbook of Chemistry and Physics* and M. C. Ball and A. H. Norbury, *Physical Data for Inorganic Chemists*, Longman, London, 1974, pp. 70–91.
Note: Temperatures enclosed in parentheses represent temperatures of sublimation.

314

As we proceed to the left in the *p*- or *d*-blocks, we see that eventually there are not enough fluorines about the central atom to satisfy its desired maximum coordination number; then fluorines are used as bridging groups, sharing their electron pairs with two (or more) different metal ions. This structure can give rise first to oligomers with moderate melting points, then to high-melting and high-boiling two- or three-dimensional polymers. Farther to the left these high-melting three-dimensional polymers grade imperceptibly into the high-melting three-dimensional ionic lattices we first noticed at the left of the table.

The break between low-melting gaseous monomeric fluorides and high-melting polymeric fluorides is quite sharp: Compare $AlF_3$ (m.p. 1272 °C) with its neighbor $SiF_4$ (m.p. $-86$ °C). This is an unusually sharp break that has nothing to do with any abrupt change in covalent versus ionic bonding but results from two factors. One is the accidental near-absence of medium-melting oligomeric structures. The more significant source of the break lies in the fact that covalent fluorides have very weak intermolecular van der Waals forces—the very electronegative fluorine atoms do not readily allow their electrons to become unevenly distributed, thus creating a temporary dipole. On the other side of the break the small fluoride ion gives large lattice energies, hence high melting points.

There are two significant trends that appear among different halides of the same element. If the halides being compared are *discrete covalent molecules*, the melting and boiling points may be expected to follow the trend fluoride $<$ chloride $<$ bromide $<$ iodide. This trend results from the increase in magnitude of the van der Waals forces holding individual molecules together as the number of electrons in the halogen(s) increases. As a consequence, *most monomeric covalent fluorides are gases at room temperature, most such chlorides are liquids, whereas most such iodides are low-melting solids.*

On the other hand, if the halides being compared are *ionic lattice compounds* or *polymeric covalent compounds*, the melting points may be expected to follow the opposite trend, namely, fluoride $>$ chloride $>$ bromide $>$ iodide. For ionic compounds this trend results from the decreasing lattice energies with the larger halide ions; for polymeric covalent compounds this results from decreasing covalent bond energies for the later halogens. Most or all of these compounds should be solids at room temperature.

Thus we expect (and find) the break in melting and boiling points to be much less pronounced among the chlorides ($MgCl_2$ melts at 714 °C, $AlCl_3$ melts at 192 °C under pressure, whereas the liquid $SiCl_4$ melts at $-70$ °C) and is even less pronounced among the iodides ($MgI_2$ 650°, $AlI_3$ 191°, $SiI_4$ 121°). The break also moves somewhat to the left, as larger halogens are less able sterically to bridge different metal atoms. This is illustrated in the halides of aluminum: $AlF_3$ adopts a typical ionic lattice with a coordination number of 6 for Al and 2 for Cl and has a very high melting point of 1545 °K. $AlCl_3$ also adopts an ionic lattice with the same coordination numbers but readily converts to an alternate (gaseous) form above 455 °K (at 1 atm) that is a dimeric covalent molecule $Al_2Cl_6$, as shown in Figure 9.1. Since this structure has much weaker intermolecular forces than ionic $AlCl_3$ but is not much higher in energy, the melting point of $AlCl_3$ is much lower than

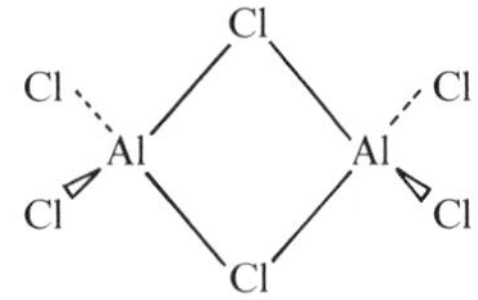

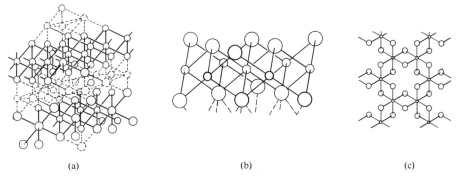

**Figure 9.1** Structure of the gas-phase dimer $Al_2Cl_6$.

**Figure 9.2** Layer structures typical of metal chlorides, bromides, and iodides having some covalent character in their bonding. Smaller circles represent metal atoms (ions); larger circles represent halogens. (a, b) The $CdI_2$ and $CdCl_2$ layer structures (coordination number of Cd = 6; Cl or I = 3): (a) The absence of alternate layers of metal cations; (b) the coordination in each layer. (c) The $CrCl_3$ layer structure (coordination number of Cr = 6; Cl = 2). From *Structural Inorganic Chemistry*, 3d ed., by A. F. Wells. Copyright © 1962 by Oxford University Press, London. Reprinted by permission.

expected for an ionic chloride. Aluminum bromide, with a larger halogen ion, consists of dimeric $Al_2Br_6$ molecules even in the solid state; hence its melting point is lower yet (371 °K). Aluminum iodide is also $Al_2I_6$ and, with its greater van der Waals forces, has a higher melting point than $Al_2Br_6$.

Many other dihalides and trihalides of moderately small, moderately electronegative metals may also vaporize (at least in part) as dimers similar to $Al_2Cl_6$ at temperatures lower than we might have expected. The bonding in such halides is polar covalent; as a result they are able to form both discrete covalent molecules and ionic lattices resembling those of highly ionic compounds. Generally, however, if the electronegativities of the metal and the halogen are within 1.8 to 2.0 units of each other, the lattices show subtle differences from those typical of purely ionic compounds (Figure 3.6). For example, many dichlorides, dibromides, and diiodides, but not difluorides, of the *d*-block metals adopt **layer-structure lattices** (Figure 9.2) instead of the fluorite or rutile lattices characteristic of difluorides. These lattices are not quite logical for purely ionic compounds, since between alternate layers of halide ions there is a complete absence of cations; the alternate layers, if completely ionic, should repel each other. But such structures are reasonable for covalent bonding involving bridging halogen atoms, since the bond angles at each halogen atom in such a layer structure are good for covalent bonding (90°, corresponding to the angle between the different unshared electron pairs in different *p* orbitals of the halogen). Only weak van der Waals attractions hold the layers together; consequently these compounds readily cleave into thin flakes.

## 9.5

## Structural Features of Selected Halides

Due to the presence of unusual oxidation states (e.g., S(II)) among the halides of the nonmetals, we present a survey of the structural features of those halides [3].

GROUP 18/VIIIA  The only reasonably stable halides of the Group 18/VIIIA elements are four fluorides, $KrF_2$, $XeF_2$, $XeF_4$, and $XeF_6$. All are volatile solids. As expected from VSEPR theory (Section 7.3), the first two of these are linear molecules, while $XeF_4$ is a square planar molecule; but the structure of $XeF_6$, expected to be distorted from an octahedron, was controversial for years. Although it is distorted in some manner from an octahedron in the gas phase, in the solid phase it consists mainly of quartets of square pyramidal $XeF_5^+$ cations bridged by $F^-$ anions.

GROUP 17/VIIA  The halogens themselves form halides that are accordingly (and commonly) called the *interhalogen* compounds. All six monohalides (ClF, BrF, IF, BrCl, ICl, IBr) exist and show properties that are in some respects intermediate

**Table 9.4**  Physical States and Structures of Halides of Elements in the Group Oxidation Number

| | | | | |
|---|---|---|---|---|
| $BF_3(g)$ | $CF_4(g)$ | | | |
| $Al^{3+}(F^-)_3(s)$ | $SiF_4(g)$ | $PF_5(g)$ | $SF_6(g)$ | |
| $Ga^{3+}(F^-)_3(s)$ | $GeF_4(g)$ | $AsF_5(g)$ | $SeF_6(g)$ | |
| $In^{3+}(F^-)_3(s)$ | $(SnF_4)_x(s)$ | $(SbF_5)_x(l)$ | $TeF_6(g)$ | $IF_7(g)$ |
| $Tl^{3+}(F^-)_3(s)$ | $(PbF_4)_x(s)$ | $(BiF_5)_x(s)$ | | |
| | | | | |
| $BCl_3(l)$ | $CCl_4(l)$ | | | |
| $Al^{3+}(Cl^-)_3(s)$ | $SiCl_4(l)$ | $PCl_4^+PCl_6^-(s)$ | | |
| $Ga_2Cl_6(s)$ | $GeCl_4(l)$ | | | |
| $In^{3+}(Cl^-)_3(s)$ | $SnCl_4(l)$ | $SbCl_5(l)$ | | |
| $Tl^{3+}(Cl^-)_3(s)$ | $PbCl_4(l)$ | | | |
| | | | | |
| $BBr_3(l)$ | $CBr_4(s)$ | | | |
| $Al_2Br_6(s)$ | $SiBr_4(l)$ | $PBr_4^+Br^-(s)$ | | |
| $Ga_2Br_6(s)$ | $GeBr_4(l)$ | | | |
| $In_2Br_6(s)$ | $SnBr_4(l)$ | | | |
| $Tl_2Br_6(s)$ | | | | |
| | | | | |
| $BI_3(s)$ | $CI_4(s)$ | | | |
| $Al_2I_6(s)$ | $SiI_4(s)$ | $PI_4^+I^-(s)$ | | |
| $Ga_2I_6(s)$ | $GeI_4(s)$ | | | |
| $In_2I_6(s)$ | $SnI_4(s)$ | | | |

NOTE: Formulas are written so as to indicate whether the structures are ionic, polymeric covalent, oligomeric covalent, or monomeric covalent; distinction between the first two of these is sometimes somewhat arbitrary.

**Table 9.5** Physical States and Structures of Halides of Elements in the Group Oxidation Number Minus 2

|  |  |  | $NF_3(g)$ |  |  |  |
|---|---|---|---|---|---|---|
|  |  |  | $PF_3(g)$ | $SF_4(g)$ | $ClF_5(g)$ |  |
|  |  | $(GeF_2)_x(s)$ | $AsF_3(l)$ | $SeF_4(l)$ | $BrF_5(l)$ |  |
|  |  | $(SnF_2)_x(s)$ | $(SbF_3)_x(s)$ | $(TeF_4)_x(s)$ | $IF_5(l)$ | $XeF_5^+F^-(s)$ |
| $Tl^+F^-(s)$ |  | $Pb^{2+}(F^-)_2(s)$ | $Bi^{3+}(F^-)_3(s)$ |  |  |  |
|  |  |  | $NCl_3(l)$ |  |  |  |
|  |  |  | $PCl_3(s)$ | $SCl_3^+Cl^-$ (dec) |  |  |
|  |  | $(GeCl_2)_x(s)$ | $AsCl_3(l)$ | $Se_4Cl_{16}(s)$ |  |  |
| $In^+Cl^-(s)$ |  | $(SnCl_2)_x(s)$ | $(SbCl_3)_x(s)$ | $Te_4Cl_{16}(s)$ |  |  |
| $Tl^+Cl^-(s)$ |  | $Pb^{2+}(Cl^-)_2(s)$ | $(BiCl_3)_x(s)$ | $PoCl_4^*(s)$ |  |  |
|  |  |  | $PBr_3(l)$ |  |  |  |
|  |  | $(GeBr_2)_x(s)$ | $AsBr_3(s)$ | $Se_4Br_{16}(s)$ |  |  |
| $In^+Br^-(s)$ |  | $(SnBr_2)_x(s)$ | $(SbBr_3)_x(s)$ | $Te_4Br_{16}(s)$ |  |  |
| $Tl^+Br^-(s)$ |  | $Pb^{2+}(Br^-)_2(s)$ | $(BiBr_3)_x(s)$ | $PoBr_4^*(s)$ |  |  |
|  |  |  | $PI_3(s)$ |  |  |  |
|  |  | $(GeI_2)_x(s)$ | $(AsI_3)_x(s)$ |  |  |  |
| $In^+I^-(s)$ |  | $(SnI_2)_x(s)$ | $(SbI_3)_x(s)$ | $Te_4I_{16}(s)$ |  |  |
| $Tl^+I^-(s)$ |  | $Pb^{2+}(I^-)_2(s)$ | $(BiI_3)_x(s)$ | $PoI_4^*(s)$ |  |  |

NOTE: Formulas are written so as to indicate whether the structures are ionic, polymeric covalent, oligomeric covalent, or monomeric covalent; distinction between the first two of these is sometimes somewhat arbitrary. (*) means the structure is unknown, (s) = solid at room temperature, (l) = liquid, (g) = gas, and (dec) = decomposed at room temperature.

between those of the pure halogens but that can also be unique. Among higher halides, all six tri- and pentafluorides ($ClF_3$, $ClF_5$, $BrF_3$, $BrF_5$, $IF_3$, $IF_5$) exist; the trifluorides have the expected T-shaped structure, while the pentafluorides have the expected square pyramidal structures. In addition to these there is $ICl_3$, which is actually the dimer $I_2Cl_6$ containing bridging chlorines as in $Al_2Cl_6$ (Figure 9.1) but which is planar at the iodine atoms (rather than tetrahedral at the aluminum atoms). One heptafluoride, $IF_7$, is known, which has a pentagonal bipyramidal structure. Among these formulas we note two expected trends: The very electronegative fluorine serves best as outer atoms in the higher halides; the less electronegative iodine atom can act as the central atom in more kinds of higher halides than the others.

GROUP 16/VIA   The hexafluorides of S, Se, and Te (and their greatly varying tendencies to hydrolyze) have already been mentioned in Section 2.4; there are also mixed hexahalides such as $SF_5Cl$. The same elements also form tetrafluorides that are, as expected, seesaw-shaped molecules (Figure 7.6), although the larger tellurium atom expands its coordination number in solid $TeF_4$, with fluorine bridges giving rise to a one-dimensional polymer (Figure 9.3a). Like sulfur, selenium forms no stable binary iodide, and sulfur does not form a tetrabromide, but otherwise all the tetrachlorides, tetrabromides, and tetraiodides of these elements (and polonium)

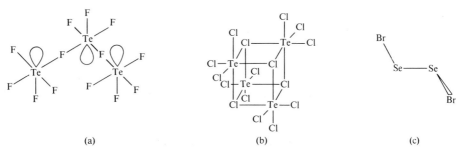

**Figure 9.3** (a) Part of the chain polymeric structure of solid $TeF_4$. (b) The "cubane" structure of $Te_4Cl_{16}$ and related compounds. (c) The geometry of $Se_2Br_2$ and related compounds.

exist. But their structures differ from that of the tetrafluorides: $SCl_4$ (unstable above $-30\,^{\circ}C$) seems to exist as $SCl_3^+$ and $Cl^-$ ions, while the larger Group 16/VIA atoms increase their coordination numbers to 6 by forming $M_4X_{16}$ with the "cubane" structure illustrated in Figure 9.3b. The octahedral coordination about the Group 16/VIA atom is distorted by the unshared electron pair on the central atoms, however. Dihalides are less abundant: $OF_2$, $SF_2$, $SCl_2$, and some polonium dihalides, $PoCl_2$ and $PoBr_2$.

In Section 6.7 we noted that these elements (especially sulfur) have strong tendencies to catenate. Accordingly there are many halides containing chains of Group 16/VIA atoms. Those of oxygen, $O_2F_2$ and $O_4F_2$, are expectedly rather unstable. Related to the hexahalides are the compounds $S_2F_{10}$ and $Te_2F_{10}$. In lower oxidation states we find $S_2F_2$, $S_2Cl_2$, $S_2Br_2$, $Se_2Cl_2$, and $Se_2Br_2$, with the geometry shown in Figure 9.3c. $S_2F_2$ also has an isomer, S—$SF_2$. The catenating tendency of sulfur is best shown in the dihalosulfanes, $S_nX_2$, which retain the chain structure found in one form of sulfur, terminated by halogen atoms. These compounds are well characterized for X = Cl and Br and $n = 2$ to 8, but dichlorosulfanes up to about $S_{100}Cl_2$ apparently exist in mixtures. Tellurium also forms polymeric halides with halogen atoms also attached along the chain.

GROUP 15/VA    Some interesting trends are found in the pentahalides of the Group 15/VA elements. Not unexpectedly for steric reasons, nitrogen does not form pentahalides (but does form the ion $NF_4^+$); the remainder of the elements do form pentafluorides. The only stable chlorides are $PCl_5$ and $SbCl_5$; $AsCl_5$ decomposes above $-50\,^{\circ}C$. Only phosphorus forms a pentabromide and pentaiodide. These trends are in accord with the trend in stability of the $+5$ oxidation states of these elements, as we saw previously in the oxo anions (Section 5.1). Structurally these halides are diverse. $PF_5$ is a trigonal bipyramidal monomer (in which axial and equatorial fluorines interchange positions rapidly). $PCl_5$ has a similar structure as a gas or liquid but as a solid is ionic, $[PCl_4]^+[PCl_6]^-$. With the larger bromide and iodide ions a different ionic structure is favored: $[PX_4]^+X^-$ (X = Br, I). $SbCl_5$ is also a trigonal bipyramidal molecule, but with the smaller fluorine atom, antimony achieves 6-coordination, forming a one-dimensional polymer with

319

bridging fluorines not unlike that in $TeF_4$ (Figure 9.3). $SbF_5$ is nonetheless a liquid at room temperature, but its polymeric nature causes it to be an extremely thick, viscous liquid.

Nearly all the trihalides of these elements are well known; $NI_3$ is unknown (although $NI_3 \cdot NH_3$ is a famous explosive), $NBr_3$ explodes above $-100\,°C$, and $NCl_3$ is explosive unless diluted with other gases. $NF_3$, the phosphorus trihalides, $AsF_3$, $AsCl_3$, and $AsBr_3$ are all volatile monomeric compounds having pyramidal structures as expected from VSEPR. $BiF_3$ is a typically ionic compound in which the large $Bi^{3+}$ ion has a coordination number of 9. The solid intermediate trihalides contain recognizable pyramidal $MX_3$ molecules, but the large central atom can accommodate a total coordination number larger than 4 (one unshared $p$-electron pair plus three halogens). The repulsion of the unshared electron pair seemingly influences the structures of these solids: They adopt higher coordination numbers utilizing bridging halogen atoms, although the bonds to the bridging halogen atoms are much longer and weaker than the bonds to the atom's "own" halogens. This type of bonding is known as *secondary bonding* [4] and is also found in the lower halides of Se, Te, and in the halogens and interhalogens.

Catenated halides are not so abundant in Group 15/VA as in 16/VIA. Nitrogen forms $N_2F_4$, a gas that, like $N_2O_4$, readily dissociates above room temperature to give colored molecules containing unpaired electrons ($NF_2$, dark blue). There is also $FN{=}NF$, in both cis and trans isomers. Phosphorus forms $P_2X_4$, which is most stable for $X = I$; arsenic forms $As_2I_4$.

GROUP 14/IVA    All the tetrahalides of the Group 14/IVA elements exist except $PbI_4$, although $PbCl_4$ and $PbBr_4$ are not stable at room temperature. All are monomeric tetrahedral molecules except $SnF_4$ and $PbF_4$, which have two-dimensional polymeric structures featuring bridging by half of the fluorine atoms. Structurally more interesting are the lower halides $GeX_2$, $SnX_2$, and $PbX_2$, in which, in order to provide the coordination numbers best for the central atoms, more extensive bridging occurs. Spiral, chain, or layer polymers shaped prominently by the unshared pairs of electrons result for the lighter dihalides. Higher coordination numbers and less influence of the unshared pair of electrons are found in the structures of $SnI_2$ and $PbX_2$.

GROUP 13/IIIA    The boron trihalides are volatile monomeric molecular compounds. The simple Lewis structure of these species gives less than an octet of electrons to boron unless pi bonding is included; the observed short B—X bond distances indicate that there is some degree of pi bonding in these halides. We have already referred (Section 9.4) to the variety of structure types in the aluminum halides; this variety continues to show up in the trihalides of the heavier Group 13/IIIA metals. Although the compound $TlI_3$ exists, it consists (in the solid state) of $Tl^+$ and $I_3^-$ (triiodide) ions.

There are many catenated halides of boron, including $B_2X_4$, for all halogen atoms. Heating the bromide and chloride produces a series of $B_nX_n$ ($n = 4$, 7 to 12) compounds with polyhedral clusters of boron atoms similar to those found in the boron hydrides (Chapter 10). Monohalides are known for Ga, In, and Tl; these are more stable than the trihalides for Tl. The monohalides tend to have ionic-type crystal structures (e.g., NaCl) but with some distortions. Compounds of stoichiometry $MX_2$ ($M = Ga, In, Tl$) are actually $M^+[MX_4]^-$, although the anion $[Cl_3Ga\!-\!GaCl_3]^{2-}$ is catenated.

METAL HALIDES   Little else will be said about the vast number of metal halides; much can be deduced from the principles given earlier in this chapter. Of some interest are structures that result with metal ions that take low coordination numbers. $BeF_2$ adopts several crystal structures that are analogous to those formed by $SiO_2$ (Chapter 4). $BeCl_2$ also has 4-coordinate beryllium and 2-coordinate halide, but the bridging is such as to produce a linear polymer. Although $HgF_2$ has the ionic $TiO_2$ structure (coordination number of $Hg = 8$), $HgCl_2$ has an unusual structure in which only two chlorines are within the (probable) sum of van der Waals radii of Hg and Cl, so it is essentially a molecular substance, with an apparent coordination number of only 2 for Hg! Secondary bonding is more evident in $HgBr_2$, which has two close but four more distant bromines around each Hg. The bridging in the stable (red) form of $HgI_2$ is normal; each Hg has a coordination number of 4 to produce a linear polymer like that in $BeCl_2$. There are also unusual structures for some of the chlorides, bromides, and iodides of some of the other soft-acid metal ions, in which the very large metal ions have surprisingly low coordination numbers.

## 9.6

## Reactions and Uses of Halides

We saw in Chapter 8 that polar covalent bonds tend to be reactive: They readily undergo acid-base reactions that produce ionic salts and products containing nonpolar covalent bonds. Many of the halides of the elements contain polar covalent bonds and are thus quite reactive in acid-base and Lewis acid-base reactions. Thus they are frequently used as starting materials in syntheses of other compounds. Their physical properties are also sometimes of direct use, and they are sometimes used as oxidizing or reducing agents. In this section we summarize some of the important types of reactions of the halides of the elements.

PHYSICAL PROPERTIES   The higher halides of some of the elements are often useful because they have a volatility that most derivatives of those elements may lack. Thus they can readily be purified in chemical plants by fractional distillation, then converted to other, less volatile forms that would be difficult to purify directly.

One example of this type of purification is the use of $UF_6$ in the preparation of uranium metal (or oxide) that is enriched in the fissionable isotope $^{235}U$. $UF_6$ is one of the very few volatile compounds of uranium. Advantage is taken of this property in reprocessing spent nuclear fuel, which contains neptunium, plutonium, and many fission products. This fuel is treated with $ClF_3$ or $BrF_3$ to oxidize the uranium to $UF_6$, which can be swept away from the nonvolatile lower fluorides of most other metals. Enrichment of the isotopes takes advantage of the slightly lower molecular weight of $^{235}UF_6$ as compared with $^{238}UF_6$; the much scarcer but lighter molecules of the former move slightly more rapidly than do the latter (recall Graham's law from general chemistry). $UF_6$ gas is pumped through thousands of porous membranes until the lighter $^{235}UF_6$ molecules are enriched enough for the uranium subsequently produced from it to be fissionable. The process is very expensive and has a very high energy demand, since this process must be carried out so many times on a very reactive gas.

Other examples include the conversion of silicon compounds to $SiCl_4$, which can be purified and freed from boron, arsenic, etc., before being reduced to the high-purity elemental Si required for making semiconductors. Another example comes in the Kroll process for producing titanium metal from the ore ilmenite, $FeTiO_3$. Heating with carbon and chlorine converts both metals to their chlorides, but the molecular $TiCl_4$ can be fractionally distilled away from the less volatile polymeric $FeCl_3$. The halide is then reduced to the metal with magnesium. A useful procedure for subsequent purification of such metals (for example, zirconium) is the van Arkel–de Boer process, in which the impure metal is heated in a vacuum with a little iodine vapor to a temperature at which the iodide of the desired metal vaporizes. The vapor then passes to an electrically heated wire. At the high temperature of the wire the relatively weak metal-iodine bonds are broken and pure metal is deposited; the iodine diffuses back to continue the process.

USE OF HALIDES AS OXIDIZING AGENTS  As we mentioned earlier in the section on the synthesis of halides, many higher halides are good oxidizing agents. They are thus useful not only in making other inorganic halides but also in preparing other inorganic and organic substances by oxidation. For example, $XeF_2$ can be used to oxidize bromate to perbromate:

$$BrO_3^- + XeF_2 + H_2O \rightarrow BrO_4^- + Xe + 2\,HF \qquad (9.18)$$

$XeF_2$ is a useful oxidizing agent in part because the byproducts are volatile and thus easily removed.

We have already mentioned the use of $ClF_3$ and $BrF_3$ as oxidizing agents in the purification of uranium from its fission products:

$$U(s) + 3\,ClF_3(g) \rightarrow UF_6(g) + 3\,ClF(g) \qquad (9.19)$$

They can also be used to analyze many compounds for their percent oxygen content. (This is probably the most difficult element to determine analytically.)

$$3\,SiO_2 + 4\,BrF_3 \rightarrow 3\,SiF_4 + 2\,Br_2 + 3\,O_2 \qquad (9.20)$$

$ClF_3$ is such a powerful oxidizing agent that even asbestos, wood, and other building materials catch fire in it.

$PtF_6$ is such a powerful oxidizing agent that it oxidizes oxygen itself:

$$O_2 + PtF_6 \rightarrow O_2^+[PtF_6]^- \qquad (9.21)$$

Upon discovering this reaction, the chemist Bartlett realized that an oxidizing agent that could oxidize $O_2$ could probably (from consideration of their similar ionization energies) oxidize xenon. $PtF_6$ did, and the myth of the complete chemical unreactivity of the Group 18/VIIIA ("inert") gases died. Possibly related to its oxidizing properties is the large-scale use of the (safely diluted) gas $NCl_3$ as a bleach and sterilant for flour.

USE OF HALIDES AS REDUCING AGENTS    Halides (especially iodides) of elements in low oxidation states can be used as reducing agents, although this is probably a less extensive use. $SnCl_2$ is widely used as a mild reducing agent; $PI_3$ is useful for removing oxo groups from some kinds of organic compounds (e.g., converting $R_2S{=}O$ to $R_2S$). Protection from the oxygen of the air is thus advisable in handling some of the halides, especially the iodides.

LEWIS ACID-BASE PROPERTIES OF THE HALIDES    The Lewis acid-base properties are probably the most important ones of the halides, being based on their polar covalent element-halogen bonds. We will divide these into four subcategories here and amplify some of the reaction types further in the remainder of this chapter and in the next chapter.

HYDROLYSIS OF THE HALIDES    Hydrolysis reactions, which result from the Lewis acidity of cations, have already been covered in detail earlier in this book. They, of course, usually give the hydrogen halide and the oxide or hydroxide of the element. As we also mentioned earlier, partial hydrolysis can produce oxo cations such as $UO_2{}^{2+}$ or their nonmetal analogues such as $POCl_3$, which we will discuss in the next section. In the case of $NCl_3$, in which the nitrogen is more electronegative than the chlorine, the $H_2O$ is attracted to the *chlorine* as a center of positive charge and produces $ClOH$ (hypochlorous acid) and $NH_3$. The prevalence of hydrolysis reactions among so many halides means that these compounds must be handled in such a way as to prevent their contact with not only liquid water but the water vapor in the air. Techniques used to handle compounds (such as reactive halides) with rigorous exclusion of air involve the use of the *glove bag*, the *dry box*, *Schlenk glassware*, and the *vacuum line* for volatile compounds.

SOLVOLYSIS REACTIONS OF HALIDES    Solvolysis reactions give rise to some important categories of inorganic compounds, to which we will devote the rest of this chapter. These are reactions that are similar to hydrolysis but involve protic solvents other than water that (like water) can eliminate the hydrogen halides. For example, an alcohol such as $CH_3OH$ may be substituted for water:

$$POCl_3 + 3\,CH_3OH \rightarrow 3\,HCl + PO(OCH_3)_3 \tag{9.22}$$

The resulting products are known as *inorganic esters* (e.g., trimethyl phosphate) or as *alkoxides* of the elements. By analogy with the reaction with water, we may suppose that the Lewis base $CH_3OH$ first is coordinated to the very acidic $P^{5+}$ center. The acidity of the $CH_3OH$ is then strongly enhanced, and its proton may then be lost (perhaps first to solvent methanol molecules but ultimately to a chloride group to give HCl).

Similar solvolysis reactions involve organic amines as Lewis bases:

$$POCl_3 + 3(CH_3)_2NH \rightarrow 3\,HCl + PO[N(CH_3)_2]_3 \tag{9.23}$$

The resulting products are known as *amides* of the elements. Use of an organic amine with one fewer R group, or of ammonia itself (or ammonium chloride) allows the loss of a second or third acidic proton from the coordinated amine or ammonia. This then permits the formation of a double or triple bond from the acidic center to nitrogen or allows the nitrogen to bridge different atoms, or both, as in this example:

$$3\,PCl_5 + NH_3 \rightarrow 3\,HCl + [Cl_3P{=}N{=}PCl_3]^+[PCl_6]^- \tag{9.24}$$

LEWIS ACID PROPERTIES OF HALIDES    As we saw in Chapter 7, the presence of water or of a protic solvent such as an alcohol or an amine is not necessary for the manifestation of acidic reactions with cations or their compounds (such as halides); these properties inhere in the cation as an electron-pair acceptor. We saw earlier in this chapter that if the number of halogen atoms around a central atom is not sufficient to achieve the expected total coordination number for the central atom, some halogens may donate additional electron pairs to neighboring metal atoms and thus become bridging halogens. But a halide ion that is already sharing one of its electron pairs is a relatively poor donor of additional electron pairs. Thus such halides still can function as Lewis acids and often react with stronger Lewis bases to give adducts or coordination compounds.

One such ligand is the free halide ion; in this case the product is a *halo anion* such as we discussed in Section 7.4. Halides that have already reached their maximum total coordination number in a monomeric halide molecule ($CCl_4$, $NF_3$, $NCl_3$, $SF_6$) have no ability to act as Lewis acids, but most halides of weakly acidic cations through very strongly acidic cations can act as Lewis acids and form halo anions. The relative stability of these tend to follow the HSAB Principle; thus hard acids such as $Be^{2+}$, $B^{3+}$, $Al^{3+}$, $Si^{4+}$, and $P^{5+}$ tend to form very stable fluoro anions such as $BeF_4{}^{2-}$, $BF_4{}^-$, $AlF_6{}^{3-}$, $SiF_6{}^{2-}$, and $PF_6{}^-$. These may be produced even during hydrolysis reactions:

$$3\,SiF_4 + 2\,H_2O \rightarrow SiO_2 + 4\,H^+ + 2\,SiF_6{}^{2-} \tag{9.25}$$

Softer acids, of course, more frequently are observed to give chloro, bromo, or iodo anions.

Some strongly Lewis-acidic halides such as $BF_3$ and $AlCl_3$ are widely used for their ability to remove halide ions from reactants. The organic Friedel-Crafts reaction involves reactions of halides ($CH_3COCl$, $CH_3Cl$, $Cl_2$, and so on) with benzene ($C_6H_6$) to give substituted benzenes ($CH_3COC_6H_5$, $CH_3C_6H_5$, $ClC_6H_5$) and HCl. In Figure 9.4 we sketch a reasonable mechanism by which such a reaction may occur. (a) We may suppose that the strong Lewis acid first coordinates to the halogen of the organic halide, forming, for example, $[CH_3Cl:AlCl_3]$. (b) This then

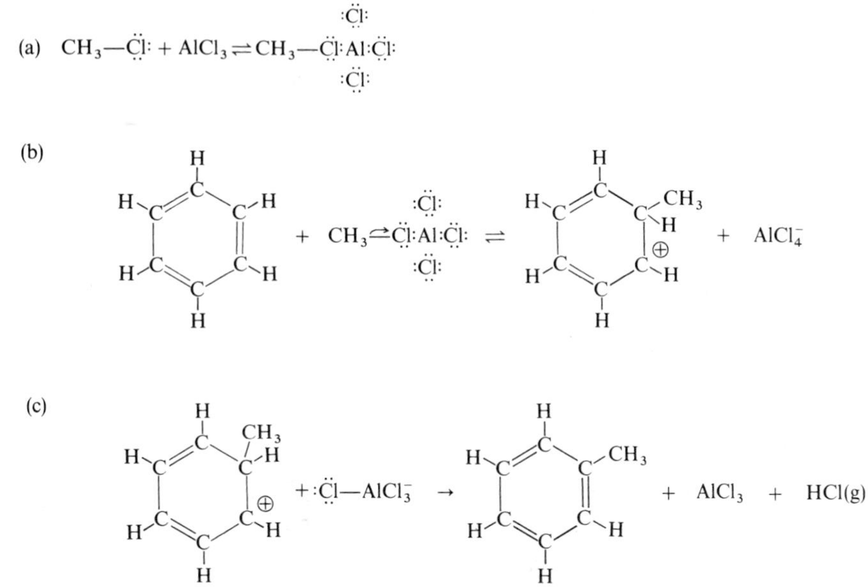

(a) $CH_3-\ddot{\underset{..}{Cl}}: + AlCl_3 \rightleftharpoons CH_3-\ddot{\underset{..}{Cl}}:Al:\ddot{\underset{..}{Cl}}:$ (with $:\ddot{Cl}:$ above and $:\ddot{Cl}:$ below Al)

(b) benzene $+\ CH_3 \rightleftharpoons \ddot{\underset{..}{Cl}}:Al:\ddot{\underset{..}{Cl}}: \rightleftharpoons$ [methylbenzenium cation] $+\ AlCl_4^-$

(c) [methylbenzenium cation] $+ :\ddot{\underset{..}{Cl}}-AlCl_3^- \rightarrow$ toluene $+\ AlCl_3\ +\ HCl(g)$

**Figure 9.4**  Possible mechanism for the Friedel-Crafts reaction of $CH_3Cl$ with benzene, catalyzed by a Lewis acid, to give $CH_3C_6H_5$ and $HCl(g)$.

may enhance the acidity of the carbon atom, allowing it to attract the pi-electron pair of the benzene ring, giving an organic cation such as $[CH_3C_6H_6]^+$ and the $AlCl_4^-$ anion. (c) Finally, the acidic organic cation transfers its proton to a chlorine of the $AlCl_4^-$ ion, giving back the catalytic halide ($AlCl_3$) and producing HCl.

We may also look at the halogenation reactions in Section 9.3, in which different halide ions are exchanged, as examples of Lewis-acid reactions of halides of the elements. Exchange of oxygen for halogen is also of this type:

$$(CH_3)_2C=O + SF_4 \rightarrow [(CH_3)_2C=O:SF_4] \rightarrow (CH_3)_2CF_2 + SOF_2 \qquad (9.26)$$

Soft-acid metal ions are often used to carry out the reverse exchange on organic halides as a method of dehalogenation:

$$CH_3I + Ag^+ \rightleftharpoons [CH_3I:Ag]^+ + OH^- \rightarrow CH_3OH + AgI(s) \qquad (9.27)$$

Another type of Lewis acid-base exchange reaction involves the swapping of organic anions or hydride ions for halide ions:

$$CH_3Li + HgCl_2 \rightarrow CH_3HgCl + LiCl \qquad (9.28)$$

These organometallic compounds and hydrides of the elements often feature low oxidation numbers for the elements and often can be made by redox reactions, so they will be considered in the next chapter.

Even a simple property such as *solubility* of a metal halide in an organic solvent is really a chemical (Lewis acid-base) reaction. Highly ionic compounds such as the fluorides of the less acidic cations are not usually soluble in organic solvents, since the lattice energies are high in such compounds. Chlorides, bromides, and iodides of the metals tend to be progressively more soluble in organic solvents, partly because their lattice energies become smaller. To be soluble at all, there must be a solvation energy (analogous to the hydration energy in the thermochemical cycle of Figure 3.4) sufficient to overcome the lattice energy. The organic solvents that can provide such a solvation energy are normally Lewis bases that form coordinate covalent bonds to the metal ions. Such solvents include alcohols, ketones, and even (for soft-acid metal ions) pi-electron donors such as benzene. Thus, for example, $NaI$ is soluble in acetone, while $NaCl$ is not; so $NaI$ can be used to exchange I for Cl in organic chlorides.

LEWIS BASE REACTIONS OF THE HALIDES  Metal halides in which the cation is nonacidic or feebly acidic do not show any tendency to form halo anions; instead their acid-base chemistry is dominated by the halide ion itself, a Lewis base; hence, for example, $CsF$ is useful in forming fluoro anions from Lewis-acid halides, since the resulting large fluoro anion is stabilized in a lattice by the large $Cs^+$ cation:

$$CsF(s) + SF_4(g) \rightleftharpoons Cs^+[SF_5]^-(s) \tag{9.29}$$

Some nonmetal halides have modest abilities to act as Lewis bases. Those that have excessive numbers of halogens (e.g., $XeF_6$) may donate one to a good Lewis acid to form a salt:

$$XeF_6 + AsF_5 \rightleftharpoons [XeF_5]^+[AsF_6]^- \tag{9.30}$$

A few nonmetal halides in lower oxidation states may be able to act as ligands by donating their unshared pairs of electrons, although the high electronegativity of all the halogens tends to discourage such donation. $PF_3$ and $PCl_3$ can act as soft bases, donating the unshared electron pairs on phosphorus. The complexes of $PF_3$ resemble those of CO (Chapter 10); $PF_3$ is similarly toxic. The chloro anion $SnCl_3^-$ is a significant soft base that readily forms complexes with the soft-acid metal ions that feature donation of the unshared electron pair on tin:

$$[PtCl_4]^{2-} + Cl^- + 5\,SnCl_2 \rightarrow [Pt(SnCl_3)_5]^{3-} \tag{9.31}$$

## 9.7

## Solvolysis with Oxygen Solvents: Inorganic Esters, Alkoxides, Silicones

A number of important compounds of the nonmetals can be thought of as solvolysis products of the halides of the elements. Partial solvolysis of halides with water (hydrolysis) produces some important oxo halides and halo oxo acids that we will discuss here, even though partial hydrolysis is often difficult to control and is not the most practical way to produce these compounds. Some products of partial solvolyses with alcohols will also be considered.

Phosgene, $COCl_2$, was mentioned in Chapter 2 as a very toxic high-temperature hydrolysis product of $CCl_4$, but it is made industrially by oxidizing CO with $Cl_2$ in the presence of activated carbon. It is used in a solvolysis reaction to be discussed later in this chapter. Combining NO with fluorine, chlorine, or bromine gives the *nitrosyl halides*, NOX, which are reactive gases; the two *nitryl halides* $NO_2F$ and $NO_2Cl$ also exist. The *phosphoryl halides*, $POX_3$, are gases or volatile liquids having tetrahedral structures with short P—O bonds, suggesting partial double-bond character; they are mostly made by the action of $O_2$ on $PX_3$. They are also used for their solvolysis reactions.

Sulfur forms the *thionyl halides* $SOF_2$, $SOCl_2$, and $SOBr_2$ and the *sulfuryl halides* $SO_2F_2$ and $SO_2Cl_2$. Thionyl chloride is made by the reaction of $SCl_2$ and $SO_3$:

$$SO_3 + SCl_2 \rightarrow SO_2 + O{=}SCl_2 \tag{9.32}$$

We have already mentioned the use of thionyl chloride as a dehydrating agent. Sulfuryl chloride is made by the combination of $SO_2$ and $Cl_2$. In concept, further partial hydrolysis of the sulfuryl halides leads to the important acids *chlorosulfuric acid*, $ClSO_2(OH)$, and *fluorosulfuric acid*, $FSO_2(OH)$. In practice these acids are made by the direct reaction of $SO_3$ with anhydrous HCl or HF. The former is used in organic syntheses and the latter is a component of "superacids" (Section 7.8).

SILICONE POLYMERS  Most of the familiar polymers in our lives are organic molecules, but polymer chemists leave no stones unturned in their search for new materials having desired but difficult-to-obtain properties, so more and more research is going into the syntheses and characterization of inorganic polymers [5]. The earliest polymers that were (substantially) inorganic were the **silicones**, obtained by hydrolysis of the (organometallic) silicon chlorides such as $R_2SiCl_2$. This reaction produces relatively unstable hydroxides that readily lose water to give cyclosiloxanes, $[R_2SiO]_n$ ($n = 3$ to 6), which have rings with a backbone of —Si—O—Si—O— linkages. Ring opening and polymerization can be induced by KOH or $H_2SO_4$ to produce polymers of the same simplest composition but with linear backbones and values of $n$ on the order of 100,000. Chemical cross-linking of the chains every 100 to 1000 silicon atoms produces silicone elastomers or rubbers that are superior to other rubbers in their inertness, elasticity, strength, and flexibility over a wide temperature range. Inclusion of some of the siloxane $[R_3Si]_2O$ in the polymerization mixture gives rise to shorter chains, $R_3Si$—O—$[R_2Si$—O—$]_nSiR_3$, the silicone oils. Inclusion of both $R_2SiCl_2$ and $RSiCl_3$ in the original hydrolysis reaction gives rise to polymers still having —OH groups attached to the Si atoms in the polymer. Heat treatment in the presence of a catalyst eliminates water between neighboring chains, cross-linking them to give three-dimensional polymeric silicone resins useful in electronics and in high-temperature paints.

SOLVOLYSIS WITH ALCOHOLS  The alkoxide ion, $RO^-$, mentioned in Section 5.5, can easily replace halide ions to form new classes of derivatives of the elements, the **inorganic esters** and the **metal alkoxides**. Alcohols are Lewis bases that presumably coordinate to strongly Lewis acidic chlorides via their oxygen atoms. Then

327

both the C—O and H—O bonds are polarized toward the Lewis acid. In some cases both the C—O and the H—O bonds are broken, and the organic group is transferred to the halogen:

$$PCl_5 + ROH \rightarrow RCl + HCl + POCl_3 \qquad (9.33)$$

Usually only the H—O bond is broken, and (especially in the presence of a basic amine to remove the proton from the coordinated alcohol) an inorganic ester is produced:

$$PCl_3 + 3\,ROH + 3\,R_3N \rightarrow P(OR)_3 + 3\,[R_3NH]^+Cl^- \qquad (9.34)$$

This product is known as a trialkyl phosphite and is a useful soft base due to the unshared pair of electrons on the phosphorus. The corresponding trialkyl phosphates can be made by related routes:

$$6\,(C_2H_5)_2O + P_4O_{10} \rightarrow 4\,(C_2H_5O)_3PO \qquad (9.35)$$

Tributyl phosphate is an important solvent for separating lanthanides. These phosphates, and especially the corresponding sulfates such as dimethyl sulfate, $(CH_3)_2SO_4$, still have polarizable C—O bonds and hence are powerful *alkylating agents*. As such compounds can alkylate DNA, they may be powerful mutagens and/or carcinogens and can be quite toxic.

A series of organophosphates (and thiophosphates) is very important in modern agriculture as the *organophosphate* pesticides, which interfere with nerve transmission in insects (and mammals) by inhibiting cholinesterase. An especially toxic ester is tetraethyl pyrophosphate, $(C_2H_5O)_2PO$—O—$PO(OC_2H_5)_2$. Some common pesticides contain a thio (S$=$) group in place of the oxo group; e.g., parathion, S$=$P$(OC_2H_5)_2(OC_6H_4NO_2)$. Also related are some chemical warfare agents, the nerve gases, such as Sarin or GB, $CH_3PF(=O)(OCH(CH_3)_2)$, the lethal dose of which may be 1 mg. Although very toxic, these pesticides have the advantage that they undergo further hydrolysis to the starting alcohols and phosphoric acid, so they do not persist in the environment. Of course, not all organophosphates are toxic: DNA, RNA, ATP, and ADP are all organophosphates.

Other relatively common inorganic esters include nitroglycerin (the trinitrate of the organic alcohol glycerin); amyl nitrite, $C_5H_{11}ONO$ (an antidote for cyanide poisoning); organic silicates, $(RO)_4Si$, and "titanates," $(RO)_4Ti$, which are useful for their slow hydrolysis to $SiO_2$ and $TiO_2$ (the latter in heat-resistant paints). The surfaces of glass and inorganic silicates also contain —OH groups that can react with a silicon chloride (such as $(CH_3)_3SiCl$) to cover the surface with water-repellent $(CH_3)_3Si$ groups linked via oxygen to silicon. An aluminum alkoxide is a useful reducing agent in organic chemistry; sodium and potassium alkoxides are useful very strong bases in nonaqueous solvents. Alkoxides of active metals such as Al, Na, and K cannot be made by reaction (9.34) but instead are produced by the redox reaction of the metal with the alcohol:

$$2\,Na + C_2H_5OH \rightarrow 2\,C_2H_5O^-Na^+ + H_2 \qquad (9.36)$$

This reaction is analogous to that of sodium with water in the activity series of metals (Chapter 5) but is somewhat less violent.

Structural features of many of these alkoxides and esters are that (excepting the Na and K) they are volatile covalent molecules that are readily soluble in nonpolar solvents. Like the halide ions, the alkoxide ions have extra unshared pairs of electrons that are often used to bridge different metal ions. Often oligomers rather than polymers result, which helps maintain volatility and solubility.

## 9.8

## Nitrogen Compounds of the Elements

Ammonia, $NH_3$, has three hydrogen atoms that can undergo solvolysis with metal halides to produce three general types of derivatives, the amides ($MNH_2$), the imides ($M_2NH$), and the nitrides ($M_3N$). In many of the amides or imides to be discussed, the remaining hydrogen atoms have first been replaced with organic groups, which can help facilitate the synthesis by preventing further solvolysis.

AMIDES    As an example of a metal amide we mention "sodamide," $Na^+NH_2^-$, which is prepared by the reaction of sodium metal with liquid ammonia in the presence of a catalytic amount of $Fe^{3+}$. (Recall that without the catalyst, this reaction gives gold or deep blue solutions containing $Na^-$ ions or solvated electrons, Section 7.8.) This compound is used in synthesis and as a very strong base. At the other extreme is the reaction of the strongly acidic nonmetal oxide $SO_3$ with $NH_3$ to give *sulfamic acid*, commonly written as $NH_2SO_2OH$ but actually having a solid-state structure closer to $H_3NSO_3$. This is a useful strong acid in analytical chemistry since it is a readily purified nondeliquescent solid that can be weighed out accurately, in contrast to strong oxo acids, which are deliquescent liquids.

Two substituted amides of phosphorus are of importance: *hexamethylphosphoramide*, $[(CH_3)_2N]_3P=O$, and tris(dimethylamino)phosphine, $[(CH_3)_2N]_3P$. These are made from dimethylamine and $POCl_3$ or $PCl_3$. The former is a useful aprotic solvent that is a good Lewis base; the latter is used as a ligand and in syntheses.

IMIDES    Two steps of solvolysis of ammonia or of a secondary organic amine, $RNH_2$, can give rise to an element-nitrogen double bond if this is favorable or, since it usually is not, to nitrogen atoms bridging between pairs of atoms of the element to give ring oligomers or chain polymers. One well-known example is the ring compound *borazine* (borazole), $B_3H_3N_3H_3$ (Figure 9.5a). Borazine is made by the reaction of $BCl_3$ with $NH_4Cl$:

$$3\,BCl_3 + 3\,NH_4Cl + heat \rightarrow 9\,HCl + (ClB{-}NH)_3 \qquad (9.37)$$

The intermediate, B-trichloroborazine, is then reduced with $NaBH_4$ to replace the chlorines with hydrogen atoms. Analogues with organic groups on N can be made

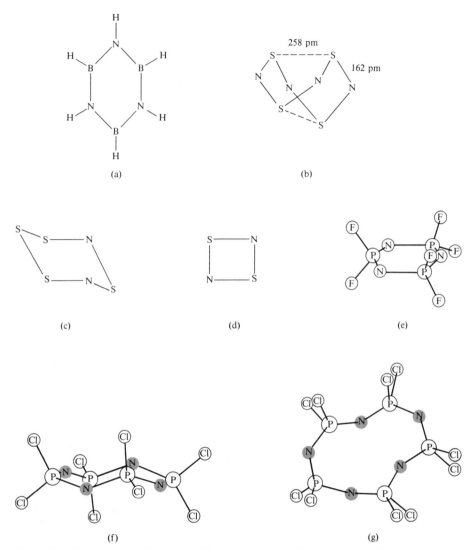

**Figure 9.5** The structure of (a) borazine; (b) tetrasulfur tetranitride; (c) tetrasulfur dinitride; (d) disulfur dinitride; (e) the trimer of $NPF_2$; (f) the tetramer of $NPCl_2$; (g) the pentamer of $NPCl_2$. All B—N, S—N, and P—N bonds have some double-bond character for which various resonance structures can be drawn.

by starting with, for example, $CH_3NH_2$ and base instead of ammonium chloride. Borazine has been called "inorganic benzene," because its physical properties are very similar to those of $C_6H_6$. Although a B and a N atom are isoelectronic to two C atoms, the (partial) B=N double bond in borazole is not equally shared as in benzene but is clearly polarized toward the more electronegative nitrogen atom, resulting in much greater chemical reactivity for borazine than for benzene.

A similar industrially important reaction is that of phosgene with secondary organic amines. After attachment of the lone pair of nitrogen at the (partially positive) carbon atom, HCl is eliminated to give an *organic isocyanate*:

$$R-NH_2 + C(=O)Cl_2 \rightarrow R-N=C=O + 2\,HCl \tag{9.38}$$

Such organic isocyanates are still quite reactive and are used to synthesize another important class of pesticides, the carbamate pesticides, by reaction with alcohols:

$$R-N=C=O + R'-OH \rightarrow R-NH-C(=O)-OR' \tag{9.39}$$

Similar reactions lead to a class of polymers known as the *polyurethanes*. The reactivity of organic isocyanates manifested itself particularly tragically in the explosion in Bhopal, India.

Solvolysis of sulfur halides (e.g., $S_2Cl_2$) in polar solvents leads to a variety of imides of sulfur, the best known of which is heptasulfur imide, $S_7NH$. This compound has an eight-membered ring structure very much like that of elemental sulfur as $S_8$. The hydrogen atom is weakly acidic, so salts of the $S_7N^-$ anion are readily made.

NITRIDES   The nitrides of the elements are commonly grouped into four categories: saltlike or ionic, *interstitial* or metallic, *covalent*, and *diamondlike*, although there are generally not clearcut differences in properties and bonding types between end members of each classification. The saltlike nitrides behave approximately as salts of the $N^{3-}$ (nitride) ion and include $Li_3N_2$ and $M_3N_2$ (M = Group 2 or 12 metal). The most active metals (K, and so on), however, do not form nitrides. The nitrides of Li and Mg result when these metals are ignited in a nitrogen or ammonia atmosphere or even in air, or they can be made by thermal decomposition of the amides.

Interstitial nitrides occur for the *d*- and *f*-block metals and have stoichiometries such as MN, $M_2N$, and $M_4N$. Nitrogen atoms occupy some or all of the interstices in cubic or hexagonal close-packed metal lattices (Sections 3.9, 6.6) in these compounds. These materials are usually very hard, chemically inert materials with metallic luster and conductivity, and they find use in crucibles, thermocouple sheaths, and the like.

Among the discrete molecular covalent nitrides are the *d*-block complexes of the nitride ion, in which $N^{3-}$ may be bridging but usually is a terminal ligand forming a triple bond to the *d*-block metal (as in the anions $[OsO_3N]^-$, $[VCl_3N]^-$, and $[WCl_5N]^{2-}$. The organosilicon compounds $[R_3Si]_3N$ (R = H, $CH_3$) are of interest because (in contrast to the predictions of VSEPR theory) they are not pyramidal but are planar or nearly planar at the nitrogen atom.

There is a very important series of molecular nitrides of phosphorus, the **cyclopolyphosphazenes**, $[-N=PX_2-]_n$ ($n = 3, 4, 5; X = F, Cl, Br$). The $-N=PX_2-$ unit is isomeric with the basic unit of silicones, $-O-SiR_2-$, and this unit also gives rise to important polymers (to be discussed later). The chlorocyclophosphazenes are readily made by the reaction

$$n \, PCl_5 + n \, NH_4Cl + \text{heat} \rightarrow [NPCl_2]_n + 4n \, HCl \qquad (9.40)$$

Structures of the trimer, tetramer, and pentamer are shown in Figure 9.5e, f, and g. Although there is double bonding between the P and N atoms in these compounds, the nature of this bonding has been a subject of controversy.

There are several significant molecular nitrides of sulfur, the most important of which is *tetrasulfur tetranitride*, $S_4N_4$. Like $S_7NH$, this can be made by the reaction of $S_2Cl_2$ and $NH_3$, but this time in a nonpolar solvent ($CCl_4$ or benzene):

$$6 \, S_2Cl_2 + 16 \, NH_3 \rightarrow S_4N_4 + S_8 + 12 \, NH_4Cl \qquad (9.41)$$

The orange-yellow crystals of $S_4N_4$ are stable to air but may explode on being heated or struck. The structure of $S_4N_4$ (Figure 9.5b) involves secondary sulfur-sulfur bonding and $S{=}N$ double bonds, but many resonance structures are needed to account for its properties. Reaction of $S_4N_4$ with sulfur gives a second nitride, $S_4N_2$ (Figure 9.5c), whereas heating $S_4N_4$ vapor over silver wool depolymerizes it to give the simplest (but also explosive) nitride, $S_2N_2$ (Figure 9.5d). This compound undergoes a very important polymerization reaction that will be discussed shortly.

The diamondlike nitrides are polymeric covalent molecules of formula MN (M = Group 13/IIIA elements except Tl). They are isoelectronic with and structurally related to diamond and graphite. The reaction of $BCl_3$ with excess $NH_3$ at 750 °C produces the slippery white hexagonal boron nitride, BN, with a layer structure much like that of graphite (Figure 6.2f). At high temperature and pressure this is converted to cubic BN with the diamond structure (Figure 6.2g), a material so hard that it will scratch diamond. The polarity of the B—N bonds makes these materials less inert than carbon, however—BN is slowly hydrolyzed by water. The nitrides of Al, Ga, and In are similar to cubic BN.

Although not isoelectronic with or structurally similar to diamond, *silicon nitride*, $Si_3N_4$, is also a polymeric covalent substance with properties that are of great interest to materials scientists and ceramics chemists. $Si_3N_4$ is made by heating $SiO_2$ and coke (C) in $N_2$ and $H_2$ at 1500 °C. It is almost completely inert chemically, is very hard and dense, has high resistance to thermal shock, and retains its high strength, shape, and resistance to corrosion even above 1000 °C. Unfortunately it is extremely difficult to fabricate and sinter components in the shape desired, so research continues on making composites of it with other materials such as $Al_2O_3$.

A polymeric nitride of historical interest is *Millon's base*, hydrated $[Hg_{2n}N_n]^{n+}$ $(OH^-)_n$, which has a polymeric cation in which each nitrogen is attached to four mercury atoms and each mercury to two nitrogens; the external anion undergoes ready ion exchange.

Polymeric nitrides of phosphorus and sulfur are of great practical interest. Of great practical use are the polyphosphazene polymers, synthesized by first heating carefully purified hexachlorocyclotriphosphazene to 250 °C whereupon the ring opens up and gives a polymer $[—N{=}PCl_2—]_n$ in which $n$ is about 10,000. Subsequent reaction of the reactive P—Cl bonds of this polymer with various reactants gives a variety of useful polymers [5]. Most of those in current use are alkoxides, $[—N{=}P(OR)_2—]_n$, resulting from reactions of the chloro polymer with sodium alkoxides. Elastomers result if the side group is methoxide or ethoxide;

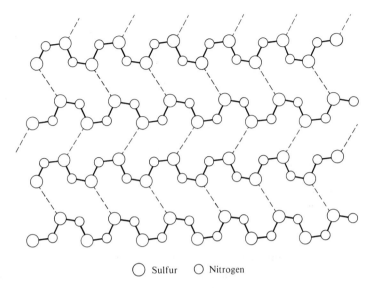

**Figure 9.6** (SN)$_x$ chains in polythiazyl. Reprinted with permission from "Inorganic Compounds with Unusual Properties," by A. G. McDiarmid et al. In R. B. King (ed.), *Advances in Chemistry Series*, No. 150, 1976. Copyright © 1976 American Chemical Society.

⬤ Sulfur ◯ Nitrogen

fluoroalkoxides impart fire resistance and water repellence. Polyphosphazene polymers tend to resist burning or oxidative breakdown better than linear organic polymers. They have many uses, such as in O-rings, gaskets, shock absorbers, and nonburning insulating materials. Solvolysis of the chloro polymers with methylamine, amino acids, or drugs containing amino groups produces water-soluble polymers with possible biomedical applications.

A polymer with quite different properties results when disulfur dinitride is simply allowed to sit at 0 °C for several days. The resulting polymer, *polythiazyl*, (SN)$_x$, is a bronze-colored solid with metallic luster; it conducts electricity about as well as mercury, and at very low temperature it is a superconductor (it conducts electricity with no resistance). In short, it appears to be a metal! Polythiazyl is a one-dimensional polymer (Figure 9.6), for each chain of which it is impossible to draw satisfactory Lewis structures. The best structures have nine electrons on each sulfur; one must be in an antibonding molecular orbital on sulfur. These apparently overlap with corresponding antibonding orbitals on adjacent polymer chains to generate a half-filled conduction band. Consistent with this, the metallic conductivity of (SN)$_x$ is in one direction only; it is a "one-dimensional metal."

## 9.9

## Sulfides of the Elements

Since the sulfide ion is the archetype of soft bases, while the oxide ion is the archetype of hard bases, there is little if any resemblance in structure, chemistry, or uses of corresponding element sulfides and oxides. Since the sulfide ion is a much better reducing agent than the oxide ion, metal sulfides are far more often colored—even black—due to charge-transfer absorptions. Metal sulfides, of course, tend to be

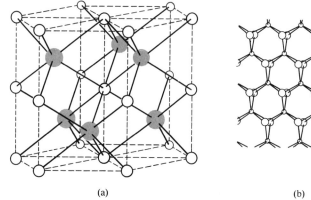

**Figure 9.7** (a) The nickel arsenide structure; (b) the molybdenum disulfide structure. From *Structural Inorganic Chemistry*, 3d ed., by A. F. Wells. Copyright 1962 by Oxford University Press. Reprinted by permission.

(a)                    (b)

much more covalent than oxides and thus more frequently adopt covalent types of lattices, such as the zinc blende and wurtzite (ZnS) lattices for MS. Many metal monosulfides adopt a peculiar structure, the nickel arsenide (NiAs) structure (Figure 9.7a). In this structure each nonmetal atom is coordinated to six metal atoms, while each metal atom is surrounded by six sulfur and two other metal atoms close enough to each other to allow bonding. These sulfides and arsenides consequently show metallic luster, and some metal sulfides show superconducting properties at very low temperatures. The NiAs structure is very unlikely for an ionic compound, as it would require close approach of cations to each other.

Many *d*-block metal disulfides adopt the $CdI_2$ layer structure, which often results in easy cleavage and lubricating properties for these compounds. Also, it is sometimes possible to insert other atoms or flat molecules between the layers of some of these compounds (e.g., $NbS_2$, BN, or graphite); such materials are known as **intercalation complexes**. This insertion is often accompanied by a partial redox reaction, giving rise to formulas such as $M_{1/4}NbS_2$, $C_8K$, or $C_8Br$; in other cases halides may intercalate without such a reaction (e.g., $Mo_2Cl_{10}$ molecules intercalated in graphite). Sometimes intercalation improves the electrical conductivity of the layer compound: The $AsF_5$–graphite intercalation complex has a conductivity comparable with that of copper.

A slightly different layer structure (Figure 9.7b) is found for the important sulfide $MoS_2$, which is widely used as a solid lubricant and as a catalyst. The Mo atom in $MoS_2$ is 6-coordinate, but its geometry is that of a trigonal prism rather than an octahedron. Such a geometry, frequently found with sulfur ligands, is unusual by VSEPR theory because it brings sulfur atoms much closer together than does the octahedron. This implies an unexpected lack of repulsion of negative ions; the S—S distance in such compounds is often short enough to suggest secondary bonding, which often occurs in sulfur and its compounds.

Many metal sulfides show extensive ranges of nonstoichiometry (Section 4.6). Zinc sulfide, ZnS, is white when pure and is used in white paint in a mixture with $BaSO_4$ known as *lithopone*, which is obtained by the reaction

$$ZnSO_4(aq) + BaS(aq) \rightarrow ZnS(s) + BaSO_4(s) \qquad (9.42)$$

ZnS is also widely used in cathode-ray tubes and radar screens, since it fluoresces upon being struck by X-rays, cathode rays, or radiation. The color of light emitted may be varied by partially substituting such ingredients as Cd in place of Zn and Se in place of S. CdS is an important yellow pigment and a useful phosphor, and CdTe may have potential as a semiconductor.

Compounds such as ZnS, CdSe, and CdTe are isoelectronic and isostructural with the semiconductors silicon and germanium and are known as "II-VI" compounds (from their traditional group numbers). Similar in properties and structure are the "III-V" compounds such as gallium arsenide, GaAs, and the other arsenides, phosphides, and antimonides of the Group 13/IIIA metals. Their conducting properties improve as the electronegativities of the elements involved approach 1.75 (Figure 9.8). The III-V compounds GaAs and GaP are now widely used as light-emitting diodes (LED's) in pocket calculators, wrist watches, scientific instrument displays, and the like, in which electrical energy is converted to light. The color of light emitted depends on the energy gap (Figure 9.8); GaAs itself, with an energy gap of 138 kJ, emits infrared light, whereas GaP (218 kJ) emits green light; a mixture with 40% P and 60% As emits red light. Doping Zn in place of Ga creates a *p*-type semiconductor; doping Te in place of As gives an *n*-type semiconductor. Recently semiconducting lasers have been designed based on these compounds.

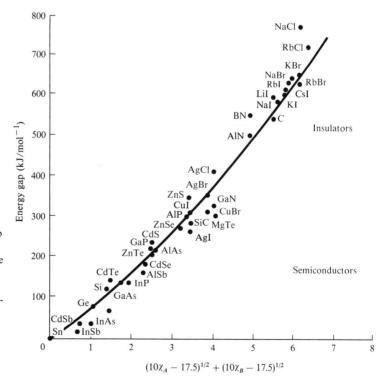

**Figure 9.8** Empirical relationship of the energy gap important in semiconduction (Section 6.6) to the electronegativities of the elements present in a I-VII, II-VI, or III-V compound or a Group 14/IVA element. From N. B. Hannay, *Solid-state Chemistry*, © 1967, p. 36. Reprinted by permission of Prentice-Hall, Englewood Cliffs, New Jersey.

These compounds are also useful in the reverse process of converting light energy into electrical energy; InSb is used as a photoconductive detector of infrared light. Several of these compounds are being investigated for use in the direct conversion of sunlight to electricity; at the present time $Cu_2S$–CdS and silicon are commercially preferred. Semiconducting properties are also found in the sulfides, selenides, and arsenides, $M_2S_3$, of arsenic, antimony, and bismuth.

A large variety of molecular sulfides of phosphorus and arsenic are known; the sulfides $P_4S_3$, $P_4S_{10}$, and $Sb_2S_3$ are used in matches as a mixture with an oxidant (e.g., $KClO_3$) that is ignited by friction. The best-known and most important molecular sulfide may be carbon disulfide, $CS_2$, which has double bonds like those of carbon dioxide. This foul-smelling liquid, however, is very volatile and extremely flammable; it may catch fire even in the absence of a flame or spark if heated to $100\,°C$. Some important soft-base ligands are produced by the reaction of $CS_2$ with bases: Amines, $R_2NH$, give *dithiocarbamate* ions, $R_2N$—$C(=S)S^-$, while alkoxides give *xanthate* ions, $RO$—$C(=S)S^-$.

## 9.10

## Study Objectives

1. Tell whether a halide of a given formula is likely to be nonexistent.
2. Tell whether a halide of a given formula is likely to be intensely colored.
3. Choose an appropriate type of synthesis for a given halide.
4. Tell whether a given halide is likely to be a gas, a liquid, a low-melting solid, or a high-melting solid at room temperature; relate this to the type of structure (monomeric covalent, ionic, and so on) that the halide is likely to have.
5. Classify a given halide as predominantly a Lewis acid or predominantly a Lewis base, and pick appropriate halides with which it might react.
6. Give examples of halides that are likely to be good oxidizing agents; good reducing agents. Write balanced chemical equations showing such functions.
7. (Review) Apply the VSEPR theory to explain or predict the geometry of a given halide.
8. Describe layer structures, know some physical properties associated with them, and tell the types of halides or chalcogenides in which they are likely to occur.
9. Give reasonable methods of preparation of the following types of compounds of the elements: (a) alkoxides or esters; (b) amides; (c) imides; (d) nitrides.
10. Know the practical uses of the types of polymers or macromolecules discussed in this chapter.

## 9.11

Exercises

†1. Which of the following hypothetical halides probably do not actually exist? $LaF_3$, $SF_6$, $CI_4$, $BiI_5$, $CuBr_2$, $XeBr_4$, $S_2I_2$, $WF_6$, $FeI_3$, $MnF_7$, $AuI_3$, $IF_7$, $FeBr_3$.

†**2.** Among the halides from the previous list that do exist, pick the ones that are most likely to be intensely colored.

**3.** Which of the following halides of bismuth is most intensely colored? $BiF_3$, $BiCl_3$, $BiBr_3$, or $BiI_3$.

†**4.** For each of the halides listed, choose one of the following four general synthetic routes as the most appropriate (workable yet simple) method of synthesis of the anhydrous halide: Method 1, anhydrous salt results from evaporating the reaction mixture of the metal hydroxide plus the hydrohalic acid; Method 2, the salt precipitates in anhydrous form upon mixing solutions containing the cation and the anion; Method 3, the hydrated halide can be dehydrated by an appropriate chemical dehydrating agent; Method 4, a strictly anhydrous method of synthesis (halogen exchange or redox) is required. Halides: **a.** $CF_4$;   **b.** $AgI$;   **c.** $KF$;   **d.** $SiCl_4$;   **e.** $BiCl_3$; **f.** $SiI_4$;   **g.** $AuI$;   **h.** $RbBr$;   **i.** $NiCl_2$;   **j.** $BiBr_3$;   **k.** $LaF_3$; **l.** $SnCl_2$;   **m.** $PrCl_3$;   **n.** $WF_6$.

**5.** For any of the above halides assigned, complete and balance an equation showing a plausible synthesis of that halide in anhydrous form.

**6.** Consider the synthesis of one of the halides listed at the end of this question. (1) Discuss the applicability of each of the seven synthetic methods listed in Section 9.3 to this halide. (2) Look up the synthesis of the halide in an advanced text, such as [3] or Cotton and Wilkinson, *Comprehensive Inorganic Chemistry*, 4th ed., Wiley-Interscience, New York, 1980. What methods are given there? Do they agree with your prediction? (3) Look up the actual details of the synthesis in the reference cited in the advanced text, in an inorganic lab manual, in *Inorganic Syntheses*, in [2], or in *Chemical Abstracts*. Are there important details (such as carrying out the synthesis in a dry or nonoxidizing atmosphere) that you would have expected from the principles of this book? Are there important details that you could not have anticipated? Halides: **a.** $MoCl_5$;   **b.** $[Mo_6Cl_8]Cl_4$;   **c.** $VCl_4$; **d.** $TiBr_4$;   **e.** $CrBr_3$;   **f.** $CrI_3$.

†**7.** Choose one of the following three structural classifications (monomeric or dimeric molecular, macromolecular, or ionic) and one of the following four categories of physical state at room temperature (gas, liquid, solid with a low melting point, or solid with a high melting point) for each of the following halides: **a.** $CF_4$;   **b.** $AgI$;   **c.** $KF$;   **d.** $SiCl_4$;   **e.** $BiCl_3$;   **f.** $SiI_4$; **g.** $AuI$;   **h.** $RbBr$;   **i.** $NiCl_2$;   **j.** $BiBr_3$;   **k.** $LaF_3$;   **l.** $SnCl_2$; **m.** $PrCl_3$;   **n.** $WF_6$.

**8.** Give actual examples of catenated halides from each of the following groups: **a.** 16/VIA;   **b.** 15/VA;   **c.** 14/IVA;   **d.** 13/IIIA.

**9.** Give the structural types of the following halides, in all of which the central atom has the group oxidation number minus two: $InI$, $SnCl_2$, $SbBr_3$, $TeCl_4$, $IF_5$. Discuss the reasons for the trends in this series.

337

**10.** From each of the following series of halides, pick the one that should have the lowest melting point, and explain why:

    **10.1** IF, BrF, ClF, $F_2$

    **10.2** $MgF_2$, $MgCl_2$, $MgBr_2$, $MgI_2$

    **10.3** InI, $SnCl_2$, $SbBr_3$, $TeCl_4$, $IF_5$

    **10.4** $SiF_4$, $SiCl_4$, $SiBr_4$, $SiI_4$

**\*11.** $GeCl_4$ is a liquid with a melting point of $223\,°K$; $GeCl_2$ is a solid with a melting point of about $400\,°K$. Is this the normal trend to be expected for two monomeric covalent compounds of these formulas? Is this the normal trend to be expected for two ionic compounds of these stoichiometries? If neither, explain why the melting points fall in this order.

**\*12.** On the bench in a professor's laboratory are six bottles containing the products of the syntheses of the six halides $CrCl_3$, $HgI_2$, $NF_3$, $NI_3$, $SnCl_2$, and $SnCl_4$. Unfortunately the labels have come loose. Examine the bottles, and reattach the labels correctly. Bottle A contains a white solid; bottle B contains an intensely red solid; bottle C contains a colorless gas; bottle D contains light violet flakes; bottle E contains a colorless liquid; and bottle F consists of broken glass—all that remains after an explosive attempted preparation of its would-be contents.

**13.** Examine a copy of Table A, the Pauling electronegativity table. Assuming that metal-halogen bonding will be ionic for Pauling electronegativity differences in excess of 1.8 but polar covalent for differences less than 1.8, identify all metals that you would expect to have ionic fluorides but polar covalent chlorides. Taking into account the likely formulas of these halides, describe how these fluorides and chlorides will differ in terms of structural types, volatility, and solubility in water and in nonaqueous solvents.

**†14.** Classify each of the following halides as: (1) predominantly a Lewis acid; (2) predominantly a Lewis base; (3) neither: **a.** $CF_4$;    **b.** AgI;    **c.** KF;    **d.** $SiCl_4$;    **e.** $BiCl_3$;    **f.** $SiF_4$;    **g.** $HgI_2$;    **h.** RbBr;    **i.** $BaCl_2$;    **j.** $BiBr_3$;    **k.** $CeF_4$;    **l.** $PF_3$;    **m.** $FeCl_3$;    **n.** $SF_6$;    **o.** $AsF_5$;    **p.** $IF_7$;    **q.** $XeF_6$;    **r.** $PI_5$.

**15.** Complete and balance plausible Lewis acid-base reactions between appropriate pairs of halides in the previous question.

**16.** List four halides that would likely be strong oxidizing agents; four that would likely be useful reducing agents. Complete and balance plausible redox reactions between appropriate pairs of these halides.

**†17.** Complete and balance equations showing the products from the following pairs of reactants, or write "NR" if no reaction will occur: **a.** $PF_5 + CsF$;    **b.** $XeF_6 + CF_4$;    **c.** $RbBr + BaBr_2$;    **d.** $XeF_6 + BF_3$;    **e.** $NaF + AlF_3$;    **f.** $SnCl_2 + PbCl_4$;    **g.** $GeCl_3^- + IrCl_6^{3-}$;    **h.** $ClF_3 + UO_3$;    **i.** $PI_5 + ZnI_2$;    **j.** $CH_3I + NaCl +$ acetone solvent.

**18.** (1) Which of the following lattice types are layer structures? **a.** fluorite; **b.** rock salt; **c.** cadmium iodide; **d.** chromium chloride; **e.** rutile; **f.** nickel arsenide; **g.** graphite; **h.** diamond.

(2) Tell for each of the following types of halide or chalcogenide which of the above lattice types is most likely, assuming an appropriate radius ratio: **a.** MO; **b.** MS; **c.** $MO_2$; **d.** $MS_2$; **e.** MTe; **f.** MF; **g.** $MF_2$; **h.** $MI_2$.

**19.** Predict the structure of each of the following halides or derivatives using VSEPR theory, and compare your prediction to the actual structure. Discuss the reasons for any discrepancies: **a.** $SF_6$; **b.** $XeF_6$; **c.** $SF_4$; **d.** $TeF_4$; **e.** $TeCl_4$; **f.** $S_2Cl_2$; **g.** $SO_2Cl_2$; **h.** $SOCl_2$; **i.** $P(OCH_3)_3$; **j.** $PCl_3$; **k.** $SbCl_3$; **l.** $IF_7$; **m.** $PI_5$.

**20.** Write balanced chemical equations illustrating reasonable methods of preparation of the following compounds: **a.** $KOC_2H_5$; **b.** $As(OCH_3)_3$; **c.** $VO[N(CH_3)_2]_3$; **d.** $B_3Cl_3N_3(CH_3)_3$; **e.** $S_4N_4$; **f.** $[N{=}P(OCH_3)_2]_3$; **g.** ZnS; **h.** CaS.

**21.** Suggest an inorganic polymer or macromolecule that might be suitable if you needed the following properties: **a.** a semiconductor that would convert electrical energy to ultraviolet light; **b.** a polymer that would exhibit metallic conduction in one dimension only; **c.** a macromolecule that would exhibit metallic conduction in two dimensions only; **d.** a polymer that might slowly hydrolyze to release the anticancer drug cisplatin in the blood; **e.** a ceramic that would maintain its strength to 1000 °C; **f.** a lubricating oil of low viscosity; **g.** a solid lubricant.

---

**Notes**

[1] Downs, A. J., and C. J. Adams, in *Comprehensive Inorganic Chemistry*, J. C. Bailar, Jr., *et al.*, ed., Pergamon Press, Oxford, 1973, vol. 2, pp. 1253–1257.

[2] Brauer, Georg, *Handbook of Preparative Inorganic Chemistry*, 2 vols., 2d ed., Academic Press, New York, 1963.

[3] Greenwood, N. N., and A. Earnshaw, *Chemistry of the Elements*, Pergamon Press, Oxford, 1984.

[4] Alcock, N. W., *Adv. Inorg. Chem. Radiochem.*, 15, 1 (1972).

[5] Allcock, H. R., "Inorganic Macromolecules," *Chem. Eng. News*, March 18, 1985, pp. 22–36.

# The Hydrides and Organometallic Derivatives of the Elements

In this last chapter of descriptive chemistry we will study some of the chemistry of the hydrides [1] and the organometallic compounds of the elements. These two classes of compounds have grown greatly in importance since the beginning of the Atomic Age, due to the importance of hydrogen in nuclear fusion and energy storage and the importance of organometallic compounds in modern industrial catalysis. Knowledge of the chemistry of these compounds has grown tremendously in recent decades; consequently we can give only an overview in this chapter, leaving further amplification for a more specialized inorganic course.

## 10.1

## Hydrides: Their Formulas, Stabilities, and Classification

The element hydrogen is uniquely difficult to locate in the periodic table. Although it is conventionally located in Group 1 and, like the other Group 1 members, predominantly exhibits a $+1$ oxidation state, it differs in lacking core electrons. It is considerably more electronegative than the other members of the group and consequently does not form the simple cation $H^+$ unless strongly hydrogen-bonded to a Lewis base such as water. Also, under normal conditions it is not a metal.

An alternative location for hydrogen in the periodic table is in Group 17/VIIA; it is a nonmetal that forms a diatomic molecule like the other members of this group. Although it does not use $p$ electrons, hydrogen is one electron shy of a noble-gas electron configuration and does (sometimes) form the hydride ion, $H^-$, which resembles the halide ions. The hydride ion in ionic hydrides is similar in size to the fluoride ion, although its radius is quite variable. But the resemblance of $H^-$ to the halide ions is limited: The electron affinity of H (Table 6.9) is much less than that of the other halogen atoms. Consequently $H^-$ is a much stronger reducing agent than the other halide ions.

A third location that has been proposed is above carbon. Although H, of course, does not have four valence electrons, its Pauling electronegativity is reasonably close to that of carbon, and there is a fair degree of resemblance between the chemical properties of the hydrides of many of the $s$- and $p$-block elements and the organometallic compounds of the same elements. Without necessarily advocating this periodic-table position, we have grouped in this chapter the hydrides and the organometallic derivatives of the elements, due to this similarity. We also briefly mention the carbides in this chapter.

Because the electronegativity of hydrogen, 2.2, is about in the middle of the Pauling scale, its compounds include polar covalent compounds in which the hydrogen is partially positively charged, ionic and polar covalent compounds in which it is partially negatively charged, and some in which there is little or no polarity. This is a contrast to the halides, in which the halogen atoms are usually partially negatively charged.

There are, of course, no fast dividing lines between these categories, so we will divide the hydrides into classifications based more on physical properties that on bond types; these are shown in Table 10.1. The *molecular* (or covalent) hydrides are those in which discrete small molecules are formed; some have partially positive hydrogen and some have partially negative hydrogen. Since each hydrogen atom contributes only one electron to these molecules, these substances have weak van der Waals forces between them and are mostly gases at room temperature. Recall that the nomenclature of the molecular hydrides was covered in Section 5.5 and Figure 5.8c; you should review this now.

Although they are fundamentally members of the same class, for convenience we have listed the molecular hydrides that involve *catenation* of the other element separately (Table 10.1b). Except for some of the lightest of these, there are enough additional electrons in the catenated molecular hydrides for these to be liquids at room temperature.

At the left of the periodic table, we find solid hydrides. These include the *saltlike* or saline hydrides, which seem to be salts of the $H^-$ ion, and those in which the hydrogen seems to serve as a bridging atom, giving *polymeric covalent* or borderline hydrides. (The division between saltlike and polymeric covalent hydrides is somewhat arbitrary but is usually made above and to the right of $MgH_2$.) The bridging function of hydride ions in polymeric covalent hydrides is best explained using the molecular orbital theory, since hydride ions do not have two unshared pairs of electrons to use in bridging.

Finally there is a category of *metal-like* or interstitial hydrides of many of the $d$- and $f$-block elements. These are also solids, but they differ from the saltlike and polymeric hydrides in that they retain metallic conductivity.

We note that there are a number of later $d$-block metals and some heavier $p$-block metals that do not form stable hydrides; furthermore, many of the hydrides of elements that are in the center of the periodic table are thermodynamically and kinetically unstable (Table 10.2). One explanation for the relative dearth of stable hydrides of the elements may be found in the Pauling electronegativity equation (8.8). By this equation, hydrides should be more stable than their components (i.e., metals or nonmetals and hydrogen) to the degree that the electronegativity of hydrogen and of the metal and the nonmetal differ. At the far left and the far right

**Table 10.1**  Hydrides of the Elements

**A. Molecular (Covalent) Hydrides**
All are gases at room temperature unless otherwise noted.

| $B_2H_6$ | $CH_4$ | $NH_3$ | $H_2O(l)$ | $HF(l)$ |
|---|---|---|---|---|
| | $SiH_4$ | $PH_3$ | $H_2S$ | $HCl$ |
| | $GeH_4$ | $AsH_3$ | $H_2Se$ | $HBr$ |
| | $SnH_4$ | $SbH_3$ | $H_2Te$ | $HI$ |

**B. Catenated Molecular Hydrides**
Liquids unless otherwise noted.

$B_nH_{n+6}$ ($n = 4$–$10$)  $C_nH_{2n+2}$ (no limit to $n$)  $N_2H_4$  $H_2O_2$
$B_nH_{n+4}$ ($n = 5$–$20$)  $C_nH_{2n}$ (no limit to $n$)
(solid for $n > 9$)  (gas for $n < 5$)
(solid for large $n$)
$Si_nH_{2n+2}$ ($n$ up to 8)  $P_2H_4$  $H_2S_n$ ($n$ up to 8)
(gas for $n = 2$)
$GeH_{2n+2}$ ($n$ up to 5)

**C. Saltlike (Saline) and Polymeric (Borderline) Hydrides**
All are solids at room temperature unless otherwise noted.

| LiH | $BeH_2$ | | | | | |
|---|---|---|---|---|---|---|
| NaH | $MgH_2$ | | | | | $AlH_3$ |
| KH | $CaH_2$ | | | CuH | $ZnH_2$ | $GaH_3(l)$ |
| RbH | $SrH_2$ | | | | | |
| CsH | $BaH_2$ | $EuH_2$ | $YbH_2$ | | | |

**D. Metal-Like (Interstitial) Hydrides**
Solids at room temperature.

| $ScH_2$ | $TiH_2$ | VH | CrH | | NiH | | |
|---|---|---|---|---|---|---|---|
| | | $VH_2$ | | | | | |
| $YH_2$ | $ZrH_2$ | NbH | | | $PdH_{<1}$ | | |
| $YH_3$ | | $NbH_2$ | | | | | |
| $LuH_2$ | $HfH_2$ | TaH | | | | | |
| $LuH_3$ | | | | | | | |

4f-element $MH_2$ and $MH_3$ except for M = Eu, Yb

| $AcH_2$ | $ThH_2$ | | | $NpH_2$ | $PuH_2$ | $AmH_2$ | $CmH_2$ |
|---|---|---|---|---|---|---|---|
| | $Th_4H_{15}$ | $PaH_3$ | $UH_3$ | $NpH_3$ | $PuH_3$ | $AmH_3$ | |

of the periodic table the electronegativities do differ greatly, and the hydrides are thermodynamically stable; but in the center of the table the elements have Pauling electronegativities close to that of hydrogen.

By equation (8.8), such hydrides should have $\Delta H_f$ approximately equal to zero; in fact, many of the hydrides have very endothermic heats of formation. The Pauling equation does not take size explicitly into account, and the very small size of the hydrogen atom results in the H—H bond being much the strongest of any single homopolar (element-element) bond (Table 6.5); hence the decomposition of many

**Table 10.2**  Standard Enthalpies of Formation (kJ/mol) and Stability to Oxidation of Some Hydrides

| | | | | | | |
|---|---|---|---|---|---|---|
| LiH −91 wat | BeH$_2$ wat | B$_2$H$_6$ +32 flam | CH$_4$ −75 stab | NH$_3$ −46 stab | H$_2$O −286 stab | HF −269 stab |
| NaH −56 wat | MgH$_2$ −76 wat | AlH$_3$ −46 | SiH$_4$ +31 flam | PH$_3$ +5 flam | H$_2$S −20 stab | HCl −92 stab |
| KH −58 wat | CaH$_2$ −174 wat | | GeH$_4$ +90 flam | AsH$_3$ +67 therm | H$_2$Se +86 stab | HBr −36 stab |
| RbH −54 wat, flam | SrH$_2$ −177 wat | | SnH$_4$ +163 therm | SbH$_3$ +145 therm | H$_2$Te +154 therm | HI +26 stab |
| CsH −56 wat, flam | BaH$_2$ −171 wat | | PbH$_4$ +250 therm | BiH$_3$ +278 therm | | |

SOURCES: Data from J. C. Bailar et al., *Comprehensive Inorganic Chemistry*, Pergamon Press, Oxford, 1973, Ch. 2; and M. C. Ball and A. H. Norbury, *Physical Data for Inorganic Chemists*, Longman, London, 1974, p. 92.
NOTE: Abbreviations: stab = (kinetically) stable to air oxidation; flam = spontaneously flammable in air; therm = thermally unstable at or near room temperature; wat = oxidized by water with release of H$_2$.

hydrides to the element and H$_2$ is thermodynamically favorable, and such reactions are often relatively rapid as well.

The hydride ion is a very good reducing agent (Section 5.5); hence we often fail to find hydrides corresponding to some of the higher group oxidation numbers. Thus PH$_5$ and SH$_6$ do not exist, and many of the *d*-block elements have surprisingly low oxidation numbers in their hydrides. The reducing property of the hydride ion persists in most of the hydrides that are thermodynamically stable, such as the saltlike hydrides. Many of these will catch fire in air or will release hydrogen on contact with water (Table 10.2):

$$CaH_2 + 2 H_2O \rightarrow Ca(OH)_2 + 2 H_2 \qquad (10.1)$$

In some cases (e.g., the carbon hydrides or *hydrocarbons*) oxidation is thermodynamically favored but is very slow due to the very small size of the central atom. Nonetheless the combustion of hydrocarbons once initiated is very exothermic, as anyone who drives a car should realize. Among the hydrides only H$_2$O and HF are completely resistant to air oxidation; even HCl can be oxidized to Cl$_2$ and H$_2$O in the presence of a catalyst.

## 10.2

## Structures of the Hydrides

MOLECULAR HYDRIDES  The structures [2, 3] of the molecular hydrides of the nonmetals other than boron are essentially as predicted by VSEPR. The simple boron hydride BH$_3$, although it forms stable adducts with Lewis bases such as H$_3$B:CO, rapidly dimerizes to give *diborane*, B$_2$H$_6$, which has two bridging hydrogen atoms (Figure 10.1). Since hydrogen, even as the hydride ion, does not

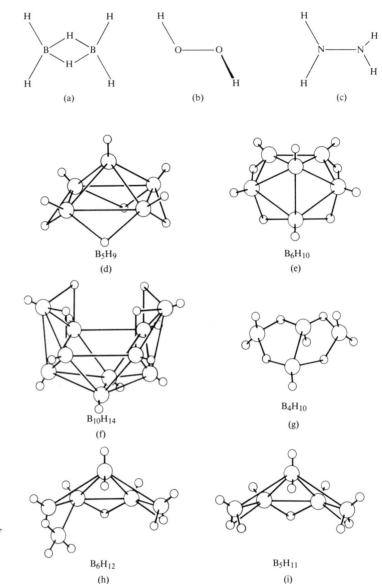

**Figure 10.1** The structures of
(a) diborane, $B_2H_6$; (b) hydrogen
peroxide, $H_2O_2$; (c) hydrazine,
$N_2H_4$; (d) to (f) the nido-boranes
$B_5H_9$, $B_6H_{10}$, and $B_{10}H_{14}$; (g)
to (i) the arachno-boranes $B_4H_{10}$,
$B_5H_{11}$, and $B_6H_{12}$. In (d) to (i)
the larger circles represent boron
atoms, and the smaller circles
represent hydrogen atoms.

have two pairs of electrons to donate, this is an example of a bonding molecular
orbital (Section 6.6), in which one electron pair of $H^-$ is shared with two boron
atoms (thus, a *three-center, two-electron* bond).

The catenated hydrides have additional structural features of interest. In order
to avoid repulsions of the electrons on neighboring nitrogen or oxygen atoms, the
hydrides **hydrogen peroxide** ($H_2O_2$) and **hydrazine** ($N_2H_4$) adopt the less symmetrical conformations shown in Figure 10.1. Related to these two compounds is

another important compound, **hydroxylamine**, $NH_2$—OH. The catenated hydrides of carbon, a very diverse group, form the backbone of the field of organic chemistry and cannot easily be summarized here. The corresponding series of hydrides of silicon and germanium are less extensive, due to their greater tendency to decompose thermally and/or catch fire.

There are about 25 catenated hydrides of boron. Although many or most of these catch fire in air, they have sufficient thermal stability that a very extensive chemistry of them is known, which can only be hinted at in this chapter. Their structures are rather complex, and all involve three-center, two-electron bonds with bridging hydrogens in addition to catenated boron-boron bonds. There are two major categories, the *nido-boranes*, $B_nH_{n+4}$, represented in Fig. 10.1 by the structures of $B_5H_9$, $B_6H_{10}$, and $B_{10}H_{14}$, and the *arachno-boranes*, $B_nH_{n+6}$, represented by the structures of $B_4H_{10}$, $B_5H_{11}$, and $B_6H_{12}$. The common nomenclature for these compounds distinguishes the nido- and arachno-boranes by enclosing the (differing) numbers of hydrogen atoms in parentheses. Thus $B_5H_9$ is known as pentaborane(9) and $B_5H_{11}$ is known as pentaborane(11). The arachno-boranes are characterized by more "open" structures than the nido-boranes. These boranes are liquids unless they contain more than nine boron atoms. The two most important and stable of these catenated boranes are pentaborane(9) and decaborane(14).

The hydrides of the three most electronegative small nonmetal atoms, fluorine, oxygen, and nitrogen, show elevated melting and boiling points due to the presence of hydrogen bonding between molecules. Consequently the hydrides HF, $H_2O$, $H_2O_2$, and $N_2H_4$ are liquids at room temperature despite their low molecular weights; $NH_2OH$ is a solid. The association of water molecules has already been discussed (Section 3.3). Crystalline HF contains zigzag chains of hydrogen-bonded HF molecules (Figure 10.2); the hydrogen bond between H and F is so strong that it persists partially in the vapor state below 80 °C in a ring hexamer, $(HF)_6$. These hydrides are also the only ones that ionize appreciably as pure liquids:

$$2\,HF \rightleftharpoons H_2F^+ + F^- \qquad K = 10^{-10} \tag{10.2}$$

$$F^- + HF \rightleftharpoons FHF^- \tag{10.3}$$

$$2\,H_2O \rightleftharpoons H_3O^+ + OH^- \qquad K = 10^{-14} \tag{10.4}$$

$$2\,NH_3 \rightleftharpoons NH_4^+ + NH_2^- \qquad K = 10^{-30} \tag{10.5}$$

SALTLIKE AND POLYMERIC HYDRIDES  Some of the saltlike and polymeric solid hydrides are too unstable to have had crystal structures determined, but the ones that have show typical lattice structures: the NaCl type for the Group 1 hydrides,

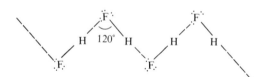

**Figure 10.2**  Hydrogen-bonded structure of crystalline HF.

a $BaCl_2$ or a rutile structure for the hydrides of Mg through Ba, a ZnS lattice for CuH, and an $AlF_3$–type structure (with hydrogen bridges) for $AlH_3$.

METALLIC HYDRIDES   The metallic hydrides are characterized by the fact that they retain metallic conductivity. They have perhaps provoked the most controversy, in part because they are often nonstoichiometric (Section 4.6). (They generally can be made stoichiometric under sufficiently forcing conditions, but there is a significant favorable entropy effect resulting from a deficiency of hydrogen atoms randomly distributed among excess sites.) This nonstoichiometry led to an original classification of these as *interstitial* hydrides, in which tiny hydrogen atoms fit in holes in metallic lattices without altering the lattice types, but it is now clear that the lattice types are altered. ($UH_3$, for example, has a lattice in which each hydrogen is surrounded by four uranium atoms, and each uranium atom has a coordination number of 12!) The hydride of palladium appears to be unique, however: The lattice type is not altered, and it is never stoichiometric, with no more than 0.8 hydrogen atom incorporated per palladium atom.

There is controversy about the nature of the hydrogen in these species; some models suppose that the hydrogen is present at $H^+$, while other models postulate the presence of $H^-$ ions. It is known that the metallic conduction band (Section 6.6) among $f$-block metals should accommodate six valence electrons per atom and that the $f$-block trihydrides are substantially poorer conductors than the $f$-block dihydrides. The reduced conductivity of $MH_3$, however, can be explained either by the presence of both $M^{3+}$ and $3H^+$ ions filling the conduction band with six electrons, or by $M^{3+}$ and $3H^-$ ions leaving no electrons to go into the conduction band.

## 10.3

## Hydride Cations and Anions

The acid-base chemistry of the hydrides is somewhat complex, since both Lewis-acid and Lewis-base properties can be found. Depending on the reaction partner of the hydride, additional significant classes of *hydride cations* and *hydride anions* can arise; hence we will subdivide these reactions into four types: (a) proton-accepting basicity, (b) electron-pair–donating basicity (other than that of donation to protons), (c) proton-donating acidity, and (d) electron-pair–accepting acidity.

(a) *Proton-accepting basicity* results in the formation of a *hydride cation*:

$$: PH_3 + H^+ \rightleftharpoons PH_4{}^+ \tag{10.6}$$

This reaction is most familiar with $NH_3$, which produces the stable *ammonium ion*, $NH_4{}^+$, which in size and chemistry resembles the potassium ion. The $PH_4{}^+$ (*phosphonium*) ion is seldom encountered, as it is readily decomposed by water:

$$PH_4{}^+ + H_2O \rightleftharpoons PH_3 + H_3O^+ \qquad K = 2.5 \times 10^{13} \tag{10.7}$$

The ammonium ion also undergoes the corresponding reaction to an extent sufficient to make its solution slightly acidic, but not to result in complete decomposition:

$$NH_4^+ + H_2O \rightleftharpoons NH_3 + H_3O^+ \qquad K = 5.5 \times 10^{-10} \tag{10.8}$$

The only other important hydride cations are the substituted ammonium ions, the hydroxylammonium cation ($NH_3OH^+$), the hydrazinium cations $N_2H_5^+$ and $N_2H_6^{2+}$, and $H_3O^+$ and $H_2F^+$. All these are more acidic than the ammonium ion. The heavier hydrides in the periodic table (e.g., $H_2S$, HCl) do not readily form hydride cations; their basicity is weaker. Recall that in Section 5.5 we observed that the basicity of a monoatomic nonmetal anion decreases down a group. Thus the decreasing basicity of the hydrides of these nonmetals down a group follows the same trend.

(b) *Electron-pair–donating basicity.* Many of the same hydrides ($NH_3$, $N_2H_4$, $NH_2OH$, $PH_3$, and $H_2O$) also function as Lewis bases by forming complexes with metals; those with $NH_3$ are called ammine complexes and those with $H_2O$ are called hydrated ions, of course. Other Group 15/VA, 16/VIA, and 17/VIIA hydrides also have unshared pairs of electrons and presumably can coordinate to metals, but they would then normally be deprotonated, since (1) these hydrides are more acidic than the ones above them (Figure 5.8), and (2) coordination enhances the acidity of a ligand (Section 7.2). Thus bubbling $H_2S$ through a solution of a soft-metal ion does not give $[M(H_2S)_n]^{x+}$; instead, deprotonation occurs to give the sulfide of the metal.

(c) *Proton-donating acidity.* The trends in the tendency of the protons of $p$-block molecular hydrides to ionize have already been covered in Figure 5.8d and should be reviewed now.

(d) *Electron-pair–accepting acidity.* The hydrides of the Group 13/IIIA elements, like the halides, have less than an octet of electrons (if regarded as $MH_3$) or have to make do with a poor donor, bridging hydrogen (if regarded as $B_2H_6$, etc.). They form some important *hydride anions*, which are more stable than the corresponding hydrides. Aluminum forms both the $AlH_4^-$ and the $AlH_6^{3-}$ ions; the former is much more important. $LiGaH_4$ is stable but unimportant; $LiInH_4$ and $LiTlH_4$ exist below $0\,°C$. Boron forms the fairly water-stable $BH_4^-$ ion (common name borohydride, more formal name tetrahydroborate). The $BH_4^-$ ion itself can act as a ligand, using bridging hydrogen atoms as three-center, two-electron donor atoms in forming complexes such as $Be(BH_4)_2$, $Al(BH_4)_3$, and $U(BH_4)_4$. (The latter complex is polymeric, involving bridging chelating borohydride groups around uranium atoms with coordination numbers of 14!)

Boron also forms some series of *borohydride* ions, of which we will mention three. The $B_nH_n^{2-}$ ions ($n = 6$ to 12) have regular polyhedral *closo-borane* structures (Figure 10.3) and are (especially for $n = 10$ and $n = 12$) very stable. The $B_nH_{n+3}^-$ ions have structures derived from those of the nido-boranes, $B_nH_{n+4}$, by removal of one bridging proton. The $B_nH_{n+5}^-$ ions ($n = 2, 3, 5, 9, 10$) have structures derived from those of the arachno-boranes.

Finally, two $d$-block metals that do not form hydrides, Tc and Re, form the stable hydride anions $TcH_9^{2-}$ and $ReH_9^{2-}$. $ReH_9^{2-}$ has the tricapped trigonal prismatic structure expected from VSEPR theory.

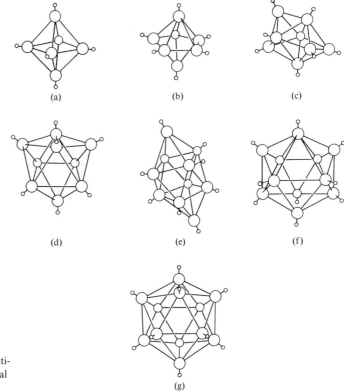

**Figure 10.3** Structures of the closo-borane, $B_nH_n^{2-}$, ions. (a) The octahedral cluster $B_6H_6^{2-}$; (b) the pentagonal bipyramidal $B_7H_7^{2-}$; (c) the dodecahedral $B_8H_8^{2-}$; (d) the tricapped trigonal prismatic $B_9H_9^{2-}$; (e) the bicapped square antiprismatic $B_{10}H_{10}^{2-}$; (f) the octadecahedral $B_{11}H_{11}^{2-}$; (g) the icosahedral $B_{12}H_{12}^{2-}$.

## 10.4

## Syntheses of the Hydrides and Hydride Anions of the Elements

The *saltlike* and *metallic* hydrides are prepared by direct combination of the metal and $H_2$. Due to the strong H—H bond and the low electron affinity of the hydrogen atom, this reaction is not nearly so easy as the reaction of these active metals with halogens; high temperatures (e.g., 300 °C) are required. This reaction is also used in the syntheses of $LiAlH_4$ and $NaAlH_4$ (lithium and sodium aluminum hydrides):

$$M + Al + 2H_2 \rightarrow MAlH_4 \qquad (10.9)$$

As we mentioned earlier, the metallic hydrides normally are produced in a nonstoichiometric form; excess pressure is needed to get fairly stoichiometric or higher hydrides.

For *molecular hydrides*, the direct combination of hydrogen and a nonmetal is seldom the method of choice: With fluorine the reaction is explosive, while with many of the nonmetals the reaction is strongly endothermic and does not go. This is, however, one route for the industrial preparation of HCl and HBr and is the basis for the extremely important *Haber-Bosch* process for the production of ammonia. In the industrial synthesis of $NH_3$, which on a mole scale is produced in larger quantities than any other industrial chemical, hydrogen is first produced by the re-forming of methane from natural gas with steam over a nickel catalyst at $750\,°C$:

$$CH_4 + 2\,H_2O \rightleftharpoons CO_2 + 4\,H_2 \tag{10.10}$$

Air is then injected; some of the hydrogen burns in the oxygen. The resulting water and the $CO_2$, CO, etc., are removed in a scrubber, leaving behind a *synthesis gas* containing the $N_2$ from the air in a 1 : 3 mixture with $H_2$. The mixture is then passed over a promoted iron catalyst at $400\,°C$ and 200 atm pressure to give about a 15% yield of ammonia:

$$N_2 + 3\,H_2 \rightleftharpoons 2\,NH_3 \tag{10.11}$$

The ammonia is condensed by refrigeration (b.p. $-34\,°C$), and the synthesis gas is recycled.

A fairly general approach to many of the molecular hydrides is the treatment of a salt of the anion of the element with strong acid (if the anion is not strongly basic—e.g., $F^-$, $Cl^-$, $S^{2-}$, $Se^{2-}$, and $Te^{2-}$) or with water if the anion is very strongly basic.

$$CaF_2 + H_2SO_4 \rightarrow CaSO_4 + 2\,HF \tag{10.12}$$

$$FeS + 2\,HCl(aq) \rightarrow Fe^{2+}(aq) + 2\,Cl^-(aq) + H_2S(g) \tag{10.13}$$

$$Ca_3P_2 + 6\,H_2O \rightarrow 3\,Ca(OH)_2 + 2\,PH_3 \tag{10.14}$$

A third general route, suitable especially for the elements toward the center of the periodic table, is the reaction of NaH, $LiAlH_4$, or $NaBH_4$ with a halide of the element, typically in ether solution, although $NaBH_4$ can be used in aqueous solution. The reactions can be regarded, according to one's preference, either as reductions of the halide of the element or as acid-base reactions in which the $:H^-$ ion displaces halide ions.

$$2\,BF_3(g) + 6\,NaH(s) \rightarrow B_2H_6(g) + 6\,NaF(s) \tag{10.15}$$

$$SbCl_3 + NaBH_4 \rightarrow SbH_3(g) + NaB H_x Cl_{4-x} \tag{10.16}$$

$$Si_3Cl_8 + 2\,LiAlH_4 \rightarrow Si_3H_8 + 2\,LiCl + 2\,AlCl_3 \tag{10.17}$$

$$Cu^{2+}(aq) + NaBH_4 \rightarrow CuH(s) \tag{10.18}$$

$$3\,LiAlH_4 + AlCl_3 + ether \rightarrow 3\,LiCl + 4\,AlH_3\!:\!ether \tag{10.19}$$

Catenated covalent hydrides are often made from the noncatenated hydrides by pyrolysis, silent electric discharge, or mercury-photosensitized irradiation. These processes eliminate hydrogen:

$$2\,B_2H_6 \rightarrow B_4H_{10} + H_2 \quad \text{(25\,°C, 10 days)} \tag{10.20}$$

$$2\,GeH_4 \rightarrow Ge_2H_6 + H_2 \quad \text{(electric discharge)} \tag{10.21}$$

349

A few special syntheses also merit mention. Arsenic may be detected at very low levels via synthesis of arsine in the Marsh test, in which an arsenic-containing sample is reduced with zinc and acid to arsine, which is then allowed to contact a hot glass plate, where it decomposes to give a silvery mirror of arsenic. Phosphine is made industrially by the disproportionation of white phosphorus in base:

$$P_4 + 3\,KOH + 3\,H_2O \rightarrow PH_3 + 3\,KH_2PO_2 \tag{10.22}$$

Hydrogen peroxide is made industrially by reducing the oxygen of the air with a hydrogen-rich organic compound, 2-ethylanthraquinol, which is then regenerated with $H_2$. Polysulfanes, $H_2S_n$, can be made by the condensation of the appropriate sulfur chloride with hydrogen sulfide:

$$S_nCl_2 + 2\,H_2S \rightarrow H_2S_{n+2} + 2\,HCl \tag{10.23}$$

Hydrazine is made by the *Raschig process*, in which ammonia and basic sodium hypochlorite solutions are combined in the presence of glue or gelatin:

$$NH_3 + OCl^- \rightarrow NH_2Cl + OH^- \tag{10.24}$$

$$NH_2Cl + NH_3 \rightarrow NH_2NH_3{}^+ + Cl^- \tag{10.25}$$

$$NH_2NH_3{}^+ + OH^- \rightarrow H_2O + N_2H_4 \tag{10.26}$$

A number of complex routes are available for synthesizing hydroxylamine.

## 10.5

## Reactions and Uses of the Hydrides

The most widely used hydrides are water and the hydrocarbons, of course; otherwise they are those of the other most electronegative nonmetals. Ammonia is used in enormous quantities in fertilizers as anhydrous (liquid) ammonia or as derivatives (ammonium salts, urea); it is also used in several major chemical processes: the manufacture of plastics based on amines, refrigeration, and the manufacture of nitrogen compounds in general. Some is oxidized over a Pt catalyst at 800 °C to give nitric oxide:

$$4\,NH_3 + 5\,O_2 \rightarrow 4\,NO + 6\,H_2O \tag{10.27}$$

The NO is subsequently oxidized by air to $NO_2$, which is converted to $HNO_3$ (equation (4.14)) and thence to nitrogen-based explosives.

Hydrogen fluoride is used in the manufacture of Freons (chlorofluorocarbons) for aerosols and in the manufacture of synthetic cryolite ($Na_3AlF_6$) used in making aluminum metal (Section 5.7); it is also a useful solvent for many biochemicals. It attacks glass, producing $H_2SiF_6$, and hence must be used in plastic containers. Anhydrous HCl is used in the manufacture of anhydrous metal halides and some metals (Mg, Al, Ti, etc.); aqueous hydrochloric acid has many uses, the largest of which is the pickling of steel (removal of surface oxides). The largest single use of hydrogen peroxide is as an industrial bleach.

Some of the hydrides of the less active nonmetals find limited uses that take advantage of their reactivity. Arsine and stibine react with metals to give arsenides and antimonides, which are useful dopants in semiconductors. The boron hydrides release enormous amounts of energy per gram upon combustion and were tested as rocket fuels; organic derivatives of hydrazine are currently used for that purpose. Hydrazine, hydroxylamine, and phosphine react with certain classes of organic compounds (e.g., ketones) to give useful derivatives (the hydrazones and oximes and the compound $[P(CH_2OH)_4]Cl$, used in flameproofing cotton cloth). $B_2H_6$ and silicon hydride derivatives react with organic compounds containing $C=C$ double bonds to form C—H and C—B or C—Si bonds; these processes are called hydroboration and hydrosilation; they are important in certain organic and organometallic syntheses. $NaBH_4$ and $LiAlH_4$ are useful reducing agents, not only in inorganic synthesis, but also in organic chemistry.

The metallic hydrides have been much investigated for possible use as an energy source in fuel cells, since they are a much more convenient form in which to store hydrogen than as gaseous $H_2$. They have also been tested as catalysts and as reactive starting materials for synthesis. $UH_3$, for example, forms as a powder that is more conveniently reactive than uranium metal and hence is used in the synthesis of some uranium compounds. Two saline hydrides find important uses: NaH as a powerful reducing agent and strong base in nonaqueous solvents, and $CaH_2$ (which is substantially less reactive) as a drying agent for organic solvents, since it reacts with water to give $Ca(OH)_2$ and $H_2$.

TOXICITY  Many of the molecular hydrides are extremely toxic materials. HF burns the skin and precipitates calcium ions in the tissues, producing excruciating pain. It is a particularly dangerous substance in part because it fails to produce pain at first, so that one may not notice that contact of skin and acid has occurred. The greatest caution should be exercised to prevent fluorides from contacting acids or HF from contacting the skin. In the event that contact does occur, not only should the burn be washed with copious amounts of water, but prompt medical attention must be obtained—otherwise surgical removal of the affected area may be needed! The other hydrogen halides are "merely" corrosive substances whose fumes are choking and should not be inhaled.

$H_2S$, $H_2Se$, and $H_2Te$ are sources of their soft-base anions and hence are very poisonous; their revolting odors usually alert us to the hazard, but our ability to smell $H_2S$ can be numbed, creating a dangerous situation. Ammonia, hydrazine, and hydroxylamine are all toxic; the latter two also can pose explosion hazards. Phosphine, arsine, and stibine also can act as soft bases and are among the most toxic gases known; they also have revolting smells. Diborane, stannane, and silane are also toxic (the latter probably due to the finely divided silica formed in the lungs upon oxidation). Although germane is not known to be toxic, it is germane to note the precedent of all its neighboring hydrides. You may also recall from Section 5.3 that some of these compounds can form in natural waters under conditions of very low pE and can cause untoward environmental consequences.

You should also recall that many of these compounds either inflame spontaneously (sometimes explosively) in air or ignite readily. And the hydrides of many of the metals less electronegative than hydrogen react with water to give oxides or hydroxides and $H_2$, which itself can inflame. (The boundary to this property seems to come near silicon: Neither carbon nor germanium hydrides react with water, but the hydrides of the less electronegative Si do.) The art of chemical synthesis had to make great strides in experimental sophistication in order to handle compounds such as these. Greaseless vacuum-line techniques (rigid exclusion of air and water) were invented by Alfred Stock and co-workers early in the twentieth century to make the study of the boranes and silanes possible.

## 10.6

## Carbides of the Elements

Before discussing the chemistry of organometallic compounds, we will briefly look at the chemistry of the carbides [4] of the elements that are less electronegative than C. (Although both classes of compounds contain metal-carbon bonds, the carbides lack any other element attached to the carbon.) These resemble the previously discussed metallic hydrides and nitrides, as well as the metallic borides and silicides, which will not be discussed. They are generally prepared either (1) by direct combination of the elements at over $2000 \, °C$, (2) by reduction of the metal oxide with carbon at high temperatures, or (3) by reaction of the metal (or metal ion) with an appropriate hydrocarbon. They may be divided into three classes: (1) saltlike or ionic carbides, (2) metal-like or interstitial carbides, and (3) polymeric covalent carbides.

SALTLIKE OR IONIC CARBIDES    Saltlike or ionic carbides are colorless ionic salts containing, in effect, one of three very basic carbon anions: $C^{4-}$, the triple-bonded $C_2{}^{2-}$ ion (acetylide ion), or the $C_3{}^{4-}$ ion. The $C^{4-}$ ion occurs in $Be_2C$ and $Al_4C_3$, which react with water to give $CH_4$. The $C_3{}^{4-}$ ion occurs in $Mg_2C_3$ and $Li_4C_3$, which hydrolyze to give $CH_3C{\equiv}CH$. Most saltlike carbides act as salts of the acetylide ion and hydrolyze to give acetylene, $HC{\equiv}CH$. Many of these can be prepared by reacting the metals with acetylene; the Group 1 metals give $M_2C_2$, the Group 2 and 12 metals give $MC_2$. $CaC_2$ is produced by heating $CaO$ and $C$ at about $2000 \, °C$:

$$CaO + 3C \rightarrow CaC_2 + CO \qquad (10.28)$$

Calcium carbide is important as a source of acetylene and for its reaction with $N_2$ at $1000 \, °C$ to give calcium cyanamide, $CaCN_2$, used ultimately in producing polymers:

$$CaC_2 + N_2 \rightarrow CaCN_2 + C \qquad (10.29)$$

Soft-acid "acetylides" can be precipitated from aqueous solutions containing ammonia by the action of acetylene, but these acetylides ($Cu_2C_2$, $Ag_2C_2$, $Au_2C_2$, $Hg_2C_2$, $HgC_2$) are very explosive.

METAL-LIKE OR INTERSTITIAL CARBIDES   The $f$-block metals also form carbides of formula $MC_2$, which contain $C_2$ groups but also $M^{3+}$ ions instead of $M^{2+}$ ions; these are metallic conductors. They hydrolyze to give a mixture of hydrocarbon products. The metal-like carbides of the $d$-block metals do not contain $C_2$ units and have typical stoichiometries of MC, $M_2C$, and $M_3C$. These conducting compounds are stable to water and are very high-melting and hard; TaC and WC are widely used in the production of high-speed cutting tools. $Fe_3C$, cementite, is an important constituent of steel, being formed during the reduction of iron ores by carbon.

POLYMERIC COVALENT CARBIDES   Two nonmetal carbides are important: boron carbide, $B_4C$, and silicon carbide or carborundum, SiC. $B_4C$ and SiC are prepared by heating $B_2O_3$ or $SiO_2$ with carbon in an electric furnace and are used as refractories and as abrasives; SiC is extremely hard. The structure of SiC is related to that of diamond, while the structure of $B_4C$ contains $B_{12}$ icosahedra linked directly to each other and also linked via $C_3$ bridges. Both are chemically quite unreactive.

## 10.7

## Methyl and Phenyl Compounds of the Elements: Structures and Stability

The study of organometallic chemistry is normally divided into the organometallic chemistry of the representative ($s$- and $p$-block) elements and the chemistry of the transition ($d$-block) elements, since different types of organometallic compounds are characteristic of these two regions of the periodic table [5]. Methyl and phenyl derivatives are far more characteristic of the $s$- and $p$-block elements, while carbonyls and complexes in which the ligand donates pi-bond electrons are far more characteristic of the $d$-block elements. We will examine the former first.

Methyl and phenyl compounds of the elements are related to the hydrides of the elements and are named by modifying the names of the corresponding hydrides (this nomenclature, given in Section 7.6, should be reviewed now). Characteristic methyl compounds (or phenyl compounds if the methyl compound is unstable) are listed in Table 10.3. (In this table we follow the common convention among organic chemists of using Me as a symbol for the methyl group and Ph as a symbol for the phenyl group.) These methyl compounds can be put into categories similar to those of the hydrides. The formulas are also similar, although in Group 15/VA there exist pentamethyl or pentaphenyl compounds that have no hydride analogues.

Since there are more electrons in the methyl group than in hydrogen, there are stronger van der Waals forces among molecular methyls than among hydrides and still stronger forces among molecular phenyls. Thus most of the molecular methyls listed in Table 10.3 are liquids, while the corresponding molecular phenyls are generally low-melting solids.

The saltlike or ionic and the polymeric covalent methyls and phenyls are solids, as we would expect, except for the dimeric $(AlMe_3)_2$, which has a structure like that of diborane (Figure 10.1) but with bridging methyl groups. Polymeric chains

**Table 10.3**  Methyl Derivatives of the Elements

**A. Molecular (Covalent) Methyls**
All are liquids at room temperature unless otherwise noted.

| | $Me_3B(g)$ | $Me_4C(g)$ | | $Me_3N(g)$ | $Me_2O(g)$ | $MeF(g)$ |
|---|---|---|---|---|---|---|
| | | $Me_4Si$ | $Ph_5P(s)$ | $Me_3P$ | $Me_2S$ | $MeCl(g)$ |
| $Me_2Zn$ | $Me_3Ga$ | $Me_4Ge$ | $Me_5As$ | $Me_3As$ | $Me_2Se$ | $MeBr$ |
| $Me_2Cd$ | $Me_3In(s)$ | $Me_4Sn$ | $Me_5Sb$ | $Me_3Sb$ | $Me_2Te$ | $MeI$ |
| $Me_2Hg$ | $Me_3Tl(s)$ | $Me_4Pb$ | $Ph_5Bi(s)$ | $Ph_3Bi$ | | |
| | | | $Me_5Ta(s)$ | | $Me_6W(s)$ | $Me_6Re(s)$ |

**B. Saltlike (Ionic) and Polymeric Covalent Methyls**
All are solids at room temperature unless otherwise noted.

| | | |
|---|---|---|
| $(LiMe)_4$ | $(BeMe_2)_n$ | |
| $(NaMe)_4$ | $(MgMe_2)_n$ | $(AlMe_3)_2(l)$ |
| $KMe$ | $CaMe_2$ | $(CuMe)_n$ |
| $RbMe$ | $SrMe_2$ | |
| $CsMe$ | $BaMe_2$ | |

**C. Representative Catenated Molecular Methyls**

| | | |
|---|---|---|
| $(Me_2Si)_6$ | $(MeP)_5$ | $Me_2S_2$ |
| $(Ph_2Ge)_6$ | $(MeAs)_5$ | $Me_2Se_2$ |
| $(Me_2Sn)_6$ | $Me_4Sb_2$ | $Ph_2Te_2$ |
| $Pb(PbPh_3)_4$ | $Me_4Bi_2$ | |

NOTE: The symbol Me is used for the $CH_3$ group; Ph is used for the $C_6H_5$ group. Phenyl derivatives are shown if the corresponding methyl is unknown.

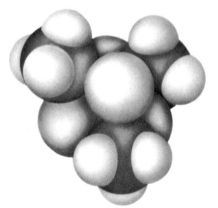

**Figure 10.4**  The structure of the methyllithium tetramer: Each methyl group bridges three lithium atoms, and each lithium atom bridges three methyl groups (one methyl group is obscured from view.) From E. Weiss and E. A. C. Lucken, *J. Organometal. Chem.*, 2, 197 (1964). Reprinted with permission of Elsevier Science Publishers.

are formed via bridging methyl groups in methylcopper, dimethylberyllium, and dimethylmagnesium, while the bridging methyl groups are used to create tetramers in methyllithium and methylsodium (Figure 10.4). In contrast, methylpotassium and methylrubidium adopt the three-dimensional lattice structure of NiAs (Figure 9.7a); these are considered ionic compounds. Nothing is known of the structures of dimethylcalcium, -strontium, and -barium.

Since organic groups such as $CH_3$ and $C_6H_5$ are larger than the hydride group, they more effectively hinder the approach of reactive substances such as $H_2O$ and $O_2$. Thus the methyls and phenyls of many elements are considerably more stable to water and air than the corresponding hydrides. Also, since their decomposition does not readily produce the very stable H—H bond, the methyls and phenyls are often more stable thermally than the corresponding hydrides. For example, while silane and germane are spontaneously flammable in air, and stannane and plumbane decompose even in the absence of air at room temperature, the corresponding tetramethyls are thermally stable at room temperature and do not catch fire in air (in the absence of a flame or spark).

But trimethyl-phosphine, -arsine, -stibine, and -bismuthine retain a point of vulnerability—their unshared pairs of electrons—and catch fire or oxidize readily in air to give species such as trimethylphosphine oxide, $Me_3P{=}O$ (halogens produce $Me_3PX_2$). Similarly, the Group 13/IIIA methyls (including $Me_6Al_2$) are spontaneously flammable or very readily oxidized. The phenyls of these elements have considerably improved stability: Triphenylphosphine is unaffected by air. The strongly reducing saltlike and polymeric covalent methyls and phenyls are oxidized readily by air, however, and also react very violently with water.

There is an interesting gradation in stability among the Group 12 methyls. Dimethylzinc is decomposed by water and is spontaneously flammable; dimethylcadmium is decomposed by water but is less quickly oxidized by air, whereas dimethylmercury is unaffected either by water or by air. This, of course, means that dimethylmercury is among the most persistent of the methyls in the environment. It is interesting that in these three species the central atom has a coordination number of only 2—there is no tendency of the methyl groups to bridge.

The tendency of methyls to be more stable than hydrides is especially predominant among the catenated hydrides and organometallics. For example, the high flammability and other forms of reactivity of the polysilanes, $Si_nH_{2n+2}$, has limited the value of $n$ to date to about 8, but there seems to be no limit at all to the extent of catenation in the fully methylated polysilanes. Table 10.3C cannot list the numerous catenated methyls and phenyls that exist, particularly for Si, Ge, and Sn. Not only are there linear methylated polysilanes, $Si_n(CH_3)_{2n+2}$, but there are also several methylated cyclic polysilanes, $(Me_2Si)_n$, and even a variety of methylated bicyclic polysilanes. Methylated polysilane polymers are now being investigated. Thus given the protection afforded by large substitutents with nonpolar bonds to silicon, it would seem that silicon chemistry could achieve at least a fair share of the variety found in carbon (organic) chemistry.

Phosphorus and arsenic form catenated compounds of two types: dimers of the type $Me_4P_2$ and $Me_4As_2$ and several compounds containing rings of P or As atoms of various sizes (e.g., $(MeP)_5$ and $(MeAs)_5$). Antimony and bismuth form only the former type of compound, but these products show the interesting property of **thermochromism**: Their colors change with temperature. Thus tetramethyldistibine is pale yellow at $-180\,°C$ but darkens as it warms, becoming bright red at $-17\,°C$. It then melts to give a pale yellow liquid. Tetramethyldibismuthine is a violet-blue solid that melts to give a red-yellow liquid. Recent structural studies have shown that the bright colors are associated with the presence of secondary bonds weakly linking antimony or bismuth atoms of different molecules into a type of polymer in the solid state only [6].

Of course, the larger size of the methyl or phenyl group is a disadvantage if the central atom of the molecule is very small. Thus there are no fully methylated analogues of the numerous catenated boron hydrides, and perphenylated organic compounds are limited: The compound hexaphenylethane, $(C_6H_5)_6C_2$, is so sterically strained that it tends to break apart in solution to give the triphenylmethyl free radical, $(C_6H_5)_3C\cdot$.

Finally, we mention that although pure methyls and phenyls of the $d$-block metals are seldom stable, methyl, phenyl, and other organic groups often occur in $d$-block organometallic compounds also containing other ligands. The preparation of pure organic derivatives of $d$-block metals is possible if still bulkier organic groups are used that also lack any hydrogens or halogens at the second carbon atom ($d$-block metal ions generally catalyze the removal of such groups). Thus there are many neopentyl derivatives of $d$-block metals, such as $Cr(CH_2CMe_3)_4$ [7].

## 10.8

## Lewis Acid-Base Properties of the Methyls and Phenyls

An important part of the chemistry of the methyls and phenyls of the $s$- and $p$-blocks is their Lewis acid-base chemistry. Characteristically, the metal atom of the saltlike and polymeric covalent methyls is much less electronegative than the carbon atom, so these methyls (and those of zinc and cadmium) either contain the methide ion, $^-CH_3$, or may act as sources of that soft base. Organolithium compounds are extensively used in organic chemistry as a source of such carbon anions ("carbanions") for the construction of more extensive carbon chains:

$$Li^+Me^- + Me_2C{=}O \rightarrow Me_3CO^-Li^+ \tag{10.30}$$

Even more well known for this use in organic chemistry are the *Grignard reagents* such as $CH_3MgBr$. The diorganozinc and diorganocadmium compounds, which contain more electronegative metals and hence less partially negatively charged organo groups, are also used for this purpose when a mild reagent is needed. You should note that these methyls, etc., are not only very strong bases but are also powerful reducing agents that can transfer electrons to an appropriate reactant. It is sometimes difficult to predict in advance which function (Lewis base or reducing agent) will predominate in a given situation.

The methyls and phenyls of the elements of Groups 15/VA, 16/VIA, and 17/VIIA can act as Lewis bases in an entirely different manner, by donating unshared electron pairs from the central atom. This function is most significant in Group 15/VA and only occasionally manifests itself in Group 17/VIIA. Thus trimethyl- and triphenylphosphine and -arsine are important soft-base ligands, forming numerous complexes with soft or borderline-acid metal ions. Fewer complexes of trimethyl- or triphenylstibine and -bismuthine are known, due perhaps to the lower covalent bond energies. Similarly, organic sulfides and selenides are soft bases, forming numerous complexes. Organic ethers, $R_2O$, and organic amines, $R_3N$, act as hard and borderline bases in forming complexes; the latter can experience steric problems in complex formation.

Another manifestation of Lewis basicity of these methyls occurs when they react with methyl halides to displace the halide ion:

$$Me_3P: + MeI \rightarrow Me_4P^+ + I^- \qquad (10.31)$$

The cation produced is analogous to the phosphonium ion mentioned previously but is more stable since it cannot readily transfer a proton and lose its positive charge. Such ions are named using the suffix -onium: $Me_4P^+$ is the tetramethyl-phosphonium ion, $Ph_4As^+$ is the tetraphenylarsonium ion, $Me_3S^+$ is the trimethyl-sulfonium ion, and $Ph_2I^+$ is the diphenyliodonium ion. These are very large nonacidic ions, and are often used to precipitate large nonbasic anions. Those from Groups 16/VIA and 17/VIIA readily transfer their methyl groups to more basic substances, however, and thus are alkylating agents that may be mutagenic (Section 9.7).

The methyls and phenyls of Group 13/IIIA lack an octet of electrons and hence are important Lewis acids. They readily form adducts with Lewis bases:

$$Me_3B + :NMe_3 \rightleftharpoons Me_3B:NMe_3 \qquad (10.32a)$$

They can, of course, react with the $Ph^-$ Lewis base to give complex anions:

$$Ph_3B + Na^+Ph^- \rightarrow Na^+ + [BPh_4]^- \qquad (10.32b)$$

This compound (sodium tetraphenylborate) is an important source of a large nonbasic anion and is used to precipitate large nonacidic cations. Similarly, copper is usually encountered, not as its polymeric methyl compound, but as lithium dimethylcuprate or related organocopper compounds, which find extensive use in organic synthesis. Some transition metal ions that do not form stable neutral methyls do give stable anions (in which more coordination sites are blocked), such as $[ReMe_8]^{2-}$ and $[PtMe_6]^{2-}$.

## 10.9

### Methods of Synthesis of Methyls and Phenyls

One of the most important methods of producing methyl and phenyl compounds of the metals and metalloids [5] is by the *direct reaction of a methyl or phenyl halide with the element.*

$$4\,CH_3Cl + 8\,Li \rightarrow [CH_3Li]_4 + 4\,LiCl \qquad (10.33)$$

With a divalent metal such as magnesium, the product is not normally the pure dimethyl compound plus $MgCl_2$ but is instead the mixed complex commonly known as a **Grignard reagent**:

$$CH_3Br + Mg \rightarrow CH_3MgBr \qquad (10.34)$$

This synthesis is generally done in a Lewis-base solvent such as an ether, which coordinates tenaciously to the magnesium atom. The reaction requires the exclusion of air and the presence of a dry solvent, and often a little iodine is also needed to clean off the surface coating of MgO that impedes this exothermic reaction. Organozinc compounds are prepared in a similar manner. The organosilicon halides used to make silicone polymers are prepared by passing gaseous $CH_3Cl$ over heated silicon mixed with copper as a catalyst:

$$2\,CH_3Cl + Si\,(300\,°C) \rightarrow (CH_3)_2SiCl_2 \tag{10.35}$$

The reaction also produces a few percent of byproducts such as $Me_3SiCl$, $MeSiCl_3$, $MeHSiCl_2$, $SiCl_4$, and disilanes.

This reaction is an oxidation-reduction reaction that the least active metals often do not undergo. These metals may be alloyed with an active metal such as Na to promote the reaction, as in the production of tetraethyllead for use in leaded gasoline:

$$4\,C_2H_5Cl + 4\,PbNa \rightarrow (C_2H_5)_4Pb + 3\,Pb + 4\,NaCl \tag{10.36}$$

Organomercury compounds can also be prepared in this manner, using the sodium-mercury alloy (sodium amalgam).

Methyl and phenyl derivatives of a more active metal can often be prepared by the *reaction of the more active metal with the organometallic derivative of a less active metal*, such as an organomercury compound. This reaction is useful for preparing organometallic compounds with no coordinated ether solvent molecules:

$$(CH_3)_2Hg + excess\,Na \rightarrow 2\,RNa + Na\,amalgam \tag{10.37}$$

The *reaction of organometallic compounds with metal halides* provides (at least formally) a Lewis acid-base exchange of the methyl anion for a halide ion: In accord with the HSAB Principle, the soft organic anion ends up coordinated with the more electronegative (soft) metal:

$$2\,CH_3MgX + CdCl_2 \rightleftharpoons (CH_3)_2Cd + MgX_2 + MgCl_2 \tag{10.38}$$

The *metallation* reaction (metal-hydrogen exchange reaction) also appears to be an acid-base (proton transfer) reaction, although the strongly reducing nature of organometallic compounds sometimes causes these reactions to proceed via electron-transfer mechanisms. Thus an organolithium compound, especially in the presence of a chelating Lewis-base ligand such as $Me_2N—CH_2CH_2—NMe_2$, will react with a more acidic organic compound to produce a less acidic organic compound and another organolithium compound:

$$C_4H_9Li + C_6H_6 \rightarrow C_4H_{10} + C_6H_5Li \tag{10.39}$$

Hard-base salts of the $Hg^{2+}$, $Tl^{3+}$, and $Pd^{2+}$ ions will often carry out metallation under quite mild conditions, readily giving products containing the soft-acid–soft-base $Hg—C$, $Tl—C$, and $Pd—C$ bonds:

$$Hg^{2+} + C_6H_6 \rightarrow C_6H_5Hg^+ + H^+ \tag{10.40}$$

Such reactions are commonly known as *mercuration, thallation,* and *palladation* reactions.

The *reaction of unsaturated organic compounds with a metal hydride or with a metal plus hydrogen* is important for the elements boron, aluminum, and silicon (plus many of the nonmetals). The *hydroboration* reaction is extremely important in organic chemistry, since the organoboranes produced can be converted to many important organic chemicals:

$$B_2H_6 + 6\,H_2C=CH_2 \rightarrow 2\,(C_2H_5)_3B \qquad (10.41)$$

A similar reaction of silicon hydrides is important for producing silanes for use in making silicones.

Organoaluminum compounds, which are very important (along with titanium compounds) in the Ziegler-Natta catalysts for producing organic polymers such as polyethylene, are produced by the direct reaction of aluminum metal, the organic compound, and hydrogen:

$$2\,Al + 3\,H_2 + 6\,CH_2=CH_2 \rightarrow 2\,(C_2H_5)_3Al \qquad (10.42)$$

The *reactions* of organometallic compounds are too many and diverse for us to cover here. The numerous applications of organometallic compounds in organic synthesis will be covered in the organic chemistry course, while the applications in organometallic chemistry may be covered in a more specialized inorganic course.

## 10.10

## Organometallic Compounds of the *d*-Block Metals: Simple Carbonyls and the 18-Electron Rule

Although *d*-block elements do form compounds containing methyl and phenyl groups, different types of ligands are more characteristic of the *d*-block organometallic compounds. These are ligands that not only can function as donors of unshared pairs of electrons but also can form partial *pi* bonds with metals. This pi bond involves partial overlap of *d* orbitals on the metal with additional orbitals on the ligand. Often the most important type of pi bond results from overlap of a filled metal *d* orbital with an empty antibonding orbital on the ligand. This type of orbital is characteristically found in *some* soft bases.

Unfortunately, without going into molecular orbital calculations it is very difficult to predict which soft-base ligands will have empty antibonding orbitals of suitable energy and which will not. Many of the suitable ligands are organic compounds, but organophosphorus compounds (using P donor atoms) are also good for this purpose, while the methyl and phenyl groups have no suitable antibonding orbitals. Since many students using this text may not have had organic chemistry, we will confine our attention to three ligands: *carbon monoxide* [8], *benzene* ($C_6H_6$), and the *cyclopentadienide anion* ($C_5H_5^-$).

As soft bases, these species characteristically form organometallic compounds with soft-acid *d*-block metal atoms. The suitable soft-acid metal atoms include not

**Table 10.4**   Carbonyls of the *d*-Block Elements

**A. Mononuclear Carbonyls**

| | | | | | | | |
|---|---|---|---|---|---|---|---|
| $V(CO)_6(s)$ | | | | | | | |
| $V(CO)_6^-$ | $Cr(CO)_6(s)$ | $Mn(CO)_6^+$ | $Mn(CO)_5^-$ | $Fe(CO)_5(l)$ | $Co(CO)_5^+$ | $Co(CO)_4^-$ | $Ni(CO)_4(l)$ |
| $Nb(CO)_6^-$ | $Mo(CO)_6(s)$ | $Tc(CO)_6^+$ | $Tc(CO)_5^-$ | $Ru(CO)_5(l)$ | | | |
| $Ta(CO)_6^-$ | $W(CO)_6(s)$ | $Re(CO)_6^+$ | $Re(CO)_5^-$ | $Os(CO)_5(l)$ | | | |

**B. Binuclear Carbonyls**

| | | |
|---|---|---|
| $Mn_2(CO)_{10}(s)$ | $Fe_2(CO)_9(s)$ | $Co_2(CO)_8(s)$ |
| $Tc_2(CO)_{10}(s)$ | $Ru_2(CO)_9(s)$ | $Rh_2(CO)_8(s)$ |
| $Re_2(CO)_{10}(s)$ | $Os_2(CO)_9(s)$ | $Ir_2(CO)_8(s)$ |

only our traditional soft acids from Chapter 8 but also most of the other *d*-block elements in *very low oxidation states*, such as $+1$, 0, or even $-1$ [9]. Table 10.4 shows the noncatenated **carbonyls** of the *d*-block elements in these three oxidation states. The carbonyls with an overall negative charge are called *carbonylate anions*; those with an overall positive charge are called *carbonyl cations*.

You may notice a periodic regularity in the formulas of these carbonyls. A full Lewis dot structure of any of the carbonyls in Table 10.4A, except $V(CO)_6$, would have 18 valence electrons about the central atom, which would bring the electron configuration of the central atom to that of the next noble gas. Corresponding to the frequent occurrence of octets in compounds of *p*-block elements, *stable organometallic compounds* (not coordination compounds) *of the d-block elements tend to have 18 valence electrons about the metal atom.* But just as there are exceptions to the octet rule among *p*-block elements, there are also some organometallic compounds with other than 18 valence electrons about the *d*-block metal atom. The largest group of these are compounds in which the metal atom is 4-coordinate in a square planar (not tetrahedral) geometry: These compounds tend to have 16 valence electrons.

Drawing Lewis structures of *d*-block compounds to check the count of 18 valence electrons is impractical; instead we generally dissect the compound into ligands and the central atom and add up the contribution for each. Thus for $V(CO)_6^-$, $Cr(CO)_6$, and $Mn(CO)_6^+$, each of the six $:CO$ molecules is a two-electron donor, accounting for 12 electrons about the metal atom or ion. We then note the number of valence electrons of the central species: in this case, 6 for $V^+$, Cr, and $Mn^-$ to give totals of 18 for each of these isoelectronic species. Neutral $V(CO)_6$ turns out to be a 17-electron species.

The 18-electron rule holds well enough that it can be used to predict the stoichiometries of *d*-block organometallics. Thus in a titanium carbonyl, $[Ti(CO)_n]^{x+}$, neutral Ti has 4 valence electrons, the *n* carbonyls contribute $2n$ electrons, and the $+x$ positive charge contributes $-x$ electrons. We expect the sum, $4 + 2n - x$, to equal 18. If we take *x* equal to zero, we find that the uncharged carbonyl should have seven CO ligands, which exceeds the maximum coordination number for Ti; hence we are not surprised to find that this compound does not exist. (Although an anion $[Ti(CO)_6]^{2-}$ could be predicted, it is not known to exist.)

We may also use the rule to predict the results of reactions in which the carbonyl ligand is replaced by another ligand, such as a halide ion, $:X^-$; the hydride ion, $H^-$; or the methyl or phenyl anions. These ions contribute as many electrons as the CO but differ in that they neutralize one unit of positive charge. Thus we might expect reactions such as the following to occur:

$$Mn(CO)_6^+ + X^- \rightleftharpoons Mn(CO)_5X + CO \tag{10.43}$$

The predicted halo-, methyl-, and phenylpentacarbonylmanganese(I) compounds are well known. Similar substitution products are known corresponding to many of the other carbonyls listed in Table 10.4A and even for some metals not forming simple carbonyls: Pd, Pt, Cu, Ag, and Au form carbonyl halides.

Similarly, $d$-block organometallic hydrides can be envisioned as resulting from replacement of a $:CO$ (or other neutral ligand) by the hydride ion, $:H^-$. Alternatively, such a compound can arise from the neutralization of a carbonylate anion by a proton, $H^+$. Either process predicts, in the case of cobalt, that its carbonyl hydride should have the formula $HCo(CO)_4$, which it does. But it is hard to predict whether this hydrogen will act more like $H^+$ or $H^-$, especially since the electronegativity of hydrogen is even closer to those of the $d$-block metals than is the electronegativity of carbon. For purposes of nomenclature and the assignment of oxidation numbers, hydrogen in $d$-block organometallic compounds should be assigned the oxidation number $-1$ and called "hydrido."

Neutral carbonyls containing only one metal atom do not arise in the odd-numbered groups of the $d$-block, with the exception of $V(CO)_6$, since these must have odd numbers of valence electrons. However, two such unstable fragments, each having 17 electrons, can share their unpaired electrons to form a metal-metal bond, producing catenated binuclear metal carbonyls in which each metal atom is surrounded by 18 valence electrons. These are shown in Table 10.4B, along with corresponding 18-electron binuclear metal carbonyls from the even groups. Evidently $V(CO)_6$ fails to form a binuclear carbonyl for steric reasons: Counting the M—M bond, such a carbonyl would require a coordination number of 7 for the vanadium [10].

When we count valence electron totals in catenated organometallic compounds, each M—M bond contributes one electron to the count for each metal atom. Thus the count for *each* cobalt atom in $(CO)_4Co—Co(CO)_4$ includes nine electrons from the cobalt atom, four times two electrons from the four carbonyl groups, and one electron from the other metal atom in the M—M bond.

Numerous even more fully catenated metal carbonyls are also known. These carbonyls are found especially for the $d$-block metals of the fifth and sixth periods. They contain from three to dozens of metal atoms, are generally not linear chains or even rings, but contain the metal atoms in clusters so as not to sacrifice the substantial metal-metal atomization energies found among such metals (Section 6.1;

see Figure 6.9f for an example). These do not always obey the 18-electron rules; as the clusters become more and more like small chunks of the metal itself different electron-counting systems are called for [11].

## 10.11

## Syntheses and Properties of Metal Carbonyls

The metal carbonyls, as molecular solids with a fairly large number of electrons, are volatile liquids or solids. (The catenated polynuclear carbonyls are predictably less volatile solids.) The M—C bonds in these compounds are not especially strong, and they readily lose carbon monoxide, for example, on heating in reactions that generate polynuclear carbonyls:

$$3\,Fe_2(CO)_9 + heat \rightarrow 2\,Fe_3(CO)_{12} + 3\,CO \tag{10.44}$$

Thus these compounds should be handled in hoods to avoid carbon monoxide poisoning. Some of them are even more toxic than carbon monoxide, however. Nickel carbonyl, for example, is 100 times as toxic as CO; it apparently decomposes in the lungs, in effect nickel-plating them.

Some metal carbonyls can be prepared by the direct reaction of the metal with carbon monoxide. This is not generally as easy a reaction as one may hope, however, since the atomization energies of many of the metals are so high. With the exception of nickel, high temperatures, long reaction times, and high pressures are required. Nickel reacts directly with CO at $30\,°C$ and 1 atm pressure to give the carbonyl, which decomposes back to nickel metal and CO at somewhat higher temperatures; this reaction was the basis of the old Mond process for purifying nickel. Synthesizing other carbonyls such as those of iron, cobalt, molybdenum, and ruthenium requires temperatures of hundreds of degrees and pressures of hundreds of atmospheres for hours or days.

Alternatively, many carbonyls can be made from CO, a metal halide, and a reducing agent (an active metal, an organometallic compound, or $H_2$, or even CO gas itself):

$$VCl_3 + 6\,CO + 4\,Na + 2\,MeO(CH_2CH_2O)_2Me \rightarrow \tag{10.45}$$
$$3\,NaCl + [Na(MeO(CH_2CH_2O)_2Me)_2]^+[V(CO)_6]^-$$

An interesting property of some of the carbonylate anions is that they act as Lewis bases, using the $d$-block metal atoms as *donor* atoms. Toward protons they vary considerably in basicity: $Co(CO)_4^-$ is nonbasic, $Mn(CO)_5^-$ and $[HFe(CO)_4]^-$ are weakly basic, while $Fe(CO)_4^{2-}$ is strongly basic. But all are very soft bases (as befits species with donor atoms of such low electronegativity) and readily form complexes with metal ions such as $Hg^{2+}$ that have Hg-metal bonds:

$$2\,[Co(CO)_4]^- + Hg^{2+} \rightarrow Hg[Co(CO)_4]_2 \tag{10.46}$$

The carbonyls are useful as starting materials for the production of other organometallic compounds, since the gas CO is relatively readily lost from them.

Many of them are also important reagents in organic synthesis and industrial catalysts (Section 7.2). An extremely important advantage of such catalysts is that they often give product yields of close to 100% with no byproducts. Older catalysts (such as strong acids) often produced many byproducts that had to be discarded, which created a serious hazardous-waste disposal problem. From the chemical industry's point of view, the best solution to the problem of disposing of hazardous byproduct wastes is not to generate them in the first place; hence research into catalysis of major industrial organic reactions by transition-metal organometallic catalysts has received top priority in funding among inorganic-chemistry research projects in the United States over the last several years.

Several of the catalytically important organometallic compounds are 16-electron, 4-coordinate square planar species: for example, Wilkinson's catalyst, $[RhCl(PPh_3)_3]$; Vaska's compound, $[Ir(CO)(PPh_3)_2Cl]$; and the catalyst in the Monsanto acetic acid process, $[Rh(CO)_2I_2]^-$. Although it is beyond the scope of this book to go into the mechanisms of action of such catalysts, the vacant coordination site present in such 16-electron species is important in these mechanisms.

## 10.12

## Metal Complexes of Pi-Donor Ligands

A property that is almost confined to the $d$-block metals is the ability to form complexes with organic compounds having *no* unshared pair of electrons but having instead one or more pi bonds. In complexes of such *pi-donor* ligands the pi-bond electron pairs are donated to $d$ orbitals of the metal; there is also some degree of donation of electrons back from other $d$ orbitals to the antibonding pi orbitals of the organic ligand. Numerous pi-donor ligands are known, starting with the simplest, ethylene ($CH_2{=}CH_2$), but we will mainly use as examples benzene ($C_6H_6$) and the cyclopentadienide anion ($C_5H_5^-$). Each has three pairs of electrons from pi-bonding orbitals available to donate (Figure 10.5).

With pi-donor ligands the counting of coordination numbers and the description of geometries become more complex, since there is no equivalence between the number of donor atoms present and the number of electron pairs being donated. In the typical compound bis(benzene)chromium (Figure 10.5c) there are 12 carbon atoms equidistant from the chromium atom, but one is reluctant to assign a coordination number of 12 to the chromium atom, since this violates so grossly the expected maximum coordination number of chromium. It is perhaps more usual in such a case to take the contribution to the coordination number from pi-donor ligands as equal to the number of pi-electron pairs donated. For $Cr(C_6H_6)_2$ this would be 2 times 3 pairs from each benzene ring, or a reasonable total of 6. For the similar molecule $Fe(C_5H_5)_2$, *ferrocene*, this would be 10 if one counted donor atoms but also a reasonable 6 if one counted pi-electron pairs donated.

The description of the geometry of such compounds is also ambiguous. The resemblance of these two molecules to octahedra is limited. They are commonly called *sandwich* molecules, since the metal atoms are sandwiched between two flat

363

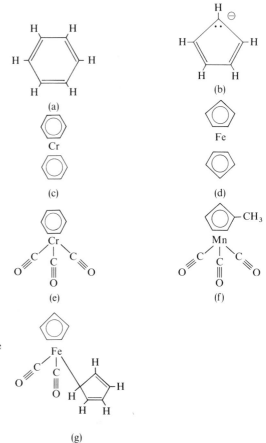

**Figure 10.5** Pi-donor ligands and their organometallic derivatives: (a) Benzene; (b) the cyclopentadienide anion, showing only one resonance structure for each to emphasize the presence of three pairs of electrons in pi-bonding orbitals. The organometallic derivatives illustrated include (c) bis(benzene)chromium, (d) ferrocene, (e) benzene-chromium tricarbonyl, (f) (methylcyclopentadienyl)-manganese tricarbonyl (MMC), and (g) (*pentahapto*-cyclopentadienyl) (*monohapto*cyclopentadienyl)-dicarbonyliron(II).

organic molecules as if they were two slices of bread. The two sandwich molecules illustrated are very important in the history of organometallic and inorganic chemistry, since they were among the very first stable *d*-block organometallic compounds prepared (excepting the carbonyls and complexes of the cyanide ion, which historically had been considered inorganic species).

Bis(benzene)chromium was discovered by E. O. Fischer as the product of a reduction of $CrCl_3$ in the presence of benzene as a pi-donor ligand and $AlCl_3$ as a Lewis acid (chloride-ion acceptor):

$$3\,CrCl_3 + 6\,C_6H_6 + AlCl_3 + 2\,Al \rightarrow 3\,[Cr(C_6H_6)_2][AlCl_4] \tag{10.47}$$

This cation was then reduced in aqueous solution to give the product, a black solid. Just before this (in 1951) Kealy and Pauson, and Miller and co-workers independently reported the discovery of bis(cyclopentadienyl)iron or ferrocene. It can be prepared from $FeCl_2$ and metal or ammonium salts of the $C_5H_5^-$ ion:

$$FeCl_2 + 2\,Tl^+C_5H_5^- \rightarrow 2\,TlCl(s) + Fe(C_5H_5)_2 \tag{10.48}$$

It is an orange solid of melting point 173 °C that is stable to air and to heating to 400 °C. This property contrasted so markedly with that of the known very unstable $d$-block methyl compounds that a flurry of research in a practically new field, transition-metal organometallic chemistry, was initiated.

Electron-counting in these species is relatively straightforward. Both $C_6H_6$ and $C_5H_5^-$ can be counted as 6-electron donors. The neutral Cr atom has 6 electrons, so bis(benzene)chromium obeys the 18-electron rule. If the cyclopentadienyl species is counted as an anion as suggested, the iron in ferrocene should be counted as the cation $Fe^{2+}$ with 6 valence electrons. Thus ferrocene is also an 18-electron species. (Note that by proper nomenclature ferrocene should be called bis(cyclopentadienyl)iron(II).)

Mixed organometallic compounds containing both carbonyl ligands and pi-donor ligands are readily available, such as the important benzenechromium tricarbonyl (Figure 10.5e) and cyclopentadienylmanganese tricarbonyl, commonly called cymantrene; both are 18-electron species containing Cr in oxidation state 0 and Mn in oxidation state +1. The closely related (methylcyclopentadienyl)-manganese tricarbonyl (MMC, Figure 10.5f) has been investigated as a possible replacement for organolead compounds in high-octane gasolines.

The nomenclature that we have just used is a common type, based on that for covalent nonmetal compounds; the justification is presumably that these organometallic compounds are also covalent. In such nomenclature, for example, $Co_2(CO)_8$ is dicobalt octacarbonyl just as $S_2Cl_2$ is disulfur dichloride. The more formal nomenclature is that previously outlined in Section 7.6, by which $Co_2(CO)_8$ is called octacarbonyldicobalt(0).

The 18-electron rule is not inviolate among organometallic compounds of pi-donor ligands, especially among those of the cyclopentadienide anion. It forms sandwich molecules not only with iron, rhodium, and osmium, but also with (among others) the $d$-block metals V, Cr, Co, and Ni. Remarkably, these so-called metallocenes (e.g., vanadocene, chromocene, cobaltocene, and nickelocene) all have melting points very close to 173 °C: The slightly differing metal atoms are evidently so buried within the sandwiches that they have no effect on the intermolecular forces. The non-18 electron metallocenes are much more reactive than ferrocene, however; thus cobaltocene is about as good a reducing agent as sodium metal and is readily oxidized to the stable 18-electron *cobalticenium ion*, $[Co(C_5H_5)_2]^+$.

Although $C_6H_6$ and $C_5H_5^-$ are the most important carbon-ring ligands, it is also possible to form sandwich or other organometallic compounds from other carbon-ring compounds, $C_nH_n^{\pm x}$, or from mixtures of two of these ring compounds. Thus complexes are known containing the two-electron donor $C_3H_3^+$ (cyclopropenium ion), the four-electron donor $C_4H_4$ (cyclobutadiene, stable only in complexes), the six-electron donor $C_7H_7^+$ (tropylium ion), and the ten-electron donor $C_8H_8^{2-}$ (cyclooctatetraenide ion). The latter is too large to form a good sandwich with most $d$-block metal ions, but it does form the most important $f$-block organometallic compounds, such as $U(C_8H_8)_2$, commonly called uranocene.

These carbon ring compounds do not necessarily have to donate all their available pi-electron pairs. For example, the compound $(C_5H_5)_2Fe(CO)_2$ is known;

this formula would appear to be that of a 22-electron molecule. But its structure shows that one ring is bonded flat over the iron with all five carbons essentially equidistant from the iron, while the other ring is attached to the iron at only one carbon atom, in the same manner as a methyl or a phenyl group attaches (Figure 10.5g). (Two double bonds of the second ring are far from the iron atom and are not bonding to it.) The two rings have quite different chemical and spectroscopic properties and are distinguished in naming by use of the prefix *pentahapto-* to indicate a ring attached via five carbon atoms and *monohapto-* to indicate a ring attached via one carbon atom. (In chemical formulas these are indicated as $\eta^5$-$C_5H_5$ and $\eta^1$-$C_5H_5$, respectively.)

Since $\eta^1$-$C_5H_5$ rings are anions that bond to a metal via only one pair of electrons, they are fairly similar to methyl and phenyl groups and form derivatives with *p*-block elements too (example: $Me_3Si$—$\eta^1$-$C_5H_5$). Although the compound $Hg(C_5H_5)_2$ has a formula much like that of ferrocene, we have repeatedly seen a preference by mercury for low coordination numbers. In this compound both rings are monohapto, giving Hg the same coordination number of 2 found in dimethylmercury and diphenylmercury.

## 10.13

## Study Objectives

1. Tell whether a given hydride is likely to have an endothermic or an exothermic heat of formation or whether it is nonexistent.
2. Tell whether a given hydride is likely to be a gas, a liquid, or a solid at room temperature; relate this to the type of structure the hydride has.
3. Tell whether a given hydride is likely to be flammable, toxic, strongly reducing, acidic or basic in a proton-transfer sense, or acidic or basic in a Lewis sense only.
4. Give a reasonable synthesis of a given hydride or hydride anion.
5. (Review) Tell the name of a hydride or organometallic compound given its formula, or vice versa.
6. (Review) Assign oxidation numbers in a given hydride or organometallic compound.
7. Classify the structure of a given borane as closo-, nido-, or arachno-.
8. Classify a given carbide as saltlike, metal-like, or polymeric covalent, and give its likely reaction products with water (if any).
9. Compare and explain the relative stability of hydrides, methyls, phenyls, and carbonyls of the same element.
10. Give the formula of the methyl or phenyl compound of a given element, classify it as molecular covalent, saltlike, or polymeric covalent, or catenated molecular, and describe its physical state.
11. For a given methyl or phenyl compound, tell its susceptibility to water and to air, and describe any Lewis acid or base properties that it has.
12. Give a reasonable synthesis of a given methyl or phenyl compound.

13. Check the formula of a *d*-block organometallic for compliance with the 18-electron rule; use this rule to predict possible formulas of organometallic compounds, cations, or anions involving specified metals and ligands.

14. Give possible syntheses and Lewis-base reactions (if any) of specified metal carbonyl compounds or anions or of sandwich compounds.

## 10.14

### Exercises

1. Tell which of the following elements form (1) no hydride; (2) thermodynamically stable hydride(s); (3) thermodynamically unstable hydride(s): **a.** C; **b.** As; **c.** O; **d.** Br; **e.** Xe; **f.** Sr; **g.** Nd; **h.** Hf; **i.** Fe; **j.** Os; **k.** Ge; **l.** B; **m.** Tl.

2. Give the formula of one hydride for each of the elements from the above list that does form hydrides.

†3. Tell whether each of the following hydrides is a gas, liquid, or solid, and explain this fact in terms of the molecular, catenated molecular, saltlike, polymeric, or metal-like structure of the hydride: **a.** NaH; **b.** HF; **c.** HCl; **d.** $H_2S_5$; **e.** $Na^+B_3H_8^-$; **f.** $SnH_4$; **g.** $N_2H_4$; **h.** CuH; **i.** $PuH_2$; **j.** $SiH_4$; **k.** $Si_8H_{18}$; **l.** $B_2H_6$; **m.** $NaBH_4$.

†4. Write three equations showing reasonable examples of three hydrides showing each of the following kinds of behavior: **a.** proton-donating acidity; **b.** proton-accepting basicity; **c.** electron-pair–accepting acidity; **d.** electron-pair–donating basicity (other than proton accepting); **e.** spontaneous flammability; **f.** decomposition by water; **g.** reducing ability.

5. Rank each of the following series in order of increasing proton-donating acidity: **a.** HF, HCl, HBr, HI; **b.** $SiH_4$, $PH_3$, $SH_2$, ClH; **c.** $[Cr(SH_2)_6]^{3+}$, $[Cr(NH_3)_6]^{3+}$, $[Cr(OH_2)_6]^{3+}$; **d.** $N_2H_6^{2+}$, $N_2H_4$, $N_2H_5^+$.

*6. Rank each of the following series in order of increasing proton-accepting basicity: **a.** $CH_4$, $NH_3$, $AsH_3$, $PH_3$; **b.** $NH_3$, $NH_2OH$, $NH_2NH_2$.

7. Only one of the hydrides discussed in this chapter shows any marked properties as an oxidizing agent. Which one is this likely to be?

†8. Complete and balance an equation showing a reasonable synthesis of each of the following: **a.** $LiAlH_4$; **b.** HF; **c.** $NH_3$; **d.** $N_2H_4$; **e.** $H_2S_3$; **f.** $H_2S_5$; **g.** $Si_2H_6$; **h.** $B_2H_6$; **i.** $NaBH_4$; **j.** $AsH_3$; **k.** $SnH_4$; **l.** $PH_3$; **m.** $B_4H_{10}$; **n.** $UH_3$; **o.** $TiH_2$.

9. Name each of the hydrides in the previous question.

10. Write the formulas of the following hydrides: **a.** bismuthine; **b.** stibine; **c.** plumbane; **d.** hydrazine; **e.** hydroxylamine; **f.** nonaborane-15; **g.** pentagermane; **h.** diborane.

367

11. Classify each of the following boranes or boron hydride anions as either closo-, nido-, or arachno-, and tell what that implies about their structures:
    **a.** $B_7H_7^{2-}$;   **b.** $B_9H_{15}$;   **c.** $B_3H_8^-$;   **d.** $B_2H_6$;   **e.** $B_6H_{10}$;
    **f.** $B_6H_{12}$;   **g.** $B_5H_8^-$.

12. Assign oxidation numbers to the central or metal atom in the following compounds: **a.** FH;   **b.** $AsH_3$;   **c.** $N_2H_4$;   **d.** $NH_2OH$;   **e.** $CH_4$;
    **f.** $MgH_2$;   **g.** $H_2O_2$;   **h.** $BiPh_3$;   **i.** $BiPh_5$;   **j.** $(MeP)_5$;
    **k.** $Me_4Bi_2$;   **l.** $C_5H_5Mn(CO)_3$;   **m.** $C_6H_6Cr(CO)_3$;
    **n.** $[Cr(C_6H_6)_2]^+$;   **o.** $Co(CO)_5^+$;   **p.** $Co(CO)_4^-$;   **q.** $[Fe(C_5H_5)_2]^+$;
    **r.** $MnCl(CO)_5$.

13. Name each of the compounds in the previous question (except for (j)).

14. Write the formulas of the following compounds: **a.** germane;
    **b.** digermane;   **c.** hexamethyldialane;   **d.** dodecamethylcyclohexasilane
    (*cyclo-* means a ring compound);   **e.** triphenylstibine;
    **f.** tetraphenylbismuthonium tetracarbonylferrate(-II);
    **g.** cobalt pentacarbonyl cation;   **h.** ruthenocene;   **i.** rhodicinium ion;
    **j.** diosmium nonacarbonyl;   **k.** benzene(cyclopentadienyl)iron(I)
    tetrachloroaluminate

*15. Classify the following carbides as saltlike, metal-like, or polymeric covalent, and give their likely reaction products with water (if any): **a.** $B_4C$;
    **b.** $Al_4C_3$;   **c.** $Li_4C_3$;   **d.** $Li_2C_2$;   **e.** $Fe_3C$;   **f.** WC.

16. From each pair of compounds given, pick the one that will be most stable (thermally and/or to oxidation) and explain why: **a.** $Si_3H_8$ or $Si_3Me_8$;
    **b.** $AsMe_3$ or $AsPh_3$;   **c.** $GeMe_4$ or $AsMe_3$;   **d.** $Cr(CH_2CMe_3)_4$ or
    $CrMe_4$;   **e.** $Fe(C_5H_5)_2$ or $Ni(C_5H_5)_2$;   **f.** $C_2Me_6$ or $C_2Ph_6$;
    **g.** $PbH_4$ or $PbMe_4$;   **h.** $FeMe_5$ or $Fe(CO)_5$;   **i.** $SbMe_5$ or $Sb(CO)_5$.

*17. Give the chemical formula of: **a.** a gaseous Group 13/IIIA methyl compound;
    **b.** a cyclic catenated Group 14/IVA methyl compound; **c.** a cyclic catenated Group 15/VA methyl compound; **d.** a solid Group 16/VIA methyl or phenyl compound; **e.** a covalent methyl compound that has fewer than eight valence electrons about the central atom; **f.** a covalent methyl compound that has more than eight valence electrons about the central atom; **g.** any polymeric covalent methyl compound; **h.** any ionic methyl compound.

18. Complete the following equations to show the products expected (if any) from (Lewis) acid-base reactions of the following pairs of substances:
    **a.** $Me_3Tl + Me_3As$;   **b.** $Me_3Tl + Me_4Sn$;   **c.** $Me_3Tl + Me_2Se$;
    **d.** $Me_6Al_2 + H_2O$;   **e.** $Me_6Al_2 + Me_2O$;   **f.** $Me_3As + MeI$;
    **g.** $MeLi + Me_4Pb$;   **h.** $MeLi + Me_3In$;   **i.** $Me_5As + Me_3Ga$;
    **j.** $MeLi + HgCl_2$.

19. Give reasonable syntheses of the following organometallic compounds:
    **a.** MeMgI;   **b.** MeK;   **c.** ether-free $Me_2Mg$;   **d.** $Me_2SiCl_2$;

**e.** $(C_2H_5)_3B$;    **f.** $Me_6Al_2$;    **g.** $Me_6Si_2$;    **h.** $Nb(CO)_6^-$;
**i.** $Fe(CO)_5$;    **j.** $Co_2(CO)_8$;    **k.** $Os(C_5H_5)_2$;    **l.** $Ti(C_5H_5)_2Cl_2$;
**m.** $Mo(C_6H_6)_2$.

**20.** (Review) Give an example of a water-soluble and a water-insoluble salt for each of the following cations and anions; explain your choices:
**a.** $Ph_4P^+$;    **b.** $Ph_4B^-$;    **c.** $[Co(C_5H_5)_2]^+$;    **d.** $[Cr(C_6H_6)_2]^+$;
**e.** $Co(CO)_4^-$;    **f.** $Co(CO)_5^+$;    **g.** $Fe(CO)_4^{2-}$.
**h.** Explain the role of the $MeO(CH_2CH_2O)_2Me$ in equation (10.45).

**\*21.** Which of the following organometallic species are and which are not 18-electron species? **a.** $Pd(CO)_4$;    **b.** $Ni(PPh_3)_4$;    **c.** $Ni(PPh_3)_3$;
**d.** $Fe(CO)_4^{2-}$;    **e.** $Fe_2(CO)_9$;    **f.** $Ni(PPh_3)_2Cl_2$;    **g.** $Ni(C_5H_5)_2$;
**h.** $Fe(C_6H_6)_2$;    **i.** $PdCl_2(CO)_2$;    **j.** $[PtMe_6]^{2-}$.

**22.** Predict the formulas of the: **a.** simplest neutral carbonyl of Rh; **b.** simplest carbonyl anion of Cr; **c.** simplest carbonyl halide of Fe; **d.** cyclopentadienyl carbonyl of Co; **e.** most stable sandwich species ($C_5H_5$ rings) for Ir; **f.** most stable sandwich species (benzene rings) for Mn.

**†23.** In principle, what metals should form the most stable neutral sandwich molecules employing: **a.** two benzene rings; **b.** two cyclopentadienyl rings; **c.** one benzene and one cyclopentadienyl ring; **d.** two cyclobutadiene rings; **e.** one tropylium and one benzene ring; **f.** one cyclopropenium and one cyclopentadienide ring?

**24.** **a.** How many two-electron-donor ligands and what valence-electron configuration of metal ion are necessary in order to have a square planar 16-electron *d*-block organometallic compound? **b.** List all the qualifying *d*-block neutral metal atoms; $+1$ ions; $+2$ ions; $+3$ ions. **c.** Write the formulas of the square planar carbonyl chlorides of these elements and ions.

Notes

[1] Mackay, K. M., "Hydrides," in *Comprehensive Inorganic Chemistry*, vol. 1, J. C. Bailar, Jr. *et al.*, ed., Pergamon Press, Oxford, 1973.

[2] Greenwood, N. N., and A. Earnshaw, *Chemistry of the Elements*, Pergamon Press, Oxford, 1984.

[3] Cotton, F. A., and G. Wilkinson, *Advanced Inorganic Chemistry: A Comprehensive Text*, 4th ed., Wiley-Interscience, New York, 1980.

[4] Holliday, A. K., G. Hughes, and S. M. Walker, in *Comprehensive Inorganic Chemistry*, vol. 1, J. C. Bailar, Jr., *et al.*, ed., Pergamon Press, Oxford, 1973, pp. 1203–1213.

[5] Coates, G. E., M. L. H. Green, P. Powell, and K. Wade, *Principles of Organometallic Chemistry*, Methuen, London, 1971.

[6] Ashe, A. A., *Organometallics*, 1, 1408 (1982), and references cited therein.

[7] Davidson, P. J., M. F. Lappert, and R. Pearce, *Acct. Chem. Res.*, 7, 209 (1974); *Chem. Rev.*, 76, 214 (1976).

[8] By historical tradition, carbon monoxide is considered an inorganic compound, so metal compounds of it are sometimes not considered organometallic compounds (for instance in *Chemical Abstracts*).

[9] The cyclopentadienide ion is more compatible with higher oxidation states of the metal.

[10] Some of these carbonyls have structures complicated by the presence of bridging carbonyl ligands; for details see C. M. Lukehart, *Fundamental Transition-Metal Organometallic Chemistry*, Brooks/Cole, Monterey, Calif., 1985, pp. 23–36.

[11] For example, K. Wade, *Adv. Inorg. Chem. Radiochem.*, 18, 1 (1976).

# The Underlying Reasons for Periodic Trends

So far in this book we have explored ways in which the chemical reactions and physical properties of the elements and their compounds are related to some more fundamental properties of the atoms of the elements: their radii, electronegativities, charges, covalent-bond types and energies, and periodic table positions. But in chemistry we also seek to explain periodic variations in atomic radii, electronegativities, charges, and covalent bond types in terms of our theory of the electronic structure of the atom itself. This is the job of the field of *theoretical chemistry*.

A full exploration of the quantum mechanical theory of the atom, how it arose, and its mathematical methods and results is beyond the scope of this book and will likely be encountered later in the career of a professional chemist. The courses that you will take later on that deal with this theory will probably not have the time to deal at length with how the results obtained from this theory relate to the observed chemical reactions and physical properties of the elements and their compounds. Consequently such courses often give students the mistaken impression that atomic theory is completely unrelated to chemistry in the "real world." In this chapter we will preview the results of quantum mechanics (some of which you have already seen in general chemistry) and try to show their relationship to the descriptive and applied chemistry we have been studying in this book.

## 11.1

### Review of Overall Periodic Trends and Anomalies

Up to this point we have been taking the overall periodic trends in atomic properties as given. Now it is appropriate to ask *why* these trends exist and why there are countertrends or anomalies. We have even taken the order of filling of the orbitals, and the shape of the periodic table itself, for granted. But why is the order of filling orbitals as it is? It does seem irregular ($4s$ before $3d$ in neutral atoms but not in ions, and so on). Why are $3d$ electrons valence electrons in Cu and Zn but not in Ga and Ge, whereas $4s$ electrons are valence electrons in all four adjacent elements?

Perhaps our most important atomic variable has been the *electronegativity* of the atom. *Why* do Pauling electronegativities tend to increase from left to right in the periodic table, and from bottom to top? And why are those trends reversed in much of the *d*-block and to some extent in the *f*-block? Why do the electronegativities alter in a seemingly irregular manner at the left of the *p*-block (B > Al < Ga > In < Tl)?

We have also made much use of variations in atomic size, as measured by various types of radii (cationic, anionic, covalent, metallic, van der Waals). In order to have the most comparable data, we will focus on trends in metallic and covalent radii. Why do atoms (in metals and covalent compounds) get smaller as we go from left to right in the periodic table, and from the bottom to the top? Why do these trends vanish in much of the *d*-block, and at the left of the *p*-block (B < Al = Ga < In = Tl)? These trends are clearly related to those in electronegativities, but they are not precisely the same.

Another important atomic variable is the *possible ionic charge(s)* or *oxidation number(s)* of elements. Why do these fall short of the group number at the right of blocks of the periodic table? In the *p*-block, why are the possible oxidation numbers highest in the middle period (the third)? In the *d*- and *f*-blocks, why are the possible oxidation numbers highest in the bottom period (the sixth)? Why are oxidation numbers so much more variable and irregular in the *d*-block than in the *p*-block or the *f*-block?

There are interesting questions dealing with *covalent-bond types and energies*. Why can only certain types of orbitals be used in forming pi and delta bonds (e.g., why can't carbon form quadruple bonds while chromium can)? Why are pi bonds weaker than sigma bonds, and why are delta bonds weaker than pi bonds? Why does a given type of bond (sigma, for instance) get weaker down a group in the *p*-block, while delta bonds seem to get stronger down the *d*-block?

If we can answer these questions, then perhaps we can answer some of the "real" chemical questions that may have occurred to you during this course. Why is gold the softest and most electronegative metal when it is neither the metal farthest to the right nor uppermost in the periodic table? Why is mercury a volatile liquid metal? Why do mercury and some of its neighbors have such low coordination numbers in their compounds? Why is $AsCl_5$ unstable when $PCl_5$ and $SbCl_5$ are stable?

We could add many questions to this list. In the space remaining we cannot answer all of these, but we will survey some results of quantum-mechanical calculations that should help us not only to answer many of these questions but also to predict the shape of parts of the periodic table yet to be filled in and the properties of yet-undiscovered elements.

## 11.2

## Angular Part of the Electronic Wave Function

Chemical change occurs among the electrons of atoms; the study of chemical change focuses on the study of electrons in orbitals. Unlike planets around the sun, electrons do not follow unvarying two-dimensional orbits around the nucleus. Their paths involve three dimensions, and (thinking of electrons as particles) they vary in distance from the nucleus over time. It is often useful to think of an electron as a wave located in a particular wave pattern around the nucleus, which has its crests and troughs (points of maximum intensity) and its level or calm points (in which the wave has zero intensity). A mathematical equation that specifies this wave pattern can (in principle and sometimes in practice) be derived from the Schrödinger wave equation; this equation is called the **wave function, $\Psi$**.

It is more difficult to describe the wave pattern or particle path of an electron in terms of Cartesian coordinates $(x, y, z)$ than in terms of spherical polar coordinates $(r, \theta, \phi)$. In spherical polar coordinates $r$ is the distance from the center of the atom (the nucleus), while $\theta$ and $\phi$ give the angles of the point in question away from a reference vertical plane and the polar axis through the nucleus, respectively. The great advantage of this system of coordinates is that it is possible to separate the description of the distance from the nucleus of the electron (or of its crest, trough, etc.) from the description of its angular distance from the "north pole" and the reference vertical plane of the atom.

Let us first illustrate this distinction using the orbits of the planets as illustrations. The *angular part* of any planet's orbit is substantially the same. All move in the equatorial plane of the solar system, so that $\phi = 90°$ ($\pi/2$ radians); the orbit is substantially unaltered as the planet sweeps the angle $\theta$ around the sun. These angular parts of the planetary orbits define the *shape* and the *orientation* of the planetary orbits: All are circular, and all are in the equatorial plane of the solar system. (Of course, comets move in orbits of different shapes and orientations.) The planetary orbits differ only in their *radial parts*: Each planet moves in an orbit of a different radius.

We will first describe the *angular* parts of the possible wave functions for electrons in a (hydrogenlike) atom. These angular wave functions result (after some mathematical manipulation to remove imaginary numbers) in the familiar shapes of orbitals. The s orbitals are spherical. In terms of a particle model of the electron, it is as likely to be at one angle from the north pole or from the reference vertical plane as another. In terms of the wave model, the wave path has only a crest; at no angle does the amplitude of the wave drop back to zero.

The p orbitals have the familiar dumbbell shape (Figure 11.1), with two **lobes**, one on each side of the nucleus at the center. The mathematical angular wave function for a p orbital involves some constants times $\cos \theta$, which goes from $+1$ to 0 to $-1$ and back to 0 as we move from the north pole to the equator to the south pole and back to the other side of the atom. Hence in the "north" hemisphere of the atom the wave function has a positive sign; in the wave model we can think of this as a crest. In the "south" hemisphere the wave function has a negative sign; this is a trough. At the "equator" the wave is at "sea level" and has zero amplitude. The amplitude of the wave disturbance is proportional to the square of the wave function, so it is equally positive in both lobes; this amplitude corresponds to the probability of finding the electron at that spot. The shaded plane in Figure 11.1

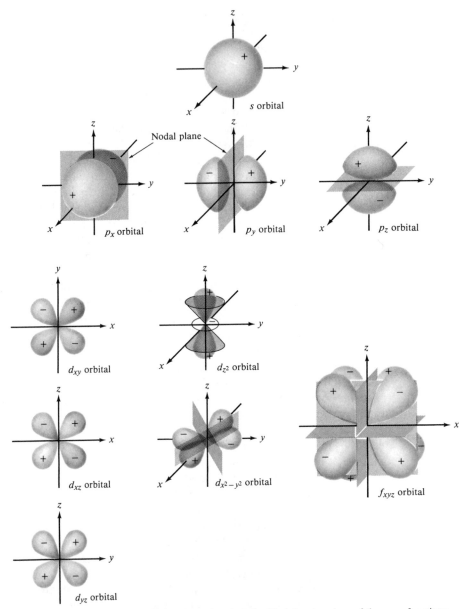

**Figure 11.1** Angular wave functions of *s*, *p*, *d*, and an *f* orbital showing signs of the wave functions in different lobes and indicating nodal planes with dark planes or solid lines. From *Chemical Structure and Bonding* by R. L. DeKock and H. B. Gray. Copyright © 1980 by Benjamin/Cummings Publishing Company. Reprinted by permission.

through the nucleus and the "equator" of the $p$ orbital is the **nodal plane** of the $p$ orbital, the plane in which the wave amplitude is zero and the electron cannot be found [1].

A $d$ orbital generally has the double-dumbbell shape indicated in Figure 11.1, with four lobes and two nodal planes. The four lobes alternately correspond to crests and troughs, but all, of course, have positive wave amplitudes and correspond to regions in which there are positive probabilities of finding the electron.

A typical $f$ orbital has the quadruple-dumbbell shape shown in Figure 11.1. Other $f$ orbitals in a given atom have a somewhat different appearance due to the different mathematical operations used to eliminate the imaginary numbers from the wave functions, but all of them share the characteristic of having *three* nodal surfaces passing through the nucleus.

You may recall from general chemistry that the shape of an orbital is associated with a particular number in the wave equation called the *secondary quantum number* and given the symbol $l$. Thus for any $s$ orbital the secondary quantum number has the value $l = 0$; for any $p$ orbital $l = 1$; for any $d$ orbital $l = 2$; for any $f$ orbital $l = 3$; and for any $g$ orbital that will be used in a yet-undiscovered atom $l$ will be 4. Note that *the number of nodal surfaces passing through the nucleus is equal to the secondary quantum number.*

The $p$, $d$, and $f$ orbitals not only have characteristic shapes but also are oriented in certain directions. The $p$ orbital we described earlier has its crest at its north pole and its trough at its south pole and is thus oriented along this polar axis, which we label the $z$-axis. A given atom also has two otherwise-identical $p$ orbitals aligned at right angles to the first one and to each other along an $x$ and a $y$ axis; we conventionally label these three orbitals as the $p_x$, $p_y$, and $p_z$ orbitals. These are also shown in Figure 11.1.

The quantum number found in the wave function that is associated with orientation is known as the *magnetic quantum number*, $m_l$. For the $p_z$ orbital $m_l = 0$. The $p_x$ and $p_y$ orbitals are associated with different combinations of $m_l = +1$ and $m_l = -1$. (Since $m$ occurs in the wave function associated with the imaginary number $i$ (the square root of $-1$), combinations are needed to produce equations that contain only real numbers.)

In a given atom there are five otherwise-identical $d$ orbitals, seven otherwise-identical $f$ orbitals, and nine otherwise-identical $g$ orbitals that differ in their orientations. These orientations are shown in Figure 11.1 for the $d$ orbitals only. The difference in appearance of the last of these, the $d_{z^2}$, is an artifact of mathematical operations; like the others it has two nodal surfaces through the nuclei, but these surfaces happen to be planes wrapped around to form conical surfaces. The $d_{z^2}$ orbital is also associated with the magnetic quantum number $m_l = 0$; the $d_{xz}$ and $d_{yz}$ orbitals are associated with $m_l = +1$ and $-1$; the $d_{xy}$ and $d_{x^2-y^2}$ correspond to $m_l = +2$ and $-2$. When discussing bonding in $d$-block complexes and organometallic compounds you must know the designations, shapes, and orientations of these five orbitals.

These factors will also be important in your future study of the molecular orbital covalent bonding theory, since an important restriction exists on the forma-

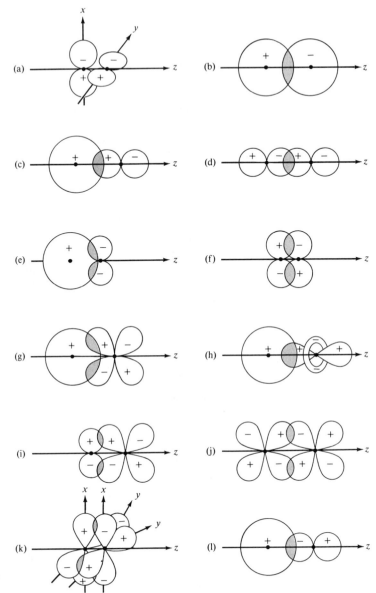

**Figure 11.2** Combinations of atomic orbitals to be tested for positive, negative, and zero overlap (see the exercises).

tion of a covalent bond between orbitals on two different atoms: *Two atomic orbitals must have positive overlap in order to form a bonding orbital.* Positive overlap means that the *signs* of the two orbitals match in the region of overlap, as in Figure 11.2c. *No* bonding interaction can result if equal amounts of positive and negative overlap occur. Suppose that we were to bring two atoms together along their mutual $z$-axes and try to overlap the $p_x$ orbital of one atom with the $p_y$ orbital of the other (Figure

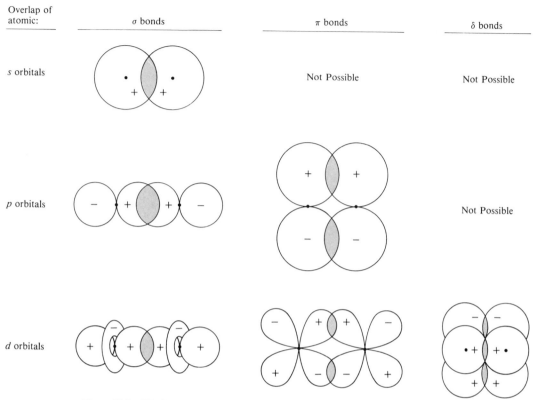

**Figure 11.3** (Similar to figure 6.7). Overlap of atomic orbitals of two atoms, the nuclei of which are represented by dots, to give covalent bonds. Regions of positive orbital overlap are indicated by shading; the $z$-axis for each atom is taken as the internuclear axis.

11.2a). No bond would result, since the positive lobe of the $p_x$ orbital on the one atom would have equal overlap with the positive and negative lobes of the $p_y$ orbital on the other atom, and the same would hold for the negative lobe of the $p_x$ orbital on the first atom. If the signs of two overlapping orbitals *oppose* each other in the region of overlap, an *antibonding* interaction results, as in Figure 11.2b.

Bonding molecular orbitals that result from positive overlap of two atomic orbitals are commonly classified as **sigma bonds**, **pi bonds**, and **delta bonds**. (Additional types are possible—in theory, at least.) If the molecular orbital has *no* nodal plane through *both* atomic nuclei, it is called a sigma bond. As shown on the left side of Figure 11.3, sigma bonds can be formed from two atomic $s$, $p_z$, or $d_{z^2}$ orbitals; each combination overlaps in one shaded region to give a sigma bond. (Although not pictured, a sigma bond can also arise from the overlap of an $s$ orbital on one atom with a $p_z$ orbital on the other atom, and so forth.)

If the molecular orbital formed has *one* nodal plane passing through both atomic nuclei, it is called a *pi* bond (center of Figure 11.3). Two given $p$-block atoms can form two such pi bonds, one using their valence $p_x$ orbitals and one using their valence $p_y$ orbitals. Two $d$-block atoms can also form two pi bonds, one with their $d_{xz}$ orbitals and one with their $d_{yz}$ orbitals. (Of course, mixed pi bonds involving a valence $d_{xz}$ or $d_{yz}$ orbital on one atom and a valence $p_x$ or $p_y$ orbital on the other are also possible, but they are important mainly between atoms of different blocks of the periodic table.)

If the molecular orbital formed has *two* nodal planes passing through both atomic nuclei, it is a *delta* bond (right side of Figure 11.3; one region of orbital overlap is obscured in this view). Only $d$- (or $f$-) block elements can form delta bonds. Two delta bonds can be formed per pair of atoms, one using two $d_{xy}$ and the other using two $d_{x^2-y^2}$ orbitals, which overlap in four regions (one hidden from view). (A $p$-block element cannot form a delta bond and hence cannot form a quadruple bond, although the Lewis structure for one looks reasonable in the case of hypothetical C≡C.)

Although Figure 11.3 does not give a computationally precise representation of overlap of atomic orbitals, it does suggest that at a given atomic internuclear distance and using orbitals of the same shape and principal quantum number, *pi bonds are weaker than sigma bonds* and *delta bonds are weaker than pi bonds*. Furthermore, what overlap there is in the pi and delta bonds is more sensitive to a small lengthening of the bond (as on going to a larger atom down a group of the periodic table) than is the sigma bond.

## 11.3

## Radial Part of the Wave Function: Shielding

As you no doubt recall from general chemistry, orbitals differ not only in shape and orientation but also in size. Some $s$ (spherical) orbitals extend farther from the nucleus than other $s$ orbitals, for example: In a given atom, an electron in a $6s$ orbital is, on the average, farther from the nucleus than one in a $2s$ orbital. Orbitals are not so completely differentiated by radius as are orbits of different planets, however: Being a wave, an electron has a distance from the nucleus of maximum amplitude or probability, but it also has amplitudes or probabilities of being found at other distances from the nucleus.

The relative size of an orbital (i.e., the most probable single radius at which the electron will be found) is a function of its *principal quantum number*, $n$. This $n$ can have whole-number values from one to infinity; it is the first number in the orbital designation (e.g., $n = 2$ in a $2s$ orbital).

In the hydrogen atom (or other one-electron species) the energy with which the one electron in the orbital is held depends only on the principal quantum number:

$$E = -\frac{2\pi^2 m Z^2 e^4}{n^2 h^2} \tag{11.1}$$

where $m$ is the mass of the electron, $Z$ is the nuclear charge, $e$ is the charge of the electron, $h$ is Planck's constant, and $n$ is the principal quantum number. Although the one electron of such an atom or ion is normally in the $1s$ orbital, it can be excited to any other orbital; this equation implies, for example, that the electron is equally stable in the $4s$, any $4p$, any $4d$, or any $4f$ orbital. This is no longer the case even in the helium atom (say, with an electron configuration of $1s^1 4x^1$), in which the $4s$, $4p$, $4d$, and $4f$ orbitals are no longer equally favorable. The reason for this is that the two electrons repel each other, and they repel each other differently depending on the structures of the orbitals in which they are located.

Thus in multielectron atoms the energies of electrons in orbitals depend on both the principal and the secondary quantum numbers of the orbitals. The splitting of the energies of electrons in differently shaped orbitals of the same principal quantum number increases with increasing atomic number (Figure 11.4) until

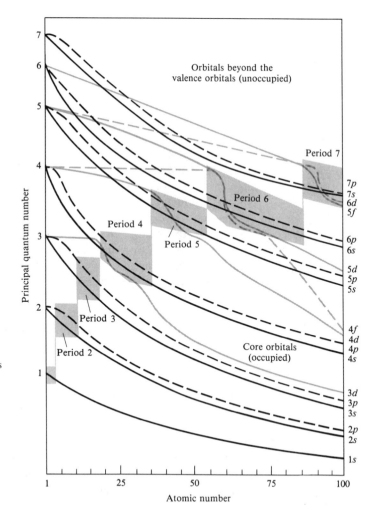

**Figure 11.4** The variation in energies of electrons in atomic orbitals with increasing atomic number in neutral atoms (energies not strictly to scale). The shaded boxes enclose the valence orbitals of the atom with the given atomic number. Adapted from *Basic Inorganic Chemistry* by F. A. Cotton and G. Wilkinson. Copyright © 1976 by John Wiley & Sons. Reprinted by permission.

approximately the point at which these orbitals actually are chosen in the ground state of the atom. At this point, it is sometimes more favorable energetically to use a spherical $s$ orbital of slightly higher principal quantum number in preference to some of the less spherical ($d$, $f$, etc.) orbitals of lower principal quantum number. This difference in energy results in the familiar order of filling of orbitals in neutral atoms:

$$1s < 2s < 2p < 3s < 3p < 4s \leq 3d < 4p < 5s \leq 4d < 5p < 6s \leq 4f \leq 5d$$
$$< 6p < 7s \leq 5f \leq 6d < \cdots$$

This is therefore the order in which the orbitals are encountered in the periodic table as one scans it from low to high atomic number (Table 1.1). Since the energetic advantages of occupying, for example, the $7s$, $5f$, and $6d$ orbitals are about the same at the time of filling, troublesome anomalous electron configurations result for many neutral atoms in the $d$- and $f$-blocks of the periodic table.

But soon after the orbitals in question have been filled and they have all become core orbitals, their order of energy again changes and becomes rather like that of the hydrogen atom (Figure 11.4), although now, of course, the orbitals are all at much lower absolute energies due to the higher nuclear charge. Consequently once the core orbitals are filled, the orbitals with high values of their secondary quantum number, which had delayed filling as long as possible, now must suddenly drop in energy very rapidly; hence once the $d$ and $f$ orbitals are completely filled and the atom in question is in the next block of the table, the $d$ and $f$ electrons rapidly become so much lower in energy that they cannot be used in bonding and can no longer be considered valence electrons. But the quickly filling $(n + 1)s$ or $(n + 2)s$ orbital electrons remain high in energy and available as valence electrons for a while.

There is a similar shift in energy of orbitals when an atom ceases being neutral and becomes positively charged: The late-filling $d$ or $f$ orbitals, under the influence of the extra positive charge, drop in energy relative to the $(n + 1)s$ or $(n + 2)s$ orbitals. Consequently the order of filling in *positive ions* of the $d$-block is also more hydrogenlike, in that the $nd$ or $nf$ orbitals are used rather than the $(n + 1)s$ or $(n + 2)s$. As a result there are no anomalous electron configurations among the chemically important positive ions.

Orbitals also have structure (lobes and nodal surfaces) outward from the nucleus. This structure can be seen in the radial portion of the wave function for an electron, which passes from crests through zero levels to troughs and back again for most types of orbitals. There are many ways that we can plot the radial wave function; in Figure 11.5 we plot what is known as the radial probability function, which involves squaring the original wave function, so that negative parts of the wave function (troughs) become positive wave amplitudes or probabilities of finding the electrons. The radial probability function also involves multiplying the squared function by $4\pi r^2$, so that the result is not just the probability of finding the electron at *one* point at a distance $r$ from the nucleus but is also the probability of finding the electron *at any point* in the spherical shell that is at a distance $r$ from the nucleus.

Figure 11.5 clearly shows that, in a given atom, the most probable distance

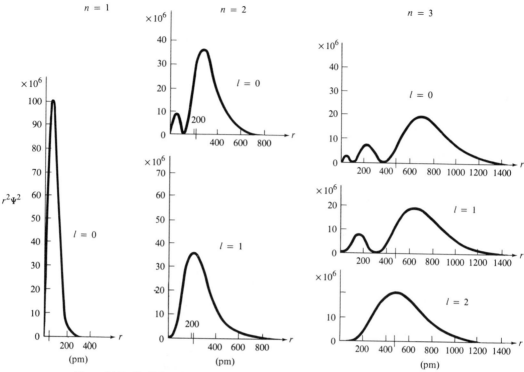

$n = 1$

$n = 2$

$n = 3$

**Figure 11.5** Radial probability functions plotted against distance from the nucleus for $n = 1, 2$, and 3 for the hydrogen atom. From *Atomic Spectra and Atomic Structure*, by Gerhard Herzberg. Copyright © 1944 by Dover Publications, Inc. Reprinted by permission.

from the nucleus for the electron increases with increasing values of the principal quantum number, $n$. But it also shows that there are some differences for different shapes of orbitals. The $1s$, $2p$, $3d$, $4f$, ... orbitals have the simplest radial probability functions, with one maximum and no radius at which there is zero probability of finding the electron (other than at the nucleus for $2p$, $3d$, $4f$, and so on). But $2s$, $3p$, $4d$, ... orbitals all have one such distance, which corresponds to a *nodal sphere*: a nodal surface that does *not* pass through the nucleus. On one side of this surface, the wave function has a positive sign (crest); on the other side, it has a negative sign (trough). (For purposes of overlapping orbitals of other atoms and forming covalent bonds, only the sign of the outermost part of the wave function is important, since overlap never extends farther into the orbital.) The $3s$ and other orbitals have two nodal spheres; the $4s$ and others have three nodal spheres. An orbital with a given principal quantum number $n$ and a secondary quantum number $l$ has a total of $n - l - 1$ nodal spheres in its radial wave function plus $l$ nodal surfaces through the nucleus in its angular wave functions, a total of $n - 1$ nodal surfaces.

It is possible and useful to combine the radial and angular wave function probability plots to give a total picture of where in the atom the electron in a given orbital is likely to be found. Without a laser holograph it is difficult to do this in three dimensions, but we can draw cross sections of orbitals. This is done for selected orbitals in Figure 11.6, which also contains (as a result of detailed calculations) contours of equal probability of finding the electron, much like geological topographic maps. Note that nodal spheres are indicated by dashed circles, while nodal planes are indicated by dashed lines.

EXAMPLE   Sketch cross sections in the $x$-$y$ plane (without density contours) of the following orbitals: (a) $4d_{x^2-y^2}$ (b) $5d_{xy}$ (c) $5s$. Indicate nodal planes and spheres by dashed lines; show positive and negative signs of the wave function in the different lobes.

Solution   ■ It is best to start by counting up the number of nodal spheres (equal to $n - l - 1$) and nodal surfaces through the nucleus (equal to $l$), then drawing these before attempting to fill in the actual lobes of the orbital. For the $4d$ orbital there will be $4 - 2 - 1 = 1$ nodal sphere; draw a dashed circle out a distance from the nucleus. For the $5d$ orbital there will be two concentric nodal spheres. The $5s$ orbital will have $5 - 0 - 1 = 4$ nodal spheres to be drawn concentrically about the nucleus. There will be two nodal surfaces through the nucleus for the two $d$ orbitals; draw these as dashed lines through the nucleus. For the $d_{x^2-y^2}$ orbital the lobes fall along the $x$ and $y$ axes, so draw the nodal planes (lines) midway between the axes; the reverse is true for the $d_{xy}$ orbital.

■ Now shade in the lobes, which will fall in between the nodal lines and circles; the last lobe(s) will fall outside the outer nodal sphere. The inner lobes in a cross section will look approximately like circles; the outer ones can be drawn as shields or curved disks, or as concentric circles.

■ Finally fill in plus and minus signs to indicate the signs of the wave function. Start anywhere with either a plus or minus sign (the first choice is completely arbitrary), but thereafter any time you cross a nodal surface you must change signs. If this procedure results in any contradictions a mistake has been made. The result is shown in Figure 11.7.   □

CONSEQUENCES OF INNER NODES OF ORBITALS   Equation (11.1) indicates that the energies of electrons in orbitals become more negative very rapidly with increasing nuclear charge ($E$ is proportional to $Z^2$). This would lead us to expect that the Li atom, for example, would be much harder to ionize than the H atom, whereas the converse is true. Evidently the outermost ($2s$) electron of the Li atom does not experience the full attractive power of all three Li nuclear protons. Equation (11.1) does not consider the repulsion of the inner ($1s^2$) electrons on the outer ($2s^1$) Li electron, which makes that electron much easier to ionize.

According to classical electrostatics, if the $2s$ electron of Li is completely outside both the $+3$-charged nucleus and the two $1s$ electrons ($-2$ charge), it is the same as if it were outside a nucleus of $+3 - 2 = +1$ charge. We may say that the $2s$ Li electron feels an **effective nuclear charge** that is less than the true nuclear charge due

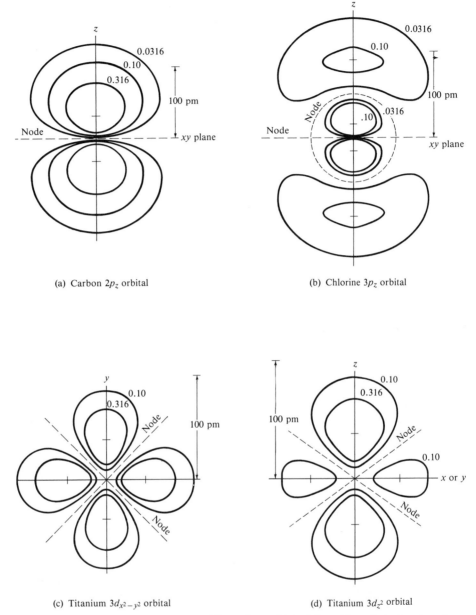

(a) Carbon $2p_z$ orbital

(b) Chlorine $3p_z$ orbital

(c) Titanium $3d_{x^2-y^2}$ orbital

(d) Titanium $3d_{z^2}$ orbital

**Figure 11.6**   Electron-density contour maps for various orbitals. Adapted from E. A. Ogryzlo and G. B. Porter, *J. Chem. Educ.*, 40, 256 (1963). Reprinted with permission.

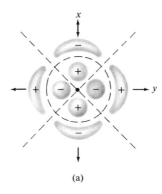

(a)

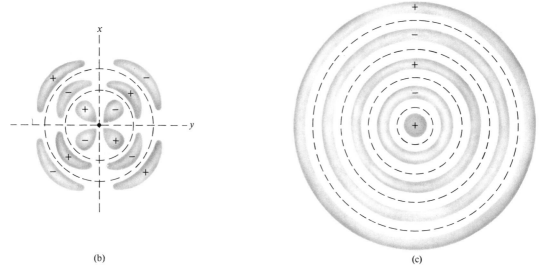

(b)                                                    (c)

**Figure 11.7**   Cross-section sketches in the $x$-$y$ plane of (a) the $4d_{x^2-y^2}$ orbital; (b) the $5d_{xy}$ orbital; (c) the $5s$ orbital.

to the **shielding** ability of the core $1s$ electrons. Thus we might want to use in equation (11.1), not $Z$, the true nuclear charge, but an effective nuclear charge, $Z^*$, that is less than $Z$ by the sum of the *screening constants* of the inner orbitals:

$$Z^* = Z - \sum S \tag{11.2}$$

But the $2s$ electron of lithium does not behave as if it were experiencing exactly a $+1$ charge either: Its properties are more consistent with those of an electron feeling an effective nuclear charge of $+1.279$. This difference can be explained by the fact that the $2s$ orbital has *two* maxima in its radial probability function (Figure 11.5) and by the fact that the lesser of these *penetrates* within the maximum of the inner $1s$ orbital; hence, although most of the time the $2s$ electron is in the outer lobe of that orbital, feeling a $Z^*$ of $+1$, for some of the time it is inside the $1s$ orbital, experiencing the full nuclear charge of $+3$. This penetration of inner orbitals is obviously going to affect the chemical properties of that valence Li electron: It will affect the ease of ionization of that electron, the size of its orbital, the electronegativity of the atom, and so forth.

The $2p$ orbital of Li lacks the penetrating inner lobe of the $2s$ orbital. Consequently the valence electron of Li (and other atoms) prefers the $2s$ orbital and fills it first, as we have seen earlier. Likewise, we can see (from the right side of Figure 11.5) the superior penetrating power of the $3s$ orbital as compared with the $3p$ orbital, which is better than the $3d$. Thus electrons in orbitals of common $n$ but increasing $l$ have progressively poorer penetration of core electrons, so these orbitals are filled later.

A second consequence is that when the $3s$, $3p$, and $3d$ orbitals (for example) are core electrons, it will be easier for valence electrons to penetrate the electrons of the $3d$ orbitals, which have little likelihood of being near the nucleus, than to penetrate the $3p$ or especially the $3s$ orbital. Consequently electrons in orbitals of common $n$ but increasing $l$ exhibit progressively poorer shielding of outer electrons and should have smaller shielding constants.

## 11.4

## Slater's Rules

In order to make use of equations such as (11.1) to predict trends in ionization energies, electronegativities, radii, and the like, it is useful to have some rules by which we can estimate the degree to which electrons in the various types of orbitals shield other electrons from the nucleus. With such rules we can estimate the effective nuclear charges experienced by other electrons. Based on calculations done in 1930 and before, Slater [2] proposed some tentative and approximate rules for making such estimations. As we will see, these do not give accurate ionization energies, radii, or electronegativities, but they do generally predict trends well, and they allow us to correlate the trends with the shielding and penetrating powers of the electrons in the orbitals involved; hence they are still useful guides to understanding the ultimate "whys" of descriptive inorganic chemistry.

Slater's rules apply to an electron in a *specified orbital* in an atom with a specified nuclear charge $Z$. Rules 2 to 5 compute the shielding constants of the other electrons for the particular electron in question; rule 6 simply applies equation (11.2) to compute the effective nuclear charge for the specified electron. Rule 7 is a completely separate calculation of an effective principal quantum number $n^*$ for the electron in question, which is needed for some applications.

**1.** Write out the complete electron configuration of the element *in the following order*. Group $ns$ and $np$ orbitals together (their shielding properties are roughly equal); group all other types of orbitals separately (their shielding properties are dissimilar). Order:

$$(1s)(2s, 2p)(3s, 3p)(3d)(4s, 4p)(4d)(4f)(5s, 5p)\cdots$$

**2.** Electrons to the *right* of the group of electrons in question contribute *nothing* to the shielding of that group of electrons. (This is an approximation that cannot always be true, since outer orbitals may have penetrating ability.)

**3.** All *other* electrons in the same group (enclosed in the same parentheses) as the electron in question shield that electron to an extent of 0.35 unit of nuclear charge each. (Some of the time the other $3p$ electrons will be between the nucleus and the given $3p$ electron, for example.)

**4.** *If the electron in question is an s or p electron*: (a) All electrons with principal quantum number one less than the electron in question shield it to an extent of 0.85 unit of nuclear charge each. (b) All electrons with principal quantum number two or more less than the electron in question shield it completely—i.e., to an extent of 1.00 unit of nuclear charge each.

**5.** *If the electron in question is a d or f electron*: *All* electrons to the *left* of the group of the electron in question shield the $d$ or $f$ electron completely—i.e., to an extent of 1.00 unit of nuclear charge each. (This is a manifestation of the poor penetrating power of the $d$ or $f$ electron.)

**6.** Sum the shielding constants from steps 2 through 5, and subtract them from the true nuclear charge $Z$ of the atom in question to obtain the effective nuclear charge $Z^*$ felt by the electron in question.

**7.** Values of $n^*$ are not the same as the true principal quantum number for higher values of $n$:

| $n$ | 1 | 2 | 3 | 4 | 5 | 6 |
|-----|---|---|---|-----|-----|-----|
| $n^*$ | 1 | 2 | 3 | 3.7 | 4.0 | 4.2 |

EXAMPLE    Calculate $Z^*$ for the $6s$ and $4f$ valence electrons of Ce.

Solution    ■ In step 1 we write the full electron configuration of Ce as follows:

$$(1s)^2(2s, 2p)^8(3s, 3p)^8(3d)^{10}(4s, 4p)^8(4d)^{10}(4f)^2(5s, 5p)^8(6s)^2$$

Proceeding with the calculation of $Z^*$ for the $6s$ electron, we find that step 2 does not apply. In step 3 the shielding constant from the *other* $6s$ electron is $1 \times 0.35 = 0.35$ nuclear charge unit. In step 4a we obtain $8 \times 0.85 = 6.80$ units; in step 4b we get $48 \times 1.00 = 48.00$ units. Step 5 does not apply; in step 6 we add these shielding constants to get 55.15 nuclear charge units; we subtract this value from the atomic number of Ce, 58 (nuclear charge units), to obtain an effective nuclear charge $Z^*$ of 2.85 nuclear charge units.

■ For the $4f$ electron, step 2 does apply: We ignore the $5s$, $5p$, and $6s$ electrons in the calculation. In step 3 the other $4f$ electron shields by $1 \times 0.35$ unit. Step 4 does not apply; in step 5 the shielding constant total is $46 \times 1.00 = 46.00$ units. In step 6 we calculate $Z^* = 58 - (0.35 + 46.00) = 11.65$. □

The calculations suggest that the $4f$ electrons, which are closer to the nucleus than the other valence electrons, feel a higher effective nuclear charge, despite the poor penetrating power of $f$ electrons. This result would have been difficult to predict with qualitative arguments only.

EXAMPLE     Calculate $Z^*$ for the last valence electron in each atom of the second period.

Solution    In step 1 we can write the electron configurations of these atoms in the general form $(1s)^2(2s, 2p)^n$, in which $n$ is the number of valence electrons. Step 2 does not apply. In step 3 ·we obtain a total shielding constant of $(n - 1) \times 0.35 = -0.35 + 0.35n$. In step 4 we obtain $2 \times 0.85 = 1.70$. In step 6 we add the shielding constants to obtain $1.35 + 0.35n$, which we subtract from the atomic number of $(2 + n)$ to obtain $Z^* = 0.65 + 0.65n$. □

The results from this example are compared in Table 11.1 with the values obtained for lighter atoms by self-consistent field (SCF) quantum-mechanical calculations [3]: As we can see, the agreement is quite good, although it is not quite so good at the left side of the third or in the fourth period. What is more important for our purposes is how well these functions work in predicting and thus explaining the periodic trends in the sizes of atoms, ionization energies, and electronegativities.

The calculation of $Z^*$ for a vertical group of atoms by Slater's rules gives interesting results: For Group 2, for example, we obtain $Z^*(\text{Be}) = 1.95$, $Z^*(\text{Mg}) = 2.85$, $Z^*(\text{Ca}) = 2.85$, $Z^*(\text{Sr}) = 2.85$, $Z^*(\text{Ba}) = 2.85$. Unfortunately, SCF calculations are not available for the heavier atoms for comparison.

It is also useful to calculate $Z^*$ across a series of $d$- and $f$-block elements.

**Table 11.1**  Comparison of Effective Nuclear Charges Calculated by Slater's Rules and from Self-Consistent Field Quantum-Mechanical Calculations

| Atom | Li | Be | B | C | N | O | F | Ne |
|---|---|---|---|---|---|---|---|---|
| $Z^*$(Slater) | 1.30 | 1.95 | 2.60 | 3.25 | 3.90 | 4.55 | 5.20 | 5.85 |
| $Z^*$(SCF) | 1.279 | 1.912 | 2.421 | 3.136 | 3.834 | 4.453 | 5.100 | 5.758 |
| Atom | Na | Mg | Al | Si | P | S | Cl | Ar |
| $Z^*$(Slater) | 2.20 | 2.85 | 3.50 | 4.15 | 4.80 | 5.45 | 6.10 | 6.75 |
| $Z^*$(SCF) | 2.507 | 3.308 | 4.066 | 4.285 | 4.886 | 5.482 | 6.116 | 6.764 |
| Atom | K | Ca | Ga | Ge | As | Se | Br | Kr |
| $Z^*$(Slater) | 2.20 | 2.85 | 5.00 | 5.65 | 6.30 | 6.95 | 7.60 | 8.25 |
| $Z^*$(SCF) | 3.495 | 4.398 | 6.222 | 6.780 | 7.449 | 8.287 | 9.028 | 9.769 |

SOURCE: The SCF values are from J. E. Huheey, *Inorganic Chemistry: Principles of Structure and Reactivity*, 3d ed., Harper and Row, New York, 1983, p. 39.

NOTE: Calculations are for the outermost valence electron; the self-consistent field (SCF) values for the outermost $s$ electron of $p$-block elements are somewhat higher than the values listed, which are for the $p$ electron.

Calculation of $Z^*$ for the $4s$ electron of the $d$-block elements having the characteristic electron configuration $1s^2 2s^2 2p^6 3s^2 3p^6 3d^n 4s^2$ gives us $Z^* = 2.85 + 0.15n$. Thus $Z^*$ values for valence $s$ electrons increase more slowly (0.15 per element) on crossing the $d$-block than on crossing the $p$-block (0.65 per element). This is because the $(n - 1)d$ electrons being added across the $d$-block are by and large inside the $ns$ electrons, while the $np$ electrons being added across the $p$-block are not inside the $ns$ electrons; hence the $(n - 1)d$ electrons do a better job of shielding.

Even though $Z^*$ does increase only slowly across the $d$-block, it does increase *ten* times; consequently the $Z^*$ values of the elements beyond the $d$-block (Ga through Kr) are 1.50 higher than the elements just above them. Were it not for the insertion of the $d$-block, the two periods of elements would have the same calculated $Z^*$ values, as do Na and K and as do Ca and Mg. We will shortly see that certain anomalies spring from this increase in effective nuclear charge among elements following the first filling of $d$ orbitals.

The computation of $Z^*$ for the outermost $s$ orbital of the $f$-block elements produces a very interesting result: $Z^*$ is 2.85 regardless of how many $f$ electrons are present in the element! This is a consequence of the fact that the $(n - 2)f$ orbitals are *two* shells below the $s$ electrons; hence, according to rule 4b, the deeply buried $(n - 2)f$ electrons completely shield the $ns$ electrons.

## 11.5

## Radii of Atoms and Trends in Covalent-Bonding Ability

The old Bohr model of the atom, which puts electrons in orbits, allows an easy calculation of the radius of the orbit of an electron:

$$r = \frac{a_0 (n^*)^2}{Z^*} \tag{11.3}$$

This equation includes the radius of the Bohr hydrogen atom, $a_0 = 52.9$ pm, and indicates, as we might expect, that atoms should get larger as the effective principal quantum number $n^*$ for the outermost electron increases but that they should get smaller as the effective nuclear charge for that electron increases. Of course, the Bohr model is no longer accepted, but the equation can still be used to give us a rough calculation of $\langle r_{max} \rangle$, the expected distance from the nucleus to the maximum in the radial probability function for the atom.

The calculation for Li, for example, gives $\langle r_{max} \rangle = 52.9 \times 2^2/1.30 = 162$ pm. Since $\langle r_{max} \rangle$ is not exactly the same as any of the types of radii we have examined, it is not surprising that our result differs from the covalent radius of Li, 134 pm (Table 6.4, also in Appendix C) [4]. The usefulness of $\langle r_{max} \rangle$ lies in looking at its periodic trends and seeing whether we can duplicate the periodic trends in other types of radii (we will use covalent and metallic radii). Table 11.2 presents calculations across some periods and down some groups and compares the results with the observed covalent (or metallic) radii from Table 6.4.

**Table 11.2**  Calculated Atomic Radii for Selected Atoms

| Atom | Li | Be | B | C | N | O | F | Ne | | |
|---|---|---|---|---|---|---|---|---|---|---|
| $\langle r_{max} \rangle$ | 162 | 108 | 81 | 63 | 54 | 46 | 41 | 36 | | |
| $r_{cov}$ | 134 | 125 | 90 | 77 | 75 | 73 | 71 | — | | |
| Atom | Be | Mg | Ca | Sr | Ba | | | | | |
| $\langle r_{max} \rangle$ | 108 | 167 | 254 | 297 | 327 | | | | | |
| $r_{met}$ | 89 | 136 | 174 | 191 | 198 | | | | | |
| Atom | Sc | Ti | V | Cr | Mn | Fe | Co | Ni | Cu | Zn |
| $\langle r_{max} \rangle$ | 241 | 230 | 219 | 210 | 201 | 193 | 186 | 179 | 172 | 166 |
| $r_{met}$ | 144 | 132 | 122 | 119 | 118 | 117 | 116 | 115 | 118 | 121 |
| Atom | C | Si | Ge | Sn | Pb | | | | | |
| $\langle r_{max} \rangle$ | 65 | 115 | 128 | 150 | 165 | | | | | |
| $r_{cov}$ | 77 | 118 | 122 | 140 | 146 | | | | | |
| Atom | Zn | Cd | Hg | | | | | | | |
| $\langle r_{max} \rangle$ | 166 | 195 | 214 | | | | | | | |
| $r_{cov}$ | 121 | 138 | 139 | | | | | | | |

We note first that the atoms of the second period are calculated to contract across the *s*- and *p*-blocks, which is indeed what they do. According to equation (11.3) the radius decreases because of the increase in $Z^*$, which suggests that the reason that atoms decrease in size across the *s*- and *p*-blocks is that since additional *s* and *p* electrons of the same principal quantum number do a poor job of shielding each other, the additional protons in each nucleus across the period can pull in *all* the valence electrons more strongly.

Second, we note that atoms of Group 2 are calculated to increase in size down the group, which is in fact observed. (The calculated values do seriously exaggerate the trend, however.) From the calculations this trend must simply be a consequence of increasing the (effective) principal quantum number of the valence orbital: Higher *n* means a larger orbital.

The third row of Table 11.2 shows a calculation across the *d*-block. Atoms are calculated to decrease in size here too due to the increases in $Z^*$. Since the increases in $Z^*$ are smaller in the *d*- than the *p*-block, we calculate smaller changes in size per element. By and large this is found, although we have not accounted for the increase in size at Cu and Zn.

The fourth row shows a calculation down the *p*-block (Group 14/IVA). As expected, the atoms increase in size, but there is a smaller-than-expected increase between Si and Ge (or, in Group 13/IIIA, between Al and Ga). This is found even more emphatically in actual covalent radii. Between these two periods, $n^*$ is increasing as usual, but $Z^*$ partly counteracts this effect by increasing 1.50 more units than expected due to the first insertion of the *d*-block elements, as we discussed in the last section. This effect is sometimes known as the **scandide contraction** of the elements Ga through Kr, since it is caused by the appearance of a new orbital type, the 3*d*, first found in scandium. Ironically, it is not really a contraction, since the expanding effect of the increase in $n^*$ is still larger than the effect of the increase in

$Z^*$ ($n^*$ is squared in equation (11.3), whereas $Z^*$ is not); but it is a contraction relative to normal trends. There is no contraction between Ge and Sn because both have $d$ orbitals and equally increased $Z^*$ values.

The scandide contraction has real consequences. In Chapter 2 we found that the maximum total coordination numbers for elements in Periods 3 and 4 are the same, whereas we might expect these to increase if the atoms get larger. This affects the formulas of compounds: $CO_3^{2-}$ in Period 2, versus $SiO_4^{4-}$ and $GeO_4^{4-}$ in Periods 3 and 4, versus $SnO_6^{8-}$ in Period 5. Chapter 2 showed that this in turn has major effects on the basicities of these compounds. In general, for properties depending on size (such as isomorphous substitution, Chapter 4), we expect unusual resemblance of the corresponding third- and fourth-period elements *beyond the d-block*.

A calculation not shown in Table 11.2 is that for crossing the $f$-block, for the simple reason that $\langle r_{max} \rangle$ is 327 pm for the whole series of elements La through Yb. (Neither $n^*$ nor $Z^*$ changes in this series of elements.) This trend is almost, but not quite, observed in the metallic radii of these elements (Table 6.4), which decrease slowly from 169 pm for La to 158 pm at Tm; (Eu and Yb are anomalies). This suggests that there really is some increase in $Z^*$ across this series, contrary to Slater's rules, and that the $6s$ orbital really does penetrate the $4f$ orbital to some extent. Since we know that $s$ orbitals are good penetrators and $f$ orbitals are poor shielders, this is not very surprising. Slater's rules, designed for simplicity, are just not subtle enough to catch this effect.

The last calculation shown in Table 11.2 is for the Group 12 elements. These are calculated as expanding normally on going down from Zn to Cd to Hg, but they do not in fact on going from Cd to Hg. Evidently there is a **lanthanide contraction** occurring in the elements after the first filling of $f$ orbitals (which starts at La), which prevents the post–$f$-block sixth-period elements from being larger (or much larger) than the corresponding fifth-period elements. This is confirmed in the metallic radii of Table 6.4. Real chemical consequences of this show up in the similar expected maximum coordination numbers of these two periods of elements (the corresponding oxo anions have the same formulas and basicities, for example). You may also recall that the elements Hf and Zr are always found isomorphously substituted in nature and are the two elements most difficult to separate.

One property related to size is that of covalent-bonding ability and covalent-bond energies; we have noted that these decrease (both for sigma and pi bonds) down the $p$-block but apparently not down the $d$-block (the related atomization energies, Table 6.1, generally increase). An important factor in the $p$-block trend is that orbitals at the bottom of the table have more nodal spheres and hence inner lobes than the same shape of orbital at the top. The valence orbitals of the heavier atoms are said to be more **diffuse**: The electrons spend more time in inner lobes that cannot overlap with orbitals of other atoms; hence the covalent bonds of these atoms are weaker.

The trend in the $d$- and $f$-blocks is more subtle. The $(n-1)d$ and $(n-2)f$ orbitals have their maxima inside the maxima of the core $(n-1)s$ and $(n-1)p$ and the valence $ns$ orbitals. Since the electrons in the filled core $(n-1)s$ and $(n-1)p$

orbitals repel other atoms, it is possible that the $(n-1)d$ or $(n-2)f$ orbitals are unable to overlap the orbitals of the other atom to form covalent bonds. Evidently this does occur for some but not all the atoms of the $d$- and $f$-blocks.

Most evidence indicates that the $f$-block atoms of the sixth period cannot involve themselves in covalent bonding using their "buried" $4f$ orbitals: These elements form only highly ionic, reactive organometallic compounds, for example. This is apparently not so true for the earlier $f$-block elements of the seventh period, which form covalent compounds such as uranocene (Chapter 10) and $UCl_6$.

Although $d$-block atoms generally can form covalent bonds involving their $d$ orbitals, there is evidence that this happens more extensively in the fifth and sixth periods than in the fourth. In Chapter 6 we saw that atomization energies are higher in the fifth and sixth periods, for example. Why this is so is not apparent from our crude calculations of atomic radius, however.

## 11.6

## Explanation of Electronegativity Trends: Allred-Rochow Electronegativities

As you should recall from Chapter 8, Pauling defined electronegativity as the ability of an atom in a molecule to attract the bond electrons to itself. Allred and Rochow interpreted this definition as corresponding to the force exerted by an atom on its ($s$ and $p$) valence electrons:

$$F = \frac{e^2 Z^*}{r^2} \tag{11.4}$$

For the charge, they used the effective nuclear charge $Z^*$ from Slater's rules, and for the radius they used the covalent radius. ($\langle r_{max} \rangle$ is clearly too inaccurate to be used here.) They added certain parameters so that the range of their numbers would correspond to the Pauling scale of electronegativities and obtained the following equation:

$$\chi_{AR} = 3590 \frac{(Z^* - 0.35)}{r^2_{cov}} + 0.74 \tag{11.5}$$

The resulting Allred-Rochow electronegativities, $\chi_{AR}$, of the elements (obtained using covalent radii available at the time) are shown in Table 11.3.

The computed Allred-Rochow and experimental Pauling electronegativities of the elements are in excellent agreement in the second, third, and fourth periods, and below these in the $s$-blocks; hence the Allred-Rochow electronegativity calculations provide us with explanations of the main periodic trends in Pauling electronegativities. Electronegativities increase from left to right because $Z^*$ increases, according to rule 3 of Slater's rules: As we add additional electrons on crossing a period, we are also adding additional protons, which are only 35% shielded by the additional electrons; hence all valence electrons (including those in the bond) are more strongly attracted to the nucleus. Down Groups 1 and 2, $Z^*$ often remains constant, so the

**Table 11.3**   Allred-Rochow Electronegativities of the Elements

| | 1 | 2 | 3 | 4 | 5 | 6 | 7 | 8 | 9 | 10 | 11 | 12 | 13/IIIA | 14/IVA | 15/VA | 16/VIA | 17/VIIA | 18/VIIIA |
|---|---|---|---|---|---|---|---|---|---|---|---|---|---|---|---|---|---|---|
| 1 | H 2.20 | | | | | | | | | | | | | | | | | He 5.50 |
| 2 | Li 0.97 | Be 1.47 | | | | | | | | | | | B 2.01 | C 2.50 | N 3.07 | O 3.50 | F 4.10 | Ne 4.84 |
| 3 | Na 1.01 | Mg 1.23 | | | | | | | | | | | Al 1.47 | Si 1.74 | P 2.06 | S 2.44 | Cl 2.83 | Ar 3.20 |
| 4 | K 0.91 | Ca 1.04 | Sc 1.20 | Ti 1.32 | V 1.45 | Cr 1.56 | Mn 1.60 | Fe 1.64 | Co 1.70 | Ni 1.75 | Cu 1.75 | Zn 1.66 | Ga 1.82 | Ge 2.02 | As 2.20 | Se 2.48 | Br 2.74 | Kr 2.94 |
| 5 | Rb 0.89 | Sr 0.99 | Y 1.11 | Zr 1.22 | Nb 1.23 | Mo 1.30 | Tc 1.36 | Ru 1.42 | Rh 1.45 | Pd 1.35 | Ag 1.42 | Cd 1.46 | In 1.49 | Sn 1.72 | Sb 1.82 | Te 2.01 | I 2.21 | Xe 2.40 |
| 6 | Cs 0.86 | Ba 0.97 | Lu 1.14 | Hf 1.23 | Ta 1.33 | W 1.40 | Re 1.46 | Os 1.52 | Ir 1.55 | Pt 1.44 | Au 1.42 | Hg 1.44 | Tl 1.44 | Pb 1.55 | Bi 1.67 | Po 1.76 | At 1.90 | Rn 2.06 |
| 7 | Fr 0.86 | Ra 0.97 | | | | | | | | | | | | | | | | |

| | 3f | 4f | 5f | 6f | 7f | 8f | 9f | 10f | 11f | 12f | 13f | 14f | 15f | 16f |
|---|---|---|---|---|---|---|---|---|---|---|---|---|---|---|
| 6 | La 1.08 | Ce 1.08 | Pr 1.07 | Nd 1.07 | Pm 1.07 | Sm 1.07 | Eu 1.01 | Gd 1.11 | Tb 1.10 | Dy 1.10 | Ho 1.10 | Er 1.11 | Tm 1.11 | Yb 1.06 |
| 7 | Ac 1.00 | Th 1.11 | Pa 1.14 | U 1.22 | Np 1.22 | Pu 1.22 | Am | Cm | Bk | Cf | Es | Fm | Md | No |

SOURCES: Values from A. L. Allred and E. G. Rochow, *J. Inorg. Nucl. Chem.*, 5, 264 (1958); E. J. Little and M. M. Jones, *J. Chem. Educ.*, 37, 231 (1960); and L. C. Allen and J. E. Huheey, *J. Inorg. Nucl. Chem.*, 42, 1523 (1980).

small decline in electronegativity is due to the larger size: The bond electrons are farther from the nucleus, hence somewhat more weakly attracted to it.

The Allred-Rochow electronegativities also duplicate the "anomalously" high electronegativities of Ga ($>$ Al) and Ge ($>$ Si). This unusual trend is due to the increase in $Z^*$ that results from the first filling of the poorly shielding $d$ orbitals—the same factor responsible for the scandide contraction discussed in the last section. Even though Slater's rules (and hence the Allred-Rochow electronegativities) do not predict the lanthanide contraction, we know it is there and expect that it too should cause the Pauling electronegativities of the post–$f$-block sixth-period elements to be higher than those of the corresponding elements of the fifth period. This is, of course, often the case: We find the electronegativity relationships Au $>$ Ag, Hg $>$ Cd, Tl $>$ In, Pb $>$ Sn in the Pauling but not the Allred-Rochow scales.

But not every anomaly is explained: In the Pauling scale the fifth-period elements of the $d$-block are strikingly more electronegative than the corresponding fourth-period elements and not so much less electronegative than the sixth-period elements, except at gold. The lanthanide contraction cannot explain that, nor can it explain the fact that the deviation between the Pauling and the Allred-Rochow scales reaches its maximum at gold, in the *middle* of the sixth period and yet far beyond the $f$-block.

## 11.7

## Relativistic Effects on Orbitals

The orbital wave functions we have been discussing up to this point do not incorporate any effects of relativity theory. But solutions of the Schrödinger wave equation that do incorporate relativity theory show that both the radial and the angular parts of the wave functions for the heavier atoms in the periodic table are appreciably altered by relativistic effects [5,6].

The radial effect is known as *relativistic contraction*. We can think of the electron as a particle that is accelerated to a certain radial velocity by the attraction of the nucleus; as the nuclear charge builds up this radial velocity builds up too, until the radial velocity may begin to approach the speed of light, which can be expressed in units of atomic charge as 137.036 atomic units. The average radial velocity of the $1s$ electrons of an atom is equal to the nuclear charge of that atom, which for the $1s$ electrons of a mercury atom is $80/137 = 0.58$ times the speed of light. According to Einstein's theory of relativity, the mass, $m$, of a particle increases over its rest mass, $m_0$, when its speed, $v$, approaches the speed of light, $c$:

$$m = \frac{m_0}{\sqrt{1 - (v/c)^2}} \tag{11.6}$$

hence the mass of a $1s$ electron in mercury is about 1.2 times its rest mass. But the radius of the Bohr orbit of an electron is inversely proportional to the mass of the electron, so we expect the radius of the $1s$ orbital to be about 20% less than otherwise expected. This is the relativistic contraction. It also affects the 2s and higher s orbitals roughly as much because of their inner lobes, which are close to the highly charged nucleus.

The effect is present to a lesser extent among p orbitals and is nearly absent among d and f orbitals, which have fewer lobes near the nucleus. Indeed, the d and f orbitals are more effectively screened out by the contracted s (and p) orbitals, so they undergo *relativistic expansion*. This expansion is proposed as the explanation of the increase in maximum oxidation numbers and (in addition to the lanthanide contraction) the increase in Pauling electronegativities down a group in the d- and f-blocks of elements. The relativistic expansion of the 5d orbitals allows them to reach out to the 6s electrons and hence overlap orbitals of other atoms, forming additional and stronger covalent bonds in the sixth period, while the 3d orbitals cannot do this in the fourth period; hence we find tungsten forming many more stable covalent compounds in its $+6$ oxidation state (e.g., $WF_6$ and $WCl_6$) than does chromium. Similarly, in the f-block uranium forms stable covalent compounds in the same oxidation state ($UF_6$, $UCl_6$), whereas neodymium does not exhibit this oxidation state at all.

While the 5d orbitals are filling and expanding, they are doing their usual poor job of shielding the 6s orbital from the nucleus. Once the 5d orbitals are filled, they are no longer valence orbitals, and the Pauling electronegativity must be determined by the 6s orbitals in Group 11 and beyond. These orbitals are contracted strongly in gold and mercury (Figure 11.8). Beyond mercury the less strongly affected and better-shielding p orbitals are valence orbitals, so the effect diminishes somewhat. Eventually, however, the increase in nuclear charge again causes atomic contraction.

Since relativistic contraction has its maximum effect in the sixth period at the 6s orbital of gold and mercury, anomalous effects appear for these elements. One of these is the enhanced electronegativity of gold and mercury as compared with

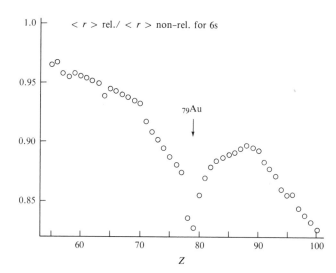

**Figure 11.8** The relativistic contraction of the 6s orbital in the elements Cs ($Z = 55$) to Fm ($Z = 100$). Reprinted with permission from P. Pyykko and J.-P. Desclaux, *Accounts Chem. Res.*, 12, 276 (1979). Copyright © 1979 American Chemical Society.

the elements above them, silver and cadmium. (Although the lanthanide contraction is partially responsible, it cannot account for the fact that enhancement is at a maximum in this part of the period.) This, of course, is related to the softness of the cations of these elements (Chapter 8). Since the $6s$ orbital is contracted and bound more tightly, its separation in energy from the $6p$ orbitals is anomalously large, so much so that the $6p$ orbitals are almost not valence orbitals anymore. Because by this time the $5d$ orbitals are also no longer effective valence orbitals, mercury, with a $6s^2$ valence electron configuration has essentially filled all its valence orbitals. This makes it very near to being a noble gas (or rather a noble liquid) and accounts for its exceptionally low atomization energy (Table 6.1) and high first-ionization energy (Table 6.6). The low atomization energy in turn explains why mercury is a liquid metal at room temperature, with a significant vapor pressure.

Gold, with a $6s^1$ valence electron configuration, is then one electron short of a noble-liquid electron configuration and thus might be expected to show some resemblance to a halogen such as iodine. Indeed, gold has the highest electron affinity of any atom other than the halogens and forms a stable $Au^-$ ion in the compounds RbAu and CsAu (which have the CsCl crystal structure and do not behave as alloys). The gaseous diatomic molecule $Au_2$ is known and has a covalent bond energy of 229 kJ/mol, which is quite high relative to most bond energies (Table 6.5). Isoelectronic with $Au_2$ is the catenated mercury(I) ion, $Hg_2^{2+}$, which has a stability unparalleled in cadmium chemistry and unknown in zinc chemistry.

Beyond gold and mercury, the $6s$ electrons are still difficult to ionize, which accounts for the low stability of the group oxidation state in the subsequent elements. These elements prefer the oxidation state equal to the group number minus two; this is sometimes known as the *inert pair* (of $6s$ electrons) *effect*. Thus Tl normally occurs as $Tl^+$, Pb as $Pb^{2+}$, and Bi as $Bi^{3+}$.

In addition to the relativistic contraction of $s$ orbitals and the relativistic expansion of $d$ and $f$ orbitals, there is another relativistic effect in heavy atoms known as **spin-orbit coupling**. The relativistic treatment of the Schrödinger wave equation was the first to reveal that electrons in atoms also have spin and need a fourth quantum number, $m_s$, which can have the value of either $+\frac{1}{2}$ or $-\frac{1}{2}$. This number corresponds to angular rotation of the electron in either a clockwise or a counterclockwise direction. This angular rotation, of course, occurs even in the lightest atoms. But in heavy atoms the angular motion due to the spin of the electron about its own axis begins to mix with the angular motion of the electron among the lobes of the nonspherical orbitals. These two types of motion may either reinforce each other or cancel each other out in part. (We say that the electron spin angular momentum and the electron orbital angular momentum are coupled with each other.)

Consequently in atoms heavier than about bromine, we observe that the $p$-, $d$-, or $f$-orbital electrons ionize at either of two energies, depending on whether the two types of angular motion are in harmony or are opposed. In the heaviest atoms, the difference between these two energies may be hundreds of kilojoules per mole, which is as large as or larger than their covalent bond energies. Consequently their bonding abilities are affected, and it becomes necessary to speak not just of a $6p$ electron but to distinguish which type of $6p$ electron is meant. This is done by specifying an additional quantum number $j$, which can have either of two values: $j = l + \frac{1}{2}$ or $j = l - \frac{1}{2}$. This number is attached as a subscript on the right of the orbital designation: Thus we find two types of $6p$ orbitals, designated the $6p_{3/2}$ and the $6p_{1/2}$ orbitals. (The $d$ orbitals are also split into $nd_{5/2}$ and $nd_{3/2}$ orbitals.) Each set of orbitals can hold $2j + 1$ electrons; thus there can be four $6p_{3/2}$ electrons and two $6p_{1/2}$ electrons. The $6p_{1/2}$ orbitals penetrate the core better and experience relativistic contraction; hence they are filled first and are more stable than the $6p_{3/2}$ orbitals by many kilojoules per mole.

Consequently in the sixth period of the $p$-block the $6p_{1/2}$ orbital is a valence orbital, while the $6p_{3/2}$ orbitals are enough higher in energy that they are not necessarily used in stable compounds. This tendency is not very pronounced, but the ion $Bi^+$ (valence-electron configuration $6s^2 6p_{1/2}{}^2$) is known in a solid compound. Some ions in the later sixth period tend to use only the $6s$ and $6p_{1/2}$ orbitals to accept electrons in forming coordination compounds, which accounts for the tendency toward the very low coordination number of 2 in $Au^+$ and $Hg^{2+}$ complexes. One familiar consequence of this low coordination number is the exceptional stability to hydrolysis of dimethylmercury and other organomercury compounds: They show little tendency to coordinate a water molecule, the first step in an expected hydrolysis reaction.

## 11.8

# Predicting the Chemistry of Superheavy Elements

At the time of this writing 109 elements have been reported; although next to nothing is known of the chemistry of the atoms beyond No, it is presumed that these elements are filling in the seventh period of the periodic table (the $d$-block through element 112, then the $p$-block through element 118). These elements should show enhanced relativistic properties over those of the sixth period, hence clarifying the nature of these effects. In element 112, the element below mercury, the stability of the $7s^2$ electron configuration might be great enough to cause it to be the next noble gas (actually liquid, since the van der Waals forces of 112 electrons will be considerable). Possibly element 114, the element below lead with the favored $7s^2 7p_{1/2}{}^2$ valence-electron configuration, will also be a noble liquid.

Even without relativistic considerations, the yet-undiscovered elements of the eighth period should be very interesting to inorganic chemists. As we can see in the longest form of the periodic table (Table 1.1), in every other period a new shape of orbital is first filled. The eighth period is expected to be one of those, in which a new type of orbital, the $g$ orbital, is first occupied. Thus a new $g$-block, 18 elements wide, is expected between the $s$-block and the $f$-block. The $5g$ orbitals will surely not extend out to reach the $8s$ orbitals and should, by Slater's rules and indeed perhaps in fact, have shielding constants of 1. Consequently these 18 elements may have identical radii, electronegativities, and chemistry and will perhaps be a nightmare for the analytical chemist to separate but a delight for the geochemist or solid-state chemist to substitute isomorphously. Following the filling of the $8s$ and the $5g$ orbitals, we expect to fill the $6f$, $7d$, and $8p$ orbitals, reaching the next noble liquid at element 168, two periods below radon.

But if we add in relativistic effects, the periodic table begins to look different from Table 1.1. In particular, calculations predict that the order of filling may not be the expected $8s < 5g < 6f < 7d < 8p$. Instead, spin-orbit coupling effects may be so great in the $8p$ orbitals that the $8p_{1/2}$ orbital may fill after $8s$, while the $8p_{3/2}$ orbital may not fill at all in this period [6]. Consequently the eighth period may end with the filling of the $7d$ orbitals, in element 164, which we may place two periods below mercury. The properties of these elements may be quite different: The first element, number 119 below francium, is predicted to have a stable oxidation state of $+4$ due to the ease of ionization of its no-longer-core $7p_{3/2}$ electrons [8].

## 11.9

## Study Objectives

1. *Without* the use of the tables in the earlier chapters, describe the main and the anomalous periodic trends in the following atomic variables: Pauling electronegativity; covalent/metallic radius; the most stable oxidation numbers or ionic charges of the elements; and covalent single- and multiple-bond energies.

2. Draw the angular part of the wave function for $s$, $p$, $d$, and $f$ orbitals.

3. Identify cases of positive, negative, and zero overlap between specified $s$, $p$, $d$, or $f$ orbitals on two different atoms. Identify cases of sigma, pi, and delta bonds resulting from such overlap. Rationalize periodic trends in sigma-, pi-, and delta-bond energies.

4. Include the radial part of the wave function in drawing specified $s$, $p$, or $d$ orbitals; identify nodal planes and surfaces and the signs of the wave functions.

5. Explain the concept of penetration or shielding of two different orbitals on the same atom and how this gives rise to an effective nuclear charge that differs from the actual nuclear charge.

6. Use Slater's rules to calculate $Z^*$ for a given atom.

7. Use $Z^*$ to calculate the radius $\langle r_{max} \rangle$ for a given atom; explain periodic trends in atomic radii. Explain the lanthanide and scandide contractions.

8. Given the covalent radius of an element, use $Z^*$ to calculate the Allred-Rochow electronegativity of an element.

9. Explain periodic trends in electronegativities in terms of periodic trends in $Z^*$. Explain discrepancies between Allred-Rochow and Pauling electronegativities.

10. Know that relativistic and spin-orbit coupling effects alter the shielding of outer electrons in very heavy atoms; know how this affects their sizes, electronegativities, and likely oxidation numbers.

## 11.10

**Exercises**

1. Describe the main periodic trends of the following from left to right in the periodic table: **a.** electronegativity; **b.** radius; **c.** covalent-bond energy; **d.** common oxidation numbers. Tell in what blocks of the periodic table (if any) you find "anomalous" countertrends.

2. Describe the main periodic trends of the following from top to bottom in the periodic table: **a.** electronegativity; **b.** radius; **c.** covalent-bond energy; **d.** common oxidation numbers. Tell in what blocks of the periodic table (if any) you find anomalous countertrends.

3. Go back to Chapter 1 and answer (or reanswer) questions 12 and 13.

†4. Without the use of tables, arrange each of the following sets in order of increasing size: **a.** Li, C, F, Ne;     **b.** Be, Ca, Ba, Ra; **c.** B, Al, Ga, In, Tl;     **d.** V, Nb, Ta, element 105.

5. Without the use of tables, arrange each of the following sets in order of increasing Pauling electronegativity: **a.** Be, Mg, Ca, Sr, Ba; **b.** Na, Al, P, Cl, Ar;     **c.** C, Si, Ge, Sn, Pb; **d.** Cu, Ag, Au, element below Au.

†6. Without the use of tables, arrange each of the following sets in order of increasing covalent-bond energies (increasing stability of covalent bonds). **6.1** The following single bonds: C—C, Si—Si, Ge—Ge, Sn—Sn, Pb—Pb. **6.2** The following single bonds: Li—Li, Be—Be, B—B, C—C. **6.3** The following Mo—Mo bonds: sigma, delta, pi.

7. Sketch $xy$-plane cross-section diagrams for the following orbitals: **a.** $4p_x$; **b.** $3s$; **c.** $4d_{xy}$; **d.** a pi bond formed from $3d_{xy}$ orbitals on two atoms; **e.** two $3d_{xy}$ orbitals on two atoms overlapping to give negative pi overlap (an antibonding $\pi^*$ orbital). Indicate regions of high electron probability by blue or black shading, nodal planes and spheres by red or penciled dashed lines, and show the positive and negative signs for the wave function. Show the $x$ and $y$ axes.

†8. Rank the $5s$, $5p$, $5f$, $5d$, and $5g$ orbitals in order of increasing: **a.** penetrating ability; **b.** shielding ability.

9. In Figure 11.2 are drawn different pairs of atoms (A on the left and B on the right) with orbitals overlapping. In each case tell whether there is positive, negative, or zero overlap between the orbitals of A and of B.

*10. In Figure 11.9 are drawn different pairs of atoms (A on the left and B on the right) with orbitals overlapping. **a.** Identify the specific atomic orbital used by each atom in each case. **b.** In each case tell whether there is positive, negative, or zero overlap between the orbitals of A and of B. **c.** For each case in which there is positive overlap, identify the bond that is formed as sigma, delta, or pi.

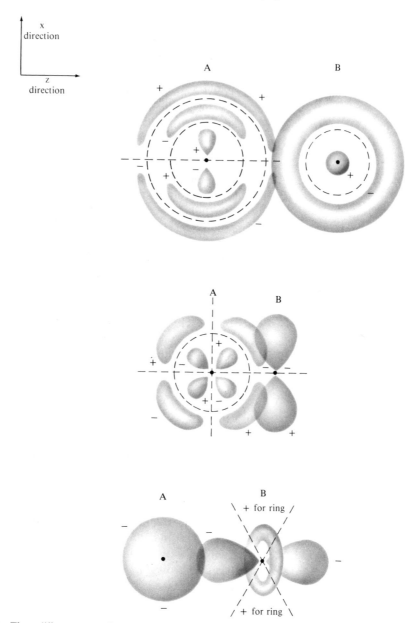

**Figure 11.9**  Three different cases of overlapping atomic orbitals on atoms A and B.

11.  **a.** Using Slater's rules, calculate $Z^*$ for a $4s$ electron of Zn. **b.** Using Slater's rules, calculate $Z^*$ for a $3d$ electron of Zn. **c.** Calculate $\langle r_{max} \rangle$ for these two orbitals of zinc. **d.** Using $Z^*$ for the $4s$ electron and given that the covalent radius of zinc is 121 pm, calculate the Allred-Rochow electronegativity of Zn. How do you think this value compares with the Pauling value for Zn? If you

think that it is either substantially ($> 0.5$ units) higher or substantially lower, explain why this is the case.

**12.** Identify all elements in the periodic table for which the Pauling and Allred-Rochow electronegativities differ by more than 0.5. In what parts of the table do these occur? For the element for which the discrepancy is *greatest*, use the Pauling electronegativity and its known covalent radius to calculate its "real" $Z^*$. Compare this with $Z^*$ computed using Slater's rules. Are the two values close?

†**13.** Let us suppose that you have discovered a new element of atomic number 162, which you choose to name Khalidium (Kh). Assume that the normal order of filling of orbitals still prevails in the eighth period.

**13.1** Write the full electron configuration of this element.

**13.2** Write the valence-electron configuration of this element and of its $+2$ ion.

**13.3** Note that a new type of orbital is used in Kh that is not used in any known atom. How many nodal planes will there be in a sketch of this type of orbital? How many nodal spheres will there be? Will the electrons in this type of orbital be good or poor at shielding the valence electrons of Kh? Will electrons in this type of orbital be good or poor at penetrating the inner orbitals of Kh?

**13.4** Calculate $Z^*$ for the outermost valence electron of Kh.

**13.5** Using an estimated covalent radius of 150 pm, calculate the Allred-Rochow electronegativity of Kh. Do you think that the Pauling electronegativity of Kh will be higher, lower, or about the same as this value?

**14.** Suppose that some all-powerful being suddenly decided to change the Schrödinger wave equation. After this change, the energies of electrons in all atoms would depend on the principal quantum number *only*. Everything else would remain the same.

**14.1** Using the new rules, list the order of filling of the $s$, $p$, and so forth orbitals for the first four values of the principal quantum number.

**14.2** Write the electron configuration of nickel under the new rules.

**14.3** What ions, if any, would nickel now tend to form? Could it achieve a noble-gas electron configuration by covalent bonding? What would be the nearest noble-gas configuration? Classify nickel now as a metal or nonmetal.

**14.4** What would happen to the nickels in your pocket when this change was made? What effect would this have on the economy?

**15.** Suppose that the solution to the Schrödinger wave equation were to be changed in one way only: The secondary quantum number $l$ would be allowed to go to *as high a value* as the principal quantum number. **a.** Write the order of filling of the eight lowest types of orbitals in an atom. **b.** Draw a new periodic table showing the arrangement of the first 36 elements. **c.** Assuming that $ns^2np^6$ remains a noble-gas electron configuration, list the noble gases in your new periodic table. **d.** If the changeover were made at the stroke of 12 midnight, discuss what would then happen to the water in the world.

16. Suppose that the inner lobes of orbitals were to be abolished, so that the $2s$ orbitals, for example, would lie completely outside the $1s$ orbitals of the same atom. **a.** Explain how this would alter the shielding of orbitals. **b.** Revise Slater's rules to accommodate this change. **c.** Use the equations for $\langle r_{max} \rangle$ and Allred-Rochow electronegativities to reevaluate these quantities for the second- and third-period elements. **d.** Discuss any changes in periodic trends that would result.

17. Explain why relativistic effects can cause gold to have both **a.** a lower oxidation number than either Ag or Cu ($-1$ in CsAu), and **b.** a higher oxidation number than either Ag or Cu ($+5$ in $AuF_5$).

18. Keeping in mind the effects of relativistic contraction and expansion of orbitals and spin-orbit coupling of $p$ orbitals, list some plausible common oxidation numbers for the elements of the seventh period beyond number 102. What element in this period has the best possibility of showing the hitherto-unknown oxidation number of $+9$, and what factor might make this possible?

Notes

[1] In relativistic quantum-mechanical calculations, the probability here is not precisely zero but is at a very low minimum.

[2] Slater, J. S., *Phys. Rev.*, 36, 57 (1930).

[3] Huheey, J. E., *Inorganic Chemistry: Principles of Structure and Reactivity*, 3d ed., Harper & Row, New York, 1983, p. 39.

[4] For the convenience of students studying Chapter 11 immediately following Chapter 1, the tables cited from Chapter 6 are also available in Appendix C at the end of the book.

[5] Pitzer, K. S., *Accounts Chem. Res.*, 12, 271 (1979).

[6] Pyykko, P., and J.-P. Desclaux, *Accounts Chem. Res.*, 12, 276 (1979).

[7] Fricke, B., and J. McMinn, *Naturwissenschaft*, 63, 162 (1976).

[8] "Extrapolation of Periodic Table Could Err," *Chem. Eng. News*, p. 27 (Sept. 3, 1973).

# Summary: Applying Theory to Chemical Reality

We have almost completed our study of the basic theories of inorganic chemistry and how these can be related to the known physical and chemical properties of the elements and their compounds. This brief chapter contains two remaining sections. The first deals with a topic—the abundance and nuclear stabilities of the elements—that does not fit well elsewhere since it involves *nuclear* properties of the elements rather than their chemical or the more common type of physical properties. Abundance of the elements relates to a very important *economic* property of the elements: their costs.

The second section is designed to help you tie together what you have learned throughout the semester by applying the concepts of the whole book to the case of a single element, trying to predict all that you can about its chemistry. You may then want to check out your predictions by looking up the chemistry of that element in an advanced textbook of descriptive inorganic chemistry. Becoming acquainted with these texts is worthwhile in itself, since you will discover that our theoretical knowledge is imperfect—we cannot predict all that we may want to know about the physical and chemical properties of the elements and their compounds.

## 12.1

## Abundance and Nuclear Stabilities of the Elements

One final property of an element that profoundly affects its practical uses is its abundance, which in part depends on the stability of the nuclei of its isotopes [1]. Figure 12.1 gives the estimated relative cosmic abundance of the elements (per each $10^6$ atoms of silicon). Some trends are apparent:

■ Hydrogen and helium, the primordial nuclear fuel and its fusion product, constitute over 99% of the atoms in the universe. Abundances generally drop off after them, although the next three, Li, Be, and B, which are readily transmuted by nuclear bombardment, are exceptionally low in abundance.

■ Iron is unexpectedly high in abundance. The iron nucleus has the greatest stability of any nuclear particle. Nuclei lighter than those of iron can (in principle)

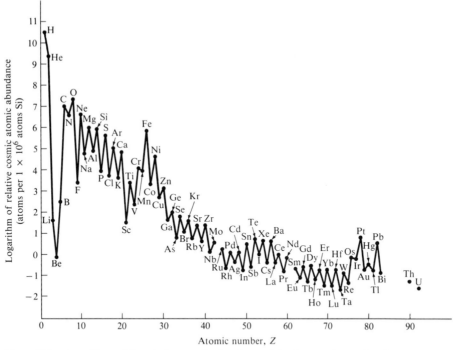

**Figure 12.1**  Logarithms of the relative cosmic abundance of the elements (per each $10^6$ atoms of silicon) versus atomic numbers. From *Inorganic Chemistry: A Modern Introduction*, by T. Moeller. Copyright © 1982 by Wiley-Interscience. Reprinted by permission of John Wiley & Sons, Inc.

release energy by combining in the process of **nuclear fusion**, whereas those heavier than iron can (in principle) release energy by breaking apart into smaller nuclei by the process of **nuclear fission**.

■ Elements with even atomic numbers are more abundant and have more stable isotopes than those with odd atomic numbers.

Neutrons are important to the stability of nuclei: They act as a "glue" to bind protons together in the nucleus despite their mutual electrostatic repulsion. Nuclei with 20 or fewer protons are effectively bound together by an approximately equal number of neutrons, but as the number of protons grows to 83, it takes a proportionately larger number of neutrons (ultimately about 5 neutrons for each 3 protons) to hold the nucleus together. Figure 12.2 indicates the combinations of neutrons and protons that form stable nuclei.

Nuclei that have insufficient numbers of neutrons to bind together their protons are called *proton-rich*; they undergo either of two types of radioactive decay that convert protons to neutrons: capture of a core electron to combine with the proton or emission of a positron (positive electron). Nuclei that have too many neutrons are called *neutron-rich*, but they can seldom emit the excess neutrons. Instead, each

403

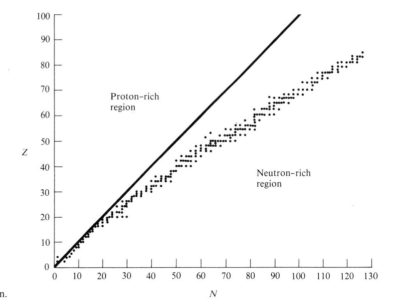

**Figure 12.2** Combinations of neutrons and protons that form naturally occurring stable nuclei. From *Nuclear and Radiochemistry*, 2d ed., by G. Friedlander, J. W. Kennedy, and J. M. Miller. Copyright © 1964 by John Wiley and Sons. Reprinted by permission.

neutron in effect converts to a proton and an electron; the latter is emitted as a *beta particle.*

No naturally occurring element of atomic number greater than 83 is stable; there is apparently no number of neutrons sufficient to bind these protons together in a stable nucleus. These "overweight" atoms often emit *alpha particles*, which are $^4$He nuclei—i.e., two protons and two neutrons. Many especially heavy atoms decay by *nuclear fission*: They break apart into two much smaller nuclei plus some spare neutrons. Atoms heavier than uranium break apart by one of these decay mechanisms fast enough that essentially none of them are left on the earth: Their *half-lives* are much shorter than the estimated age of the earth, 4.5 billion years. As the atoms get heavier, their half-lives get shorter (Table 12.1), and the available world supply (from nuclear reactions) gets smaller. At the extreme, element 109 was produced in 1982 in the quantity of one atom, which lasted for 0.005 s! (It was identified from its decay products [2].)

Given these trends, it would seem impossible to verify the interesting chemical predictions made for superheavy elements in the last section of Chapter 11. But the study of the occurrence of stable isotopes and the binding energies within them reveals that there are nuclear energy levels somewhat comparable to those of electrons, and that especial nuclear stability is reached with certain "magic numbers" of neutrons and protons, corresponding to noble-gas electron configurations among electrons. *Even* numbers of neutrons and of protons are favored. Beyond that, filled nuclear energy levels (hence extra stability) seem to result with 2, 8, 20, 28, 50, and 82 protons (i.e., with helium, oxygen, calcium, tin, and lead); calculations suggest that 114 and 164 protons might fill the next two proton energy levels. Similarly, filled nuclear energy levels result from 2, 8, 20, 28, 50, 82, and 126 neutrons; cal-

**Table 12.1**  Principal Isotopes of Transuranium Elements

| Isotope | Half-life | Quantities Available |
|---|---|---|
| $^{237}$Np | 2,200,000 years | many kilograms |
| $^{239}$Pu | 24,360 years | many kilograms |
| $^{244}$Pu | 82,800,000 years | > 1 milligram |
| $^{243}$Am | 7,650 years | > 100 grams |
| $^{244}$Cm | 18.12 years | > 100 grams |
| $^{247}$Cm | 16,000,000 years | traces |
| $^{247}$Bk | 1,400 years | traces |
| $^{249}$Bk | 314 days | > 1 milligram |
| $^{251}$Cf | 800 years | traces |
| $^{252}$Cf | 2.57 years | > 1 milligram |
| $^{254}$Es | 276 days | > 1 milligram |
| $^{257}$Fm | 94 days | > 0.001 milligram |
| $^{258}$Md | 53 days | traces |
| $^{255}$No | 3 minutes | traces |
| $^{256}$Lr | 45 seconds | |
| $^{261}$Rf | 70 seconds | |

SOURCES: Data from F. A. Cotton and G. Wilkinson, *Advanced Inorganic Chemistry: A Comprehensive Text*, 4th ed., Wiley-Interscience, New York, 1980; and the *Handbook of Chemistry and Physics*, 50th ed., Chemical Rubber Co., Cleveland, 1969, pp. B-267 to B-561.

culations suggest that 184 and 196 neutrons might fill the next two energy levels; hence the hope exists that, although the elements from 93 to 109 protons lie far from the regions of nuclear stability, there might be one or more "islands of stability" for superheavy elements.

Combining the proton magic number of 114 with the neutron magic number of 184, one might hope that extraordinary stability would be found for the isotope of mass number 298 of the element of atomic number 114. Calculations indicate that this isotope ought not to decay by beta particle emission, and that its half-life with respect to spontaneous nuclear fission ought to be very long—$10^{16}$ years! Unfortunately the conditions for stability against alpha particle emission, beta particle emission, and nuclear fission differ somewhat; the half-life of this isotope with regard to alpha particle emission is expected to be only 1 to 1000 years. The longest half-lives have been calculated for the isotope of superheavy element 110 with 184 neutrons; this might have a half-life of $10^5$ years. If this isotope could be made and if it had this kind of half-life, much chemistry could be done with it and practical uses could perhaps be found for it.

Synthesizing such nuclei provides a formidable challenge, however. The early transuranic elements were synthesized by bombarding uranium (and later, heavier) atoms with neutrons or alpha particles. But this method can raise the atomic numbers only by two at most; the necessary target nuclei are not available for elements 114 or 164. Instead it is proposed to use heavier bombarding particles, such as $^{58}$Fe nuclei, on reasonably heavy targets such as $^{209}$Bi (this reaction actually led to element 109). In order to overcome electron-electron repulsions, however, *all* the electrons must be stripped from the bombarding particle. Removal of all of the electrons from iron to produce a $Fe^{26+}$ ion requires the use of an extremely high energy particle accelerator. The ion must also be given enough kinetic energy to overcome the repulsion of the $Fe^{26+}$ nucleus and the $Bi^{83+}$ nucleus but not so much energy as to blast the target to pieces! Finally, there is the serious problem that most feasible projectiles do not have the 5 : 3 ratio of neutrons to protons that will be needed in a stable superheavy nucleus [3,4]. At the time of this writing the theoretical island of stability of long-lived superheavy elements has not yet been reached.

Let us return now to consideration of the stable, naturally occurring elements. The abundances of the elements in the crust of the earth are not always similar to those in the cosmos, affected as they are by the primary and secondary differentiation of the elements. For example, elements present as light gaseous molecules or noble gases tend, over time, to escape from the earth's atmosphere when they exceed escape velocity from the earth. Metals less active than iron have tended to concentrate in the core of the earth, dissolved in the iron, and thus are inaccessible to us. On the other hand, some rare elements that do not readily substitute isomorphously for common ones did not crystallize out during the solidification of the mantle and are especially abundant in the crust of the earth. The estimated crustal abundances of the more common elements are indicated in Table 12.2.

Most important for the overall usefulness of an element is its cost. Although related to its abundance, the cost is also related to how many other demands there are for the element, whether it is obtained as a byproduct of other more abundant elements, whether it is widely dispersed in many minerals (due to isomorphous substitution) or concentrated in convenient ore deposits, and the political situations in the countries in which it is found, if it is rare. In Table 12.3 we give costs of the elements (in elemental form). These costs are for small quantities of high-purity, research-grade samples, and would be much lower on an industrial scale, although the relative trends among elements ought to be reasonably similar. The costs, of course, are also affected by the relative activity of the different elements: More active ones are more difficult to produce (sodium is much cheaper as sodium ion in rock salt than as the metal).

Many of the highly industrialized countries of the world depend almost completely on imports for their supplies of many strategic metals and have become accustomed to their ready availability in the marketplace. Many technologies have been developed based on the elements with the optimal combination of cost and physical and/or chemical properties; the telephone handset, for example, contains more than 40 different elements. Some of these elements are mined in only a few countries, which may be politically unstable; for example, most of the world's cobalt comes from the Katanga province of Zaire, which has been shut off from the rest of the world more than once by civil wars or invasions. Consequently the price of cobalt has fluctuated wildly during these times.

Even in the absence of such sudden interruptions, many people expect producer countries to attempt to put together cartels in the future to attempt to increase the prices for these critical materials [5]. Then the problem of substitution of materials

**Table 12.2** Estimated Crustal Abundances of the Elements

| 1 | H | | | | | | | | | | | | | | | | | He |
|---|---|---|---|---|---|---|---|---|---|---|---|---|---|---|---|---|---|---|
| 2 | Li | Be | | | | | | | | | | | B | C | N | O | F | Ne |
| 3 | Na | Mg | | | | | | | | | | | Al | Si | P | S | Cl | Ar |
| 4 | K | Ca | Sc | Ti | V | Cr | Mn | Fe | Co | Ni | Cu | Zn | Ga | Ge | As | Se | Br | Kr |
| 5 | Rb | Sr | Y | Zr | Nb | Mo | | Ru | Rh | Pd | Ag | Cd | In | Sn | Sb | Te | I | Xe |
| 6 | Cs | Ba | * | Lu | Hf | Ta | W | Re | Os | Ir | Pt | Au | Hg | Tl | Pb | Bi | Po | At | Rn |
| 7 | Fr | Ra | † | | | | | | | | | | | | | | | |

| * | La | Ce | Pr | Nd | | Sm | Eu | Gd | Tb | Dy | Ho | Er | Tm | Yb |
|---|---|---|---|---|---|---|---|---|---|---|---|---|---|---|
| † | Ac | Th | Pa | U | Np | Pu | | | | | | | | |

Greater than 1000 ppm (0.1%) of the earth's crust

99 to 1000 ppm

10 to 99 ppm

5 to 10 ppm

less than 5 ppm

Data from T. Moeller, *Inorganic Chemistry: A Modern Introduction*, John Wiley, New York; 1982, p. 24.

in industrial processes may suddenly become critical for inorganic chemists and materials scientists. Finding substitutes becomes a problem, too, when a material such as mercury or asbestos is found to cause such severe environmental problems that it must be replaced. We suggest an exercise at the end of this chapter in which you try to invent appropriate and cost-effective substitutes for industrial materials; this is not an easy task but it may be a realistic one.

It is also worth noting that the problem of cost is much more critical for an element and its compounds if it is used in very large quantities in the production of basic industrial inorganic chemicals (e.g., those in the table included in the exercises); there are often not many economically viable possibilities for substitution. But the chemical industry is also involved in the production of *specialty chemicals*, which are produced in small quantities at high prices for uses the importance of which cannot be measured by tons of output. (The solid-state semi-

**Table 12.3** Costs of the Elements

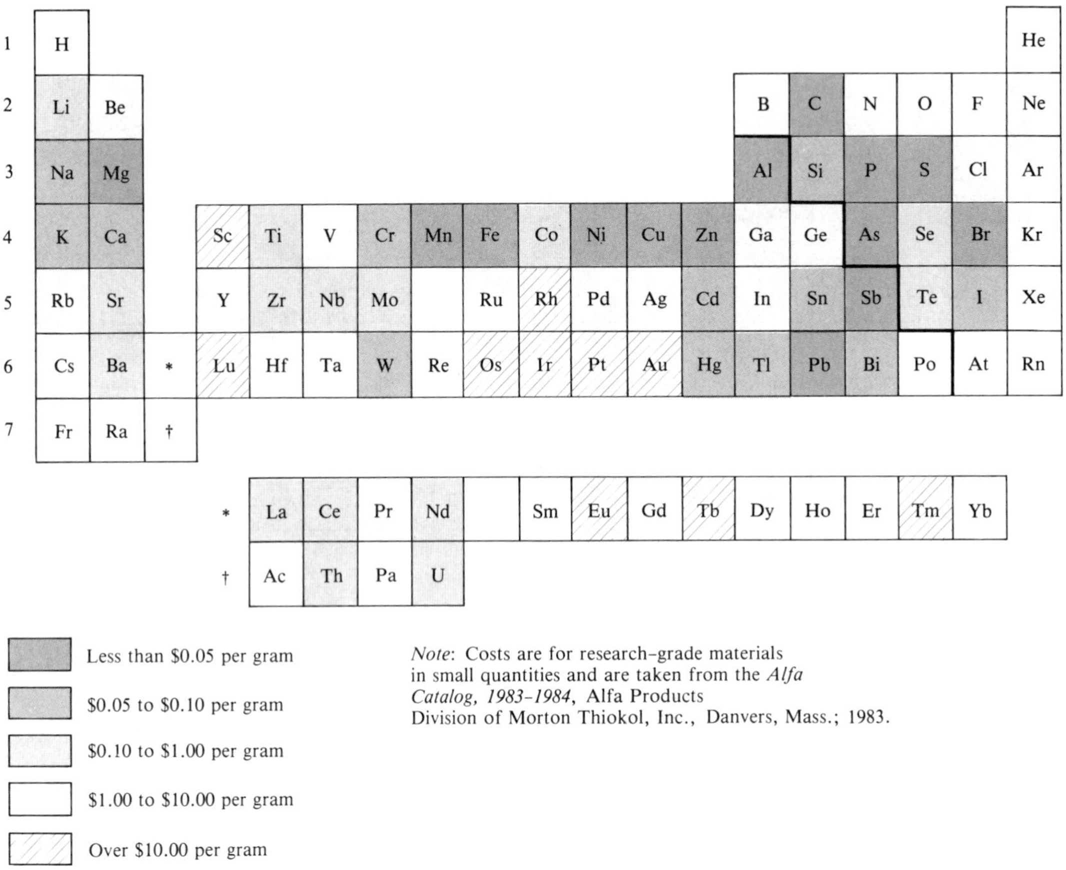

| 1 | H | | | | | | | | | | | | | | | | | | He |
|---|---|---|---|---|---|---|---|---|---|---|---|---|---|---|---|---|---|---|---|
| 2 | Li | Be | | | | | | | | | | | B | C | N | O | F | | Ne |
| 3 | Na | Mg | | | | | | | | | | | Al | Si | P | S | Cl | | Ar |
| 4 | K | Ca | | Sc | Ti | V | Cr | Mn | Fe | Co | Ni | Cu | Zn | Ga | Ge | As | Se | Br | Kr |
| 5 | Rb | Sr | | Y | Zr | Nb | Mo | | Ru | Rh | Pd | Ag | Cd | In | Sn | Sb | Te | I | Xe |
| 6 | Cs | Ba | * | Lu | Hf | Ta | W | Re | Os | Ir | Pt | Au | Hg | Tl | Pb | Bi | Po | At | Rn |
| 7 | Fr | Ra | † | | | | | | | | | | | | | | | | |

| * | La | Ce | Pr | Nd | | Sm | Eu | Gd | Tb | Dy | Ho | Er | Tm | Yb |
|---|---|---|---|---|---|---|---|---|---|---|---|---|---|---|
| † | Ac | Th | Pa | U | | | | | | | | | | |

Less than $0.05 per gram

$0.05 to $0.10 per gram

$0.10 to $1.00 per gram

$1.00 to $10.00 per gram

Over $10.00 per gram

*Note*: Costs are for research-grade materials in small quantities and are taken from the *Alfa Catalog, 1983–1984*, Alfa Products Division of Morton Thiokol, Inc., Danvers, Mass.; 1983.

conductor industry is much more important than would be guessed by the number of tons of microchips produced.) So the chemistry of rare elements can indeed be very important. As an admittedly extreme example, the unstable radioactive element technetium, which virtually does not even exist naturally on the earth, is nonetheless important in the radiopharmaceutical industry.

## 12.2

## Final Review

In this book we have tried to present (or let you discover) principles that can be used to predict the physical properties and chemical reactions of simple inorganic compounds with reasonable accuracy and minimal complexity. As a means of reviewing these trends before you take your final exam, we suggest you try Exercise 1, which asks you to tell all you can about the chemistry of (any element whatsoever).

EXAMPLE    Tell all you can about the chemistry of gallium.

Solution    ■ To answer a question such as this you may want to think through the different chapters of the book and apply the general concepts of each chapter to the specific case of gallium.

■ Chapter 1 deals with the fundamental properties of the atoms. You have been using tables to make predictions up to this point, but perhaps your instructor would now want you to make your predictions *without* the tables. If so, you could calculate the Allred-Rochow electronegativitity of Ga using Slater's rules and equation (11.5) (it is approximately 1.82). Then recalling that Ga will be affected by the scandide contraction but that Slater's rules include this effect, you could take this number to be equal to the Pauling electronegativity. Your instructor would need to provide you with radii, since radii calculated from Slater's rules are too inaccurate; your instructor might give you the covalent and ionic radii of Al, in which case you could reason that Ga should have almost the same radii (covalent 120 pm, cationic 76 pm) due to the scandide contraction. Finally, you could predict the most common positive oxidation number of $+3$.

■ Using the concepts of Chapter 2, you would begin predicting the acid-base chemistry of Ga. The $Z^2/r$ ratio for $Ga^{+3}$ is 0.118, enough to classify it as a moderately acidic cation. (Its electronegativity value of 1.81 would be barely enough to raise its category to strongly acidic according to our approximate rules, but detailed calculations would leave it in the moderately acidic category.) Thus recalling Experiment 1, you would expect its chloride to hiss upon contact with water, giving a very acidic solution that would deposit insoluble $Ga(OH)_3$ or $Ga_2O_3$ unless its pH was kept below roughly 2 to 4. Ga would also form an oxo anion, $GaO_4{}^{5-}$, but only in nonaqueous reactions, since this oxo anion is very strongly basic and would react with water to give the hydroxo anion $Ga(OH)_4{}^-$ in basic aqueous solutions.

■ From Chapter 3 you could predict that since $Ga^{3+}$ is moderately acidic, its salts with basic oxo anions (i.e., most oxo anions) would be insoluble, whereas its perchlorate, nitrate, sulfate, permanganate, salts of fluoro anions such as $BF_4{}^-$, etc. would be soluble in water. The radius ratio of $Ga^{3+}$ to oxo anions would be such that these soluble salts would most often be found as hydrated salts; the anhydrous salts would be good drying agents.

■ From Chapter 4 you would predict that $Ga_2O_3$ would be a solid with very high melting and boiling points and that it might substitute isomorphously for the very abundant $Al^{3+}$ in spinels, silicates, and aluminosilicates. Its acid-base chemistry in natural waters would feature these insoluble minerals or insoluble $Ga_2O_3$ or $Ga(OH)_3$; it would be very unlikely to be a significant component in solution in natural waters unless acid-rain pollution was present.

■ In Chapter 5 we began looking at the oxidation-reduction reactions of the elements. From gallium's electronegativity of 1.81 you would locate it among the electropositive metals. Perhaps you would note that due to the poor shielding effects of the scandide $3d$ orbitals, its electronegativity is significantly higher than that of Al, so its position in the activity series of metals ought to be significantly lower. It might not be necessary to produce gallium by electrolysis or by reduction with a more active metal; reduction of $Ga_2O_3$ with carbon might suffice. Being in the $p$-block of elements, Ga could have a $+1$ oxidation state, but being in the middle of the $p$-block this would be much less stable than the $+3$ oxidation state, so it would be difficult to produce.

■ Using the principles of Chapter 6, you would recognize Ga as a metal with three valence electrons to contribute to its conduction band; it should be a good conductor with reasonably high melting and boiling points. (But see ahead concerning these predictions.) No great number of catenated Ga compounds would be expected.

■ Chapters 7 and 8 deal with principles for predicting coordination compounds of the elements. You could apply VSEPR theory to predict the geometries of the halides or complex ions of Ga as you encounter them. From Chapter 8 you would classify Ga as a borderline acid on the basis of its electronegativity and proximity to gold, but its common oxidation number of $+3$ would tend to emphasize hard rather than soft properties. Its halides consequently would be soluble in water; its sulfide, selenide, telluride, arsenide, and the like would probably be somewhat insoluble. It would precipitate in the qualitative analysis scheme in Group III, probably as a sulfide rather than as a hydroxide or oxide. As a borderline-to-hard cation $Ga^{3+}$ might have some toxicity but would probably not be as toxic as the soft acids; it might cause some difficulties by substituting for $Fe^{3+}$ in key enzymes. As a borderline-to-hard metal ion $Ga^{3+}$ might form many coordination complexes in which its maximum coordination number would be 6.

■ In Chapters 9 and 10 the principles of the previous chapters are applied to the remaining major classes of inorganic compounds. The fluoride $GaF_3$ would be expected to adopt an ionic lattice with a very high melting point; the other halides ($GaCl_3$, $GaBr_3$, and $GaI_3$) would show a decreasing tendency to do this and at some point would form 4-coordinate dimers, $Ga_2X_6$, which would be solids with relatively low melting points. These should be Lewis-acidic halides, readily forming the halo anions $GaX_6^{3-}$ or $GaX_4^{-}$. The halides would hydrolyze too readily to form by evaporation of a solution of $Ga_2O_3$ in HX, but they could be made by direct combination of the elements or, for example, by heating the oxide with carbon and chlorine. These halides would probably be colorless, since the $Ga^{3+}$ ion would be a relatively poor oxidizing agent, unable to give charge-transfer bands (except possibly with $I^-$).

■ An alkoxide such as $Ga(OCH_3)_3$ could perhaps be made by reacting $GaCl_3$ with methanol and trimethylamine, but since it had not reached its maximum coordination number it might be difficult to prevent it from combining with excess trimethylamine or methanol to give a coordination compound. Ga would form "diamondlike" nitride, phosphide, and arsenide compounds, which could have considerable importance as semiconductors; its sulfide ($Ga_2S_3$), selenide, and telluride would not be isoelectronic with important semiconductors and might be of less interest.

■ The hydride gallane, $(GaH_3)_x$, would be expected to be of very low stability, although the $GaH_4^-$ ion might be of some interest. The compound $Ga(CH_3)_3$ would be a Lewis-acidic liquid methyl that would be quite reactive with oxygen and air; the phenyl compound would be a solid that would be less reactive; these would function as borderline acids (softened by the soft donor atoms attached to Ga). The metal would not, of course, form carbonyl complexes.

■ From this chapter we might note that Ga is rather scarce and expensive; it is unlikely to be found among the top 50 industrial chemicals. Relativistic effects (Chapter 11) should be insignificant in its chemistry.                                      □

Having carried out such an exercise in prediction, you might then be instructed to check your predictions against the facts. Useful moderately sized comprehensive works on descriptive inorganic chemistry include the works by Cotton and Wilkinson [6], Greenwood and Earnshaw [7], and the five-volume *Comprehensive Inorganic Chemistry* [8]. These books contain descriptive chemistry organized the old way, element by element or group by group, but this organization does make them very useful reference works for details of descriptive chemistry that may not be covered by our principles or that may not be predicted quite accurately. Having the principles of descriptive inorganic chemistry in your mind should make these works easier to read and digest. As you read the sections on gallium, you may often say to yourself, "That makes sense," or "I could have guessed that," and sometimes, "Now that is surprising—I wonder why that is?" The principles we have developed are designed to be used quickly and to be widely applicable, but they may, of course, be inaccurate in certain specific situations. For example, our prediction that Ga metal should have a relatively high melting point turns out to be completely wrong—it melts in the palm of the hand. But it *does* have a relatively high boiling point; it is a liquid over a range of over 2000 °C. No one can explain this; it is one of the many mysteries remaining to tantalize the imagination and rational powers of the inorganic chemist.

## 12.3

## Study Objectives

1. Explain the difficulties in synthesizing superheavy nuclei, and predict some such nuclei that ought to be relatively stable.
2. Predict some facts ("descriptive chemistry") about any element that your instructor names or about an as-yet-undiscovered superheavy element.
3. Select likely substitutes for a given element in some of its technological uses.

## 12.4

Exercises

1. Describe the chemistry of the following elements: **a.** Re;    **b.** At;    **c.** Be; **d.** the element below Au;    **e.** the element below Hg.

2. Check your predictions from the previous question in a standard comprehensive text of descriptive inorganic chemistry and in the *Handbook of Chemistry and Physics* if necessary. Which of your predictions were confirmed? Which were erroneous? What should you have been able to predict but overlooked? What facets of the chemistry fall outside the scope of predictability using the principles of this book?

3. In Table 12.4 are listed the inorganic chemicals from the Chemical and Engineering top 50 list in terms of total tonnage. For each one of them that you can, give a likely important industrial use and list another chemical that could best substitute for that chemical should it, for some reason, be unavailable.

**Table 12.4**    Inorganic Chemicals among the Top 50 in U.S. Production in 1984

| Rank | Chemical Substance | Production (millions of tons) |
|---|---|---|
| 1 | Sulfuric acid | 79.37 |
| 2 | Nitrogen | 43.41 |
| 3 | Ammonia | 32.41 |
| 4 | Calcium oxide (lime) | 32.20 |
| 6 | Oxygen | 31.04 |
| 7 | Sodium hydroxide | 22.45 |
| 8 | Phosphoric acid | 22.22 |
| 9 | Chlorine | 21.45 |
| 10 | Sodium carbonate | 17.02 |
| 11 | Nitric acid | 16.08 |
| 13 | Urea | 14.30 |
| 14 | Ammonium nitrate | 14.01 |
| 19 | Carbon dioxide | 7.80 |
| 25 | Hydrochloric acid | 5.72 |
| 30 | Ammonium sulfate | 4.13 |
| 31 | Potassium carbonate | 3.53 |
| 33 | Carbon black | 2.89 |
| 38 | Aluminum sulfate | 2.16 |
| 40 | Calcium chloride | 2.10 |
| 44 | Sodium sulfate | 1.74 |
| 45 | Titanium dioxide | 1.60 |
| 46 | Sodium silicate | 1.50 |
| 49 | Sodium tripolyphosphate | 1.33 |
| | Total inorganics | 366.15 |
| | Total organics | 180.39 |
| | Grand total | 546.54 |

SOURCE: Data from *Chem. Eng. News*, May 6, 1985, p. 13. For data on metals and nonmetals, see Table 8.10.
NOTE: Does not include the production of metals.

(In some cases you may decide that the properties of a given compound are so unique that it will be irreplaceable; if so, justify this conclusion.)

**4.** Explain the fact that the element tin has an unusually high number of stable isotopes (ten).

**5.** The stable isotope $^{48}Ca$ is unusual among light-element isotopes in its neutron-to-proton ratio. **a.** Compute this ratio and explain why it is unusual. **b.** Explain why this isotope is stable anyway. **c.** Explain why this isotope is an attractive choice for use as a projectile in trying to create stable superheavy nuclei.

**6.** Work the following crossword puzzle.

**Figure 12.3** An inorganic chemistry crossword puzzle by Leonard F. Druding. From Leonard F. Druding, *J. Chem. Educ.*, 50, 773 (1973). Reprinted by permission.

ACROSS

2. Principal ionic constituent of igneous rocks
9. Color of liquid bromine
10. Symbol of the element for which a coin is named
11. Symbol for an element prepared by the Hall process
12. Principal source of magnesium
14. A chalcogen
15. A halide that gives a red liquid on oxidation
19. A synthetic element named after a woman
20. A pseudohalide used to prepare detonators
21. Forms 15 across
23. Where most rare gases are found
24. Main body of the hydrosphere
25. A rare gas that forms compounds
27. The most abundant alkali
29. The solid phase of an important liquid
31. Viscous, black liquid hydrocarbons
32. See 19 across
33. An alkaline earth that gives a crimson flame test
34. Prefix for eight
37. See 32 down
38. Mineral form of $Mg_2SiO_4$
41. A heavy alkali
42. An important constant
43. An unreactive rare gas
44. Condensation from the atmosphere
45. First 5f series element
46. A radioactive halogen
47. The commercially most important halogen
48. Group 17/VIIA elements
49. A multidentate complex

DOWN

1. Prefix for four
3. It forms a purple solution in $CCl_4$ on oxidation
4. A scarce Group 13/IIIA element
5. Geometrical designation of two groups on the same side
6. A silicate mineral that has lubricant properties
7. The basic kind of matter
8. The element preceding 13 down
12. Generic name for a silicon hydride

13. A rare earth used to prepare a red phosphor
16. $O_3$
17. A mineral that has a sheetlike structure
18. To remove activity
22. Symbol for prescription (pharm)
26. A metallic Group 15/VA element
28. The most abundant *metal*
30. Symbol for element with $3d^4 4s^2$ configuration
32. A Group 9 element that forms many inert $d^6$ coordination complexes
35. Obsolete name for Ce(IV)
36. $RNH_2$
37. Its mineral source is chromite
38. A semiprecious gem
39. Atomic number 3
40. A platinum metal
46. Forms photosensitive halides
47. Metallic element in limestone

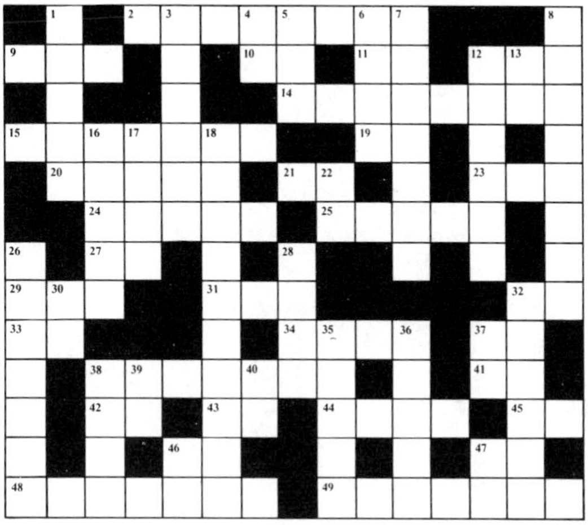

Notes

[1] Moeller, T., *Inorganic Chemistry: A Modern Introduction*, Wiley-Interscience, New York, 1982, pp. 17–33.

[2] "Element 109 Made by West German Researchers," *Chem. Eng. News*, p. 27 (Oct. 11, 1982).

[3] Seaborg, G. T., *Chem. Eng. News*, p. 46 (Apr. 16, 1979); G. T. Seaborg, W. Loveland, and D. J. Morrissey, *Science*, 203, 711 (1979).

[4] "Better Route to Neutron-rich Nuclides Found," *Chem. Eng. News*, p. 30 (Sept. 15, 1980).

[5] Lepkowski, W., "Politics and the World's Raw Materials," *Chem. Eng. News*, p. 14 (June 4, 1979).

[6] Cotton, F. A., and G. Wilkinson, *Advanced Inorganic Chemistry: A Comprehensive Text*, 4th ed., Wiley-Interscience, New York, 1980.

[7] Greenwood, N. N., and A. Earnshaw, *Chemistry of the Elements*, Pergamon Press, Oxford, 1984.

[8] Bailar, J. C., Jr., H. J. Emeleus, R. Nyholm, and A. F. Trotman-Dickinson, *Comprehensive Inorganic Chemistry*, Pergamon Press, Oxford, 1973.

# Laboratory Experiments in the Principles of Descriptive Inorganic Chemistry

The following laboratory experiments are designed to allow students to practice the scientific method by observing descriptive inorganic chemistry in the laboratory (or in a lecture demonstration). They should then draw up their own hypotheses and attempt to verify them before the theoretical principles of descriptive inorganic chemistry are presented in the textbook and in lecture. It is suggested that each experiment be performed before the chapter and section indicated below.

| Experiment No. and Name | Best done before Chapter and Section: |
|---|---|
| **1.** Some Reactions of Cations | 2.1 |
| **2.** Some Reactions of Oxo Anions | 2.6 |
| **3.** Reaction of Anions with Cations | 3.1 |
| **4.** Periodicity in the Activity (Electromotive) Series of Elements | 5.4 |
| **5.** Nonaqueous Reactions of Metal Ions and Compounds | 7.1 |
| **6.** Competitive Precipitation and Complexation Reactions | 8.1 |

Experiments 4 and 6 are related to two common general chemistry laboratory sets of experiments, those dealing with the activity series of the elements and the inorganic qualitative analysis scheme, respectively. But these experiments go beyond the general chemistry experiments in that they call for discovering periodic trends in the results. For students who have already done these general chemistry lab experiments, alternative instructions are included in Experiments 4 and 6 that call for no new experimentation but rather for further interpretation of the general chemistry lab experiments.

13.1

## Experiment 1: Some Reactions of Cations

We will begin our study of the chemistry of the elements with a laboratory investigation (or classroom demonstration and discussion) of their chemistry under a very common set of circumstances: when the elements have positive oxidation numbers and are in aqueous solution. We will start with the presumption that an element in a positive oxidation state is present as a cation. (Of course, we cannot obtain cations by themselves, so in this experiment we will use the chlorides of these elements—the chloride anion does not substantially affect the results of most of these experiments.) We will observe a most elementary reaction of these chlorides—dissolving them in water—and will find that something does indeed happen chemically when this experiment is performed. Finally, we will try to find periodic trends in the degree to which this chemical reaction occurs, by relating the tendency to react to the basic periodic properties of atoms discussed in Chapter 1.

**1.a.** Take five test tubes; to each add 2 to 3 mL of distilled water. To the first test tube add *nothing*. Measure the pH of the distilled water using first long-range and then short-range pH papers. Record the pH in the data table below. (Do not presume that the pH of the distilled water is 7.0!) Also feel the test tube to note qualitatively the temperature of the distilled water.

**1.b.** To the second test tube add LiCl (enough to cover the larger end of a standard nickel spatula). Stir until it is dissolved, then measure and record its pH as before. Also note whether there is any detectable change in temperature.

**1.c.** To the third test tube add a similar amount of dry $ZnCl_2$, stir to dissolve, measure its pH, and record any temperature change. To the fourth test tube add a similar amount of *fresh, anhydrous* $AlCl_3$, stir to dissolve, measure the pH, and record any temperature change.

**1.d. GO TO THE HOOD** to obtain a micro test tube (closed with a tight stopper or a septum) containing 0.5 mL of $TiCl_4$. **Cautiously** open the test tube and pour the contents into the (larger) fifth test tube of distilled water. Stir. Measure the pH of the solution and (using moistened pH paper) of the gas being evolved from the test tube; cautiously feel the bottom of the test tube to note any temperature change. Record your pH's and observations of temperature changes and any other visible or audible changes in the table below.

| Solute | Cation radius | Electronegativity | Charge | pH | Observations |
|---|---|---|---|---|---|
| Distilled water | — | — | — | | — |
| LiCl | 90 | 0.98 | | | |
| $ZnCl_2$ | 88 | 1.65 | | | |
| $AlCl_3$ | 67 | 1.61 | | | |
| $TiCl_4$ | 74 | 1.54 | | | |

**2.** What do you think happened in the test tubes in which reactions occurred? On a separate report sheet, write plausible chemical equations that would account for your observations.

**3.** Fill in the oxidation numbers (cation charges) of the (nonchlorine) elements in the above table, then look at the three periodic properties listed there (radius of the cation, Pauling electronegativity of the cation, and cation charge). Which one of these varies most significantly in this series of four compounds? Finally, decide how the tendency of a cation to undergo the reaction you described in part 2 depends on this periodic property. Write your conclusion on the report sheet.

**4.** Skywriting involves spraying $TiCl_4$ from an airplane into the air. Explain the chemistry of skywriting. How would you have to handle compounds like $TiCl_4$ to prevent this reaction from happening?

**5.** In the hood, carry out the same sort of experiment, adding spatula-tipfuls (or a few drops taken using a *dry* eyedropper) of the following compounds to 2 to 3 mL of distilled water. What is the significant periodic variable in this series of compounds? Can you write a conclusion relating the reaction tendency in this series of compounds to this variable?

| Solute | Cation radius | Electronegativity | Charge | pH | Observations |
|--------|--------------|-------------------|--------|-----|-------------|
| $BiCl_3$ | 117 | 2.19 | | | |
| $SbCl_3$ | 90 | 2.05 | | | |
| $PCl_3$ | (about 66) | 2.02 | | | |

**6.** Design and carry out an experiment to determine whether the Pauling electronegativity of the cation has any effect on this reaction tendency. Use some, but not all, of the chlorides from the following list: $CaCl_2$, $SrCl_2$, $MnCl_2$, $FeCl_2$, $ZnCl_2$, $SnCl_2$ (estimated cation radius of 126 pm), $Pb(NO_3)_2$, $Hg(NO_3)_2$ (chlorides not suitable here), $PrCl_3$ (or other $f$-block trichloride), $BiCl_3$. *Hint*: Rather than starting by testing all these chlorides, look at the periodic properties of the cations first, and pick out only the set or sets of compounds to test that will give you the comparison that you want.

**7.** Double-check your conclusion in part 5 by checking the results with these two Group 14/IVA chlorides: $SnCl_4$ and $CCl_4$. First use your principles to predict what will happen, then do the test. Can you explain any discrepancy between theory and observation?

**8.** A far-too-frequent experience of people who make up solutions of metal salts such as $SnCl_2$, $Hg(NO_3)_2$, and $BiCl_3$ and get a cloudy solution is to assume that their compound or water was contaminated; hence they throw out the solution and try again, only to get the same result. Looking at the equations you wrote in part 2, suggest what must be done to get clear solutions of these metal ions. Test your answer by trying it with one of the above three salts.

417

## 13.2

## Experiment 2: Some Reactions of Oxo Anions

The principles we have developed to predict the extent of hydrolysis of cations are most useful with the metallic elements when they are in not too high an oxidation state. When we deal with nonmetals or metals in high oxidation states, the notion of a "cation" of the nonmetal or metal is too far divorced from reality to enable us to make reasonably accurate predictions—for one thing, only limited numbers of "cationic" radii for such species have been tabulated. Our best approach to systematizing the chemistry of nonmetals in aqueous solution with positive oxidation numbers is to start with them in their "final" form—as **oxo anions**. In this experiment we will try to find simple but useful principles for predicting some of their properties.

**1.** Using long- and short-range pH test papers, check the pH of the distilled water. Then measure and compare the pH's of solutions of the following two salts of oxo anions: $NaClO$ and $NaClO_4$. Likewise, measure and compare the pH's of solutions of $Na_2SO_3$ and $Na_2SO_4$. All these salts contain oxo anions having the general formula $MO_x{}^{y-}$. In each pair of compounds, what component of the general formula are we varying? What effect does this variation have on the pH of the solutions? Predict which of each of the following pairs of salts will have the higher pH: $NaNO_2$ or $NaNO_3$; $Na_3AsO_3$ or $Na_3AsO_4$. Test your hypothesis by measuring and comparing the pH's within each pair of salts.

**2.** What type of reaction is occurring here to produce the pH's observed? Write an equation to illustrate this type of reaction, using the most reactive of the above oxo anions as an example.

**3.** Can you suggest any physical reason for the relationship of this reaction tendency to the structural variables you investigated in part 1?

**4.** Design an experiment to determine the effect of the *charge* $-y$ of an oxo anion on its basicity, and carry out the experiment. You may use the data from part 1; in addition, solutions of the following salts are available: $Na_3PO_4$, $Na_4SiO_4$, $NaIO_3$. (You may decide that you have enough compounds to set up two experiments; if so, set up and do both of them.) What relationship between charge and basicity do you observe? How would you explain this relationship?

**5.** Predict the trend in pH's of the following series of solutions, two of which are too intensely colored to be tested by pH paper: $K_3VO_4$, $K_2CrO_4$, $KMnO_4$. (If a pH meter is available, you may want to check the pH's of these solutions.)

**6.** Below are listed the oxo anions of the later *p*-block elements in their highest oxidation states. Note that the number of oxygen atoms (**oxo groups**) changes down a group. How would you explain the fact that the number does not remain constant?

$$CO_3{}^{2-} \qquad NO_3{}^{-}$$
$$SiO_4{}^{4-} \qquad PO_4{}^{3-} \qquad SO_4{}^{2-} \qquad ClO_4{}^{-}$$

$$\text{GeO}_4{}^{4-} \qquad \text{AsO}_4{}^{3-} \qquad \text{SeO}_4{}^{2-} \qquad \text{BrO}_4{}^{-}$$

$$\text{SnO}_6{}^{8-} \qquad \text{SbO}_6{}^{7-} \qquad \text{TeO}_6{}^{6-} \qquad \text{IO}_6{}^{5-} \qquad \text{XeO}_6{}^{4-}$$

**7.** Note that down a group, not only the number of oxo groups but also the charge on the oxo anion changes. Looking at your previous conclusions, would you expect these simultaneous changes in structure of oxo groups to have the same effects on basicity (i.e., to reinforce each other's effects)? If so, describe the basicity trend for oxo anions down a group of the periodic table. Or would you expect the two changes to have opposite effects on the basicity of the oxo anions? If the latter is the case, you need to determine which of the two effects is dominant. To do this, compile the pH's of all the oxo anions listed above for which you have data. If possible, test some additional solutions to add to your data (suggestion: $Na_2CO_3$). What is the observed basicity trend for oxo anions down a group of the periodic table?

**8.** Putting together all your trends, identify the oxo anion listed in part 6 that would most strongly undergo the type of chemical reaction you wrote in part 2.

## 13.3

## Experiment 3: Reaction of Anions with Cations

**1.** In this experiment we will investigate what happens when cations of varying acidity are combined with oxo anions of varying basicity. We will study the reactions of the following eight cations: $Cs^+$, $K^+$, $Ag^+$, $Mg^{2+}$, $Sr^{2+}$, $Hg^{2+}$, $Zn^{2+}$, $Al^{3+}$. Recalling the principles of Chapter 2, give the category of acidity of each cation. Then try to arrange the ions within each category in order of increasing acidity, so that, when you list all eight ions, they will be in order of increasing acidity. Finally check yourself by listing the $pK_a$ values for the ions (Table 2.3). (You probably won't be precisely right, since the rules of thumb in Chapter 2 are not exact, but the ordering you predict should put the $pK_a$ values reasonably close to the correct order.)

**2.** We will now study the reactions of the above eight cations with the following four oxo anions: $SiO_4{}^{4-}$, $SO_4{}^{2-}$, $PO_4{}^{3-}$, $ClO_4{}^{-}$. List these four in order of increasing basicity, and give the category of basicity and the approximate $pK_b$ for each.

**3.** Test the reactions of each of the eight cations in part 1 with each of the four anions in part 2. For each test mix equal volumes (say, one eyedropperful) of the two solutions, mix well, and allow a minute (if necessary) for the reaction to occur. Note whether the test tubes get hot or cold. Describe the reactions and list your observations in tabular form, listing the eight cations down one side of the table in a logical order, and listing the four anions across the top of the table in a logical order.

**4.** What do we call the kind of reaction that occurs in some of the test tubes? Write the formulas of some of the products.

**5.** How does the tendency for this kind of reaction appear to relate to the acidity and basicity of the cations and anions involved?

**6.** Predict which categories of metal cations would give insoluble salts with each of the following anions: selenate, permanganate, chromate, carbonate, nitrate. If a solution of one of these ions is available, test your predictions using the eight available cations. Check your answers for one of the anions that is not available by looking up the solubilities of the salts of that ion in the *Handbook of Chemistry and Physics.*

**7.** The perbromate ion was first synthesized in the 1960s in minute quantities by the radioactive decay of $*SeO_4^{2-}$. How would you go about separating the $BrO_4^-$ ion from the selenate ion?

**8.** In Section 2.3 we studied principles that could be used to predict the solubilities of metal hydroxides. Classify the basicity of the hydroxide ion. We could consider the hydroxide ion as a type of oxo anion. Are the principles in Section 2.3 consistent with the principles you derived in part 5 of this experiment?

## 13.4

## Experiment 4: Periodicity in the Activity (Electromotive) Series of the Elements

### A. Introduction

In this experiment we will study the stability of *low* oxidation states of the metallic elements—i.e., in the zero oxidation state, the metals themselves. We will examine the relative reactivity (*activity*) of different metals with the hydrogen ion and will list the different elements in order of decreasing reactivity with the hydrogen ion (such a list is called an activity or electromotive series of elements). But in contrast with the usual general-chemistry experiment, we will also attempt to discover the periodicity in such an activity series so that we can also predict the activities of other metals not yet tested and also (in the next chapter) gain some insight into the bonding that occurs in metals.

The reaction of some metals with standard 1 $M$ hydrogen ion is dangerously exothermic, so we will begin our tests by studying the reactivity of metals with pure cold water, in which the concentration of the hydrogen ion is only $10^{-7}$ $M$. In terms of predominant species such a reaction can usually be summarized as

$$M(s) + n\,H_2O \rightarrow M(OH)_n(s\ or\ aq) + \frac{n}{2}H_2(g) \tag{13.1}$$

Only very reactive metals undergo this reaction with cold water; their relative activity will be judged by the relative rate of evolution of $H_2$. If this reaction is not perceptible, we will try to speed the rate of the reaction by using hot water. Metals that show no activity to hot water will then be reacted with cold (roughly 1 $M$) hydrochloric acid:

$$M(s) + n\,H^+ \rightarrow M^{n+}(aq) + \frac{n}{2}H_2 \qquad\qquad (13.2)$$

If cold hydrochloric acid produces no reaction, then hot hydrochloric acid will be tested.

A certain number of metals fail to react with the hydrogen ion at all; these are said to be less active than hydrogen and are listed below it in an activity series. To rate their relative activity, stronger oxidizing agents than $H^+$ must be used; we will use oxidizing metal ions.

$$n\,M(s) + m\,N^{n+}(aq) \rightarrow m\,N(s) + n\,M^{m+}(aq) \qquad\qquad (13.3)$$

## B. Forming Hypotheses

There are more than 80 metals in the periodic table; testing all these would be very time-consuming and very expensive too. If we can find some form of periodicity in the tendency of metals to react with oxidizing agents, we will not need to test all 80 metals or memorize the results. Let us first form hypotheses as to how a metal's activity might relate to each of these fundamental properties of a metal: its position in the periodic table, its ionization potential, and its electronegativity.

**1.** Recalling general chemistry, describe how metallic properties of an element (such as their activities) relate to their position in the periodic table. Can you use this principle to predict the order of decreasing activity of the elements Ag, Al, Ca, Cu, Hg, Mg, Mn, Na, Ni, Pb, and Zn? Why or why not?

**2.** The *first ionization energy* (or first ionization potential) is the energy required to remove one electron from a gaseous atom of an element to produce a $+1$ ion:

$$M(g) + energy \rightarrow M^+(g) + e^- \qquad\qquad (13.4)$$

How do you think the activity of an element ought to be related to its first ionization potential? Table 6.6 (Appendix C) presents the first ionization energies of the elements. Predict a decreasing order of reactivity of the above elements based on their first ionization energies.

**3.** How do you think the *Pauling electronegativity* of an element and its activity might be related? Predict a decreasing order of reactivity of the above elements based on their Pauling electronegativities. (The elements lead and thallium go to their lower oxidation numbers in this experiment, so the Pauling electronegativities for Tl(I) and Pb(II) should be used for predictions involving these elements.)

**4.** The concept of electronegativity is complicated by the fact that there is more than one way of defining and measuring electronegativity. Allred and Rochow's scale of electronegativity is presented in Table 11.3 (Appendix C). Predict a decreasing order of reactivity of the above elements based on their Allred-Rochow electronegativities.

## C. Experimental

**1.** Heat a 400-mL beaker of water on a ringstand to near boiling. This will be used from time to time throughout the experiment.

**2.** The following metals are to be tested: Ag, Al, Ca, Cu, Hg, Mg, Mn, Na, Ni, Pb, Zn. *Certain metals require special precautions*:

■ *Sodium*—Use only a very tiny cube of metal. Do not point the test tube of water at anyone; do this reaction in the hood behind the glass shield.

■ *Aluminum* is coated with thin films of tightly adhering oxide that must be cleaned off before its reactions can be observed. To do this, put a *few* granules of the metal in a large test tube with 2 mL $H_2O$ and 2 mL 6 $M$ HCl. Heat in the hot water bath until vigorous reaction just begins (a few minutes are required), then quickly remove the test tube, dilute the acid with cold distilled water, pour off the diluted acid, half-fill the test tube with distilled water, pour this off, and again half-fill the test tube and pour off nearly all the water (leave enough to keep the metal out of contact with the air). Observe whether the metal is reacting with this cold water—if not, go immediately to step 4.

**3.** Put a few granules of each of your metals in large test tubes that contain about 5 mL of distilled water. Observe whether bubbling occurs, and if so, record which metal gives the most rapid reaction.

**4.** If you observe no reaction with cold water (or if you observe only a very faint reaction), put the test tube in your hot water bath (which should be *very hot* but *not boiling*). Observe whether bubbles of $H_2$ now form.

**5.** From any unreactive metals pour off all but about 2 mL of the water, then add 2 mL of 6 $M$ HCl. Observe over a period of a few minutes; note the relative bubbling rates.

**6.** For any metals that are still unreactive, put the test tube in the beaker of almost-boiling water and heat for a few minutes. Note relative bubbling rates.

**7.** Arrange the above eleven metals, insofar as possible, in order of decreasing reactivity (an *activity series*).

**8.** In each of six test tubes place 2 mL of 0.5 $M$ $Pb(C_2H_3O_2)_2$ solution. To each test tube add a few granules (or one drop) of the following six metals: Cu, Hg, Ag, Mg, Mn, Zn. Observe whether reaction occurs (wait 10 minutes before deciding "no reaction"). The $Pb^{2+}$ ion (in $Pb(C_2H_3O_2)_2$) is capable of oxidizing which of these metals? What is the observed product? Write balanced equations for the reactions that occur.

**9.** As a generalization, how do the activities of metals that react with the $Pb^{2+}$ ion compare with the activity of Pb itself?

**10.** Devise a series of experiments by which you could determine the relative positions of the metals Ag, Cu, and Hg in the activity series, using any of the following reactants: Ag(s), Cu(s), Hg(l), $1\,M$ AgNO$_3$, $0.5\,M$ Hg(NO$_3$)$_2$, $0.5\,M$ CuSO$_4$. Carry out and describe the experiments and your results, and complete the activity series in part 7.

## D. Conclusions

Decide which periodic property of a metallic element (periodic-table position, ionization potential, Pauling electronegativity, Allred-Rochow electronegativity) correlates most strongly with the activity of metals. Justify your choice.

## E. Applications

**1.** Based on your conclusions, predict the products of the following reactions (if they go), and describe the vigor of the expected reaction: (a) Cu + AuCl; (b) La + H$_2$O; (c) Au + HBr; (d) Ti + HCl; (e) Be + H$_2$O; (f) U + HCl; (g) Pt + Hg(NO$_3$)$_2$.

**2.** Many of the metals can be conveniently prepared by the *thermite* reaction:

$$2\,Al + (e.g.)\,M_2O_3 \rightarrow 2\,M + Al_2O_3 \tag{13.5}$$

Which of the following metals could be produced in this way? (a) Sc; (b) La; (c) U; (d) Cr; (e) Fe; (f) Ca; (g) Bi.

**3.** Many of the least active metals have been prized by humans for ages for their durability (and scarcity); these are sometimes referred to as the noble metals. Which of the metals would most likely fail to react with oxygen (under neutral conditions) and hence might be called noble metals?

## F. Alternative Version of the Experiment

This is for students who carried out an activity series experiment in general chemistry. Go through this experiment omitting part C, the experimental part; substitute your data sheet (or a textbook activity series) for this experiment. Draw up your hypotheses in part B for the elements you tested or for the elements in your textbook activity series.

## 13.5

## Experiment 5: Nonaqueous Reactions of Metal Ions and Compounds

**1.** Put 1 mL of $0.5\,M$ HCl in a test tube. Using long- and then short-range pH papers, measure the pH of this solution. Then **IN THE HOOD** add an eyedropperful of pyridine, C$_5$H$_5$N (Figure 13.1), to the test tube. Stir with a glass rod, then measure the pH of the solution again. What term do we use to describe what the pyridine has done to the HCl solution? What class of compounds reacts with acids

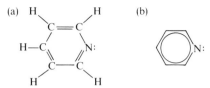

**Figure 13.1** The Lewis dot structure of pyridine, (a) completely drawn out and (b) in the type of shorthand organic chemists use, implying the presence of carbon and hydrogen atoms at the corners of the hexagon.

such as HCl in a manner similar to pyridine? Does it seem to be necessary for that class of compound to contain hydroxide or oxide ions?

**2.** Try to write an equation for the above reaction, using full Lewis dot structures, showing how the pyridine reacts with the HCl to give the observed result. **IN THE HOOD** try holding the stopper from a bottle of concentrated HCl over a watch glass in which you have put an eyedropperful of pyridine. Does it appear that this reaction can proceed even when water is completely absent? (*Note:* The water in the air and in the concentrated HCl is not involved in this reaction.)

**3.** Looking back at your results in Experiment 1, select the chlorides of three of the more acidic cations from that experiment (suggestions: $TiCl_4$, $BiCl_3$, $SnCl_4$, $SbCl_3$). If you select a liquid chloride, dissolve half an eyedropperful in 5 mL of dry toluene in a test tube. If you select a solid chloride, dissolve two large spatula-tipfuls in 2 to 3 mL of acetone. To each test tube of a chloride that you prepare in this manner, add half an eyedropperful of pyridine (**IN THE HOOD**). Describe the results.

**4.** By analogy with the reaction in part 1 of this experiment, what is probably happening in this reaction? Draw Lewis dot structures of the products you think might have been produced in this reaction. The function of the pyridine in this reaction is to act as what class of compound? The function of the metal cation (or metal chloride) is to act as what kind of compound?

**5.** In each of three test tubes put 3 mL of a 1 $M$ aqueous solution of cesium chloride, CsCl. To each add two good spatula-scoopfuls or a half-eyedropperful of one of your chlorides from the previous part of this experiment. Stir. Let the solution sit awhile if no reaction is immediately apparent. Explain your results. Draw possible structures of the products. What is the function of the cesium chloride, and which ion ($Cs^+$ or $Cl^-$) is performing this function? (*Note:* Although water is present, it is not participating in this reaction. If you are skeptical of that statement, as any good chemist ought to be, try the experiment with the three chlorides and 3 mL of water instead of the CsCl solution and see if the results are quite the same.)

**6.** Add a half-eyedropperful of $CCl_4$ to 3 mL of the 1 $M$ CsCl solution. Explain the results (or lack of results) with $CCl_4$.

**7.** Repeat part 5 using 3 mL of a saturated aqueous solution of tetramethylammonium chloride, $(CH_3)_4 N^+ Cl^-$, in place of the CsCl in part 5. Are your results similar? Write a Lewis dot structure of a plausible product for one of the reactions. Which ion (chloride or tetramethylammonium) is directly involved? Compare or contrast the behavior of the two nitrogen-containing species, pyridine and the tetramethylammonium ion, using their Lewis dot structures to explain their similarity or dissimilarity. In summary, what structural feature seems to be essential for a species to act as a base?

## 13.6

## Experiment 6: Competitive Precipitation and Complexation Reactions

**1.a.** Prepare a mixture of 10 mL of 2.0 $M$ KI solution with 10 mL of 2.0 $M$ KF solution (**AVOID CONTACT WITH ACID! HF IS A VERY DANGEROUS POISON!**). Stir well, then put 1 mL of this mixture in each of six test tubes. To each of the six test tubes add 1 mL of a different one of the following six solutions: 1 $M$ LiCl, 1 $M$ AgNO$_3$, 0.5 $M$ SrCl$_2$, 0.5 $M$ HgCl$_2$, 0.5 $M$ MgCl$_2$, 0.5 $M$ CuSO$_4$. Record your results, noting the colors of the products.

**1.b.** Design and carry out some experiments that will enable you to determine what the precipitate is in each case. Identify each precipitate in part 1.a.

**1.c.** Which metal ions have fluorides that are less soluble than their iodides? Which metal ions have iodides that are less soluble than their fluorides? Which of the fundamental atomic properties that we have used as predictors (Pauling electronegativity, ionic charge, ionic radius, etc.) seems to be most useful in predicting the relative solubility of iodides versus fluorides? Use it to predict the identities of four more insoluble fluorides; four more insoluble iodides. Verify your predictions with your instructor before proceeding.

**2.a.** Take 10 mL of a 0.5 $M$ Na$_2$S solution. Test its pH. What other anions (besides S$^{2-}$) are present? Put 1 mL in each of eight test tubes. To each of the eight add 1 mL of a different one of these eight solutions: 1 $M$ LiCl, 1 $M$ AgNO$_3$, 0.5 $M$ MgCl$_2$, 0.5 $M$ CuSO$_4$, 0.5 $M$ SrCl$_2$, 0.5 $M$ HgCl$_2$, 0.33 $M$ PrCl$_3$ (or another $f$-block +3 ion), and 0.33 $M$ BiCl$_3$. Record the results.

**2.b.** Carry out experiments that will enable you to determine the identity of the precipitate in each case.

**2.c.** Which metals form sulfides that are less soluble than their hydroxides or oxides? Which metals prefer to remain as hydrated ions, or to form hydroxides or oxides, rather than to precipitate sulfides? Which fundamental atomic property seems to work best in predicting the relative tendency to form sulfides versus hydroxides, oxides, or hydrated ions? Predict four more metals that will have sulfides more insoluble than hydroxides; four more that will have hydroxides more insoluble than sulfides. Verify with your instructor.

**3.a.** Mix 10 mL of a saturated urea [(NH$_2$)$_2$C=O] solution with 10 mL of a saturated thiourea [(NH$_2$)$_2$C=S] solution. Stir well, then put 2 mL in each of eight test tubes. To each of these test tubes add a different one of the same eight metal-ion solutions used in part 2.a. Record your results.

**3.b.** Identify your products and write plausible formulas for them.

**3.c.** Which metals react with thiourea in preference to urea? Which prefer to react with urea or to stay in hydrated form?

**4.** Do you see any relationship among the results of the experiment with the $F^-/I^-$ mixture, that with the $S^{2-}/OH^-$ mixture, and that with the urea/thiourea mixture? If so, give the relationship.

**5.a.** Prepare a mixture of 10 mL of 0.25 $M$ $Na_4SiO_4$ and 5 mL of 0.5 $M$ $Na_2S$. (The mixture is now 0.167 $M$ in each.) Put 1.5 mL of this mixture in each of eight test tubes. To each of the eight add 0.5 mL of one of the eight metal-ion solutions you have been using. Record the results and identify the products.

**5.b.** The largest number of minerals are silicates or sulfides, which presumably arise as the result of competitive precipitation experiments. Draw a periodic table in which you indicate the metals you would expect to find in nature as silicates. (Geochemists refer to these as the *lithophile* elements.) Indicate the metals that you would expect to find in nature as sulfides. (Geochemists refer to these as the *chalcophile* elements.) Also indicate any area of the tables for which you are not yet able to make predictions.

**6.** To the extent that you can, predict the results of the computer study cited on page 266: With which ligand or ligands will each of the metal ions be predominantly associated?

## Alternative Version of this Experiment

This is for students who have done the qualitative analysis scheme (qual scheme) in general chemistry.

**a.** List all the metals (in the general known sample) that precipitate as chlorides in Group I (note that these would precipitate as sulfides in Group II if you neglected to add the chloride ion). List all the metals that precipitate as sulfides in Group II *or as sulfides* in Group III. List all the metals that precipitate as *hydroxides* or *oxides* in Group III. List all the metals that precipitate as carbonates in Group IV. List all the metals that remain as hydrated ions in Group V.

**b.** Identify the donor atom of the carbonate ion (Group IV) and the water molecule (Group V). Consolidate your lists from **a.** into two larger lists—one of metal ions that prefer to precipitate with the sulfide ion (sulfur donor atom), the other of metal ions that prefer to coordinate with oxygen donor atoms.

**c.** Which of the fundamental atomic properties that we have used as predictors (Pauling electronegativity, ionic charge, ionic radius, etc.) would be most useful in predicting which metal ions prefer to coordinate with sulfur donors and which prefer to coordinate with oxygen donors?

**d.** Use your relationship to predict six more metal ions that should belong to each list. Verify these with your instructor. Can you also decide which of the six additional oxygen-preferring metal ions would fall in qual scheme Group III, which in qual scheme Group IV, and which in qual scheme Group V? (This requires principles explained much earlier in this text.)

**e.** Go back and answer questions 5b and 6, predicting the results of the experimental part of question 5.

# Answers to Selected Exercises

## A.1

### Chapter 1

1. **a.** $7s^1$  **b.** $4s^24p^3$  **c.** $6s^25d^8$  **d.** $6s^24f^{10}$  **e.** $6s^26p^6$  **f.** $6s^2$  **g.** $6s^0$  **h.** $5d^8$
   **i.** $4f^9$  **j.** $4s^24p^2$  **k.** $4s^2$  **l.** $4s^24p^6$

3. **a.** :Ö=C=Ö:  **b.** [:Ö—N̈=Ö:]⁻  **c.** [:N≡O:]⁺  **d.** H—Ö—Ö—H

            H                H                  :F̈:            :Ö:

            |                 |                  |               |

   **e.** H—C—H  **f.** :B̈r—Si—B̈r:  **g.** :F̈—I—F̈:  **h.** :Ö—Os—Ö:

            |                 |                  ⁄＼             |

            H                H              :F̈::F̈:       :Ö:

5. Answers are for question 3: **a.** C, $+4$; O, $-2$  **b.** N, $+3$; O, $-2$  **c.** N, $+3$; O, $-2$
   **d.** H, $+1$; O, $-1$  **e.** H, $+1$; C, $-4$
   **f.** H, $-1$; Br, $-1$; Si, $+4$ (most easily determined from Lewis structures)
   **g.** I, $+5$; F, $-1$  **h.** Os, $+8$; O, $-2$

7. **a.** H, $+1$; O, 0; F, $-1$  **b.** H, $-1$; Si, $+3$
   **c.** H, $-1$; terminal Si, $+3$; central Si, $+2$ (or an average of $+\frac{8}{3}$)
   **d.** C, 0; Se, 0 (This answer will vary depending on the table of electronegativities used.)
   **e.** V, $-1$; C, $+2$; O, $-2$  **f.** N, $-3$; B, $+3$; H on N, $+1$; H on B, $-1$
   **g.** H, $+1$; C, $-1$; Cl, $-1$

9. Six valence electrons: O, S, Se, Te, Po, Cr, Mo, W, Nd, U. Commonly give up or share all six: S, Se, Mo, W, U. Atomic numbers: 106 (known but chemistry not yet studied), 116, 124, 142, 156.

13.1. $8s^25g^6$; $8s^26f^6$; $8s^27d^{10}$

13.2. $> 1.38$; $> 2.54$; $< 2.2$

13.3. $+6$, $+3$, $+5$, or higher; $+2$, $+4$, or higher; $+2$

13.4. $Rn^{6+}$, $> 62$ pm; No. 121 difficult to extrapolate; eka-$Np^{+5}$, $> 89$ pm; eka-$No^{2+}$, $> 124$ pm; eka-$Os^{4+}$, $< 66$ pm; eka-$Pb^{2+}$, $> 133$ pm. But see Chapter 11 for possible revisions of some of the answers to this question and the preceding four questions.

## A.2

## Chapter 2

**2.1.** **a.** $Z^2/r = 0.077$, so weakly acidic **b.** $Z^2/r = 0.008$ and $\chi_P > 1.8$, so feebly acidic **c.** $Z^2/r = 0.272$, so very strongly acidic **d.** $Z^2/r = 0.533$ and $\chi_P > 1.8$, so very strongly acidic **e.** $Z^2/r = 0.125$ and $\chi_P > 1.8$, so strongly acidic **f.** $Z^2/r = 0.006$, so nonacidic **g.** $Z^2/r = 0.148$, so moderately acidic

**2.2.** Precipitation of oxide or hydroxide with $Pa^{5+}$, $As^{3+}$, possibly with $Th^{4+}$. This could be cleared up by adding excess acid for $As^{3+}$ and $Th^{4+}$ but not for $Pa^{5+}$.

**2.3.** $U^{3+}(aq)$; $Ag^+$; $Pa_2O_5$ or $Pa(OH)_5$; in original form such as $CCl_4$; $As(OH)_3$ or $As_2O_3$; $Tl^+(aq)$; $ThO_2$ or $Th(OH)_4$.

**3.** Calculated $pK_a$'s: 8.35; 10.79; $-8.84$; $-40.73$; $-1.63$; 13.59; 2.09. $Ag^+$ is more accurately classified as weakly acidic and $Tl^+$ as feebly acidic.

**7.** **a.** Carbonate ion, $pK_b = 6.7$, moderately basic **b.** Perbromate ion, $pK_b = 22.6$, nonbasic; **c.** Periodate ion, $pK_b = -6.8$, so strongly basic that it gives a hydroxo ion

**8.** **a.** $SiO_4^{4-}$, $pK_b = -8.0$, so very strongly basic **b.** $TeO_6^{6-}$, $pK_b = -17.0$, so very strongly basic **c.** $BrO_4^-$, $pK_b = 22.6$, so nonbasic **d.** $SO_3^{2-}$, $pK_b = 6.7$, so moderately basic

**11.** **a.** Uranyl sulfate **b.** Titanium(III) chloride **c.** Sulfuryl chloride

**12.** **a.** $BiONO_3$ **b.** $Sr(ClO_4)_2$ **c.** $EuSO_3$ **d.** $Fe_3(PO_4)_2$ **e.** $CrCO_3$

**16.** **a.** $HMnO_4$, very strongly acidic **b.** $H_2SeO_4$, strongly acidic **c.** $H_3AsO_3$, weakly acidic

**17.** **a.** $+1$, $Li^+(aq)$ **b.** $+3$, $Al(OH)_3$ or $Al_2O_3$ **c.** $+6$. There is some uncertainty as to whether the oxo anion of $W^{6+}$ would form: If it showed a maximum coordination number of 6, its oxo anion $WO_6^{6-}$ would be too strongly basic to persist, so $WO_3$ or $W(OH)_6$ would be expected; if it showed a maximum coordination number of 4, its oxo anion would be $WO_4^{2-}$, which would be feebly basic and could persist. The actual oxo anion is $WO_4^{2-}$, but at pH 7, $WO_4^{2-}$ and $WO_3$ are at equilibrium, and polynuclear tungstate anions are also likely to be present (Section 4.9).

## A.3

## Chapter 3

**1.** $CsBrO_4$, $CePO_4$, $BaSeO_4$, $Hg_5(IO_6)_2$, $TiO_2$.

**3.** **a.** All chromates are soluble except those of $Ca^{2+}$, $Sr^{2+}$, $Ba^{2+}$, $Ra^{2+}$, $Tl^+$, $Eu^{2+}$, $Tm^{2+}$, $Yb^{2+}$. (This prediction is not highly accurate. We may note that the solubilities of salts of feebly acidic cations and feebly basic anions are difficult to predict accurately. Also, our classification of $CrO_4^{2-}$ as a feebly basic anion is not realistic, since its $pK_b$ is 7.5, which makes it a weakly basic anion with quite different solubility rules. We might have predicted this if we had taken into account the effect of the comparatively low electronegativity of Cr on the basicity of its oxo anion.) **b.** Ferrates would be predicted in the same manner as chromates. (Little is known of their solubilities.) **c.** All pertechnetates are soluble except those of $Na^+$, $K^+$, $Rb^+$, $Cs^+$, and $Fr^+$.

**5.** **a.** $Rb^+$ or $(C_4H_9)_4N^+$ **b.** $Al^{3+}$ **c.** $(C_4H_9)_4N^+$ **d.** $Al^{3+}$ or $Li^+$ **e.** $Rb^+$

**8.1.** $LiF$, $-1017$ kJ/mol; $NaCl$, $-764$ kJ/mol

**9.1.** $CaSO_4$, $-1079$ kJ/mol; $SrSO_4$, $-1167$ kJ/mol; $BaSO_4$ $-1100$ kJ/mol

**11.** **a.** Radius ratio $= 0.802$; coordination number of $Ce = 8$, of $O = 4$; fluorite ($CaF_2$) lattice type **b.** $-11,035$ kJ/mol

**23.1.** $Ba^{2+}(aq) + SO_4^{2-}(aq) \rightarrow BaSO_4(s)$

**23.2.** $2\,H^+(aq) + SeO_3^{2-}(aq) \rightarrow H_2SeO_3(aq)$

**24.** **a.** Since $Co^{2+}$ is a weakly acidic cation, the soluble salt needed would be one of a nonbasic (or perhaps feebly basic) anion; $Co(NO_3)_2$ would be suitable. The *Handbook of Chemistry and Physics* shows that cobalt(II) nitrate is found as a hexahydrate, formula mass = 291.04. To make 0.2 mol of cobalt(II) vanadate, 0.6 mol or 174.6 g of $Co(NO_3)_2 \cdot 6\,H_2O$ should be dissolved in water. Since $VO_4^{3-}$ is a moderately basic anion, its salt with a nonacidic cation such as $K^+$ or $Na^+$ should be soluble. In principle, however, this nonacidic cation will give a precipitate with the nonbasic nitrate ion, contaminating the product. In practice, this does not happen with $Na^+$, which is a good choice of cation. The *Handbook of Chemistry and Physics* shows that sodium vanadate occurs as anhydrous (formula mass = 183.94), as a decahydrate (formula mass = 364.06), or as a hexadecahydrate (formula mass = 472.15). 0.4 mol of sodium vanadate should be dissolved in water, and the resulting solution should then be mixed with the cobalt(II) nitrate solution. The amount of sodium vanadate needed is 73.6 g of the anhydrous salt, or 145.6 g of the 10-hydrate, or 188.8 g of the 16-hydrate.

## A.4

## Chapter 4

**1.** **a.** Basic and soluble     **b.** Amphoteric or neutral and insoluble     **c.** Acidic and soluble
**d.** Acidic and insoluble

**2.** Small molecules: $CO_2$, $SO_2$; oligomeric molecules: $P_4O_{10}$; ionic or macromolecular: the rest

**3.1.** (Most acidic) $P_4O_{10} > CO_2 > SiO_2 > Cr_2O_3 > Na_2O$ (least acidic)

**3.2.** $CO_2 > ZrO_2 > Y_2O_3 > SrO > Rb_2O$

**4.1.** $CO_2(g) < P_4O_{10} < Na_2O < Cr_2O_3, SiO_2$

**6.1.** $Tl_2O + H_2O \rightarrow 2\,Tl^+(aq) + 2\,OH^-(aq)$

**6.2.** $I_2O_5 + H_2O \rightarrow 2\,H^+(aq) + 2\,IO_3^-(aq)$

**6.3.** $ClO_2 + 2\,OH^- \rightarrow ClO_3^-(aq) + ClO_2^-(aq) + H_2O$

**6.4.** $La_2O_3 + 6\,H^+(aq) \rightarrow 2\,La^{3+}(aq) + 3\,H_2O$

**6.5.** $B_2O_3 + 2\,OH^-(aq) + 3\,H_2O \rightarrow 2\,B(OH)_4^-(aq)$

**6.6.** $6\,FeO + P_4O_{10} \rightarrow 2\,Fe_3(PO_4)_2$

**6.7.** $MnO + 2\,H^+(aq) \rightarrow Mn^{2+}(aq) + H_2O$

**8.** **a.** $SrO + H_2O \rightarrow Sr^{2+} + 2\,OH^-$     **b.** $MoO_3 + H_2O \rightarrow 2\,H^+(aq) + MoO_4^{2-}(aq)$
**c.** $RuO_4$     **d.** $RuO_4$

**12.** $Fe_{0.95}O$, $Co_{0.95}O$, $Eu_{0.95}O$.

**13.** Spinels: $NiFe_2O_4$, $Ni_3O_4$; perovskites: $BaTiO_3$, $NaTaO_3$.

**16.** **a.** Monticellite     **b.** Wollastonite     **c.** Grunerite     **d.** Talc     **e.** Monticellite
**f.** Stishovite

## A.5

## Chapter 5

**1.** P

**3.** Br

5. N, Bi

7. 14

9.1. $FeO_4^{2-}$    9.2. $BrO_4^-$    9.3. $PoO_6^{6-}$

11. **a.** $CoO_2$    **b.** $H_2Se$    **c.** $Eu^{2+}$

14.1. $PuO_2$, in sludge

14.2. React: probable overall reaction is $Pu + 2H_2O \rightarrow PuO_2 + 2H_2$, although first step could be $2Pu + 6H^+ \rightarrow 2Pu^{3+} + 3H_2$.

14.3. Acidic; may be protected in basic solution by forming adhering coating of insoluble $Pu_2O_3$.

14.4. $Pu^{3+}$, in solution.

15. **a.** Does not exist.    **b.** Strongly basic solution.

17. **a.** Disproportionate at some pH: $Am(OH)_4$ or $Am_2O_5 \cdot H_2O$; $CrO_2$; $Mn_2O_3$, $Mn_3O_4$, or $MnO_4^{2-}$; $Bi_4O_7$ or $BiO_2$; $I_3^-$ or $I_2$.
    **b.** Oxidize water: $Am_2O_5 \cdot H_2O$ or $AmO_3$; no Cr species; $MnO_4^-$ in acid solution; $Bi_2O_5(?)$; no I species(?).

19. See Table D for list of elements; Te is not a metal. Predicted from electronegativities: Tb $\geq$ Tm > Th > Ta $\geq$ Ti > Tl > Sn > Tc > W. Experimental from Table 5.4 (no value available for Ta): Tm > Tb > Th > Ti > Tc > Tl > Sn > W.

20. **a.** Reacts with 1, 2, 3, 4    **b.** Unreactive with 1, 2, 3, perhaps 4    **c.** Reacts with 2, 3 (and 4?—may be passivated)    **d.** Reacts with 1, 2, 3, 4    **e.** Unreactive with 1, 2, 3, perhaps 4    **f.** Reacts with 1, 2, 3, 4    **g.** Reacts with 2, 3, (and 4?—may be (actually is) passivated by 4).

25. **a.** Reacts with $Ag^+$    **b.** Reacts with $I^-$    **c.** Neither    **d.** Neither

27. **a.** Unstable    **b.** Although $E^\circ_{ox} - E^\circ_{red}$ is only 0.67 V, we know from experience that this is unstable    **c.** Stable    **d.** Unstable    **e.** Unstable    **f.** Similar to **b.**; from experience, unstable

34.1. Chemical conversion    **34.2.** Reduction    **34.3.** Beneficiation

## A.6

### Chapter 6

2. **a.** C    **b.** Mo

6. **a.** All macromolecular forms (B, diamond, graphite, Si, Ge, gray Sn, black and red P, gray As, gray Sb, Bi, polymeric S, helical Se, Te). Reason: To boil, these must be converted to small molecules by rupturing covalent bonds.
    **b.** The smallest of the small molecules: He, Ne, Ar, Kr, Xe, $H_2$, $F_2$, $Cl_2$, $Br_2$, $O_2$, $O_3$, $N_2$. Reason: These are held together in the liquid or solid state only by very weak van der Waals forces.
    **c.** The larger of the small molecules: $I_2$, $S_8$, $S_6$ etc., $Se_8$, $P_4$, $As_4$, $Sb_4$. Reason: Having more electrons, these molecules are held together in the liquid or solid state by stronger van der Waals forces.

8. **a.** Largest, Li; smallest, Ne    **b.** Largest, Ra; smallest, Be    **c.** Largest, van der Waals radius of Na; smallest, cationic radius of Na.

11. In the order shown: delta, pi, sigma. Bond dissociation energies: delta < pi < sigma. (In general, a different order might result if widely varying sizes of orbitals were compared.)

12. $+210$ kJ/mol, $O_2$.

17. B.

21.1. $-2630$ kJ/mol

21.2. $-3437$ kJ/mol (Note that you need the *second* electron affinity of oxygen, given in the text.)

21.3. $CaF_2$: theoretical lattice energy $-2628$ kJ/mol; CaO: predict CsCl lattice type, for which lattice energy is $-3571$ kJ/mol; actually has NaCl lattice type, for which lattice energy is $-3540$ kJ/mol.

24. **a.** $+535$ kJ/mol  **b.** No, $-2.77$ V

---

## A.7

## Chapter 7

1. **a.** 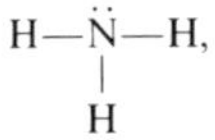 H—N̈—H, N donor atom (with H below N)  **b.** 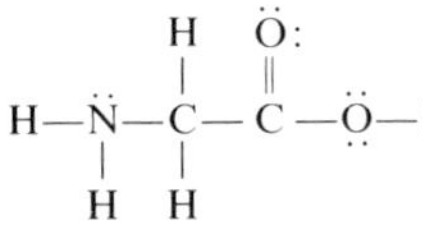 H—N̈—C—C—Ö—H, N and two O donor atoms. (with H, Ö: above; H, H below)

3. **a.** $Br^-$, $CH_3^-$, $C_6H_5^-$, and $H_2N$—$CH_2$—$CH_2$—$NH_2$ in A; $CN^-$, CO, $(CH_3)_2O$, and $SnCl_3^-$ in B.  **b.** $H_2N$—$CH_2$—$CH_2$—$NH_2$ in A; none in B  **c.** $Pb^{4+}$ in A; $Ni^{2+}$ in B  **d.** $Cl^-$ in A; none in B

8. **a.** $[:\ddot{O}=N=\ddot{O}:]^+$, 180°, linear  **b.** 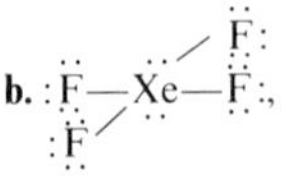 :F—Xe—F:, 90°, square planar (with :F: above and :F: below)

**c.** 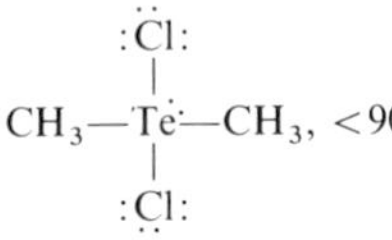 $CH_3$—Te—$CH_3$, $<90°$ and $<120°$, seesaw (with :Cl: above and :Cl: below)

**d.** 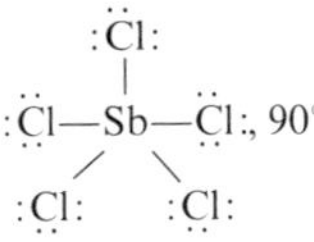 :Cl—Sb—Cl:, 90° and 120°, trigonal bipyramidal (with :Cl: above and :Cl: :Cl: below)

**e.** 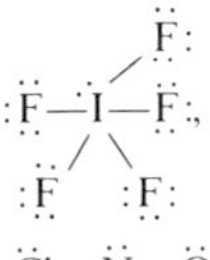 :F—I—F:, $<90°$, square pyramidal (with :F: above and :F: :F: below)

**f.** :Cl—N̈=Ö:, $<120°$, bent

10. **a.** Hg(l), HgS(s)  **b.** $HgCl_2$(aq)  **c.** $Hg(CH_3)_2$(l)  **d.** $Hg(CH_3)_2$(l)
    **e.** Hg(l), $Hg(CH_3)_2$(l)

14. **a.** $CH_3$—O—$CH_2$—$CH_2$—O—$CH_2$—$CH_2$—O—$CH_3$ (let us call this A); $CH_3^-$; $N_3^-$; $Cl^-$; $H_2O$; $NH_3$; $C_6H_5$—$CH(NH_2)$—$C(=O)$—$O^-$ (let us call this B); and the $O_2$ ligand, which could be $O_2^{2-}$ (the peroxide ion), $O_2^-$ (the superoxide ion), or $O_2$ (the dioxygen molecule). For purposes of the remaining questions we assume that it is $O_2^{2-}$.  **b.** A is tridentate; B is bidentate.  **c.** $N_3^-$, $Cl^-$, $O_2^{2-}$.  **d.** $Mo^{4+}$ and $Mo^{5+}$.  **e.** Each Mo has a coordination number of 7.

15. **a.** C  **b.** E  **c.** A, C, D, E  **d.** B  **e.** B

21. **a.** Calcium hexachlorovanadate(III)
    **b.** Tetraaquadihydroxocobalt(III) tetrachloroaluminate(III) (III optional for the anion)
    **c.** cesium tricarbonyltris(triphenylphosphine)vanadate($-I$)

432

**25.** **a.** $-305.1$ kJ/mol; $-80.2$ kJ/mol; $-128.7$ kJ/mol; and $-78.9$ kJ/mol, in the order given in the equation    **b.** $+177.7$ kJ; to the left

**29.1.** Potassium hexacyanoferrate(II); potassium hexacyanoferrate(III), iron(III) hexacyanoferrate(III), iron(II) hexacyanoferrate(II), and potassium iron(II) hexacyanoferrate(III) *or* potassium iron(III) hexacyanoferrate(II)

## A.8

## Chapter 8

**2.** For Lewis structures see question 1 in Chapter 7.    **a.** N donor atom is borderline.
**b.** N donor atom is borderline, and O donor atoms are hard.    **c.** O donor atoms are hard.
**d.** O donor atoms are hard, and S donor atom is soft.

**4.** **a.** $K^+ < Cu^+ < Ag^+ < Au^+$    **b.** $F^- < Cl^- < Br^- < I^-$

**6.1.** Products    **6.2.** Products    **6.3.** Reactants    **6.4.** Products    **6.5.** Products

**11.** Very small size, and high charge in the case of $B^{3+}$.

**13.** CdTe; AgI; $TiO_2$; $PtAs_2$. (EuSe and $TiTe_2$ should undergo hydrolysis.)

**16.** **a.** Group II, SnS    **b.** Group III, $Pr(OH)_3$ or $Pr_2O_3$
**c.** Forms complex $CuCl_2^-$ ion in Group I, so carries over to Group II as $Cu_2S$
**d.** Group IV, $SrCO_3$    **e.** Group II, $Sb_2S_3$

**17.** **a.** Answers for $Zr^{4+}$:    **17.1** Soluble    **17.2.** Insoluble, but hydrolyzed to $H_2S$ and insoluble $ZrO_2$    **17.3.** Insoluble    **17.4.** Group III    **17.5.** Yes    **17.6** Silicate

**18.1.** A, $LaPO_4$    **18.2.** B, $PtAs_2$    **18.3.** A, $ZrSiO_4$    **18.4.** B, $Sb_2S_3$    **18.5.** B, PbTe

**20.** The last ligand

**25.** 319 kJ/mol in $CCl_4$; 366 kJ/mol in $(CH_3)_3CCl$

## A.9

## Chapter 9

**1.** From among the *first five* possibilities only: **$BiI_5$**.

**2.** From among the *first five* possibilities only: $CI_4$, $CuBr_2$.

**4.** **a.** Method 3    **b.** Method 2    **c.** Method 1    **d.** Method 4    **e.** In principle Method 3, but since $BiCl_3$ can be obtained from solution in anhydrous form, Method 1 can be used *with a large excess of HCl*    **f.** Method 4

**7.** **a.** Molecular, gas    **b.** Macromolecular (or ionic), solid with high m.p.    **c.** Ionic, solid with high m.p.    **d.** Molecular, liquid    **e.** Macromolecular, solid with low m.p. (low m.p. because polymerization is by weak secondary bonds only)    **f.** Molecular, solid with low m.p.

**11.** Neither. The normal trend for ionic compounds would be for $GeCl_4$ to have the higher melting point due to the higher lattice energy with a $+4$ ion. The normal trend for monomeric covalent compounds would be for $GeCl_4$ to have the higher melting point due to stronger van der Waals forces when more electrons are present. Actually $GeCl_4$ is monomeric covalent, but $GeCl_2$ does not meet a reasonable coordination number for Ge, so it is macromolecular; hence $GeCl_2$ has the higher melting point.

**12.** A, $SnCl_2$; B, $HgI_2$; C, $NF_3$; D, $CrCl_3$; E, $SnCl_4$; F, $NI_3$

**14.** **a.** Neither    **b.** Lewis acid    **c.** Lewis base    **d.** Lewis acid    **e.** Lewis acid
**f.** Lewis acid.

**17.** **a.** $PF_5 + CsF \rightarrow Cs[PF_6]$    **b.** $XeF_6 + CF_4 \rightarrow NR$    **c.** $RbBr + BaBr_2 \rightarrow NR$
**d.** $XeF_6 + BF_3 \rightarrow [XeF_5][BF_4]$ (from information given in text about structure of $XeF_6$)
**e.** $3\,NaF + AlF_6 \rightarrow Na_3[AlF_6]$

## A.10

---

## Chapter 10

**3.** **a.** Solid, since it is saltlike    **b.** Liquid, since it is molecular with hydrogen bonding
**c.** Gaseous, since it is molecular    **d.** Liquid, since it is catenated molecular
**e.** Solid, since it is a salt    **f.** Gaseous, since it is molecular

**4.** (One example of each): **a.** $HCl + H_2O \rightarrow H_3O^+ + Cl^-$    **b.** $PH_3(g) + HI(g) \rightarrow PH_4^+I^-(s)$
**c.** $B_2H_6(g) + 2\,Na^+H^-(s) \rightarrow 2\,Na^+BH_4^-$    **d.** $CrCl_3(s) + 6\,NH_3(l) \rightarrow [Cr(NH_3)_6]Cl_3$
**f.** $Na^+H^-(s) + H_2O(l) \rightarrow Na^+(aq) + OH^-(aq) + H_2(g)$
**g.** $SiH_4(g) + 2\,O_2(g) \rightarrow SiO_2(s) + 2\,H_2O(l)$

**6.** **a.** $CH_4 < AsH_3 < PH_3 < NH_3$    **b.** $NH_2OH < NH_2NH_2 < NH_3$

**8.** **a.** $Li + Al + 2\,H_2 \rightarrow LiAlH_4$ (or other answer)    **b.** $CaF_2 + H_2SO_4 \rightarrow CaSO_4 + 2\,HF$
**c.** $N_2 + 3\,H_2 \rightarrow 2\,NH_3$    **d.** $NaOCl + 2\,NH_3 \rightarrow NaCl + H_2O + N_2H_4$
**e.** $SCl_2 + 2\,H_2S \rightarrow H_2S_3 + 2\,HCl$ (or analogous to **8f**)
**f.** $2\,Na_2S + S_8 \rightarrow 2\,Na_2S_5$; $Na_2S_5 + 2\,HCl \rightarrow 2\,NaCl + H_2S_5$ (or analogous to **8e**)
**g.** $2\,Si_2Cl_6 + 3\,LiAlH_4 \rightarrow 2\,Si_2H_6 + 3\,LiAlCl_4$ (or other reactions)

**15.** **a.** Polymeric covalent, unreactive with water    **b.** Saltlike, $Al(OH)_3 + CH_4$
**c.** Saltlike, $LiOH + CH_3CCH$    **d.** Saltlike, $LiOH + HCCH$
**e.** Metal-like, unreactive with water    **f.** Metal-like, unreactive with water

**17.** (for example) **a.** $BMe_3$    **b.** $(SiMe_2)_6$    **c.** $(PMe)_5$    **d.** $Ph_2Te_2$    **e.** $BMe_3$    **f.** $Me_5As$
**g.** $BeMe_2$    **h.** $K^+Me^-$

**21.** Are 18-electron species: a, b, d, e, j.

**23.** **a.** Cr, Mo, W    **b.** Fe, Ru, Os    **c.** Mn, Tc, Re

## A.11

---

## Chapter 11

**4.** **a.** $Ne < F < C < Li$    **b.** $Be < Ca < Ba < Ra$

**6.1.** $Pb-Pb < Sn-Sn < Ge-Ge < Si-Si < C-C$

**6.2.** $Li-Li < Be-Be$ (but $Be_2$ molecule itself is unstable) $< B-B < C-C$

**8.** **a.** $5g < 5f < 5d < 5p < 5s$

**10.** From top to bottom of figure: **a.** $A = 4p_x$, $B = 2s$; $A = 4d_{xz}$, $B = 2p_x$; $A = 1s$, $B = 3d_{z^2}$
**b.** Zero, positive, positive    **c.** (Does not apply), pi, sigma

**13.1.** $1s^2 2s^2 2p^6 3s^2 3p^6 4s^2 3d^{10} 4p^6 5s^2 4d^{10} 5p^6 6s^2 4f^{14} 5d^{10} 6p^6 7s^2 5f^{14} 6d^{10} 7p^6 8s^2 5g^{18} 6f^{14} 7d^{10}$

**13.2.** $8s^2 7d^{10}$; $7d^{10}$

**13.3.** Four nodal planes; zero nodal spheres; poor shielders; poor penetrators

**13.4.** $Z^* = 4.35$

# Glossary

Definitions are taken from the *McGraw-Hill Dictionary of Scientific and Technical Terms*, 3d ed., by S. B. Parker, (ed). Copyright © 1984 by McGraw-Hill, Inc. Copied by permission. Phrases included in brackets represent modifications or notes concerning usage in this book.

**acid**   1. Any of a class of chemical compounds whose aqueous solutions turn blue litmus paper red, react with and dissolve certain metals to form salts, and react with bases to form salts. 2. [Arrhenius, Bronsted-Lowry] A compound capable of transferring a hydrogen ion in solution. 3. [Lewis] A molecule or ion that combines with another molecule or ion by forming a covalent bond with two electrons from the other species.

**acid anhydride, [acidic oxide]**   An acid with one or more molecules of water removed; for example, $SO_3$ is the acid anhydride of $H_2SO_4$, sulfuric acid.

**activation energy**   The energy, in excess over the ground state, that must be added to an atomic or molecular system to allow a particular process to take place.

**allotropy**   The assumption by an element or other substance of two or more different forms of structures that are most frequently stable in different temperature ranges, such as different crystalline forms of carbon as charcoal or diamond.

**alloy**   Any of a large number of substances having metallic properties and consisting of two or more elements; with few exceptions, the components are usually metallic elements.

**ambident**   A reagent or substrate that can have two or more attacking sites.

**amorphous**   Pertaining to a solid that is noncrystalline, having neither definite form nor structure.

**amphiprotic, amphoteric**   Having both acidic and basic characteristics.

**anion**   An ion that is negatively charged.

**antibonding orbital**   An atomic or molecular orbital whose energy increases as atoms are brought together, indicating a net repulsion rather than a net attraction and chemical bonding.

**aprotic solvent**   A solvent that does not yield or accept a proton.

**band**   A restricted range in which the energies of electrons in solids lie or from which they are excluded, as understood in quantum-mechanical terms.

**base**   Any chemical species, ionic or molecular, capable of accepting or receiving a proton (hydrogen ion) from another substance; the other substance acts as an acid in giving of the proton; the hydroxyl ion is a base.

**basic oxide**  A metallic oxide that is a base or that forms a hydroxide when combined with water, such as sodium oxide to sodium hydroxide.

**beneficiation**  Improving the chemical or physical properties of an ore so that metal can be recovered at a profit. Also known as mineral dressing.

**bond energy**  The heat of formation of a molecule from its constituent atoms. [*Note*: Bond energies are usually tabulated as positive numbers, which represent the energy required to dissociate (break apart) a gaseous diatomic molecule into its constituent gaseous atoms, or a corresponding fraction for a larger molecule composed of two elements.]

**Born-Haber cycle**  A sequence of chemical and physical processes by means of which the cohesive energy of an ionic crystal can be deduced from experimental quantities; it leads from an initial state in which a crystal is at zero pressure and $0\,°K$ to a final state that is an infinitely dilute gas of its constituent ions, also at zero pressure and $0\,°K$.

**bridging ligand**  A ligand in which an atom or molecular species that is able to exist independently is simultaneously bonded to two or more metal atoms.

**catalyst**  Substance that alters the velocity of a chemical reaction and may be recovered essentially unaltered in form and amount at the end of the reaction.

**catenation**  The property of an element to link to itself to form molecules, as with carbon.

**chalcogen**  One of the elements that form Group 16/VIA of the periodic table; included are oxygen, sulfur, selenium, tellurium, and polonium.

**close packed**  Referring to a crystal structure in which the lattice points are centers of spheres of equal radius arranged so that the volume of the interstices between the spheres is minimal.

**conductor**  A wire, cable, or other body or medium that is suitable for carrying electric current.

**coordinate covalent bond**  A chemical bond between two atoms in which a shared pair of electrons forms the bond, the pair having been supplied by one of the two atoms.

**corrosion**  Gradual destruction of a metal or alloy due to chemical processes such as oxidation or the action of a chemical agent.

**covalent bond**  A bond in which each atom of a bound pair contributes one electron to form a pair of electrons.

**covalent radius**  The effective radius of an atom in a covalent bond.

**cracking**  A process that is used to reduce the molecular weight of hydrocarbons by breaking the molecular bonds by various thermal, catalytic, or hydrocracking methods.

**crystal defect**  Any departure from crystal symmetry caused by free surfaces, disorder, impurities, vacancies, and interstitials, dislocations, lattice vibrations, and grain boundaries. Also known as lattice defect.

**deliquescence**  The absorption of atmospheric water vapor by a crystalline solid until the crystal eventually dissolves into a saturated solution.

**desiccant**  A soluble or insoluble chemical substance that has such a great affinity for water that it will abstract water from a great many fluid materials.

**displacement reaction**  A chemical reaction in which an atom, radical, or molecule displaces and sets free an element of a compound.

**disproportionation**  The changing of a substance, usually by simultaneous oxidation and reduction, into two or more dissimilar substances.

**electrolysis**  A method by which chemical reactions are carried out by passage of electric current through a solution of an electrolyte or through a molten salt.

**electromotive force**  1. The difference in electric potential that exists between two dissimilar electrodes immersed in the same electrolyte or otherwise connected by ionic conductors. 2. The resultant of the relative electrode potential of the two dissimilar electrodes at which electrochemical reactions occur. Abbreviated emf.

**electron affinity**    The work needed in removing an electron from a negative ion, thus restoring the neutrality of an atom or molecule.

**electronegative**    Pertaining to an atom or group of atoms that has a relatively great tendency to attract to itself electrons [in a molecule].

**energy gap**    A range of forbidden energies in the band theory of solids.

**enthalpy**    The sum of the internal energy of a system plus the product of the system's volume multiplied by the pressure exerted on the system by its surroundings. Also known as heat content, sensible heat, and total heat. [*Note*: In this book (and in many other works) reactions are usually understood to occur with no change in external pressure, so the terms *heat* or *energy* are often used interchangeably with *enthalpy*, as heat of vaporization for enthalpy of vaporization, or bond energy for bond enthalpy.]

**entropy**    1. Measure of the disorder of a system, equal to the Boltzman constant times the natural logarithm of the number of microscopic states corresponding to the thermodynamic state of the system. 2. Function of the state of a thermodynamic system whose change in any differential reversible process is equal to the heat absorbed by the system from its surroundings divided by the absolute temperature of the system.

**equilibrium constant**    A constant at a given temperature such that when a reversible chemical reaction $cC + dD = gG + hH$ has reached an equilibrium, the value of this constant $K°$ is equal to

$$\frac{a_G{}^g a_H{}^h}{a_C{}^c a_D{}^d}$$

**explosion**    A chemical reaction or change of state that is effected in an exceedingly short space of time with the generation of a high temperature and generally a large quantity of gas.

**ferroelectricity**    Spontaneous electric polarization in a crystal; analogous to ferromagnetism.

**ferromagnetism**    A property, exhibited by certain metals, alloys, and compounds of the transition (iron group) rare-earth and actinide elements, in which the internal magnetic moments spontaneously organize in a common direction; gives rise to a permeability considerably greater than that of a vacuum, and to magnetic hysteresis.

**Fischer-Tropsch process**    A catalytic process to synthesize hydrocarbons and their oxygen derivatives by the controlled reaction of hydrogen and carbon monoxide.

**Gibbs free energy, free energy**    The thermodynamic function $G = H - TS$, where $H$ is enthalpy, $T$ absolute temperature, and $S$ entropy.

**glass**    A hard, amorphous, inorganic, usually transparent, brittle substance made by fusing silicates, sometimes borates and phosphates, with certain basic oxides and then rapidly cooling to prevent crystallization.

**Grignard reagent, RMgX**    The organometallic halide formed in the reaction between an alkyl or aryl halide and magnesium metal in a suitable solvent, usually ether.

**group**    A family of elements with similar chemical properties.

**halogen**    Any of the elements of the halogen family, consisting of fluorine, chlorine, bromine, iodine, and astatine.

**heteropoly compound**   Polymeric compounds of molybdates with anhydrides of other elements such as phosphorus; the yellow precipitate $(NH_4)_3[PMo_{12}O_{40}]$ is such a compound.

**hexagonal close packing**   Close-packed crystal structure characterized by the regular alteration of two layers; the atoms in each layer lie at the vertices of a series of equilateral triangles, and the atoms in one layer lie directly above the centers of the triangles in neighboring layers.

**homogeneous catalysis**   Catalysis occurring within a single phase, usually a gas or liquid.

**hydration energy, heat of hydration**   The increase in enthalpy accompanying the formation of 1 mol of a hydrate from the anhydrous form of the compound and from water at constant pressure.

**hydrolysis**   In aqueous solutions of electrolytes, the reactions of cations with water to produce a weak base or of anions to produce a weak acid.

**hydrothermal synthesis**   Mineral synthesis in the presence of heated water.

**intercalation**   A layer located between layers of different character.

**ionic bonding**   A type of chemical bonding in which one or more electrons are transferred completely from one atom to another, thus converting neutral atoms into electrically charged ions; these ions are approximately spherical and attract one another because of their opposite charge.

**isoelectronic**   Pertaining to atoms having the same number of electrons outside the nucleus of atoms.

**isomorphous, isomorphic minerals**   Any two or more crystalline mineral compounds having different chemical compositions but identical structure, such as the garnet series or the feldspar group.

**lanthanide contraction**   A phenomenon encountered in the rare-earth elements; the radii of the atoms of the members of the series decrease slightly as the atomic numbers increase; starting with element 58 in the periodic table, the balancing electron fills in an inner incomplete $4f$ shell as the charge on the nucleus increases.

**lattice energy**   The energy required to separate ions in an ionic crystal an infinite distance from each other. [*Note:* Inorganic chemists usually take the lattice energy to be the energy released when ions come together from infinite separation to form an ionic crystal; this is the usage in this book.]

**layer structure**   A crystalline structure found in substances such as graphites and clays in which the atoms are largely concentrated in a set of parallel planes, with the regions between the planes comparatively vacant.

**Madelung constant**   A dimensionless constant that determines the electrostatic energy of a three-dimensional periodic crystal lattice consisting of a large number of positive and negative point charges when the number and magnitude of the charges and the nearest-neighbor distance between them is specified.

**metallic bond**   The type of chemical bond that is present in all metals, and may be thought of as resulting from a sea of valence electrons that are free to move throughout the metal lattice.

**molecular orbital**   A wave function describing an electron in a molecule.

**native**   Pertaining to an element found in nature in a nongaseous state.

**$n$-type semiconductor**   An extrinsic semiconductor in which the conduction electron density exceeds the hole density.

**oxidation-reduction reaction**   An oxidizing chemical change, where an element's positive valence is increased (electron loss), accompanied by a simultaneous reduction of an associated element (electron gain).

**oxidation state, number**   The number of electrons to be added to (or subtracted from) an atom in a combined state to convert it to elemental form. Also known as oxidation number.

**oxidizing agent**   Compound that gives up oxygen easily, removes hydrogen from another compound, or attracts negative electrons.

**passivate**   To render passive; to reduce the reactivity of a chemically active metal surface by electrochemical polarization or by immersion in a passivating solution.

**period**   A family of elements with consecutive atomic numbers in the periodic table and with closely related properties.

**periodic table**   A table of the elements, written in sequence in the order of atomic number and arranged in horizontal rows (periods) and vertical columns (groups) to illustrate the occurrence of similarities in the properties of the elements as a periodic function of the sequence.

**photochemical reaction**   A chemical reaction influenced or initiated by light, particularly ultraviolet light, as in the chlorination of benzene to produce benzene hexachloride.

**pi bonding**   Covalent bonding in which the greatest overlap between atomic orbitals is along a plane perpendicular to the line joining the nuclei of the two atoms.

**piezoelectric**   Having the ability to generate a voltage when mechanical force is applied, or to produce a mechanical force when a voltage is applied, as in a piezoelectric crystal.

**pnicogen**   Any member of the nitrogen family of elements, Group 15/VA in the periodic table.

**polar covalent bond**   A bond in which a pair of electrons is shared in common between two atoms, but the pair is held more closely by one of the atoms.

**polymorphism**   The property of a chemical substance of crystallizing into two or more forms having different structures, such as diamond and graphite.

**$p$-type semiconductor**   An extrinsic semiconductor in which the hole density exceeds the conduction electron density.

**radiochemistry**   That area of chemistry concerned with the study of radioactive substances.

**radius ratio**   The ratio of the radius of a cation to the radius of an anion; relative ionic radii are pertinent to crystal lattice structure, particularly the determination of coordination number.

**redox system**   A chemical system in which reduction and oxidation (redox) reactions occur.

**reducing agent**   1. A material that adds hydrogen to an element or compound. 2. A material that adds an electron to an element or compound, that is, decreases the positiveness of its valence.

**reduction**   Chemical reaction in which an element gains an electron (has a decrease in positive valence).

**refractory**   A material of high melting point.

**resonance structure**   Any of two or more structures of the same compound that have identical geometry but different arrangements of their paired electrons; none of the structures has physical reality or adequately accounts for the properties of the compound, which exists as an intermediate form.

**scrubber**   A device for the removal, or washing out, of entrained liquid droplets or dust or for the removal of an undesired gas component from process gas streams.

439

**sedimentation**   The act or process of accumulating sediment in layers.

**semiconductor**   A solid crystalline material whose electrical conductivity is intermediate between that of a conductor and an insulator, ranging from about $10^5$ mho to $10^{-7}$ mho per meter, and is usually strongly temperature-dependent.

**sigma bond**   The chemical bond resulting from the formation of a molecular orbital by the end-on overlap of atomic orbitals.

**sintering**   Forming a coherent bonded mass by heating metal powders without melting; used mostly in powder metallurgy.

**slag**   A nonmetallic product resulting from the interaction of flux and impurities in the smelting and refining of metals.

**solvolysis**   A reaction in which a solvent reacts with the solute to form a new substance.

**spectrum**   A display or plot of intensity of radiation (particles, photons, or acoustic radiation) as a function of mass, momentum, wavelength, frequency, or some related quantity.

**spin-orbit coupling**   The interaction between a particle's spin and its orbital angular momentum.

**steel**   An iron-base alloy, malleable under proper conditions, containing up to about 2% carbon.

**sublimation**   The process by which solids are transformed directly to the vapor state, or vice versa, without passing through the liquid phase.

**synthesis gas**   A mixture of gases prepared as feedstock for a chemical reaction, for example, carbon monoxide and hydrogen to make hydrocarbons or organic chemicals, or hydrogen and nitrogen to make ammonia.

**valence**   A positive number that characterizes the combining power of an element for other elements, as measured by the number of bonds to other atoms that one atom of the given element forms upon chemical combination; hydrogen is assigned valence 1, and the valence is the number of hydrogen atoms, or their equivalent, with which an atom of the given element combines.

**valence electron**   An electron that belongs to the outermost shell of an atom. [*Note:* This definition is not satisfactory for *d*- or *f*-block elements.]

**water gas**   A mixture of carbon monoxide and methane produced by passing steam through deep beds of incandescent coal; used for industrial heating and as a gas engine fuel.

**wave function, Schrödinger wave function**   A function of the coordinates of the particles of a system and of time that is a solution of the Schrödinger equation and that determines the average result of every conceivable experiment on the system.

**weathering**   Physical disintegration and chemical decomposition of earthy and rocky materials on exposure to atmospheric agents, producing an in-place mantle of waste.

# Frequently Used Tables

**Table 2.1**   Hydration Enthalpies of Metal Cations (kJ/mol)

| Electronegativity $\leq$ 1.5 | | | Electronegativity $\geq$ 1.5 | | |
|---|---|---|---|---|---|
| ION | RADIUS | $\Delta H_{hyd}$ | ION | RADIUS | $\Delta H_{hyd}$ |
| **+1 Ions** | | | | | |
| Cs | 181 | $-263$ | | | |
| Rb | 166 | $-296$ | Tl | 164 | $-326$ |
| K | 152 | $-321$ | | | |
| Na | 116 | $-405$ | Ag | 129 | $-475$ |
| Li | 90 | $-515$ | Cu | 91 | $-594$ |
| H | | $-1091$ | | | |
| **+2 Ions** | | | | | |
| Ra | | $-1259$ | | | |
| Ba | 149 | $-1304$ | | | |
| Sr | 132 | $-1445$ | Pb | 133 | $-1480$ |
| No | 124 | $-1485$ | Sn | | $-1554$ |
| Ca | 114 | $-1592$ | Cd | 109 | $-1806$ |
| | | | Cr | 94 | $-1850$ |
| | | | Mn | 97 | $-1845$ |
| | | | Fe | 92 | $-1920$ |
| | | | Co | 88 | $-2054$ |
| | | | Ni | 83 | $-2106$ |
| | | | Cu | 91 | $-2100$ |
| | | | Zn | 88 | $-2044$ |
| Mg | 86 | $-1922$ | Be | 59 | $-2487$ |

*(continued)*

441

**Table 2.1** Hydration Enthalpies of Metal Cations (kJ/mol) (continued)

| Electronegativity ≤ 1.5 | | | Electronegativity ≥ 1.5 | | |
|---|---|---|---|---|---|
| ION | RADIUS | $\Delta H_{hyd}$ | ION | RADIUS | $\Delta H_{hyd}$ |
| **+3 Ions** | | | | | |
| Pu | 114 | −3441 | | | |
| La | 117 | −3283 | | | |
| Lu | 100 | −3758 | Tl | 102 | −4184 |
| Y | 104 | −3620 | In | 94 | −4109 |
| Sc | 88 | −3960 | Ga | 76 | −4685 |
| | | | Fe | 78 | −4376 |
| | | | Cr | 75 | −4402 |
| | | | Al | 67 | −4660 |
| **+4 Ions** | | | | | |
| Ce | 101 | −6489 | | | |

Ionic radii are from Table C; hydration enthalpies are taken from J. Burgess, *Metal Ions in Solution*, Ellis Horwood, Chichester, England, 1978, pp. 182–183.

**Table 2.5** Shannon-Prewitt Radii and Hydration Enthalpies (kJ/mol) of Some Anions

| Anion | Radius | Hydration Energy | Anion | Radius | Hydration Energy |
|---|---|---|---|---|---|
| $F^-$ | 119 | −497 | $OH^-$ | 119 | −453 |
| $O^{2-}$ | 126 | | $CN^-$ | 177 | −334 |
| $N^{3-}$ | 132 | | $N_3^-$ | 181 | −290 |
| $Cl^-$ | 167 | −355 | | | |
| $S^{2-}$ | 170 | −1356 | $SH^-$ | 193 | −328 |
| $Br^-$ | 182 | −328 | $BF_4^-$ | 215 | −215 |
| $Se^{2-}$ | 184 | | $ClO_4^-$ | 226 | −227 |
| $I^-$ | 206 | −287 | $I_3^-$ | | −160 |
| $Te^{2-}$ | 207 | | | | |

SOURCES: Radii for monoatomic anions are taken from R. D. Shannon and C. T. Prewitt, *Acta Crystallogr.*, B25, 925 (1969) and R. D. Shannon, ibid., A32, 751 (1976). "Radii" for polyatomic anions (which of course are not truly spherical) are thermochemical radii, taken from J. E. Huheey, *Inorganic Chemistry*, 3d ed., Harper and Row, Cambridge, 1983, p. 78.
NOTE: Hydration enthalpies are from M. C. Ball and A. H. Norbury, *Physical Data for Inorganic Chemists*, Longman, London, 1974.

**Table 3.4**  Lattice Types and Madelung Constants for Different Stoichiometries and Radius Ratios of Cations and Anions

| Radius Ratio (Cation/Anion) | Lattice Type | Coordination Number of Cation | Anion | Madelung Constant | Reduced Madelung Constant |
|---|---|---|---|---|---|
| **A. 1:1 Stoichiometry of Salt (MX)** | | | | | |
| 0.225–0.414 | Wurtzite (ZnS) | 4 | 4 | 1.63805 | 1.63805 |
| | Zinc blende (ZnS) | 4 | 4 | 1.64132 | 1.64132 |
| 0.414–0.732 | Rock salt (NaCl) | 6 | 6 | 1.74756 | 1.74756 |
| 0.732–1.000 | CsCl | 8 | 8 | 1.76267 | 1.76267 |
| **B. 1:2 Stoichiometry of Salt (MX$_2$)** | | | | | |
| 0.225–0.414 | Beta-quartz (SiO$_2$) | 4 | 2 | 2.201 | 1.467 |
| 0.414–0.732 | Rutile (TiO$_2$) | 6 | 3 | 2.408[a] | 1.605 |
| 0.732–1.000 | Fluorite (CaF$_2$) | 8 | 4 | 2.51939 | 1.6796 |
| **C. 2:3 Stoichiometry of Salt (M$_2$X$_3$)** | | | | | |
| 0.414–0.732 | Corundum (Al$_2$O$_3$) | 6 | 4 | 4.1719[a] | 1.6688 |
| **D. Other Stoichiometries and Lattice Types** | | | | | |
| Never favored | Ion pair | 1 | 1 | 1.00000 | 1.0000 |
| 0.000–0.155 | | 2 | | | |
| 0.155–0.225 | | 3 | | | |
| 0.225–0.414 | | 4 | | | |
| 0.414–0.732 | | 6 | | | |
| 0.732–1.000 | | 8 | | | |
| 1.000 | | 12 | | | |

NOTE: Reduced Madelung constant = Madelung constant × $2/p$, where $p$ = number of ions in the simplest formula of the salt.

[a] Exact value dependent on details of the structure.

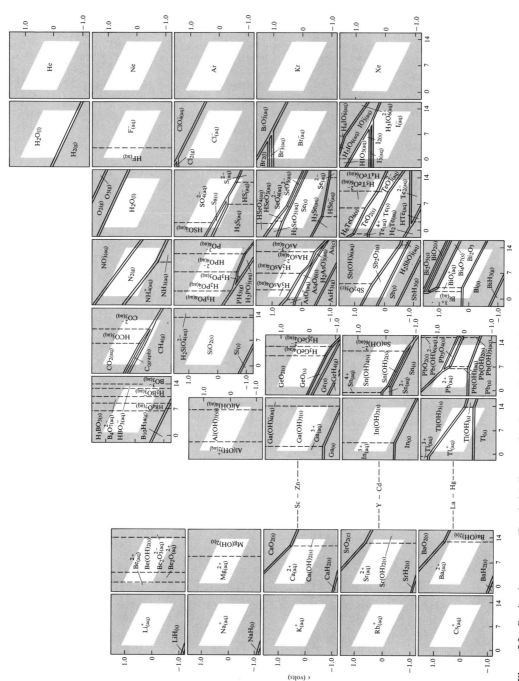

**Figure 5.2** Predominance area (Pourbaix) diagrams for the *s*- and *p*-block elements. The white "windows" are bounded on the top by the theoretical *E* for the oxidation of water, and on the bottom by the theoretical *E* for the reduction of water. From J. A. Campbell and R. A. Whiteker, "A Periodic Table Based on Potential-pH Diagrams." *J. Chem. Educ.*, 46, 92 (1969). Reprinted with permission.

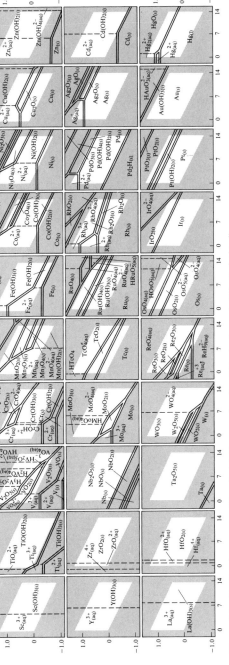

**Figure 5.3** Predominance area (Pourbaix) diagrams for the *d*-block elements. The "window" boundaries are the same as those given in Figure 5.2. From J. A. Campbell and R. A. Whiteker, "A Periodic Table Based on Potential-pH Diagrams," *J. Chem. Educ.*, 46, 92 (1969). Reprinted with permission.

**Table 6.1**  Heats of Atomization of the Elements (kJ/mol)

| 1 | 2 | 3 | 4 | 5 | 6 | 7 | 8 | 9 | 10 | 11 | 12 | 13/IIIA | 14/IVA | 15/VA | 16/VIA | 17/VIIA | 18/VIIIA |
|---|---|---|---|---|---|---|---|---|---|---|---|---|---|---|---|---|---|
| H 218 | | | | | | | | | | | | | | | | | He 0 |
| Li 159 | Be 324 | | | | | | | | | | | B 563 | C 717 | N 473 | O 249 | F 79 | Ne 0 |
| Na 107 | Mg 146 | | | | | | | | | | | Al 326 | Si 456 | P 315 | S 279 | Cl 122 | Ar 0 |
| K 89 | Ca 178 | Sc 378 | Ti 471 | V 515 | Cr 397 | Mn 283 | Fe 415 | Co 426 | Ni 431 | Cu 338 | Zn 131 | Ga 277 | Ge 377 | As 303 | Se 227 | Br 112 | Kr 0 |
| Rb 81 | Sr 165 | Y 423 | Zr 605 | Nb 733 | Mo 659 | Tc 661 | Ru 652 | Rh 556 | Pd 377 | Ag 285 | Cd 112 | In 244 | Sn 302 | Sb 262 | Te 197 | I 107 | Xe 0 |
| Cs 76 | Ba 182 | Lu $^a$414 | Hf 621 | Ta 782 | W 860 | Re 776 | Os 789 | Ir 671 | Pt 564 | Au 368 | Hg 64 | Tl 182 | Pb 195 | Bi 207 | Po 142 | At | Rn 0 |

| | 3f | 4f | 5f | 6f | 7f | 8f | 9f | 10f | 11f | 12f | 13f | 14f | 15f | 16f |
|---|---|---|---|---|---|---|---|---|---|---|---|---|---|---|
| 6 | La 423 | Ce 419 | Pr 356 | Nd 328 | Pm 301 | Sm 207 | Eu 178 | Gd 398 | Tb 389 | Dy 291 | Ho 301 | Er 317 | Tm 232 | Yb 152 |
| 7 | Ac $^a$293 | Th 575 | Pa $^a$481 | U 482 | Np $^a$337 | Pu 352 | Am $^a$239 | Cm | Bk | Cf | Es | Fm | Md | No |

SOURCES: Heats (enthalpies) of atomization of the s- and d-block elements were taken from W. L. Jolly, *Modern Inorganic Chemistry*, McGraw-Hill, New York, 1984, p. 292; those of the d-block elements were taken from W. W. Porterfield, *Inorganic Chemistry: A Unified Approach*, Addison-Wesley, Reading, Mass., 1984, p. 84; those of the f-block elements are from N. N. Greenwood and A. Earnshaw, *Chemistry of the Elements*, Pergamon, Oxford, 1984.

NOTE: Values preceded by $^a$ are enthalpies of *vaporization*, which are normally slightly less than true heats of atomization, since the metals vaporize in part as diatomic or polyatomic molecules.

446

**Table 6.4**  Covalent and Single-Bonded Metallic Radii of the Elements

| Period | 1 | 2 | 3 | 4 | 5 | 6 | 7 | 8 | 9 | 10 | 11 | 12 | 13/IIIA | 14/IVA | 15/VA | 16/VIA | 17/VIIA | 18/VIIIA |
|---|---|---|---|---|---|---|---|---|---|---|---|---|---|---|---|---|---|---|
| 1 | H 37 | | | | | | | | | | | | | | | | | He 32 |
| 2 | Li 134 / 122 | Be 125 / 89 | | | | | | | | | | | B 90 / 80 | C 77 | N 75 | O 73 | F 71 | Ne 69 |
| 3 | Na 154 / 157 | Mg 145 / 136 | | | | | | | | | | | Al 130 / 125 | Si 118 / 117 | P 110 / 110 | S 102 / 104 | Cl 99 | Ar 97 |
| 4 | K 196 / 202 | Ca 174 | Sc 144 | Ti 136 / 132 | V 122 | Cr 119 | Mn 139 / 118 | Fe 125 / 117 | Co 126 / 116 | Ni 121 / 115 | Cu 135 / 118 | Zn 120 / 121 | Ga 120 / 125 | Ge 122 / 124 | As 122 / 121 | Se 117 / 117 | Br 114 | Kr 110 |
| 5 | Rb 216 | Sr 191 | Y 162 | Zr 148 / 145 | Nb 134 | Mo 130 | Tc 127 | Ru 133 / 125 | Rh 132 / 125 | Pd 131 / 128 | Ag 152 / 134 | Cd 148 / 138 | In 144 / 142 | Sn 140 / 142 | Sb 143 / 139 | Te 135 / 137 | I 133 | Xe 130 |
| 6 | Cs 235 | Ba 198 | Lu 156 | Hf 144 | Ta 134 | W 130 | Re 128 | Os 133 / 126 | Ir 132 / 126 | Pt 131 / 129 | Au 140 / 134 | Hg 148 / 139 | Tl 147 / 144 | Pb 146 / 150 | Bi 146 / 151 | Po | At | Rn 145 |

| | 3f | 4f | 5f | 6f | 7f | 8f | 9f | 10f | 11f | 12f | 13f | 14f | 15f | 16f |
|---|---|---|---|---|---|---|---|---|---|---|---|---|---|---|
| 6 | La 169 | Ce 165 | Pr 164 | Nd 164 | Pm 163 | Sm 162 | Eu 185 | Gd 162 | Tb 161 | Dy 160 | Ho 158 | Er 158 | Tm 158 | Yb 170 |
| 7 | Ac | Th 165 | Pa | U 143 | Np | Pu | Am | Cm | Bk | Cf | Es | Fm | Md | No |

SOURCES: Covalent radii are listed in the first row under the symbol for each element; these are taken from J. E. Huheey, *Inorganic Chemistry: Principles of Structure and Reactivity*, 3d ed., Harper and Row, New York, 1983, pp. 258–259, or, if not available there, from M. C. Ball and A. H. Norbury, *Physical Data for Inorganic Chemists*, Longman, London, 1974, pp. 139–144. Single-bonded metallic radii are listed in the second row under the symbol for each element; these are taken from Ball and Norbury, op.cit.

**Table 6.5**  Element-Element Covalent Bond Energies (kJ/mol)

**(a) Single (Sigma) Bond Energies**

| | | | | | | |
|---|---|---|---|---|---|---|
| H—H | | | | | | |
| 432 | | | | | | |
| Li—Li | Be—Be | B—B | C—C | N—N | O—O | F—F |
| 105 | (208) | 293 | 346 | 167 | 142 | 155 |
| Na—Na | Mg—Mg | Al—Al | Si—Si | P—P | S—S | Cl—Cl |
| 72 | (129) | | 222 | 201 | 226 | 240 |
| K—K | Ca—Ca | Ga—Ga | Ge—Ge | As—As | Se—Se | Br—Br |
| 49 | (105) | 113 | 188 | 146 | 172 | 190 |
| Rb—Rb | Sr—Sr | In—In | Sn—Sn | Sb—Sb | Te—Te | I—I |
| 45 | (84) | 100 | 146 | 121 | 126 | 149 |
| Cs—Cs | | | | | | At—At |
| 43 | | | | | | 116 |

**(b) Pi-Bond Energies**

| | | C—C | N—N | O—O |
|---|---|---|---|---|
| | | 256 | 387 | 352 |
| | | | P—P | S—S |
| | | | 140 | 199 |
| | | Ge—Ge | As—As | Se—Se |
| | | 84 | 117 | 100 |
| | | | Sb—Sb | Te—Te |
| | | | 87 | 92 |

SOURCE: Data from J. E. Huheey, *Inorganic Chemistry: Principles of Structure and Reactivity*, 3d ed., Harper and Row, New York, 1983, Table E-1.

NOTE: Pi-bond energies are calculated as double-bond energies minus single-bond energies for the same element or (triple-bond energies minus single-bond energies) times one-half.

**Table 6.6**  First Ionization Energies of the Elements (kJ/mol)

| | $s^1$ | $s^2$ | $s^2d^1$ | $s^2d^2$ | $s^2d^3$ | $s^2d^4$ | $s^2d^5$ | $s^2d^6$ | $s^2d^7$ | $s^2d^8$ | $s^2d^9$ | $s^2d^{10}$ | $s^2p^1$ | $s^2p^2$ | $s^2p^3$ | $s^2p^4$ | $s^2p^5$ | $s^2p^6$ |
|---|---|---|---|---|---|---|---|---|---|---|---|---|---|---|---|---|---|---|
| 1 | H 1312 | | | | | | | | | | | | | | | | | He 2372 |
| 2 | Li 520 | Be 899 | | | | | | | | | | | B 801 | C 1086 | N 1402 | O 1314 | F 1681 | Ne 2081 |
| 3 | Na 496 | Mg 738 | | | | | | | | | | | Al 578 | Si 786 | P 1012 | S 1000 | Cl 1251 | Ar 1520 |
| 4 | K 419 | Ca 590 | Sc 631 | Ti 658 | V 650 | Cr 653 | Mn 717 | Fe 759 | Co 758 | Ni 737 | Cu 746 | Zn 906 | Ga 579 | Ge 762 | As 944 | Se 941 | Br 1140 | Kr 1351 |
| 5 | Rb 403 | Sr 550 | Y 616 | Zr 660 | Nb 664 | Mo 685 | Tc 702 | Ru 711 | Rh 720 | Pd 805 | Ag 731 | Cd 868 | In 558 | Sn 709 | Sb 832 | Te 869 | I 1008 | Xe 1170 |
| 6 | Cs 376 | Ba 503 | Lu 524 | Hf 654 | Ta 761 | W 770 | Re 760 | Os 840 | Ir 880 | Pt 870 | Au 890 | Hg 1007 | Tl 589 | Pb 716 | Bi 703 | Po 812 | At | Rn 1037 |
| 7 | Fr | Ra 509 | Lr | Rf | Ha | | | | | | | | | | | | | |

| | $s^2f^1$ | $s^2f^2$ | $s^2f^3$ | $s^2f^4$ | $s^2f^5$ | $s^2f^6$ | $s^2f^7$ | $s^2f^8$ | $s^2f^9$ | $s^2f^{10}$ | $s^2f^{11}$ | $s^2f^{12}$ | $s^2f^{13}$ | $s^2f^{14}$ |
|---|---|---|---|---|---|---|---|---|---|---|---|---|---|---|
| 6 | La 538 | Ce 528 | Pr 523 | Nd 530 | Pm 536 | Sm 543 | Eu 547 | Gd 592 | Tb 564 | Dy 572 | Ho 581 | Er 589 | Tm 597 | Yb 603 |
| 7 | Ac 490 | Th 590 | Pa 570 | U 590 | Np 600 | Pu 585 | Am 578 | Cm 581 | Bk 601 | Cf 608 | Es 619 | Fm 627 | Md 635 | No 642 |

SOURCE: First ionization energies, IE(1), of the elements listed are taken from James E. Huheey, *Inorganic Chemistry: Principles of Structure and Reactivity*, 3d ed., Harper and Row, New York, 1983, pp. 42–44.

NOTE: Group headings indicate the characteristic valence-electron configurations of the atoms being ionized.

449

**Table 6.7**  Second Ionization Energies of (+1 Ions of) the Elements (kJ/mol)

| Period | $s^1$ | | | | | | | | | | | $s^2$ | $s^2p^1$ | $s^2p^2$ | $s^2p^3$ | $s^2p^4$ | $s^2p^5$ | $s^2p^6$ |
|---|---|---|---|---|---|---|---|---|---|---|---|---|---|---|---|---|---|---|
| 1 | | | | | | | | | | | | | | | | | He 5250 | Li 7298 |
| 2 | Be 1757 | | | | | | | | | | | B 2427 | C 2353 | N 2856 | O 3388 | F 3374 | Ne 3952 | Na 4562 |
| 3 | Mg 1451 | | | | | | | | | | | Al 1817 | Si 1577 | P 1903 | S 2251 | Cl 2297 | Ar 2666 | K 3051 |
| 4 | Ca 1145 | Sc 1235 | Ti 1310 | V 1414 | Cr 1496 | Mn 1509 | Fe 1561 | Co 1646 | Ni 1753 | Cu 1958 | Zn 1733 | Ga 1979 | Ge 1537 | As 1798 | Se 2045 | Br 2100 | Kr 2350 | Rb 2633 |
| 5 | Sr 1064 | Y 1181 | Zr 1267 | Nb 1382 | Mo 1558 | Tc 1472 | Ru 1617 | Rh 1744 | Pd 1875 | Ag 2074 | Cd 1631 | In 1821 | Sn 1412 | Sb 1595 | Te 1790 | I 1846 | Xe 2046 | Cs 2230 |
| 6 | Ba 965 | Lu 1340 | Hf 1440 | Ta | W | Re | Os | Ir | Pt 1791 | Au 1980 | Hg 1810 | Tl 1971 | Pb 1450 | Bi 1610 | Po | At | Rn | Fr |
| 7 | Ra 979 | Lr | Rf | Ha | 106 | 107 | | 109 | | | | | | | | | | |

f-block:

| | | | | | | | | | | | $s^2$ | $s^2p^1$ | $s^2p^2$ | $s^2p^3$ |
|---|---|---|---|---|---|---|---|---|---|---|---|---|---|---|
| 6 | La 1067 | Ce 1047 | Pr 1018 | Nd 1034 | Pm 1052 | Sm 1068 | Eu 1085 | Gd 1170 | Tb 1112 | Dy 1126 | Ho 1139 | Er 1151 | Tm 1163 | Yb 1175 |
| 7 | Ac 1170 | Th 1110 | Pa | U | Np | Pu | Am | Cm | Bk | Cf | Es | Fm | Md | No |

SOURCE: Second ionization energies, IE(2), of the +1 ions of the elements listed are taken from James E. Huheey, *Inorganic Chemistry: Principles of Structure and Reactivity*, 3d ed., Harper and Row, New York, 1983, pp. 42–44.

NOTE: Group headings indicate the valence electron configurations of the ions being ionized a second time.

**Table 6.9**   Electron Affinities of the Elements (kJ/mol) (Zeroth Ionization Energies of the −1 Ions of the Elements)

| Period | $s^1$ | $s^2$ | $s^2p^1$ | | | | | | | | | | | $s^2p^2$ | $s^2p^3$ | $s^2p^4$ | $s^2p^5$ | $s^2p^6$ |
|---|---|---|---|---|---|---|---|---|---|---|---|---|---|---|---|---|---|---|
| 1 | | H 73 | | | | | | | | | | | | | | | | |
| 2 | He 0 | Li 60 | Be 0 | | | | | | | | | | | B 27 | C 122 | N −7 | O 141 | F 328 |
| 3 | Ne 0 | Na 53 | Mg 0 | | | | | | | | | | | Al 44 | Si 134 | P 72 | S 200 | Cl 349 |
| 4 | Ar 0 | K 48 | Ca 0 | Sc 0 | Ti 20 | V 50 | Cr 64 | Mn 0 | Fe 24 | Co 70 | Ni 111 | Cu 118 | Zn 0 | Ga 29 | Ge 120 | As 77 | Se 195 | Br 325 |
| 5 | Kr 0 | Rb 47 | Sr 0 | Y 0 | Zr 50 | Nb 100 | Mo 100 | Tc 70 | Ru 110 | Rh 120 | Pd 60 | Ag 126 | Cd 0 | In 29 | Sn 121 | Sb 101 | Te 190 | I 295 |
| 6 | Xe 0 | Cs 46 | Ba 0 | Lu 50 | Hf 0 | Ta 60 | W 60 | Re 15 | Os 110 | Ir 160 | Pt 205 | Au 223 | Hg 0 | Tl 30 | Pb 110 | Bi 110 | Po 180 | At 270 |

SOURCE: Electron affinities (EA) of the elements listed are taken from James E. Huheey, *Inorganic Chemistry: Principles of Structure and Reactivity*, 3d ed., Harper and Row, New York, 1983, pp. 42–44.

NOTE: Group headings indicate the characteristic valence electron configurations of the anions being ionized.

**Table 11.3**  Allred-Rochow Electronegativities of the Elements

| 1 | 2 | 3 | 4 | 5 | 6 | 7 | 8 | 9 | 10 | 11 | 12 | 13/IIIA | 14/IVA | 15/VA | 16/VIA | 17/VIIA | 18/VIIIA |
|---|---|---|---|---|---|---|---|---|---|---|---|---|---|---|---|---|---|
| 1 H 2.20 | | | | | | | | | | | | | | | | | He 5.50 |
| 2 Li 0.97 | Be 1.47 | | | | | | | | | | | B 2.01 | C 2.50 | N 3.07 | O 3.50 | F 4.10 | Ne 4.84 |
| 3 Na 1.01 | Mg 1.23 | | | | | | | | | | | Al 1.47 | Si 1.74 | P 2.06 | S 2.44 | Cl 2.83 | Ar 3.20 |
| 4 K 0.91 | Ca 1.04 | Sc 1.20 | Ti 1.32 | V 1.45 | Cr 1.56 | Mn 1.60 | Fe 1.64 | Co 1.70 | Ni 1.75 | Cu 1.75 | Zn 1.66 | Ga 1.82 | Ge 2.02 | As 2.20 | Se 2.48 | Br 2.74 | Kr 2.94 |
| 5 Rb 0.89 | Sr 0.99 | Y 1.11 | Zr 1.22 | Nb 1.23 | Mo 1.30 | Tc 1.36 | Ru 1.42 | Rh 1.45 | Pd 1.35 | Ag 1.42 | Cd 1.46 | In 1.49 | Sn 1.72 | Sb 1.82 | Te 2.01 | I 2.21 | Xe 2.40 |
| 6 Cs 0.86 | Ba 0.97 | Lu 1.14 | Hf 1.23 | Ta 1.33 | W 1.40 | Re 1.46 | Os 1.52 | Ir 1.55 | Pt 1.44 | Au 1.42 | Hg 1.44 | Tl 1.44 | Pb 1.55 | Bi 1.67 | Po 1.76 | At 1.90 | Rn 2.06 |
| 7 Fr 0.86 | Ra 0.97 | | | | | | | | | | | | | | | | |

| | 3f | 4f | 5f | 6f | 7f | 8f | 9f | 10f | 11f | 12f | 13f | 14f | 15f | 16f |
|---|---|---|---|---|---|---|---|---|---|---|---|---|---|---|
| 6 | La 1.08 | Ce 1.08 | Pr 1.07 | Nd 1.07 | Pm 1.07 | Sm 1.07 | Eu 1.01 | Gd 1.11 | Tb 1.10 | Dy 1.10 | Ho 1.10 | Er 1.11 | Tm 1.11 | Yb 1.06 |
| 7 | Ac 1.00 | Th 1.11 | Pa 1.14 | U 1.22 | Np 1.22 | Pu 1.22 | Am | Cm | Bk | Cf | Es | Fm | Md | No |

SOURCES: Values from A. L. Allred and E. G. Rochow, *J. Inorg. Nucl. Chem.*, 5, 264 (1958); E. J. Little and M. M. Jones, *J. Chem. Educ.*, 37, 231 (1960); and L. C. Allen and J. E. Huheey, *J. Inorg. Nucl. Chem.*, 42, 1523 (1980).

# Index

# Index

# Index

# Index

# Table C  Shannon–Prewitt Crystal Ionic Radii of Cations (pm)

| | 1 | 2 | 3 | 4 | 5 | 6 | 7 | 8 | 9 | 10 | 11 | 12 | 13/IIIA | 14/IVA | 15/VA | 16/VIA | 17/VIIA | 18/VIIIA |
|---|---|---|---|---|---|---|---|---|---|---|---|---|---|---|---|---|---|---|
| 1 | H | | | | | | | | | | | | | | | | | He |
| 2 | Li +1 90 | Be +2 59 | | | | | | | | | | | B +3 41 | C +4 30 | N +5 27 | O | F | Ne |
| 3 | Na +1 116 | Mg +2 86 | | | | | | | | | | | Al +3 67 | Si +4 54 | P +5 52 | S +6 43 | Cl +7 41 | Ar |
| 4 | K +1 152 | Ca +2 114 | Sc +3 88 | Ti +2 100 +3 81 +4 74 | V +2 93 +3 78 +4 72 +5 68 | Cr +2 94 +3 75 +4 69 +6 58 | Mn +2 97 +3 78 +4 67 +7 60 | Fe +2 92 +3 78 +4 72 | Co +2 88 +3 75 +4 67 | Ni +2 83 +3 74 | Cu +1 91 +2 87 | Zn +2 88 | Ga +3 76 | Ge +2 87 +4 67 | As +3 72 +5 60 | Se +4 64 +6 56 | Br +7 53 | Kr |
| 5 | Rb +1 166 | Sr +2 132 | Y +3 104 | Zr +4 86 | Nb +3 86 +4 82 +5 78 | Mo +4 79 +5 75 +6 73 | Tc +4 78 +5 74 +7 70 | Ru +3 82 +4 76 +5 70 | Rh +3 80 +4 74 +5 69 | Pd +2 100 +4 75 | Ag +1 129 +2 108 +3 89 | Cd +2 109 | In +3 94 | Sn +4 83 | Sb +3 90 +5 74 | Te +4 111 +6 70 | I +5 109 +7 69 | Xe +8 62 |
| 6 | Cs +1 181 | Ba +2 149 | Lu +3 100 | Hf +4 85 | Ta +3 86 +4 82 +5 78 | W +4 80 +5 76 +6 74 | Re +4 77 +5 72 +7 67 | Os +4 77 +5 71 +6 66 | Ir +3 82 +4 76 +5 71 | Pt +2 94 +4 76 | Au +1 151 +3 99 | Hg +2 116 | Tl +1 164 +3 102 | Pb +2 133 +4 91 | Bi +3 117 +5 90 | Po +4 108 +6 81 | At +7 76 | Rn |
| 7 | Fr +1 194 | Ra | Lr | Rf | Ha | 106 | 107 | | 109 | | | | | | | | | |

| | 3f | 4f | 5f | 6f | 7f | 8f | 9f | 10f | 11f | 12f | 13f | 14f | 15f | 16f |
|---|---|---|---|---|---|---|---|---|---|---|---|---|---|---|
| 6 | La +3 117 | Ce +3 115 +4 101 | Pr +3 113 +4 99 | Nd +3 112 | Pm +3 111 | Sm +3 110 | Eu +2 131 +3 109 | Gd +3 108 | Tb +3 106 +4 90 | Dy +2 121 +3 105 | Ho +3 104 | Er +3 103 | Tm +2 117 +3 102 | Yb +2 116 +3 101 |
| 7 | Ac +3 126 | Th +4 108 | Pa +3 118 +4 104 +5 92 | U +3 116 +4 103 +5 90 +6 87 | Np +3 115 +4 101 +5 89 +6 86 +7 85 | Pu +3 114 +4 100 +5 88 +6 85 | Am +3 111 +4 99 | Cm +3 111 +4 99 | Bk +3 110 +4 97 | Cf +3 109 +4 96 | Es | Fm | Md | No +2 124 |

Data are for cations with coordination number of six and are from R. D. Shannon and C. T. Prewitt, *Acta Cryst.*, **B25**, 925 (1969), and R. D. Shannon, *Acta Cryst.*, **A32**, 751 (1976).